Plant Genetic Resources, Inventory, Collection and Conservation

Siva Ramamoorthy • Inocencio Jr. Buot •
Rajasekaran Chandrasekaran
Editors

Plant Genetic Resources, Inventory, Collection and Conservation

Springer

Editors
Siva Ramamoorthy
Department of Biotechnology
Vellore Institute of Technology
Vellore, India

Inocencio Jr. Buot
Institute of Biological Sciences
University of the Philippines Los Baños
Laguna, Philippines

Rajasekaran Chandrasekaran
Department of Biotechnology
Vellore Institute of Technology
Vellore, India

ISBN 978-981-16-7698-7 ISBN 978-981-16-7699-4 (eBook)
https://doi.org/10.1007/978-981-16-7699-4

This Springer imprint is published by the registered company Springer Nature Singapore Pte Ltd.
The registered company address is: 152 Beach Road, #21-01/04 Gateway East, Singapore 189721, Singapore

Preface

Plant genetic resources (PGR) are the potential sources for farming, experimenting, and conserving plants. The conservation and effective utilization of PGR have been documented centuries ago. The changing socioeconomic situation in the early 1900s has increasingly put pressure on plant biodiversity, in some cases leading to a severe loss of many native species. Thus, the crisis impacted by genetic erosion instigated global conservation schemes in the form of forest laws and regulations to protect biodiversity. Still, there are several discrepancies in global- and national-level conservation policies that require the pressing attention of authorities. The fruitful employment of advanced biotechnological approaches will undeniably aid the conservation of PGR, which will be an immense treasure for the development of our society and the socioeconomic status of mankind. It is also imperative to safeguard the indigenous plant varieties and the knowledge of many tribal groups utilizing them.

The book *Plant Genetic Resources, Inventory, Collection and Conservation* is designed considering the magnitude of PGR conservation as well as the most preferable approaches in doing the same. The book discusses the global and regional plant biodiversity, in situ, ex situ conservation strategies, biotechnological advances in conservation, bioprospecting, notable plant conservation case studies, and conservation practices followed by various tribal communities, and global and national plant conservation policies. Particular significance is given to adding chapters on medicinal PGR conservation. Concurrently, the conservation, research, and utilization and natural wild varieties of rice, an important crop plant, are also included. Besides, a chapter on dye-yielding plants is also incorporated, making the book exceptional. Thus, the readers can access and analyze the status of genetic resources conservation of different genera of plants. Conservation case studies from different geographical domains like the Philippines, Thailand, Bangladesh, South Africa, and India are also included to demarcate the conservation policies in different parts of the world. A chapter solely on the plant diversity of Mount Makiling Forest Reserve in the Philippines illustrates the conservation policies practiced in vast forest ecosystems and points to the devastating human interference in the protection of natural resources. Likewise, the book includes a chapter on how the rare forest genetic resource in Bangladesh was explored, identified, multiplied, and safeguarded. Besides, exceptional case studies on various in situ and ex situ conservation strategies for PGR are also comprised as chapters. Furthermore, educating the young minds on the gravity of PGR is a requisite, and hence a critical analysis of the role of education on PGR conservation is also added as part of the book.

Advances in science and technology are well explored in conservation biology also. Therefore, a chapter on the utilization of geographic information system (GIS) on conservation and management of PGR for the identification of conservation sites, development of accession explorations and collection plans, and georeferencing of germplasm in gene banks is also added. Besides, the efficient amalgamation of advanced biotechnological approaches with orthodox conservation practices is also discussed in various chapters. Chapters critically reviewing the policy frameworks associated with PGR are another highlight of the book. Timely renewal of policies and their strict application considering the resources in developing and developed countries are a requisite for maintaining and implementing efficient PGR conservation globally.

The book is written and edited by leading experts in the area of ecology, conservation biodiversity, ethnobotanicals, and bioprospecting aspects of PGR. The book is intended for use by graduate and advanced undergraduate students and researchers in plant physiology, molecular biology, biochemistry, and agriculture and will also be extremely useful as a general reference for the conservation of flora and large-scale cultivation.

The authors are thankful to Vellore Institute of Technology management for their constant support. We sincerely hope that the book will be a pertinent reference for all those who are interested in conservation biology.

Vellore, India Siva Ramamoorthy
Laguna, Philippines Inocencio Jr. Buot
Vellore, India Rajasekaran Chandrasekaran

Contents

Editors and Contributors

About the Editors

Siva Ramamoorthy is Professor and Dean of School of Biosciences and Technology at Vellore Institute of Technology, India. His PhD thesis was on *plant genetic diversity*, which was completed at Bharathidasan University, India. He completed postdoctoral research at Ben Gurion University (Israel) and Gyeongsang National University (South Korea). For around two decades, he has carried out extensive research on plant dyes and pigments. Dr. Siva is an IUCN species commission member for medicinal plants. He was nominated as Fellow of Linnean Society (FLS), Fellow of the Royal Society of Biology (FRSB), and Fellow of the Royal Society of Chemistry (FRSC), and he has published around 150 manuscripts in reputed journals with high impact factor. He has received many awards for his research contribution. He is the associate and academic editor of various reputed international journals. Dr. Siva has been recognized as one among the top 2% scientists of the world as per survey of a leading publisher in 2021 Stanford University and Mendeley in 2020.

Inocencio Jr. Buot is Professor of Botany and curator in the Plant Biology Division Herbarium (PBDH) and current head of the Plant Systematics Laboratory at the Institute of Biological Sciences (IBS), a Center of Excellence for Biology since 1984, University of the Philippines Los Banos (UPLB). He obtained his PhD in botany (ecology) from the University of Chiba, Japan, in 1998 through a Monbusho Scholarship of the Japan Government and a postdoctoral research training in landscape ecology from 2003–2005 through the auspices of the Japan Society for the Promotion of Science (JSPS). His laboratory is now focusing on biodiversity conservation and plant systematic studies. He has published more than 150 articles in journals. He has received many prestigious awards for his research contribution.

Rajasekaran Chandrasekaran is serving as a full-time professor in the Department of Biotechnology, Vellore Institute of Technology. He graduated in botany at the University of Madras and Annamalai University. His obtained his doctoral degree from H.N.B. Grahwal University, specified with eco plant physiology. His research focuses on ecophysiology, conservation biodiversity, bioprospecting of medicinal and aromatic plants, phycoremediation, and climate change indicators. He has authored 77 manuscripts in high-impact factor journals and edited 3 books and contributed 22 book chapters. He has been nominated as a Fellow of Linnean Society (FLS), London, and elected as a Fellow of the Royal Society of Biology (UK), Fellow of Indian Society for Plant Physiology (FISPP) Fellow of Academy of Sciences Chennai (FASCh). He is serving as commission member of many theme groups of International Union for Conservation of Nature (IUCN). He is the recipient of many awards.

Contributors

Mohammed Iqram Uddin Al Amran International Union for Conservation of Nature, Dhaka, Bangladesh

Mariadoss Alphonse Department of Biotechnology, School of Bio Sciences and Technology, Vellore Institute of Technology, Vellore, Tamilnadu, India

Nestor C. Altoveros Institute of Crop Science, College of Agriculture and Food Science, University of the Philippines Los Baños, College, Los Baños, Laguna, Philippines

K. Anitha ICAR-National Bureau of Plant Genetic Resources, Regional Station, Hyderabad, India

Augusto A. Asis, Jr Puerto Princesa Subterranean River National Park Management Office, Puerto Princesa City, Palawan, Philippines

Ricardo T. Bagarinao University of the Philippines Open University, Los Baños, Laguna, Philippines

Maria Celeste N. Banaticla-Hilario International Rice Genebank, International Rice Research Institute, Los Baños, Laguna, Philippines

Ponnusamy Baskaran Biotechnology, Dami OPRS, NBPOL, Dami, Kimbe, Papua New Guinea

Janele Ann C. Belegal University of the Philippines Open University, Los Baños, Philippines

Teresita H. Borromeo Institute of Crop Science, College of Agriculture and Food Science, University of the Philippines Los Baños, College, Los Baños, Laguna, Philippines

Inocencio Jr. Buot Plant Biology Division, Institute of Biological Sciences, College of Arts and Sciences, University of the Philippines Los Baños College, Los Baños, Laguna, Philippines

Mbaoji Camillus Nigeria Natural Medicine Development Agency, Victoria Island, Lagos, Nigeria
School of Clinical Research, Texila American University (TAU), Georgetown, Guyana

Christian R. Casio Plant Biology Division, Institute of Biological Sciences, University of the Philippines Los Baños, Los Baños, Laguna, Philippines

Cleofas R. Cervancia Institute of Biological Sciences, University of the Philippines Los Baños College, Los Baños, Laguna, Philippines

Rajasekaran Chandrasekaran Department of Biotechnology, School of Bio Sciences and Technology, Vellore Institute of Technology, Vellore, Tamilnadu, India

Suchitra Changtragoon Department of National Parks, Wildlife and Plant Conservation, Forest and Plant Conservation Research Office, Bangkok, Thailand

Ajaegbu Henry Chukwudi Nigeria Natural Medicine Development Agency, Victoria Island, Lagos, Nigeria

April Karla N. Conde Plant Biology Division, Institute of Biological Sciences, University of the Philippines Los Baños, Los Baños, Laguna, Philippines

Hidelisa D. de Chavez Institute of Plant Breeding, College of Agriculture and Food Science, University of the Philippines Los Baños, College, Los Baños, Laguna, Philippines

Marjorie D. Delos Angeles Plant Biology Division, Institute of Biological Sciences, University of the Philippines Los Baños College, Los Baños, Laguna, Philippines

Jonathan C. Descalsota National Plant Genetic Resources Laboratory, Institute of Plant Breeding, College of Agriculture and Food Science, University of the Philippines, Los Baños, Laguna, Philippines

Michelle Lyka V. Descalsota National Plant Genetic Resources Laboratory, Institute of Plant Breeding, College of Agriculture and Food Science, University of the Philippines, Los Baños, Laguna, Philippines

Raymundo Dong-e Poblation, Hungduan, Philippines

C. George Priya Doss School of Bio Sciences and Technology, Vellore Institute of Technology, Vellore, India

Leizel M. Estoque Plant Biology Division, Institute of Biological Sciences, University of the Philippines Los Baños, Los Baños, Laguna, Philippines

Alexander W. Fagyan Soil Science Department, Benguet State University, La Trinidad, Benguet, Philippines

Jabez Joshua M. Flores School of Environmental Science and Management, University of the Philippines Los Baños, Los Baños, Laguna, Philippines

Adrian G. Gandecilla Plant Biology Division, Institute of Biological Sciences, University of the Philippines Los Baños, Los Baños, Laguna, Philippines

Mrinalini Goswami Centre for Ecological Economics and Natural Resources (CEENR), Institute for Social and Economic Change (ISEC), Nagarabhavi, Bengaluru, India

Lavernee S. Gueco National Plant Genetic Resources Laboratory, Institute of Plant Breeding, College of Agriculture and Food Science, University of the Philippines, Los Baños, Laguna, Philippines

Consuelo DI Habito Faculty of Management and Development Studies, University of the Philippines Open University, Los Baños, Laguna, Philippines

Annalee S. Hadsall Institute of Biological Sciences, University of the Philippines Los Baños College, Los Baños, Laguna, Philippines

Md. Akhter Hossain Institute of Forestry and Environmental Sciences, University of Chittagong, Chittagong, Bangladesh

Mohammad Mosharraf Hossain Institute of Forestry and Environmental Sciences, University of Chittagong, Chittagong, Bangladesh

Mohammed Kamal Hossain Institute of Forestry and Environmental Sciences, University of Chittagong, Chittagong, Bangladesh

Ifeoluwa Margaret Idowu Nigeria Natural Medicine Development Agency, Victoria Island, Lagos, Nigeria
Tianjin University of Traditional Chinese Medicine (TUTCM), Tianjin, China

Luke Michael Ifreke Nigeria Natural Medicine Development Agency, Victoria Island, Lagos, Nigeria

Shalan Joseph E. Kitma Biostadt Philippines, Inc., Ortigas Center, Pasig, Metro Manila, Philippines

Ivy Amor F. Lambio Institute of Biological Sciences, University of the Philippines Los Baños College, Los Baños, Laguna, Philippines

Jeoffrey M. Laruya Institute of Biological Sciences, University of the Philippines Los Baños College, Los Baños, Laguna, Philippines

Lerato N. Madike Department of Biotechnology, Vaal University of Technology, Vanderbijlpark, South Africa

Damasa B. Magcale-Macandog Institute of Biological Sciences, University of the Philippines Los Baños College, Los Baños, Laguna, Philippines

Pastor L. Malabrigo Jr Department of Forest Biological Sciences, College of Forestry and Natural Resources, University of Philippines Los Baños, Los Baños, Laguna, Philippines

Sherry B. Marasigan University of the Philippines Los Baños, Los Baños, Philippines

Rosalie C. Mendoza Department of Forest Products and Paper Science, College of Forestry and Natural Resources, University of the Philippines Los Baños, Los Baños, Laguna, Philippines

Mohammed Danesh Miah Institute of Forestry and Environmental Sciences, University of Chittagong, Chittagong, Bangladesh

Roger Vincent B. Nabua Plant Biology Division, Institute of Biological Sciences, University of the Philippines Los Baños, Los Baños, Laguna, Philippines

Kanagam Nachiappan Department of Biotechnology, School of Bio Sciences and Technology, Vellore Institute of Technology, Vellore, Tamilnadu, India
Department of Biotechnology, Sri Venkateswara College of Engineering, Sriperumbudur, Kanchipuram, Tamilnadu, India

Subashen Naidu Microscopy and Microanalysis Unit (MMU), University of KwaZulu-Natal, Pietermaritzburg, Scottsville, South Africa

Nagaraj Nallakaruppan Department of Biotechnology, School of Bio Sciences and Technology, Vellore Institute of Technology, Vellore, Tamilnadu, India

Sunil Nautiyal Centre for Ecological Economics and Natural Resources (CEENR), Institute for Social and Economic Change (ISEC), Nagarabhavi, Bengaluru, India

Aurfeli D. Nietes Plant Biology Division, Institute of Biological Sciences, University of the Philippines Los Baños, Los Baños, Laguna, Philippines

Olaniyan Kayode Olaoluwa Nigeria Natural Medicine Development Agency, Victoria Island, Lagos, Nigeria
Department of Veterinary Parasitology and Entomology, Federal University of Agriculture Abeoakuta, Abeokuta, Nigeria

Emmanuel Adikwu Orgah Nigeria Natural Medicine Development Agency, Victoria Island, Lagos, Nigeria
School of Clinical Research, Texila American University (TAU), Georgetown, Guyana

Joyce N. Paing Biology Department, Benguet State University, La Trinidad, Benguet, Philippines

Vachel Gay V. Paller Institute of Biological Sciences, University of the Philippines Los Baños College, Los Baños, Laguna, Philippines

S. R. Pandravada ICAR-National Bureau of Plant Genetic Resources, Regional Station, Hyderabad, India

Michael Pillay Department of Biotechnology, Vaal University of Technology, Vanderbijlpark, South Africa

Abul Hayat Poyal Institute of Forestry and Environmental Sciences, University of Chittagong, Chittagong, Bangladesh

Sarena Grace L. Quiñones Institute of Biological Sciences, University of the Philippines Los Baños College, Los Baños, Laguna, Philippines

Md. Oliur Rahman Department of Botany, University of Dhaka, Dhaka, Bangladesh

Chandrasekaran Rajasekaran School of Bio Sciences and Technology, Vellore Institute of Technology, Vellore, India

Siva Ramamoorthy Department of Biotechnology, School of Bio Sciences and Technology, Vellore Institute of Technology, Vellore, Tamilnadu, India

Jananipriya Rameshbabu School of Social Sciences and Languages, Vellore Institute of Technology, Vellore, India

Disna Ratnasekera Department of Agricultural Biology, Faculty of Agriculture, University of Ruhuna, Matara, Sri Lanka

A. Gomez Romeo Jr. Biology Department, Benguet State University, La Trinidad, Benguet, Philippines
Office of the Vice President for Research and Extension, Benguet State University, La Trinidad, Benguet, Philippines

Esther Josephine D. Sagalla Agronomy Department, Benguet State University, La Trinidad, Benguet, Philippines

Mohammed Saifullah Bangladesh Agriculture Research Council, Farmgate, Dhaka, Bangladesh

Andres Godwin Sajise International Rice Genebank, International Rice Research Institute, Los Baños, Laguna, Philippines

Salinda Sandamal Department of Agricultural Biology, Faculty of Agriculture, University of Ruhuna, Matara, Sri Lanka
State Key Laboratory of Systematic and Evolutionary Botany, Institute of Botany, Chinese Academy of Sciences, Beijing, China
University of Chinese Academy of Sciences, Beijing, China

R. Seenivasan Center for Nanotechnology, Vellore Institute of Technology, Vellore, India

Manoj Sekaran Department of Biotechnology, School of Bio Sciences and Technology, Vellore Institute of Technology, Vellore, Tamilnadu, India

T. Senthilkumar Department of Plant Science, Bharathidasan University, Tiruchirappalli, India

Joane V. Serrano University of the Philippines Open University, Los Baños, Philippines

Kittiya Singthong Department of National Parks, Wildlife and Plant Conservation, Forest and Plant Conservation Research Office, Bangkok, Thailand

Lorna E. Sister Institute of Crop Science, College of Agriculture and Food Science, University of the Philippines Los Baños, College, Los Baños, Laguna, Philippines

N. Sivaraj ICAR-National Bureau of Plant Genetic Resources, Regional Station, Hyderabad, India

Patrick A. Sodusta Plant Biology Division, Institute of Biological Sciences, University of the Philippines Los Baños, Los Baños, Laguna, Philippines

Rachel C. Sotto Institute of Biological Sciences, College of Arts and Sciences, University of the Philippines, Los Baños, Laguna, Philippines

Usha Swaminathan School of Social Sciences and Languages, Vellore Institute of Technology, Vellore, India

Belinda A. Tad-awan Agronomy Department, Benguet State University, La Trinidad, Benguet, Philippines
Office for Research Services, Benguet State University, La Trinidad, Benguet, Philippines

Samkeliso Takaidza Department of Biotechnology, Vaal University of Technology, Vanderbijlpark, South Africa

Maila S. Tejano National Plant Genetic Resources Laboratory, Institute of Plant Breeding, College of Agriculture and Food Science, University of the Philippines, Los Baños, Laguna, Philippines

Mabeth F. Tejida Plant Biology Division, Institute of Biological Sciences, University of the Philippines Los Baños, Los Baños, Laguna, Philippines

Asanka Tennakoon Department of Agricultural Biology, Faculty of Agriculture, University of Ruhuna, Matara, Sri Lanka

Unisa Terblanche Department of Biotechnology, Vaal University of Technology, Vanderbijlpark, South Africa

Kalaivani Thaigarajan Department of Biotechnology, School of Bio Sciences and Technology, Vellore Institute of Technology, Vellore, Tamilnadu, India

Nerissa K. Torreta Institute of Biological Sciences, University of the Philippines Los Baños College, Los Baños, Laguna, Philippines

Chimezie Esther Uchechukwu Nigeria Natural Medicine Development Agency, Victoria Island, Lagos, Nigeria
School of Clinical Research, Texila American University (TAU), Georgetown, Guyana

Johannes Van Staden Research Centre for Plant Growth and Development, School of Life Sciences, University of KwaZulu-Natal, Pietermaritzburg, Scottsville, South Africa

Ressin Varghese School of Bio Sciences and Technology, Vellore Institute of Technology, Vellore, India

Chandra Veluchamy Department of Biotechnology, School of Bio Sciences and Technology, Vellore Institute of Technology, Vellore, Tamilnadu, India

Kamala Venkateswaran ICAR-National Bureau of Plant Genetic Resources, Regional Station, Hyderabad, India

Jose Dale L. Viacrucis III Plant Biology Division, Institute of Biological Sciences, University of the Philippines Los Baños, Los Baños, Laguna, Philippines

Parakkrama Wijerathna Department of Agricultural Biology, Faculty of Agriculture, University of Ruhuna, Matara, Sri Lanka

Lyka A. Yanos National Plant Genetic Resources Laboratory, Institute of Plant Breeding, College of Agriculture and Food Science, University of the Philippines, Los Baños, Laguna, Philippines

A Checklist of Some Economically Important Philippine Ferns (Eupolypod I)

1

Aurfeli D. Nietes, Marjorie D. Delos Angeles, Christian R. Casio, Jose Dale L. Viacrucis, III, April Karla N. Conde, Leizel M. Estoque, Mabeth F. Tejida, Adrian G. Gandecilla, Roger Vincent B. Nabua, Patrick A. Sodusta, and Inocencio Jr. Buot

Abstract

This checklist provides an updated reference for the economically important ferns that can be used by stakeholders both from the community and academe such as consumers, hobbyists, students, teachers, and researchers alike. Available comprehensive online and printed materials regarding the economic uses of ferns classified under Eupolypod I were used. Furthermore, this checklist also notes the conservation status of these select ferns as categorized by the IUCN Red List and/or DAO 2017–11. A total of 40 fern species representing 20 genera and 6 families categorized under Eupolypod I were recorded to have various economical uses. Twenty-five species were ornamentals, 19 species were used as medicinals, and 6 species were edible. Five fern species were listed as threatened/endangered in the DAO 2017–11. The Philippine endemic *Platycerium grande* is categorized as Critically Endangered due to its increased horticultural value. Data derived from this paper can be a good informative guide to stakeholders on the economic uses of some ferns and stimulate future research, conservation, and management practices towards the protection of these economically important ferns.

Keywords

Eupolypod · Economic uses · Conservation · Philippine ferns

A. D. Nietes (✉) · M. D. Delos Angeles · C. R. Casio · J. D. L. Viacrucis, III · A. K. N. Conde · L. M. Estoque · M. F. Tejida · A. G. Gandecilla · R. V. B. Nabua · P. A. Sodusta · I. J. Buot
Plant Biology Division, Institute of Biological Sciences, University of the Philippines Los Baños, Los Baños, Laguna, Philippines
e-mail: adnietes@up.edu.ph; mddelosangeles1@up.edu.ph; jlviacrucis@up.edu.ph; anconde@up.edu.ph; lmestoque@up.edu.ph; mftejida@up.edu.ph; aggandecila@up.edu.ph; rbnabua@up.edu.ph; pasodusta@up.edu.ph; iebuot@up.edu.ph

S. Ramamoorthy et al. (eds.), *Plant Genetic Resources, Inventory, Collection and Conservation*, https://doi.org/10.1007/978-981-16-7699-4_1

1.1 Introduction

Eupolypod I is the largest clade of leptosporangiate fern with approximately 4208 species classified under this taxon (Rothfels et al. 2012; PPG 2016; Tan and Buot Jr 2019). This group is made up of nine families (Didymochlaenaceae, Hypodematiaceae, Davalliaceae, Nephrolepidaceae, Lomariopsidaceae, Tectariaceae, Oleandraceae, Dryopteridaceae, and Polypodiaceae) (Smith et al. 2006; PPG 2016) with great variation among their morphologies but these were grouped together based on their shared characteristic, which is the numerous vascular bundles observed in the stipe, thus delineating it from its sister clade—Eupolypod II (Rothfels et al. 2012).

Most of the species under Eupolypod I were noted to have economic importance (de Winter and Amoroso 2003) as these ferns became a vital source of food, raw materials for handicrafts, ornamentals, and most importantly medicine (McCalla 1994; Smith et al. 2006; Schippmann et al. 2002). Over the years, ferns have provided humans with these services, thus establishing the important impact of pteridophytes in society development.

But, despite these economic values, there was scarcity of published articles made to extensively record pteridophytes for their economic importance. In the Philippines, there were only few works that featured this taxon's commercial value (Delos Angeles and Buot 2012; de Winter and Amoroso 2003; Amoroso et al. 2015; Mustacisa 2016) and the information among these articles was not consolidated; thus, the full range of the economic relevance of ferns is still hard to assess. Due to these, there is a need of a comprehensive data gathering of the economic importance of ferns and consolidation of such data. Thus, this study aims to evaluate literature and collect information on the specific uses of ferns to provide a robust checklist of economically important Eupolypod I species in the country.

1.2 Materials and Methodology

A comprehensive review of published literature online and in print was conducted regarding economic importance and the current and underlying conservation status of Philippine fern species classified under Eupolypod I. Copeland's Fern Flora of the Philippines Vol. 1, 2, and 3 (1958-1960); published monographs; and online resources such as "Co's Digital Flora of the Philippines" (Pelser et al. 2011) were consulted for the distribution and accepted names of the fern species. The endemic status of each species was verified by consulting available checklists, related publications, and online resources.

1.3 Results and Discussion

A total of 40 fern species representing 20 genera and 6 families categorized under Eupolypod I were recorded to have various economic uses (Table 1.1). The most represented fern families were Polypodiaceae (21) followed by Davalliaceae (5) and Dryopteridaceae (3). The families Hypodematiaceae, Nephrolepidaceae, and Tectariaceae are represented by three fern species.

Among these ferns, 25 species were used as ornamentals as majority of fern species were known to be of aesthetic purpose (Zamora and Co 1986; Buot Jr 1999; Banaticla and Buot Jr. 2006; Delos Angeles and Buot 2012; Shukla and Khare 2014) due to their unique frond and spore patterns. With the remarkable forms of their foliage and habit (Jones 1997; Ranil et al. 2015), and their ability to adapt to a wide variety of ecology (Fernández and Revilla 2003; Dong et al. 2009), they are considered as valuable components of urban landscaping and rehabilitation of damaged environs (Ranil et al. 2015; Dong et al. 2009). They have also become a subject of domestication as they are applied to the floriculture industry, where ferns are propagated and cultivated not just outdoor but also in indoor settings as these plants are capable of purifying the air (Ranil et al. 2015). Nineteen species were also noted to be medicinal and six were edible ferns. For many years, numerous ethnobotanical and pharmacological studies were conducted on these fern flora as they have high potential for medicinal uses (de Winter and Amoroso 2003). Almost every part of ferns is an abundant source of phytochemicals which make ferns a great prospect for drug discovery.

Some species under the family Polypodiaceae are commercially known as high-priced ornamentals such as *P. coronarium* and the Philippine endemic *P. grande* (Zamora and Co 1986; Aspiras 2010; Amoroso et al. 2015; Amoroso et al. 2015). However, these species of polypods were also assessed as threatened and even *P. coronarium* and *P. grande* were specifically recorded as critically endangered indicating the declining population of this taxon in the wild (Fernando et al. 2008; Aribal and Buot Jr 2009). The recent hype of horticultural activities also resulted for these ferns to be listed as one of the top ten commonly poached plants, regardless of its threatened status (DENR Soccsksargen 2020). This certainly calls for the need of legal protection of these traded species (Flores-Palacios and Valencia-Diaz 2007). There were policies embodied in the Philippine Republic Act No. 4167 also known as "Wildlife Resources Conservation and Protection Act" established by the government to stop illegal trading and ensure protection of these wild plants. However, such dilemma in illegal exploitation undeniably persists until now mostly caused by the lack of awareness of the local public of the population status of the species and the little to no understanding on the environmental and legal consequences entailed with these illegal acts (Dadang et al. 2020). The lack of knowledge of the status of these taxa cripples the society on taking its part in conserving the species. As a consumerist society, humans may have been reliant on plants to survive; however, humans also have to take on their role in promoting protection of the environment. It is essential to practice sustainable growth without disregarding the economy, social development, and protection of the existing natural resources (UN 2018).

Table 1.1 List of Philippine economically important Eupolypod I species (sensu PPG 2016)

Family/species	Common name	Medicinal uses				References
		Ornamental	Medicinal	Edible	Raw material	
Davalliaceae						
Davallia denticulata (Burm. f.) Mett. ex Kuhn var. elata (G. Forst) Kuhn	No recorded common name	*				Yatskievych (2021)
D. embolostegia Copel.	Hare's foot fern	*				Yatskievych (2021)
D. hymenophylloides (Blume) Kuhn	No recorded common name	*				Alcala et al. (2019), Yatskievych (2021)
D. lorrainii Hance	No recorded common name	*				Yatskievych (2021)
D. solida (G. Forst.) Sw. var. solida	Rabbit foot fern	*				Yatskievych (2021)
Dryopteridaceae						
Arachniodes amabilis (Blume)	No recorded common name	*				Ranil et al. (2015)
Bolbitis heteroclita (C. Presl)	El Niño Fern	*				Balkrishna et al. (2019)
B. rhizophylla (Kaulf.) Hennipman	Pako	*				MBCF (2012)
Cyrtomium caryotideum (wall. ex Hook. and Grev.) C. Presl	Dwarf netvein hollyfern		*			Balkrishna et al. (2019)
C. falcatum (L.f.) C. Presl	Holly fern, Japanese netvein hollyfern, Japanese holly	*	*			Gul et al. (2016)
Hypodematiaceae						
Didymochlaena truncatula (Sw.)	Tree maidenhair fern	*				Roux (2003)

Hypodematium crenatum (Forssk.)	No recorded common name		*	*		Singh and Singh (2012), Sathiyaraj et al. (2015), Gul et al. (2016)
Leucostegia truncata (D. Don) Fras-Jenk	No recorded common name		*	*		Dudani et al. (2013)
Nephrolepidaceae						
Nephrolepis biserrata (Sw.) Schott	Broad sword fernAlolokdo (Philippines)		*	*		de Winter and Amoroso (2003)
N. cordifolia (L.) C. Presl	Erect sword fern Bayabang (Philippines)	*	*			de Winter and Amoroso (2003)
N. hirsutula (G. Forst.) C. Presl	No recorded common name		*	*	*	de Winter and Amoroso (2003)
Polypodiaceae						de Winter and Amoroso (2003)
Aglaomorpha quercifolia (L.) Hovenkamp and *S. Linds*	No recorded common name		*			Costa et al. (2021)
A. rigidula (Sw.) Hovenkamp and S. Linds.	No recorded common name	*	*			de Winter and Amoroso (2003)
A. sparsisora (Desv.) Hovenkamp and S. Linds.	No recorded common name	*	*			de Winter and Amoroso (2003)
Dendroconche linguiforme (Mett.) Testo, Sundue and Field	No recorded common name	*				de Winter and Amoroso (2003)
Goniophlebium benguetense (Copel.) Copel.	(an-) anam-am; Benguet fern					Reid and Madulid (1972)
Lemmaphyllum carnosum (J. Sm. ex Hook.) C. Presl	No recorded common name		*			Mannan et al. (2008)
Leptochilus longissimus (Blume) L.Y. Kuo	No recorded common name	*				de Winter and Amoroso (2003)

(continued)

Table 1.1 (continued)

Family/species	Common name	Medicinal uses				References
		Ornamental	Medicinal	Edible	Raw material	
Loxogramme avenia (Blume.) C. Presl	No recorded common name	*				Boonkerd and De Winter (2003)
L. involuta (D. Don) C. Presl	No recorded common name				*	Boonkerd and De Winter (2003)
L. scolopendrioides (Gaudich.) C.V. Morton	No recorded common name	*			*	de Winter and Amoroso (2003)
L. parallela Copel	No recorded common name	*				Janakiram et al. (2019)
Microsorum punctatum (L.) Copel.	No recorded common name	*	*			de Winter and Amoroso (2003)
Phymatosorus commutatus (Blume) Pic.-Serm.	No recorded common name	*				de Winter and Amoroso (2003)
P. membranifolius (R.Br.) S. G. Lu	No recorded common name					de Winter and Amoroso (2003)
P. scolopendria (Burm. f.) Pic.-Serm	No recorded common name	*	*			de Winter and Amoroso (2003), Stuart (2018)
Platycerium coronarium (König ex Müller) Desv.	No recorded common name	*		*		Amoroso et al. (2015)
P. grande (A. Cunn. ex Fée) J. Sm.	No recorded common name	*				Amoroso et al. (2015)
Pyrrosia lanceolata (L.) Farw	Humang anapatpat; Holog; Apatpat an dodologapdi (Luzon, Philippines)		*			de Winter and Amoroso (2003)
Pyrrosia longifolia (Burm. f.) C.V. Morton	No recorded common name		*			de Winter and Amoroso (2003)
P. nummularifolia (Sw.) Ching	No recorded common name		*			de Winter and Amoroso (2003)

P. piloselloides (L.) M.G. Price	No recorded common name		*			de Winter and Amoroso (2003)
Tectariaceae						
Pleocnemia irregularis (C. Presl) Holttum	No recorded common name	*	*			Darnaedi and Ngatinem-Praptosuwiryo (2003)
Tectaria crenata Cav.	No recorded common name	*	*	*		Rusea and Aguilar (2003)
T. griffithii (Baker) C. Chr.	No recorded common name	*				de Winter and Amoroso (2003)

It can be noted that the IUCN and the Philippines National Red List based on DAO 2017–11 (2021) have differences on the designation of conservation status. Five were listed as threatened/endangered in the DAO 2017–11 (2021). However, these species have not been evaluated in the IUCN yet (Table 1.2). The assessment between these two units varies since the DENR is localized in the Philippines, and hence, the list would represent the status of the population of such species in the country, while that of IUCN is a global assessment that considers and depends on all available data from various countries, to assess the status of the species. Without enough data on the population of the species, the IUCN could not evaluate the conservation status of such taxa.

Several conservation efforts were done and promoted by the government, academes, and NGOs to protect and conserve these floras: from conserving its natural habitat and on-site species protection (in situ) to modern practices of ex situ conservation such as fernery establishment, spore and vegetative propagation, and tissue cultures (Barnicoat et al. 2011; Mehltreter et al. 2010). All of these are aimed at restoring fern population and avoiding its possible extinction. But among these conservation strategies, in situ also known as on-site conservation of species and habitat is highly promoted and should be the first option in the conservation of fern. This is to ensure prevention of species from becoming threatened in the wild (Mehltreter et al. 2010). With this goal in mind, this would accentuate the important role of the local people specifically indigenous people (IPs) in conservation, as they are the ones interacting with these ferns in its natural habitat. These people should be educated of the sustainable utilization of these floras and should be taught of ways on how to actively practice protection and conservation of the natural resources they have at their territory (Sobrevila 2008). The role of conservation should be a collective effort and should start from the local level as represented by the IPs. Collaboration and partnership with the IP should be promoted and stakeholders from the local government, academe, and NGOs may make it their goal to enhance capability-building efforts for the indigenous people (IPs) to help them become more engaged, informed, and equipped for biodiversity management and conservation.

1.4 Conclusion and Recommendation

Several fern species of Eupolypod I are economically important as they render great service to the society. Some of their economic values include but are not restricted to ornamentals, food and nutrition, and being great sources of plant active components essential to ethnobotany and pharmacology. Due to these human services, ferns are constantly exploited which has led to the decline of its numbers in the wild. With this, it is essential to practice conservation and start implementing guidelines and management strategies towards the protection of these fern floras. Collaborative efforts towards biodiversity management and conservation should be materialized and it is recommended that the partnership of local government, academe, and NGOs with the indigenous people (IPs) should be enhanced through education and

Table 1.2 Economically important Philippine Eupolypod I fern species with available conservation status from the IUCN and/or DAO 2017–11

Family	Scientific name	Conservation status	
		IUCN status	DAO 2017–11
Davalliaceae			
	Davallia denticulata (Burm. f.) Mett. ex Kuhn var. elata (G. Forst) Kuhn	NE	OTS
	D. embolostegia Copel.	NE	OTS
	D. hymenophylloides (Blume) Kuhn	NE	–
	D. lorrainii Hance	NE	OTS
	D. solida (G. Forst.) Sw. var. *solida*	NE	OTS
Dryopteridaceae			
	Arachniodes amabilis (Blume)	NE	–
	B. heteroclita (C. Presl)	NE	–
	B. rhizophylla (Kaulf.) Hennipman	NE	–
	Cyrtomium caryotideum (Wall. ex Hook. and Grev.) C. Presl	NE	–
	C. falcatum (L.f.) C. Presl	–	–
Hypodematiaceae			
	Didymochlaena truncatula (Sw.)	LC	–
	Hypodematium crenatum (Forssk.)	NE	–
	Leucostegia truncata (D. Don) Fras-Jenk	–	–
Nephrolepidaceae			
	Nephrolepis biserrata (Sw.) Schott	NE	–
	N. cordifolia (L.) C. Presl	NE	–
	N. hirsutula (G. Forst.) C. Presl	NE	–
Polypodiaceae			
	Aglaomorpha quercifolia (L.) Hovenkamp and *S. Linds.*	NE	–
	A. rigidula (Sw.) Hovenkamp and S. Linds.	–	–
	A. sparsisora (Desv.) Hovenkamp and S. Linds.	–	–
	Dendroconche linguiforme (Mett.) Testo, Sundue and Field	–	–
	Goniophlebium benguetense (Copel.) Copel.	NE	–
	Lemmaphyllum carnosum (J.Sm. ex Hook.) C. Presl	NE	–
	Leptochilus longissimus (Blume) L.Y. Kuo	–	–
	Loxogramme avenia (Blume.) C. Presl	–	
	L. involuta (D. Don) C. Presl	–	
	L. scolopendrioides (Gaudich.) C.V. Morton	NE	–
	L. parallela Copel.		
	Microsorum punctatum (L.) Copel.	NE	–
	Phymatosorus commutatus (Blume) Pic.-Serm., Webbia	NE	–
	P. membranifolius (R.Br.) S.G. Lu	–	–

(continued)

Table 1.2 (continued)

Family	Scientific name	Conservation status	
		IUCN status	DAO 2017–11
	P. scolopendria Pic.-Serm.	NE	–
	P. coronarium (König ex Müller) Desv.	NE	CE
	P. grande (A. Cunn. ex Fée) J.Sm.[a]	–	CE
	Pyrrosia lanceolata (L.) Farw	–	–
	P. longifolia (Burm. f.) C.V. Morton	NE	–
	P. nummularifolia (Sw.) Ching	NE	–
	P. piloselloides (L.) M.G. Price	NE	–
Tectariaceae			
	Pleocnemia irregularis (C. Presl) Holttum	NE	–
	Tectaria crenata Cav.	NE	–
	T. griffithii (Baker) C. Chr.	NE	–

Conservation status: *VU* vulnerable, *CE* critically endangered, *EN* endangered, *NT* near threatened, *OTS* other threated species, *LC* least concern, *NE* not evaluated.
[a]Philippine endemic.

capability building to ensure protection and conservation of these important floras at the local level. It is recommended however to consider a priority for in situ management strategies to prevent fern species population from declining in the wild, rather than trying to avoid species loss at the last moment. Fern species may be protected within their native range by conservation of their natural habitat (in situ) or by cultivation of individual species in botanical gardens or other horticultural institutions by either growing them from spores, vegetative propagation, or tissue culture (ex situ).

Future research can move forward with pharmacological and ethnobotanical studies on the medicinal uses of ferns and investigate aspects that a particular taxon shares. Also, the availability and conservation of edible varieties can boost food security along with other cultural benefits. It is also significant to ponder on the vulnerability and urgency of biodiversity loss affecting ferns and of deforestation in the Philippines.

Acknowledgements The authors would like to express their gratitude to the Institute of Biological Sciences of the University of the Philippines Los Baños. Furthermore, the authors would like to extend their appreciation to the Department of Science and Technology-Accelerated Science and Technology Human Resource Development Program (DOST-ASTHRDP) for awarding graduate scholarships to Aurfeli D. Nietes.

References

Alcala AA, Angeles MDD, Buot IE Jr (2019) Fern species diversity across various land use types of Mt. Makiling, Luzon Island. Philippines Biodivers: J Biol Divers 20(9):2437–2445

Amoroso VB, Chiou WL, Liu YC, Callado JRC, Briones NNG, Coritico FP (2015) Threatened endemic, rare and economic species of Philippine Pteridophytes. In: Instructional materials and development center. Central Mindanao University, Musuan, Bukidnon

Aribal LG, Buot IE Jr (2009) The threatened plant species in various regions in Mindanao Island, Philippines. J Nat Stud 8(2):23–33

Aspiras RA (2010) Sporophyte and gametophyte development of Platycerium coronarium (Koenig) Desv. and P. grande (Fee) C. Presl.(Polypodiaceae) through in vitro propagation. Saudi J Biol Sci 17(1):13–22

Balkrishna A, Joshi B, Srivastava A, Shankar R, Tiwari S, Tyagi SS, Mishra RK (2019) Some economic aspects of ferns and Fern-Allies of Seijosa Forest area of Pakke-Kessang District, Arunachal Pradesh. Int J Adv Res Bot 5(3):14–20

Banaticla MCN, Buot IE Jr (2006) Endemic ferns on the northeastern slopes of Mt. Banahaw de Lucban, Sierra Madre mountain range, southern Luzon, Philippines: potentials for domestication and landscaping. J Nat Stud 5:13–22

Barnicoat H, Cripps R, Kendon J, Sarasan V (2011) Conservation in vitro of rare and threatened ferns—case studies of biodiversity hotspot and island species. In Vitro Cellular Develop Biol-Plant 47(1):37–45

Boonkerd T, De Winter WP (2003) Loxogramme scolopendrina (Bory) C. Presl. In: de Winter WP, Amoroso VB (eds) Plant resources of South-East Asia no 15(2): cryptogams: ferns and fern allies. PROSEA Foundation, Bogor

Buot Jr IE (1999) Pteridophytes frequently sold at carbon market, Cebu [Philippines]: implications to urban horticulture and nature conservation, Philippine Scientist, Philippines. https://agris.fao.org/agris-search/search.do?recordID=PH2000101484. Accessed 17 Jan 2021

Copeland EB (1958–1960) Fern flora of the Philippines. Manila Bureau of Printing, Manila

Costa HR, Simão I, Silva H, Silveira P, Silva A, Pinto DC (2021) Aglaomorpha quercifolia (L.) Hovenkamp & S. Linds a wild fern used in Timorese cuisine. Foods 10(1):87

Dadang RJ, Casinillo NGB, Coritico FP, Simborio LT, Amoroso VB (2020) Living with endangered species: collection of tree ferns in the forest-reserve of Marilog district, Southern Philippines. Trees Forests People 2:100041

Darnaedi D, Ngatinem-Praptosuwiryo T (2003) Pleocnemia irregularis (C. Presl) Holttum. In: de Winter WP, Amoroso VB (eds) Plant resources of South-East Asia no. 15(2): cryptogams: ferns and fern allies. PROSEA Foundation, Bogor

de Winter WP, Amoroso VB (2003) Plant resources of South-East Asia no. 15(2). In: Cryptograms: fern and fern allies. Backhuys Publisher, Leiden

Delos Angeles M, Buot I (2012) Orders and families of Philippine pteridophytes. J Nat Stud 11(1&2):19–33

DENR Administrative Order (DAO) 2017–11 (2021) Updated National List of threatened Philippine plants and their categories. http://www.bmb.gov.ph/index.php/e-library/laws-and-policies/denr-administrative-orders/dao-2017-2019?download=197:denr-administrative-order-2017-11. Accessed 14 Jan 2021

DENR SOCCSKSARGEN (2020) Commonly poached threatened plants. https://www.facebook.com/denr12official/posts/2740421649527244?comment_id=2740432506192825. Accessed 17 Jan 2021

Dong L, Xu W, Yu L (2009) Investigation and evaluation of ornamental fern in Beijing area. Acta Horticul 813:351–357

Dudani SN, Mahesh MK, Mukri V, Subash MDC, Ramachandra TV (2013) An appraisal and conservation strategies for the pteridophytes of Uttara Kannada. In: CES technical report, 129. Bangalore, Energy & Wetlands Research Group Centre for Ecological Sciences, Indian Institute of Science

Fernández H, Revilla MA (2003) In vitro culture of ornamental ferns. Plant Cell Tiss Org Cult 73(1):1–13

Fernando ES, Co LC, Lagunzad DA, Gruezo WS, Barcelona JF, Madulid DA, Lapis AB, Texon GI, Manila AC, Zamora PM (2008) Threatened plants of the Philippines: a preliminary. Asia Life Sci 3:1–52

Flores-Palacios A, Valencia-Diaz S (2007) Local illegal trade reveals unknown diversity and involves a high species richness of wild vascular epiphytes. Biol Conserv 136(3):372–387

Gul A, Alam J, Ahmad H, Shah GM, Hussain M (2016) Traditional, medicinal and food uses of Pteridophytes of district Mansehra (Pakistan) and their some adjacent areas. Int J Biosci 9(5): 116–133

Janakiram T, Safeena SA, Prasad KV (2019) Status of indigenous ornamental plants in India. Indian Hort 64(4):11–20

Jones DL (1997) Encyclopaedia of ferns. In: An introduction to ferns, their structure, biology, economic importance, cultivation and propagation. Melbourne, Thomas C. Lothian Pty Ltd

Mannan MM, Maridass M, Victor B (2008) A review on the potential uses of ferns. Ethnobot Leafl 2008(1):33

McCalla AF (1994) Agriculture and food needs to 2025: why we should be concerned. Consultative group on international agricultural research, sir John Crawford memorial lecture. International centers week. https://cgspace.cgiar.org/handle/10947/159. Accessed 19 Jan 2021

Mehltreter K, Walker LR, Sharpe JM (eds) (2010) Fern ecology. Cambridge University Press, Cambridge

Mindoro Biodiversity Conservation Foundation (MBCF) (2012). http://database.mbcfi.org.ph/2012/03/08/field-survey-iv/. Accessed 19 Nov 2021

Mustacisa MM (2016) Diversity, uses and economic value of ferns: an instrument for epistemological perception. Int J Environ Sci Educ 11(18):13111–13146

Pelser PB, Barcelona JF, Nickrent DL (2011) Co's digital flora of the Philippines.-http://www.philippineplants.org. Accessed 19 Jan 2021

PPG I (2016) A community-derived classification for extant lycophytes and ferns. J Syst Evol 54(6):563–603

Ranil RHG, Beneragama CK, Pushpakumara DKNG, Wijesundara DSA (2015) Ornamental pteridophytes: an underexploited opportunity for the Sri Lankan floriculture industry. J Natl Sci Found 43(4)

Reid L, Madulid D (1972) Some comments on Bontoc ethnobotany. Phippine J Linguist:1–25

Rothfels CJ, Sundue MA, Kuo LY, Larsson A, Kato M, Schuettpelz E, Pryer KM (2012) A revised family–level classification for Eupolypod II ferns (Polypodiidae: Polypodiales). Taxon 61(3): 515–533

Roux JP (2003) Didymochlaena truncatula (Sw.) J.Sm. [family Dryopteridaceae]. https://plants.jstor.org/stable/10.5555/al.ap.flora.sffa003100371500001. Accessed 19 Jan 2021

Rusea G, Aguilar NO (2003) Tectaria Cav. In: de Winter WP, Amoroso VB (eds) Plant resources of South-East Asia no 15(2): cryptogams: ferns and fern allies. PROSEA Foundation, Bogor

Sathiyaraj G, Muthukumar T, Ravindran KC (2015) Ethnomedicinal importance of fern and fern allies traditionally used by tribal people of Palani Hills (Kodaikanal), Western Ghats, South India. J Med Herbs Ethnomed 5:4–9

Schippmann U, Leaman DJ, Cunningham AB (2002) Impact of cultivation and gathering of medicinal plants on biodiversity: global trends and issues. Inter-Department Working Group on Biology Diversity for Food and Agriculture, FAO, Rome

Shukla SP, Khare PB (2014) In vitro conservation of some threatened and economically important ferns belonging to the Indian subcontinent. J Bot 2014:1–8

Singh S, Singh R (2012) Ethnomedicinal use of Pteridophytes in reproductive health of tribal women of Pachmarhi Biosphere Reserve, Madhya Pradesh, India. Int J Pharm Sci Res 3(12): 4780

Smith AR, Pryer KM, Schuettpelz E, Korall P, Schneider H, Wolf PG (2006) A classification for extant ferns. Taxon 55(3):705–731

Sobrevila C (2008) The role of indigenous peoples in biodiversity conservation: the natural but often forgotten partners. The World Bank. https://documents.worldbank.org/en/publication/documents-reports/documentdetail/995271468177530126. 19 Jan 2021

Stuart G (2018) Rabbit foot fern. Philippine medicinal plants. http://www.stuartxchange.com/RabbitFootFern.html. Accessed 18 Jan 2021

Tan JMP, Buot IE Jr (2019) Cluster and ordination analyses of leaf architectural characters in classifying Polypodiales sensu PPG. Thailand Nat History Museum J 13(1):27–42

United Nations (2018) Achieving sustainable development and promoting development cooperation. United Nations Department of Economic and Social Affairs. United Nations Publications, New York

Yatskievych G (2021) Davalliaceae. Britannica Inc. https://www.britannica.com/plant/Davalliaceae. Accessed 17 Jan 2021

Zamora PM, Co L (1986) Guide to Philippine flora and fauna. Natural Resources Management Center, Ministry of Natural Resources and University of the Philippines, Quezon City

On-Farm Conservation of Traditional Rice Varieties in the Provinces of Quirino, Iloilo, Antique, and Davao Oriental, Philippines: Drivers and Implications for Policy

2

Lorna E. Sister, Hidelisa D. de Chavez, Shalan Joseph E. Kitma, Teresita H. Borromeo, and Nestor C. Altoveros

Abstract

An important objective of in situ on-farm conservation is the continued evolution of crop diversity in farmers' fields, in order to generate potentially useful adaptations that can help sustain the capacity of agricultural and food systems in the face of change. Aside from generating diversity in the crop, in situ on-farm conservation should also provide incentives for the community to continue their conservation and sustainable use of TRVs. This chapter describes the process of engaging communities in designing their own conservation strategy for traditional rice varieties: TRV inventories, seed preparation and storage, and community planning for in situ on-farm conservation strategy, considering local knowledge, access to resources, livelihood incentives, and sociocultural norms. In order for communities to make informed decisions, technical experts provided inputs on the principles of conservation and its various methods through Conservation Field Schools (CFS) after obtaining a good appreciation of local conservation knowledge and practices. The primary driver of in situ on-farm conservation of TRVs is household food security. Other drivers are socioeconomic factors, cultural pride and traditions, and market incentives linked to TRVs.

L. E. Sister (✉) · T. H. Borromeo · N. C. Altoveros
Institute of Crop Science, College of Agriculture and Food Science, University of the Philippines Los Baños, College, Los Baños, Laguna, Philippines
e-mail: lesister@up.edu.ph; thborromeo@up.edu.ph

H. D. de Chavez
Institute of Plant Breeding, College of Agriculture and Food Science, University of the Philippines Los Baños, College, Los Baños, Laguna, Philippines
e-mail: hddechavez@up.edu.ph

S. J. E. Kitma
Biostadt Philippines, Inc., Ortigas Center, Pasig, Metro Manila, Philippines

S. Ramamoorthy et al. (eds.), *Plant Genetic Resources, Inventory, Collection and Conservation*, https://doi.org/10.1007/978-981-16-7699-4_2

Keywords

On-farm conservation · Traditional rice varieties · Conservation field school

2.1 Introduction

2.1.1 Securing Resources for Food and Agriculture in the Face of Insecurity

We rely on genetic resources for food and agriculture for food security, now and in the future. They directly provide food, as well as raw materials for future crop improvement. Thus, losing them may not only mean food and livelihood insecurity of the producing households; it may actually mean a highly insecure future for all humankind.

Rice (*Oryza sativa* L.) is a crop that is closely associated with food security. Rice is the major staple food in the Philippines. Yet, while rice importation has flooded markets with imported and high-yielding varieties (HYVs), the remote uplands still hold traditional rice varieties (TRVs) that may not be able to compete in markets, but are culturally and economically important in the communities where they are produced.

Safekeeping of important plant genetic resources for food and agriculture including rice at the national and international gene banks with safety duplication at Svalbard Global Seed Vault is an important step in securing rice genetic resources for the future. However, ex situ approaches like these do not have significant bearing for the rural households facing food insecurity on a day-to-day basis. For instance, the loss of seed due to a crop failure in one rice season may mean the loss of that variety for the farmer and the community in the absence of any knowledge of other sources to replace the lost variety. This brings to the fore the importance of in situ on-farm conservation for local livelihood security, complementing off-site institutional conservation strategies (Sthapit et al. 2009).

Jarvis and Hodgkin (2008) mention circumstances that favor the occurrence of high levels of crop genetic diversity: a variable environment, technical difficulties of production, and where access to resources and markets is difficult.

2.1.2 In Situ On-Farm Conservation

The Convention on Biological Diversity (CBD) and the International Treaty on Plant Genetic Resources for Food and Agriculture (ITPGRFA) define in situ conservation as “the conservation of ecosystems and natural habitats and the maintenance and recovery of viable populations of species in their natural surroundings and, in the case of domesticated or cultivated plant species, in the surroundings where they have developed their distinctive properties” (UNEP 1992). Further, the CBD defines domesticated or cultivated plant species as "species in which the evolutionary process has been influenced by humans to meet their needs."

By inference, the surroundings where these crop species have developed their distinctive properties would be the farmland and its surroundings in the farming communities. The scope of the project was limited to the latter part of the definition, due to the fact that rice is a domesticated and cultivated species.

Traditional varieties and primitive forms were the types of genetic resources dealt with. Traditional varieties (also known as farmers' varieties) or landraces are unique varieties of crops that have adapted to local conditions through a process of continued farmer selection over many cropping cycles in the case of rice and cultivated root crops. Primitive forms, on the other hand, are crop varieties that have not been subjected to intensive breeding or growers' selection; they have features or traits that are similar to wild relatives as the case may be for some yams.

An important objective of in situ on-farm conservation is the continued evolution of crop diversity in farmers' fields, in order to generate potentially useful adaptations that can help sustain the capacity of agricultural and food systems in the face of change. Aside from generating diversity in the crop, in situ on-farm conservation should also provide livelihood and societal benefits to encourage farmers and local communities to continue cultivation (Bellon and Smale 1999; Brush and Meng 1998). It can also contribute to environmental services.

This chapter describes the process of engaging communities in designing their own conservation strategy for traditional rice varieties that they will, and can, commit to managing and supporting. In the end, it is the farmers/communities—the central figure in in situ on-farm conservation—who make the decisions regarding the conservation and use of on-farm diversity (Brown 1963). These decisions are a function of not only knowledge, but also physical, human, and social resources available to implement a conservation and use strategy.

In order for communities to make informed decisions, technical experts provided inputs on the principles of conservation and its various methods through Conservation Field Schools (CFS) after obtaining a good appreciation of local conservation knowledge and practices. The chapter then presents reflections on factors that either drove in situ on-farm conservation of TRVs based on farmers' and key informant accounts and on observations in communities and markets. These genetic resources were targeted for conservation in the established community-based on-farm conservation strategies.

2.2 In Situ On-Farm Conservation of Traditional Rice Varieties

2.2.1 Communities of TRV Conservation

Identification of target sites was initially done through online and secondary literature scoping and telephone interviews with local agricultural officers and local government officials from municipalities in the general area of key biodiversity areas covered by the ongoing UNDP-supported Biodiversity Partnerships Program

Table 2.1 General description of areas where traditional varieties are usually found in the focus sites

Site	Occurrence of TRVs
Quirino province (Maddela and Nagtipunan)	• Usually in the highlands: e.g., *Barangay* Villa Agullana and Jose Ancheta • In some lowland areas in Maddela: *barangay* Cofcaville and Divisoria • Areas dominated by Ifugao and Bugkalot tribes
Antique province (San Remigio)	• Mountain communities: *barangay* Timbuto, Timbabad, Nagbangi I, Bugo, Panpanan, Walker, and Aningalan (excluding 15 lowland *barangay*) • In the fringes of the NIPAS areas, away from roads
Iloilo province (Janiuay)	• Mountain communities: e.g., *barangay* Panuran, Aglobong, Monte-magapa, and Muhon
Davao Oriental province (Baganga)	• Mountainous areas: e.g., *barangay* Mahan-ub (upland *sitios*), Binondo, Mikit, Kinablangan, San Isidro, Campawan

Source: Comprehensive Land Use Plan/Municipal Profile of the Municipalities of Maddela, Nagtipunan, San Remigio, Janiuay, Baganga; key informant interviews.

(BPP) of the Biodiversity Management Bureau of the Philippines Department of Environment and Natural Resources (DENR-BMB). Final ocular surveys, face-to-face key informant interviews, and focus group discussions were conducted in the communities with the strongest potential.

Diversity in TRVs cultivated by communities is mostly reported in the margins of highland forest areas (Table 2.1).

Six *barangay* or villages were identified as project focus sites across the provinces of Quirino, Iloilo, Antique, and Davao Oriental (Fig. 2.1). These sites are located in the vicinity or at least in the same province as the selected KBAs. In these areas, production of traditional varieties of rice forms a significant part of household livelihoods (Table 2.2).

2.2.2 The Conservation Field School (CFS): Setting the Stage for On-Site Conservation

Conservation Field Schools became the vehicle for introducing technical information for improving local conservation knowledge practice as well as for validating good local conservation practices. The community field schools covered seed preparation, storage, and management in two sessions: (1) inventory of traditional varieties or landraces and associated knowledge and practices and (2) introduction of plant genetic conservation and management principles and methods. A final session was devoted to community planning for a community-based gene bank.

2.2.2.1 Inventories of Traditional Rice Varieties

The first step in the preparation of protocols for in situ on-farm conservation of TRVs involves gathering of relevant information on production to seed-keeping.

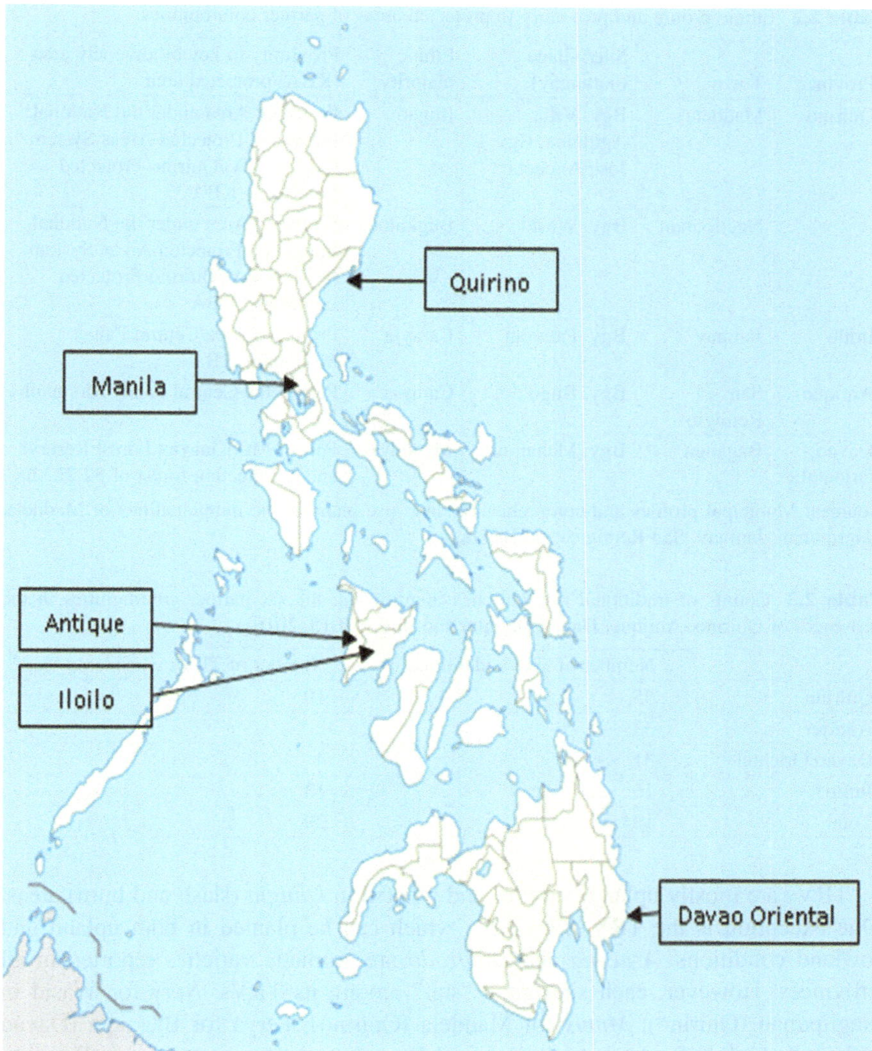

Fig. 2.1 Location of Quirino, Antique, Iloilo, and Davao Oriental provinces with reference to Manila, Philippines

While farmers do have an intimate knowledge of crop production (e.g., phenology, cultural management, pests and diseases, harvesting), conservation (seed processing and storage), products, and markets, such information is usually not documented systematically. Documentation was initiated with TRV inventories along with the sharing of associated knowledge on conservation and management. Table 2.3 shows a summary of the number of TRVs identified in the inventories.

Table 2.2 Ethnic profile and proximity to protected areas of partner communities

Province	Town	Site/village (barangay)	Ethnic majority	Proximity to key biodiversity area (KBA)/protected area
Quirino	Maddela	Bgy. Villa Agullana, Bgy. Jose Ancheta	Ifugao	Protected Area under the National Integrated Protected Areas System (NIPAS-PA)/Quirino Protected Landscape KBA
	Nagtipunan	Bgy. Wasid	Bugkalot	Protected Area under the National Integrated Protected Areas System (NIPAS-PA)/Quirino Protected Landscape KBA
Iloilo	Janiuay	Bgy. Panuran	Caray-a	Protected Area/Central Panay Mountains KBA
Antique	San Remigio	Bgy. Bugo	Caray-a	Part of the Central Panay Mountains KBA
Davao Oriental	Baganga	Bgy. Mahan-ub	Mandaya	Part of Mt. Kinayan Forest Reserve and a protection forest of 32,222 ha

Sources: Municipal profiles and comprehensive land-use plans of the municipalities of Maddela, Nagtipunan, Janiuay, San Remigio, and Baganga

Table 2.3 Counts of traditional rice varieties identified in the six partner communities in the provinces of Quirino, Antique, Davao Oriental and Iloilo (2015–2016)

	Number of TRVs identified	Number of TRVs considered "lost"
Quirino	55	10
Antique	33	2
Davao Oriental	31	4
Iloilo	15	13
Total	134	29

TRVs are mostly upland varieties and planted in *kaingin* (slash and burn) areas. One exception is the TRV *Bungkitan*, which can be planted in both upland and lowland conditions. *Azucena* and *Dinorado* are aromatic varieties reported in all provinces. However, each site has a "star" among its TRVs: *Nery* for Wasid in Nagtipunan (Quirino); *Mimis* for Maddela (Quirino); Perya for Baganga (Davao Oriental); *Malidu* for Janiuay (Iloilo); and *Dumale* and *Sampokoy* for San Remigio (Antique). Except for the red-grained *Nery*, all of these varieties have white and red variants.

2.2.2.2 Initial Assessment of Diversity

Even as farmers continue to cultivate TRVs, a narrowing of genetic diversity was noted through farmers' selection and spread of the more preferred varieties. An initial assessment of diversity involving 94 TRVs for which grain samples were available is shown in Table 2.4.

Endosperm types were almost invariably non-glutinous across all sites. This is especially true in the Iloilo site, where all the rice samples were non-glutinous and had white seed color, regardless of the high variation in awning characteristics.

Table 2.4 Summary of findings on the level of diversity of traditional rice varieties from all sites based on grain qualitative characters, and the dominant traits for each character ($n = 94$)

Grain characters	ALL sites	Quirino	Antique	Iloilo	Davao Or.
Endosperm type	Low (non-glutinous)	Moderate (non-glutinous)	Low (non-glutinous)	Monomorphic (non-glutinous)	Low (non-glutinous)
Awning	Moderate (absent)	Moderate (absent)	Low (absent)	High (absent; short; partly awned)	Moderate (absent)
Seedcoat color	Moderate (white)	Moderate (straw)	Moderate (red)	Low (white)	Moderate (white)
Awn color	Moderate (absent)	Moderate (absent)	Low (absent)	Moderate (straw)	Moderate (absent)
Lemma-palea color	Moderate (straw)	Moderate (straw)	Low (straw)	Low (straw)	Moderate (straw)
Apiculus color	Moderate (straw)	Moderate (straw)	Low (straw)	Low (straw)	Moderate (brown)
Lemma-palea pubescence	Moderate (short hairs)	Moderate (short hairs)	Low (short hairs)	Moderate (short hairs)	Moderate (short hairs)
Sterile lemma length	Moderate (medium)	Moderate (medium)	Low (medium)	Low (medium)	Moderate (medium)

Seedcoat color was usually white, although it was generally red for Antique and straw-colored for Quirino.

Since this was only a subset of the varieties named by the communities in the inventory, there were no grain samples for many other TRVs. This reflects the fact that many of these varieties are already hard to find—growers are becoming fewer, and sometimes unknown to others, or planting materials are becoming scarce. However, other plant characters and quantitative characters were not included in the analysis and it will not be safe to assume that the same trend will be observed for these set of characters.

The data seemed to support the notion that agrobiodiversity has been pushed to the fringes of forests and protected areas, particularly to very inaccessible areas. It is also in these areas where traditions and knowledge associated with the conservation of these native varieties are still vaguely remembered—and may even be practiced by some. Wherever conditions become more favorable for agriculture, mono-cropping of modern varieties in high-input, modern agricultural production systems takes over, even in those areas which are very close to the key biodiversity areas (KBAs).

2.2.2.3 Indications of Genetic Erosion of Traditional Rice Varieties

Even in the cultivation of TRVs, the diversity assessment of rice based on grain qualitative characters gives insight into erosion of diversity affecting heirloom varieties as well. Low to moderate diversity in these characters was noted for the more popularly grown or more readily available rice varieties characterized. Since only 94 of the 134 TRVs mentioned by respondents were characterized, the trend can change if the other less common varieties are included. Also, quantitative traits and other plant characters were not included in the analysis.

Ten TRVs have been reported "lost" in the province of Quirino. Among the 64 TRVs in the inventory, only 26 were mentioned for household consumption, which may indicate that many are possibly disappearing. Respondents in the predominantly Ifugao community of Villa Agullana mentioned that the native violet rice was lost along with many others when people left the area in the 1980s to 1990s due to a volatile peace and order situation. Community accounts in the Iloilo, Antique, and Davao Oriental sites mention 13, 2, and 4 TRVs, respectively, as "lost." Losses were attributed to typhoon, bird attacks, pest infestation, peace and order problems, as well as displacement due to modern/high-yielding varieties/crops in the more favorable areas.

2.2.2.4 On-Farm Conservation of TRVs

Continued planting of TRVs has been a key factor in maintaining the TRVs in the six communities. Seed is maintained in the communities since farmers do not plant simultaneously or share seed. With up to ten TRVs held by one household, asynchronous planting and the more common practice of exchanging seeds supported conservation through use. A farmer missing his planting season or suddenly facing a reduced production area can save his seeds by "lending" the seeds to another farmer.

Communities have traditional conservation practices on which a community-based gene bank can be built on. The Ifugao and the Bugkalot have traditional seed storage structures for rice, while others employ storage practices to keep seed. The role of women in TRV conservation is highlighted, particularly among the Ifugao and Bugkalot tribes in Quirino province, and the Mandaya in Davao Oriental province.

Some key highlights of the conservation of TRVs in the six partner communities, briefly described in the short case narratives that follow.

Barangay Wasid, Nagtipunan, Quirino province. Locals have been planting TRVs since the 1940s. Planting is done by the females as well as seed selection. In fact, most activities on TRV production are done by females while the men of the community are more focused on corn production.

The onset of rains signals the start of the planting season. In Wasid, this is usually in May or June, or as late as July. Areas selected for upland rice planting are those which have not been planted with any agricultural crop for the past 5 years. Once planted with rice, the area may be used for up to 3 years (three planting seasons), replacing the rice with leguminous crops after every season. After that, they allow the land to fallow and move to a new area. Whole plants are uprooted when the beans are harvested so no plant dry matter is returned to the soil.

The role of men in TRV production is often limited to field cleaning and preparation. Women take charge of all other activities. Peanuts and glutinous rice specifically are considered as "female" crops. Women are interested in glutinous rice because these can be processed into local delicacies (e.g., *tupig*, *pinipig*, and *inuruban*) that help break the monotony of steamed or boiled rice. Thus, they transform rice from being a plain-tasting staple into snack foods and meal alternatives to please the palates of household members, particularly the children.

Harvesting is done with the help of other farmers in the community—a custom called *bayanihan* or *suyuan*. With varieties that have grains which easily fall off the panicle (e.g., Palawan), harvesting is done with bare hands (*lasao/lab-o/lap-o*). Otherwise, a tool called *rakim* is used. Harvesters may reap panicles from the panicle base or with a portion of the culm attached. A collector called *men gatab* ties the panicles into bundles or transfers the contents of the harvest basket (*lukbut*) to sacks.

Harvested panicles are hung and air-dried in a covered but well-ventilated area before storage. Locals make a *begantay*—structure of wood and cogon grass (*Imperata cylindrica* L.) from which to hang the harvested bundles for air-drying. Grains meant for eating are spread over a mat or on a wide pavement and are dried under the sun. They may also be force-dried by heating them in a dry pot set over a low flame. Grains are stored in a sturdy, weather-resistant wooden container. In the olden days, the grain storage was made of a grass called *runo* (*Miscanthus sinensis* Andersson). Grains for seed are kept undisturbed at the bottom of the container while grains for consumption are placed on top. As most participants from Wasid are of the *Bugkalot* indigenous group, they have a specially designed shack for storing harvested rice grains called an *agang*.

Locals maintain their TRVs mainly because of their pleasing aroma and great eating quality. In extreme weather conditions during harvest time, no seeds are obtained. Some TRVs have disappeared because of this. Other reasons for loss of some TRVs are pest damage (rats and birds), damages due to typhoons, sharing of seeds to other farmers, and replacement by other TRVs.

The elders of the community are the ones growing TRVs. Not all are able to pass this practice on to younger generations. Thus, when they pass away, the varieties they have been growing disappear as well. Cash crops—especially corn—have been replacing subsistence crops. However, a recent reversal back to TRVs is happening with the increase in corn production-related bankruptcies.

Barangay Jose Ancheta, Maddela, Quirino province. The *Mimis* rice variety of *barangay* Jose Ancheta, a red-grained variant, has been planted since 1975. *Ultok* is another rice variety planted since the 1970s, and in the 1980s was the *Anippa* variety. There were also other varieties planted then but the locals do not know their names anymore. The *Mimis*, *Donya*, and *Denorado* varieties were most commonly planted around the first decade of 2000. TRVs in the villages vary across *purok* or household clusters. Topography, degree of sloping, water availability, and accessibility to milling facilities are some of the factors affecting which varieties are grown in each *purok*. Regardless of location, the most common reasons for maintaining a variety are its organoleptic qualities (aroma, eating quality) and yield.

Areas that are either forested or *oblag* (fallow) for 3–4 years are used for planting TRVs. The custom of *bayanihan* (*ammuyo*) is practiced in upland rice planting, with the number of helpers a farmer is entitled to depending on his participation in other farmers' *ammuyo*. The men clear the area and burn plant residues (*kaingin*) to discourage rats from sheltering there. Both men and women take part in planting. The men bore planting holes while the women sow the seeds. Seeds for planting are stored in bottles. If these seeds are not planted as scheduled, farmers do a test planting in the next cropping to check viability.

Seeds for the next season are obtained from the batch of grains that were harvested under favorable harvesting conditions (i.e., no rains and the grains are ripe). Harvesting may be done manually (*rungos*) or with the use of a *rakim*. *Rineppet* or *binettek* is the system of harvesting whole panicles and tying them in bundles. Selection of seeds may be done by either men or women. Grains for seed should be healthy-looking and dry. Unfilled, discolored, and infested grains are discarded as these will not grow well. Harvested panicles are tied into bundles as thick as a man's arm for drying under the sun for 1–3 days. In the morning, bundles are dried with the grains underneath. At noon, the bundles are reoriented such that the grains are on top.

The practice of *kirog* or drying grains for food in a pot set over a fire is done on cloudy or rainy days. Another alternative drying method is by placing the grains on top of the *insuoran* or the slatted platform above a fireplace or cooking area.

Grains for seeds are separated from those meant for food. The *kamarin* is the traditional rice storage shack. Others built inside their house a large rectangular wooden box called a *kuloung* that can hold up to 100 bundles of rice panicles. Yet, farmers no longer found the need for a storage shack or box since they do not get to harvest so much anymore. Grains are just kept inside the house in an area where they will not be spoiled by rain or other sources of moisture. Seed bags are not stacked so that rats are easy to spot. Weevils only become a problem if seeds are left hanging for 2 years. At 2 weeks before planting, seeds are threshed, winnowed, and then dried for an hour.

Barangay Villa Agullana, Maddela, Quirino province. Rice has been planted in Villa Agullana even before the 1960s. The practice of *kaingin* has just started back then and rice varieties were abundant. Villa Agullana was occupied by the Ifugao tribe. In the 1980s to the 1990s, an unstable peace and order situation prompted people to leave the area and their TRVs. When they eventually went back, many were lost. At present, both lowland and upland rice varieties are grown.

TRVs that were soft when cooked, were aromatic, and have good eating quality are maintained, but planting is dependent on the availability of seeds. Seed is farm-saved, so they can either use their own seed or ask from other farmers—for free or to be replaced after harvest without interest.

Not all farmers select their seeds for the next planting, but for those who do, men and women make the selection. It does not matter from which part of the field the seeds are taken as long as they come from healthy plants. When selecting for seeds, they pick grains that are relatively bigger and disease free. Harvested seeds are dried for 2–3 days before being stored in the traditional storage shack or *alang*. However, farmers rarely build them nowadays. Some keep grains in the bundled panicles. Some keep threshed grains stored in large biscuit cans. No chemical weevil (*bukbok*) control is employed. Rice grains for food and for seeds are kept in separate containers inside homes. When the food bins run low, the grains kept for seeds are often consumed as well.

Barangay Panuran, Janiuay, Iloilo. TRVs have been planted in the village of Panuran in Janiuay, Iloilo province, even before the 1930s. The current volume of rice production in the community, however, is just enough for household consumption. At most, farmers plant one sack of seeds or enough for about a hectare. Lack of financial capital limits production expansion.

Farmers may borrow seeds from other farmers and give back twice the amount of seeds lent to him after harvest. Choice of variety often depends on the availability of seeds, or is limited to the usual varieties planted. Glutinous (*pilit*) rice varieties are getting rare.

Areas planted with traditional rice are usually either virgin lands or those which have been left fallow for up to 10 years. The area is used for rice production for up to 2 years before moving to a new area, unless a new area cannot be found. Then, fertilizers are used.

Planting is in March to April. Locals claim that the crops will not produce good grains when planted in any other month. In June, the seedlings can be overcome by weeds. In February, the seeds will not grow. The problem, however, is that the harvest time falls during the typhoon season. Some farmers intercrop TRVs with corn while others intercrop with pigeon pea. Men prepare the land by plowing twice with the aid of a carabao and then pulverizing the soil using a hoe (*ginasadol*). Some farmers have attended a training 15 years ago which introduced the use of the A-frame and how to plant in terraces. Dibbling (*panggas*) is done when a carabao is not available to plow the land. Seeds are, however, broadcasted (*sabog*) if sown on flat land. *Paminhi* is a ritual done right before the start of planting. Here, the owner of the farm sows rice seeds one by one while holding an egg (the larger, the better so as to promote full grains), a piece of turmeric (to prevent pest and disease occurrence), and some lemongrass (to promote tillering). The sowing should be done gently, as locals believe that if they are handled roughly, they will not produce a good harvest. After the seventh seed has been sown, the helpers (*kubreros*) may now enter the field and start planting. Farmers fertilize their rice using ammonium sulfate usually at 15–30 days after emergence. They often apply one sack a hectare or a maximum of two sacks depending on the appearance of the standing crops. One of the main causes of disappearance of varieties pointed out by the locals is typhoon damage. Rat infestation is another. Birds can also be damaging pests but farmers are able to drive them away using a scarecrow and also a local version of a noisemaker made of bamboo which makes a loud clapping sound.

As long as there are plants with flowers open, a person is not allowed to pick grains for the purpose of sniffing or tasting. Otherwise, the locals say that the person will get a stomachache. On the day before harvesting, locals also have a traditional practice of placing seven *tuos* along the peripheries of the rice field. *Tuos* is the looped blade of a tall grass (*tigbaw* or *udyong*). One of the blades that will be placed will have a piece of cloth-wrapped ginger (turmeric is also acceptable) tied to it. This is usually done by, but not limited to, the husband. They do this ritual so that conditions will be favorable for the harvest the following morning. Some farmers say that the *tigbaw and udyong* grasses are able to ward off spirits who may take away a portion of the harvest. Once the *tuos* have been placed no one should walk into or walk across the rice field.

Whether planting or harvesting, both men and women work equally. On the other hand, seed selection is done by the men as they work with the crops more often. They obtain seeds from tall and sturdy plants and avoid any volunteer varieties in the field. Selective harvesting is made easy using a handheld harvesting tool called *kayog*. There are other farmers, however, who are not very particular in selecting the plants from which they will obtain the seeds. They simply take a portion of the harvest and store them separately. Harvested panicles are collected in a bamboo basket worn around the waist called *bande*. Unlike in planting wherein they work at a fast rate, farmers work at a slow pace during harvesting as if there is so much to harvest that it would take them the entire day to finish. The direction of harvesting is done perpendicular to the sun's movement for fear that the volume of harvest might go down, in synchronicity with the sun's setting. The rule of not walking through the field still applies. Also, the owner of the farm should never turn to his right while harvesting as they believe that the dragon they call *Bakunawa* awaits at his right side. This dragon is said to be able to devour the harvest if the owner comes face to face with him. After harvesting, grains are sun-dried for 2 days and then cleaned.

After harvesting, the grains are piled in one place and are split among the farm owner and the *kubreros*. Ratio of the split is always 10:2 in favor of the farm owner. Only one person should do the splitting (no preferred gender). The work should also be done with gentle movements. Whenever grains are separated from the pile, one should always take grains from the foot of the pile. Short breaks may be taken whenever needed, but once started the task should not be abandoned or given to somebody else. According to local belief, these practices are supposed to prevent the grains from becoming upset or from being frightened away. Though not particularly mentioned by the locals, it is possible that this belief is linked to spoilage during storage.

Grains are poured over a *trapal* (wide sheet of tarpaulin) or *sawali* (bamboo mat) and spread under the sun or air-dried inside the house on the occasion of rains. Ordinary straw sacks are used for storage. Separate sacks are used for food grains and those for seeds. The sacks, in turn, are kept in large baskets. A basket that can hold up to three sacks is called *lasok* while a larger basket called *tabungos* can hold as much as 15–20. These two are also the largest of a series of traditional baskets of different sizes including *bande*. The grains are kept inside their homes in a place where they are easily seen. This is so as to easily shoo away any rat or chicken and to protect in case of strong rains or any other calamity.

Rice harvesting happens only once a year and rice is always at its best when newly harvested. That is why the first meal wherein the newly harvested rice is served is always a special occasion and a considerable feast. This feast which they call *pamag-o* always goes with a dish with native chicken or pork. All members the family should be present during this meal if possible. Sometimes even relatives, friends, and neighbors are invited to join. *Panunghol* is a ritual done before starting the first meal of the new harvest as a kind of offering of newly harvested rice to the spirits. Everyone in the family does this but young children are usually the ones made to do it. The child will first take a small portion of the cooked rice with one hand, smells it, and then throws it away behind him or her. Then he or she takes another small portion with his or her other hand and does the same thing but throws forward instead. Headaches and stomach pains after the *pamag-o* are said to be prevented when this ritual is done. Other than the *panunghol*, a small amount of food (rice, chicken, wine, and whatever food is on the banquet table) is also laid in a corner of the house for a few minutes for the family members who have passed away to take part in the feast. No matter how many will take part in the *pamag-o*, a portion of rice should always be left untouched. The more leftovers, the better the celebration is for them. Locals perceive having leftover rice after the *pamag-o* as an indication of abundance while if the rice gets fully consumed, the rice of the household will most probably be short for the year. This is why a lot of rice is always cooked for this occasion.

Pinipig is a popular local delicacy similar to rice crispies. The locals, however, do not get to make so much of these for them to sell. These are used mainly as a gift item. *Ubás* is another local delicacy made from grains harvested at the soft dough stage. Again, production is limited and seasonal. From a field of about 1 hectare, just 1 *gantang* of soft dough grains is taken to be made into *ubás*. Also, the black rice variety (*tapul*) is used by medicine men in curing illnesses.

Barangay Bugo, San Remegio, Antique. The communities in San Remigio, Antique, were affected by typhoon Frank (Fengshen) in 2008, causing rivers to overflow, destroying roads and rice fields, and forcing residents along the riverside to migrate to upland *sitios*.

Traditional rice varieties or TRVs have been planted even before the 1940s. The choice of variety to plant is dependent on seed availability. Given the choice though, they would choose *varieties* with good yield and full grains, high milling recovery, early maturing, and good eating quality and that are aromatic. Upland dwellers plant TRVs for consumption while in Bugo's community center, modern rice varieties are the staple.

Farmers often exchange rice seeds with neighbors and relatives. However, when seeds are no longer available within the community, farmers have to buy from other *barangay* where TRVs are grown.

The planting season begins in April and men prepare the land. As most growing areas depend on rain for moisture, TRVs are grown only once a year. Off-season cropping is usually for seed production purposes. During droughts, TRVs are irrigated using long hoses and sprinklers. Women usually plant using a long, pointed tool called *tagad* for dibbling at the rate of 3–4 seeds per hill. The *tagad* is especially useful when the soil is hard and compact.

Over the decades, farmers have not used chemical sprays. There are areas, however, that have started applying chemical fertilizers at half the application rates for lowland rice but plants turned out lanky.

A number of rituals take place when harvest time draws near. In the few days before harvest day itself, noise is forbidden near the rice field. An offering of chicken and rice grains spread around the field is given on the eve of harvest day. Seven to ten knotted grass blades called *tuos* tied together with a chicken feather called *tigbao* are also placed in all corners and edges of the field. Some farmers even practice incensing (*usok ng kamangyan*) to drive away what locals call "unseen beings." Locals believe that these "unseen beings" may join the harvesting and spirit away a portion. These, however, are not practiced with modern rice varieties.

Rice is harvested with a traditional tool called *kayog* and panicles are collected in a basket worn around the waist (*kabig*). Culms are burned in the field after harvesting and legumes (e.g., peanuts) are planted after rice. Harvested grains are sun-dried afterwards on a large woven mat (*makan*). Solar exposure of grains for seeds is limited to 2–3 h a day for 3 days to avoid damage. On rainy days, grains are brought inside the house; spread over the *makan* or on top of a 4 ft tall, narrow-slatted bamboo platform (*giikan*); and air-dried using an electric fan. Then, farmers would still bring them out to the sun before storage. When dry, grains are threshed by dancing barefoot on the grains (*linas*). Under the *giikan*, a container catches falling grains as they pass between the bamboo slats. This practice is being slowly replaced by mechanical threshers that have become accessible to upland rice farmers.

Men select the seeds by hand. Seeds are obtained from tall, fine plants with firm culms and from panicles with full grains. Pest-infested and disease-infected plants, as well as off-types, are avoided. Weeds and other off-type seeds are also removed from the harvest. Grains are cleaned often using a hand-operated blower. Others stick to the traditional *bilao*—a flat, woven circular tray used for winnowing. Seeds are stored in a round basket called *tabungos*, but the traditional storage container, locally called *bayong*, is a length of bamboo culm with ends tightly plugged with cloth. This method lets them store their seeds for up to 6 months.

A few farmers sell TRVs at the nearby town of Aningalan, a local tourist destination. The varieties *Tapul*, *Asucena*, and *Camuros* are sought after in the market for the widely held belief in their medicinal properties. Though farmers are aware that there is a demand for traditional rice, they do not get to produce more due to the limited availability of seed.

The rule of seed trade in the area is to return 2–3 units after harvest for every unit of seed traded.

Barangay Mahan-ub, Baganga, Davao Oriental province. When super typhoon Pablo (Bopha) struck Baganga in November 2012, most farmers have planted all their seeds. All of this was wiped out and what was left of the TRVs were from the few farmers who have not planted yet, and who were able to keep their seeds safe. After this calamity, farmers found planting TRVs in large areas too risky. To increase their seeds, some tried growing TRVs in the lowlands and were more than satisfied with the outcome. Farmers then ventured into producing TRVs off-season, although crop performance was still better in-season. Crops were more prone to disease infection (leaf spots) and were less competitive with weeds in the off-season. Farmers also blamed typhoon Pablo for what they perceived as a decline in performance of crops grown in the highlands (yellowing crops) as well as an apparent degradation of the resource base (more weeds and less earthworms seen).

Limitations in available production area, seeds, labor, and capital as well as rodent infestation were perceived as the reasons for the disappearance of TRVs. Organoleptic qualities (good eating quality) are the foremost reason farmers maintain certain varieties. Also, farmers chose to plant varieties that were drought tolerant, not labor intensive, popular, and with a high market price. Others maintain varieties that are easy to pound while glutinous varieties are important for special occasions. *Perya* is deeply ingrained in the Mandaya culture of the province. This importance is highlighted by the locals' willingness to switch to corn for their staple over consuming their *Perya* seeds.

Mahan-ub's *sitio* (household cluster) Coog is the most popular place for locals to obtain traditional rice seeds through a barter system. Seeds may be exchanged for goods such as fish and other products or for services agreed upon by both parties. Seeds can also be "borrowed" (*paki*) and repaid fourfold at the end of the season with part of the harvest share earned through cooperative work or with cash. Yet, there are still some who may give seeds for free.

TRVs are planted on newly opened virgin or fallow land. An area may be used for two rice seasons with one corn season in between before it is left fallow. Men clear the area of vegetation and then burn the remnants as field rats find shelter in them. Some apply nonselective herbicides and begin manual land preparation (*hasok*) and planting after 15 days. Others allow weeds to grow over the burned field to about a foot high. Seeds are then broadcast before the young weeds are cut down (*takyab*) to serve as mulch. In November, land preparation and planting are done by men and women. Some wait for the new moon as insects are said to be fewest during this time. As a planting tradition, seeds are mixed with a gold ring (for a golden harvest), a cross, *kila na bulaklak* (the top section of a red ti plant, *Cordyline fruticosa*, for good-quality grains), and a few stalks of *tanglad* or lemongrass (*Cymbopogon citratus*). Some oversimplify the ritual by just saying a prayer before the start of planting, while others carry on without any rites or rituals.

Weeds, volunteer plants, and off-types are also removed a month after sowing to get rid of pests. Weeding is done by men and women, but men apply the fertilizers and insecticides. Complete (14-14-14) fertilizer is applied once while pesticides are sprayed as needed, usually up to three applications only for the entire season. The pesticide is a mix of a molluscicide, chili pepper, ground *hipto*, or marigold (*Tagetes* sp.). Chemical insecticides such as Cymbush and Lannate are also used by locals to control rats.

A wooden stick in the field with seven panicles tied to it and prayers are common rituals especially on the day before the harvest. Couples may offer prayers of thanksgiving together. However, many farmers nowadays are becoming indifferent when it comes to rites and rituals.

Harvest time begins in April and can last until the end of May. Men and women usually work together but older females can take over when the men become involved in other livelihood activities during this time of the year. Pregnant women and those having their monthly period are not allowed to participate as they may bring bad luck. Panicles (*uhay*) are harvested excluding the stalk using a handheld tool called *kayog* and collected in a woven basket called *kayag*. A *kayag* that is more than half filled up means the harvest is good.

Farmers are particular in selecting vigorous plants with relatively large, fully filled, and disease-free grains for seeds. Using the *kayog* allows harvesters to easily separate panicles that bear grains fit to be seed material. A few farmers do not find it important to scrutinize the harvested grains to select seeds. Seed selection is usually done by the couple. After harvesting, the panicles are spread over a sheet of sack material and threshed (*buy-ok*) usually using the feet. Threshers rub the grains off the panicles while walking or even dancing on them. Panicles for seeds are threshed first. The rest of the panicles are threshed only after the grains for seeds have been collected. Grains are then cleaned by winnowing and afterwards spread over a large sheet of tarpaulin or a woven mat for drying. These are laid over a flat, unpaved area to avoid accumulation of too much heat. Grains are exposed for a day under full sun or up to 4 days under cloudy skies, but they lose viability with longer drying. When it rains, seeds are placed in sacks until conditions improve. Grains for consumption, on the other hand, may be heated in a dry pan set over a gentle flame. After drying, grains are allowed to cool down and are winnowed before storing. If there are grains meant for commercial use, these are pounded (*bayó*) and cleaned first before they are bagged.

Grains for planting material and for household consumption are kept in separate sacks and stored in a *kurob*—a special shed of nipa palm fronds (*Nypa fruticans*) roof, slatted bamboo flooring, and wooden walls that may hold up to 20 sacks of seeds. Some keep seeds in a covered plastic pail or drum, or in a covered tin can to protect the seeds from rodents and the elements. Half a cup of dried native chili peppers (*Katumbaw* variety) is mixed with the stored grains to prevent weevils and ants from damaging the grains. Seeds remain viable in storage for a period of

6–7 months, making it necessary for them to be sown in the next cropping season. A farmer claims that she can keep seeds viable longer inside a bamboo culm.

2.2.3 Community-Led Design of On-Farm TRV Conservation

As all communities expressed interest in conserving TRVs and having obtained an appreciation for long-term conservation, community gene banks were established. The various modalities and management strategies took into consideration the technical requirements for long-term storage as discussed in the CFSs, resources immediately available to the communities, local government support, and relevant cultural traditions.

A facilitated inventory led to the identification of 134 TRVs across focus sites in the provinces of Quirino, Iloilo, Antique, and Davao Oriental. To raise awareness on conservation, improve local conservation practices, and put communities in-charge of their own rice genetic resources, community gene banks were initiated through the conduct of Conservation Field Schools.

The design and management of community gene banks combined technical recommendations with the unique institutional and sociocultural context of each community (Figs. 2.2, 2.3).

The expressed option of the communities in *Barangay* Wasid, Jose Ancheta, and Villa Agullana—all of the province of Quirino—to directly engage in the conservation of TRVs is a reflection of their cultural familiarity with collective efforts in many aspects of daily life, including the conservation and management of TRVs. The community gene banks were extensions of the Ifugao and Bugkalot tribes' conservation traditions and practices. The Ifugao that have migrated into the province have brought with them their TRVs to ensure their continued availability for food as well as for the practice of cultural traditions. Thus, they have also brought with them knowledge of constructing and managing an *alang*, the traditional rice seed storage. A similar structure is the *agang* of the Bugkalot that is central to the chosen community TRV gene banking strategy. However, the choice of the Ifugao community of Villa Agullana to repurpose an existing concrete building is prompted by their intimate knowledge of local weather conditions. It was chosen over the traditional Ifugao *alang* in order for the structure to withstand the strong winds that the community often experiences.

The choices made by communities in Janiuay, San Remigio, and Baganga to give the Municipal Agriculture Office leadership were made in consideration for accessibility of the facility to other communities and availability of resources to meet the requirements of management such as technical capabilities, appropriate structures and environmental conditions, as well as promotion of conservation of TRVs wider to other municipalities and beyond. Visitors to the local government-managed gene bank in San Remigio, Antique, received contributions from other villages as other farmers visited the facility. Visitors' interest in accessing the TRVs in the collection highlights the need to continue capacity building into the documentation and utilization of these genetic resources.

Fig. 2.2 (**a**) Barangay Wasid, Nagtipunan, managed by a women's group, and (**b**) Ifugao *alang*-inspired structure in Barangay Jose Ancheta, Maddela, managed by women in the community

Many challenges remain in terms of maintaining the desired environmental conditions; human resource capability building; and collection, characterization, regeneration, evaluation, and documentation of the collections. To provide further capacity building for the key people in local governments and in the communities actively involved in the Conservation Field Schools and in the management and monitoring of the community gene banks, a Training of Trainers for Conservation Field Schools was conducted (Fig. 2.4). Capacity building needs to be sustained, and in the current pandemic situation may be possible through remote platforms. The experience highlights the importance of weaving improvements to in situ on-farm conservation strategy into what farmers have, know, and are doing. Thus, linking it to institutional ex situ efforts requires two-way understanding.

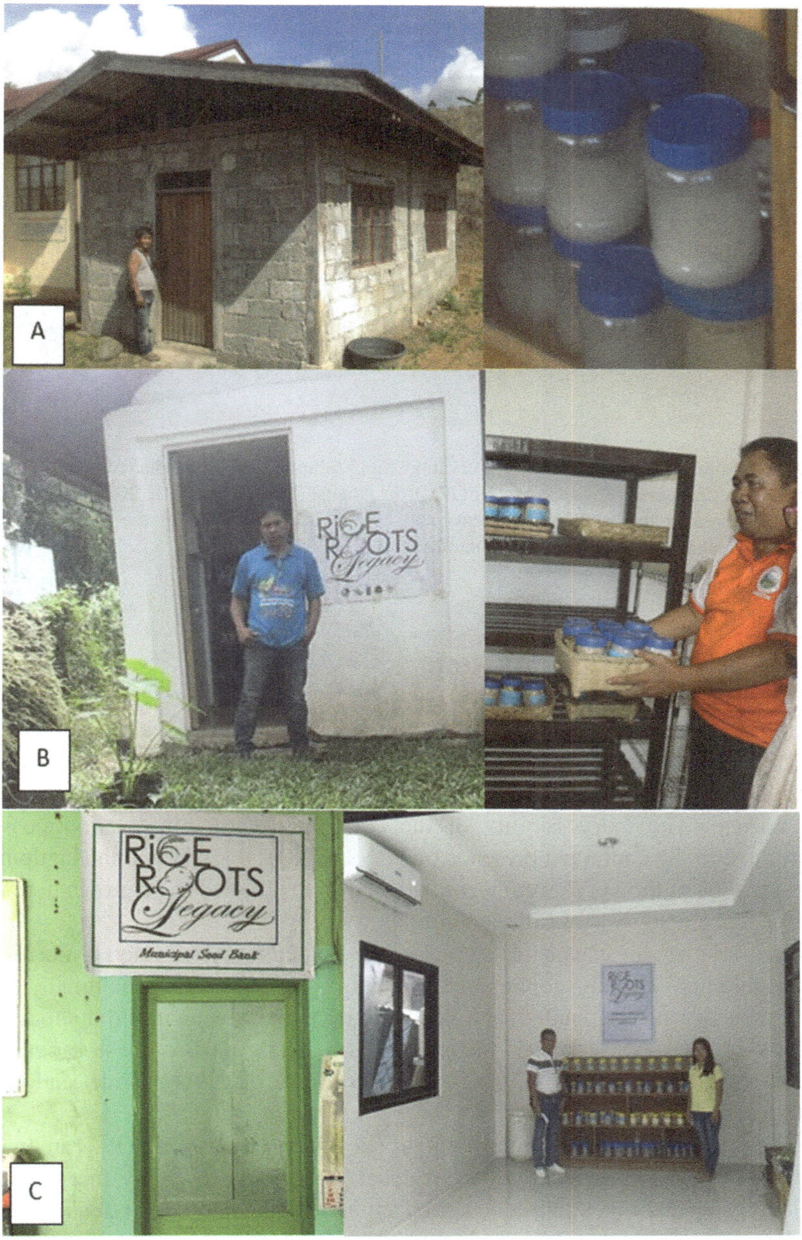

Fig. 2.3 Local government-managed community gene banks: (**a**) Retrofitted building chosen in Barangay Villa Agullana, Maddela, and Quirino province to withstand strong winds and managed by the *Barangay* local government unit; (**b**) retrofitted stockroom at the San Remigio Municipal Agriculture Office, San Remigio, Antique province; and (**c**) initial retrofitted office at the Janiuay Municipal Agriculture Office, Iloilo province, and final relocation to a new air-conditioned space in the new office building of the Municipal Agriculture Office of Janiuay, Iloilo province, in 2020. Not in photo: Retrofitted office pantry at the Baganga Municipal Agriculture Office, Davao Oriental province

Fig. 2.4 Training of Trainers for Conservation Field Schools (CFS) conducted at the National Plant Genetic Resources Laboratory (NPGRL) on August 31–September 1, 2016, for local key people involved in the local governments and communities involved in the CFS and in the management and monitoring of the community-based TRV gene banks

2.3 Drivers of Conservation

Quirino province reported the highest number of TRVs even if majority of its lands (mostly mountainous areas) have been converted into maize production areas. Most of the TRVs are being maintained by the Bugkalot tribe and to some extent by Ifugao and Ilocano. The inaccessibility of the communities of conservation possibly contributed to the continued presence of TRVs. In *Barangay* Wasid in Nagtipunan, Quirino, most of the TRVs were identified, and were reported to be grown in remote parts of this already very-far-flung community.

Rice varieties are usually produced in the uplands, where water resources are scarce and land—usually sloping and recently cleared *kaingin*—is minimally tilled. Small plots of land are planted with rice varieties that one cannot find in the regular rice markets. Yet, households in the focus sites maintain seeds of these varieties for several reasons: food security, food culture, a source of cultural pride, and adaptability to the local agroecology.

Food security appears to be the main driver for conservation, as TRVs are mainly grown for household consumption. Producer households generally face low cash availability so that this represents significant avoidance of costs for acquiring rice, which is the main staple. This cost includes the cost of transport from the markets to these remote communities. It also represents avoidance of hospitalization or illness costs, since some respondents reported health issues associated with consumption of commercial varieties. For these households, seeds of specific traditional varieties must be maintained despite the availability of cash and modern rice.

Food culture. Rice is a staple in the six villages, so that the centrality of these TRVs in the local food culture contributed to their conservation on-farm. The lowest number of TRVs was reported in Iloilo, where majority of the lands are planted with cash crops like sugarcane, peanut, and pigeon pea and modern rice varieties are widely grown. "Good eating quality" is the primary reason for the popularity of most varieties that still exist.

Most certainly, for the rice-eating population of the Philippines, not all rice is the same. Although the promotion of *Perya* is linked to its identification with the Mandaya culture, its strength lies in its highly preferred eating quality. In Quirino, *Penilisa* and *Nere* are similarly sought after for their taste. Because of these varieties, it is said that Bugkalot native rice tastes good even without a complementary viand.

In the province of Quirino, TRVs are usually mixed with the low-quality rice from the National Food Authority (NFA) to improve eating quality. The Bugkalot of *Barangay* Wasid in Nagtipunan use *Nere*, a glutinous white-grain upland TRV, while *Makaale*, a purple non-glutinous upland TRV, is used by the Ifugao of *Barangay* Villa Agullana in Maddela. In San Remigio, Antique, *Kapawod*, *Sampokoy*, *Victory*, *Malido*, and *Pinorong* are the most preferred TRVs. Although growers are becoming fewer and fewer, these keepers have strong preferences for TRVs.

Since rice is a main staple and upland rice is mainly produced for household consumption, organoleptic quality, particularly those that relate to the cooking and consumption experience, is very important. This is reflected in the varieties that farmers keep and continue to grow. Most varieties were described as having "good eating quality"—soft, aromatic when cooked, and with high volume expansion (*maalsa*). In general, TRVs are perceived to have excellent eating qualities.

Cultural pride. *Malidu* is the pride of Iloilo, while *Perya* is a highly valued icon of the Mandaya culture. *Perya* is usually given away or served as a sign of friendship or high esteem for guests. It is also an indication of the affluence of the giver—a Mandaya whose cultural traditions and symbolisms have become an established marketing brand for the province of Davao Oriental.

The *Bungkitan* variety is considered as the best for processing *tapuy*, a traditional rice wine of the Ifugao. The importance that the Ifugao place on ceremonies and celebrations provides the motivation for keeping this variety, even if they have established their communities in Quirino province, away from their home provinces. *Bungkitan* can be planted in either upland or lowland conditions and is known for better wine production volume and taste compared to other varieties. This variety is also used in the preparation of local rice-based delicacies like *suman*, *kakanin*, *tinudok*, *tambong-tambong*, and *palitaw*, among others. The Ilocano and the Ifugao share a common tradition of handing down rice seed from generation to generation to ensure that they can continue to produce the delicacies and food products that mark the affairs of the tribe.

Adaptability to upland agroecology. TRVs are known for their adaptability to less fertile soils and for their low external input requirements. Some are known to be less susceptible to lodging, or exhibit less grain shattering during strong winds. In the less favorable and highly variable conditions of the uplands, these characteristics allow farmers to grow food. For farmers, TRVs are for food security, but in more favorable areas, they plant high-yielding commercial varieties for both food and cash.

Cost avoidance. Farmers far from rice mills prefer varieties with hulls that easily separate from grains in hand pounding. Their motivation to maintain varieties with this characteristic is high to avoid the high cost of accessing rice milling services.

Niche market opportunities. Since TRVs are widely believed to help people facing major health problems, urban consumers offer a high premium for them. Kaporkas, Malido, and Tinuramis have buyers in Janiuay, Dumangas, and Iloilo City. However, access to market information by farmers in Panuran is very limited, and most notably for TRVs. Much of the market information is accessed through middlemen/assemblers locally called *bolantero*.

Urban consumers in Iloilo suffering from severe ailments have been known to seek out TRVs as a "healthy rice" alternative to commercial rice that was found to make them fall ill. Trade usually happens through direct contact by the buyer with the farmer.

Malidu, Iloilo's claim to fame, has reached higher end urban consumers. A locally famous restaurant takes pride in serving the sought-after Iloilo's *Malidu* rice.

The observations point to the importance of securing the household food first particularly in relation to the main staple, rice. Securing the household food as a primary concern reflects the maximization of short-run cash flows that describe the behavior of most of the farmers growing TRVs. They avoid cash outlays as much as possible, and growing traditional varieties that are known to thrive well in their local agroecological environment with low resource endowments is one of the ways they know best to avoid additional costs for securing household subsistence. An underlying theme is the relative remoteness of TRV production areas, and the difficulties faced by producing households in accessing markets and resources.

Related to food security is local food culture, which also dictates what people want to eat, and thus, what they grow. Thus, it is important for them to maintain the seeds of preferred TRVs. Certainly, conservation behavior also has noneconomic dimensions such as cultural pride and traditions in which TRVs play key—and even central—roles.

The chance at markets comes with the production surplus, depending on market information available to the farmer.

Figure 2.5 provides a visual representation of how demand and value addition may interact to create incentives for conservation of TRVs such as those reported or observed.

The diversity of traditional rice genetic resources in farmers' fields reflects these conservation drivers and incentives, which will also hopefully be reflected in the community-based gene banks. With these conservation structures and modalities, a wider appreciation for the value of TRVs is expected to be fostered and encourage local investment of resources in on-farm conservation. There are market incentives that can push conservation and sustainable use further. Yet, at the moment, the community gene banks that have been initiated need much work to be up to the challenge.

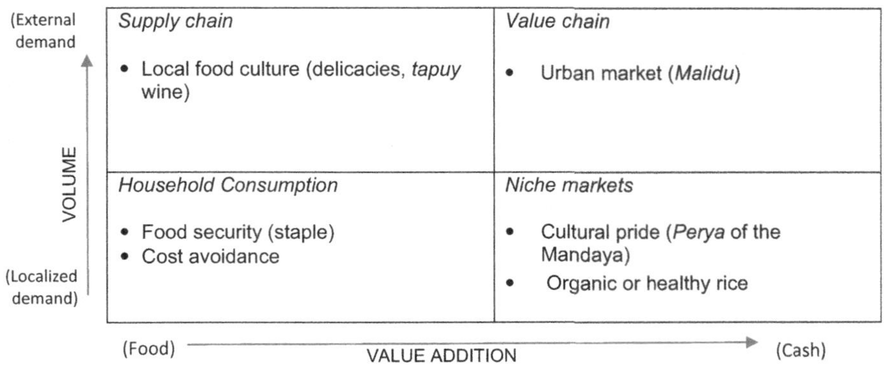

Fig. 2.5 Drivers of conservation of TRVs on a supply volume-value addition matrix

As ex situ conservation strategies struggle to cope with the task of conservation, efforts such as these should eventually be linked to institutional efforts and expertise towards a concerted conservation effort that also allows for diversity of strategies arising from the diversity of conditions faced by community stakeholders.

2.4 Policy Recommendations to Support On-Farm Conservation

During a National Consultation Meeting held in September 2015, the following policy recommendations for the sustainable community conservation of plant genetic resources for food and agriculture were put forward:

1. Provide institutional support to community gene bank (CGB) and seedbank (CSB).
   Three of the project sites had established a community gene bank and the rest have their community collections maintained and managed by the municipal agricultural offices. These gene banks will contribute to the seed/planting material conservation of the traditional varieties of rice and root crops as safeguard to their loss following natural disasters and land-use changes. This can be extended to other crop species grown in the communities. For example, communities planned to landscape the area around the community TRV gene bank with traditional root crops. These community gene banks require funds for operation and management including maintenance of the viability of the collections and documentation.
   Inventory of all the collections should be done. Section 17 of the Indigenous Peoples' Rights Act (IPRA) states that indigenous cultural communities/indigenous peoples (ICCs/IPs) may, on their own initiative, make an inventory of biological and genetic resources found inside their domains/lands, for their exclusive use. Traditional varieties in each community are products of farmers' selections over generations, and to protect them from misappropriation,

communities may opt to provide a list of their traditional varieties (community registration) to the Plant Variety Protection Office (PVPO). Section 72 of the PVP Act states that farming communities and bona fide farmers' organizations are encouraged to build an inventory of locally bred varieties as an option to protect these resources from misappropriation and unfair monopolization. CGB then empowers the farmers to control over their genetic resources (farmers' rights). Also, RA 8371 (IPRA), Part 4 Section 1, states that to ensure biological diversity, sustainable indigenous agriculture shall be encouraged while the system of monocropping shall be discouraged.

Also, support to CGB/CSB can be done through their inclusion in national and local policies and plans. The Comprehensive Land Use Plan (CLUP) serves as a guide of each municipality for future actions on land use and development in the community. It presents the vision of each municipality which may include farmland preservation and other areas of ecological, aesthetic, and cultural significance.

For indigenous cultural communities, the conservation of biological resources including plant genetic resources can be incorporated in the Ancestral Domain Sustainable Development and Protection Plan (ADSDPP). Inclusion of the sustainable conservation and production of traditional varieties in the CLUP/ADSDPP of each municipality will ensure the conservation and production of traditional varieties in the communities. It is therefore recommended that a barangay resolution be passed to provide funds for the operations and management of the CGB/CSB for the sustainable conservation and production of their traditional varieties of crops. A draft model resolution is presented in the attachments.

2. Provision of financial services and crop insurance for traditional crop varieties' production.

   It is recommended that the Department of Agriculture, through the ADCP and Philippine Crop Insurance Corporation, shall lead in the formulation of policies for provision of financial services including crop insurance for traditional crop varieties' production.

   Agricultural loans traditionally exclude farmers of traditional varieties; hence, these farmers suffer total failures when natural disasters, pests, and disease incidence hit their farms. Granting crop loans to farmers growing traditional varieties will allow farmers to pay for necessary inputs including payments for labor or even for accessing markets or seed. In turn, higher returns may be obtained from better marketability owing to stability of TRV supply, as well as better household food and nutrition security.

3. Promote conservation and sustainable use of traditional crop varieties by incorporating payment for conservation as a component of Payment for Environmental Services (PES).

   Biodiversity conservation is one of the environmental services provided by upland communities and Arocena-Francisco (2013) proposed that payments for such services could be made in the form of rewards such as provision of free planting materials, conduct of skill training, technical assistance, tenure security, or even support for community conservation strategies. Farming communities growing traditional varieties of crops continuously and employing sustainable farming practices should be rewarded for conserving many of the traditional varieties of crops. These resources are important for the development of new varieties adapted to changing climate and emerging pests.

4. Include traditional varieties of crops in official crop production statistics.
   The Philippine Statistics Authority provides production statistics of various agricultural crops in the country. However, production statistics for traditional varieties are not included, and thus no information can be obtained on traditional varieties; hence, their importance and contribution to the local food security are often not included. Production statistics should therefore include the contribution of traditional varieties of crops, or, in cases where these are lumped with the aggregate statistic, to disaggregate this and make explicit the production statistics of traditional varieties of crops.
5. Local government to support the establishment of community seed banks with a negotiated modality and set of operating guidelines.
   R.A. 7308 (Seed Development Act) consolidated the policy and structure for variety testing and recommendation, seed quality control and certification, and seed production and distribution in the public sector at the national level. At the community level, almost all seed requirements are supplied through farmer-to-farmer exchanges or the informal seed supply system. Ensuring planting materials of traditional varieties every planting season especially after the occurrence of natural calamities such as strong typhoons, extreme droughts, and pest incidences is essential for the continued growing of traditional varieties. With the support of the municipal agricultural office, the community should set up a system for a seed bank with consideration for a preferred modality by the communities, as well as negotiated guidelines for managing and monitoring, and the resource-sharing mechanisms for the operation of the community seed bank. The management guidelines should integrate technical knowledge into identified good local conservation practices.
6. Strengthen local policies that link farmers of traditional crop varieties to markets.
   The National Program on Seed Banking that included growing the best traditional rice varieties per municipality supports the conservation and production of traditional varieties. The MAO should encourage/assist communities to use traditional varieties in promoting products of the municipality under the One Town One Product (OTOP) program, and expand this to traditional varieties of other crops. Furthermore, communications infrastructure support towards the producing communities is expected to contribute to increased farmers' access to markets and value addition through availability of information on qualities of traditional varieties and their products. In light of the increasing importance of digital platforms and other creative modes of communications, farmers should be supported to take advantage of information platforms to share and access information as well as to access market networking opportunities.
7. Safeguard community collections of genetic resources through safety duplication.

To safeguard the communities' genetic resource collection from being lost in cases of natural disasters, collection of each community can be safely duplicated at the Philippine Rice Research Institute (PhilRice) for rice; at the Philippine Root Crops Research and Training Center (PhilRoots) for yam (*Dioscorea alata* L.), sweet potato (*Ipomoea batatas* (L.) Lam.), and gabi (*Colocasia esculenta* L., *Xanthosoma sagittifolium* (L.) Schott); and at the National Plant Genetic Resources Laboratory (NPGRL) for all crops except rice. Safety duplication in these gene banks can be in a black box, where access to these materials shall be under the management and control of designated member(s) of the communities. Hence, institutional arrangements that promote traditional crop varieties' production, conservation, and sustainable use should be strengthened.

8. Monitor traditional crop variety conservation.
   To raise awareness on traditional varieties' conservation among stakeholders, the Department of Agriculture-Bureau of Agricultural Research (DA-BAR) and other agencies with relevant mandates should incorporate traditional crop variety conservation in their standard monitoring systems.

Acknowledgements The authors thank the Benefit-Sharing Fund of the International Treaty of Plant Genetic Resources for Food and Agriculture (Window 1: Strategic Action Plans) for providing funding support through the project "Integrating the conservation of plant genetic resources for food and agriculture into decentralized landscape management for food security and biodiversity conservation in critical eco-regions of the Philippines" conducted in 2014–2016. We also express appreciation for the support of the people and the local governments of Maddela and Nagtipunan, Quirino province; San Remigio, Antique province; Janiuay, Iloilo province; and Baganga, Davao Oriental province, particularly the following municipal agriculture officers (MAO) and staff: Maddela, Quirino MAO Jovencio Salvador; Nagtipunan, Quirino MAO Heidi Marquez; Janiuay, Iloilo MAO Nemesio B. Adolacion and staff, Alrey Adurante and Vivian Lastimoso; San Remigio, Antique MAO Marcos Melicano, Jr. and staff Irene Cabaya-Cajeben; the National Plant Genetic Resources Laboratory (NGPGRL) based at the University of the Philippines Los Baños as venue for the Training of Trainers for Conservation Field School facilitators; the Department of Agriculture-Bureau of Agricultural Research; and Ms. Edna Mercado for the administrative support.

References

Arocena-Francisco H (2013) Environmental service "payments": experiences, constraints and potential in the Philippines. https://www.cbd.int/financial/pes/philippines-pespotential.pdf

Bellon MR, Smale M (1999) A conceptual framework for valuing on-farm genetic resources. In: Wood D, Lenne J (eds) Agrobiodiversity: characterization, utilization and management. CABI Publishing, Wallington

Brown AHD (1963) The genetic structure of crop landraces and the challenge to conserve them in situ on farms. In: Brush SB (ed) Genes in the field: on-farm conservation of crop diversity. Lewis Publishers, the International Plant Genetic Resources Institute, and the International Development Research Centre, New York, pp 29–50

Brush SB, Meng E (1998) The value of wheat genetic resources to farmers in Turkey. In: Evenson R, Gollin D, Santaniello V (eds) Agricultural values of plant genetic resources. CABI Publishing, Wallingford, pp 97–116

Jarvis DI, Hodgkin T (2008) The maintenance of crop genetic diversity on farm: supporting the convention on biological diversity's programme of work on agricultural biodiversity. Biodiversity 9:2

Sthapit, B, Padulosi S, Bhag M (2009) Roles of on-farm conservation and underutilized crops in the wake of climate change. Paper presented in the national symposium on recent global developments in the management of plant genetic resources, NBPGR, New Delhi, India, 17–18 Dec 2009, p 1

UNEP (1992) Convention on Biological Diversity. United Nations Environment Programme, Nairobi

3 Recent Developments in Wild Rice Conservation, Research, and Use

Maria Celeste N. Banaticla-Hilario and Andres Godwin Sajise

Abstract

In this chapter, we assess the current status of wild *Oryza* gene bank collections and in situ protection sites. We have identified *O. neocaledonica* and *O. schlechteri* as priority species, and Africa and South America as priority regions for ex situ and in situ conservation. We review recent genetic diversity studies and explained how information on population genetic structure, ecogeographic differentiation within species, and intra- and interspecific gene flow can be used in enhancing the effectivity and efficiency of conservation methods. We emphasize the need for further studies on species-specific dormancy-breaking treatments and on seed longevity of wild rice species under storage conditions. We discuss the potentials of molecular taxonomic markers, tissue culture, and plant-dwelling symbionts in improving gene bank practices. Finally, we present the significant contributions of wild *Oryza* species to rice improvement.

Keywords

Wild rice · *Oryza* · Conservation · Gene bank · Crop wild relative

3.1 Introduction

A crop wild relative (CWR, plural CWRs) is a noncultivated plant taxon that is genetically related, and potentially crossable to a crop. Taxonomically, CWRs correspond to species belonging to the same genus of the crop, but may also include

M. C. N. Banaticla-Hilario (✉) · A. G. Sajise
International Rice Genebank, International Rice Research Institute, Los Baños, Laguna, Philippines
e-mail: a.sajise@irri.org

S. Ramamoorthy et al. (eds.), *Plant Genetic Resources, Inventory, Collection and Conservation*, https://doi.org/10.1007/978-981-16-7699-4_3

closely related genera. CWRs possess an arsenal of traits that can be used to combat threats to crop production such as pests and diseases, declining agricultural resources, and climate change. Yet they themselves are threatened by overexploitation; habitat fragmentation, homogenization, degradation, and loss; climate change; exotic species; and natural calamities (Food and Agriculture Organization of the United Nations (FAO) 2017).

Over the past decade, great strides have been made to better understand, document, and conserve the wild relatives of crops. In 2011, the Crop Trust and the Royal Botanic Gardens, Kew, started a decade-long project called "Adapting Agriculture to Climate Change: Collecting, Protecting and Preparing Crop Wild Relatives." This project aimed to collect, conserve, and enhance the use of important CWRs (Dempewolf et al. 2014). FAO (2017) introduced guidelines that can be used for CWR conservation and utilization. Several studies raised various recommendations to improve current in situ and ex situ conservation practices (Dempewolf et al. 2017; Engels and Thorman 2020; Khoury et al. 2019; Philips et al. 2017; Riordan et al. 2019). New tools that can help countries in planning CWR conservation were developed (Brehm et al. 2019). Priority taxa were identified at the global (Castaneda-Alvarez et al. 2016), regional (Allen et al. 2019; Fitzgerald 2019), and national levels (Contreras-Toledo et al. 2018; Rahman et al. 2019). Nevertheless, the need for conservation remains urgent, as many CWR species continue to be in danger of extinction (Diederichsen and Schellenber 2018; Engels and Thorman 2020; Khoury et al. 2021; Tyack et al. 2020).

This chapter focuses on the wild relatives of Asian rice *Oryza sativa* L., a crop that sustains more than 3.5 billion people (Ricepedia 2021). Asian rice has a cultivated sibling *O. glaberrima* Steud., produced and consumed mainly in West Africa. Aside from the two mentioned cultigens, the genus *Oryza* L. comprises 25 wild species growing through a wide range of ecological and climatic conditions. There are ten closely related genera to *Oryza,* namely *Chikusichloa* Koidz, *Hygroryza* Nees, *Leersia* Soland, *Luziola* Juss., *Maltebrunia* Kunth, *Potamophila* R. Br., *Prosphytochloa* Schweickerdt, *Rhynchoryza* Baill., *Zizania* Gronov ex Linn., and *Zizaniopsis* Doell. et Aschers. However, only wild *Oryza* species will be tackled in this work.

The infrageneric classification of *Oryza* is presented in Table 3.1. There are three sections and six series that are also referred to as species complexes. Species from the same series/complex have the same genomic constitution. The series *Sativae* or *O. sativa* complex represents the primary gene pool as it contains the two cultivated species. The largest and most genomically diverse series *Latifoliae* or *O. officinalis* complex belongs to the secondary gene pool along with monospecific series *Australianses* and *Brachyanthae.* Finally, series *Meyerianae* (*O. meyeriana* complex), *Ridleyanae* (*O. ridleyi* complex), and *Schlechterianae* composed the tertiary gene pool. *O. coarctata* Roxb., the latest addition to the genus, has not yet been designated to a formal section and series. Its KKLL genome places it in the tertiary gene pool. Phylogeography of these complexes and biology of the species have been discussed by Vaughan et al. (2008) and Mondal and Henry (2018), respectively.

Table 3.1 The sections and series of genus *Oryza* and their placement in the rice gene pool

Section	Series	Genome	Gene pool	Number of species	Distribution
Oryza L.	*Sativae* Sharma et Shastry	AA	Primary	8	Pantropic
	Latifoliae Sharma et Shastry	BB, CC, BBCC, CCDD	Secondary	10	Pantropic
	Australianses Tateoka ex Sharma et Shastry	EE	Secondary	1	Tropical Australia
Brachyantha B. R. Lu	*Brachyanthae* Sharma et Shastry	FF	Secondary	1	Africa
Padia (Zoll. et Mor) Baill.	*Meyerianae* Sharma et Shastry	GG	Tertiary	3	South and Southeast Asia, Andaman and Nicobar Islands, New Caledonia
	Ridleyanae Sharma et Shastry	HHJJ	Tertiary	2	Southeast Asia and Papua New Guinea
	Schlechterianae Sharma et Shastry	HHKK	Tertiary	1	Irian Jaya and Papua New Guinea

In this chapter, we assess the current status of wild *Oryza* ex situ and in situ conservation and identify gaps in terms of species and geography. We review recent genetic diversity and ecogeographic studies and discuss the implications of the results to wild rice conservation and use. We also look into other research results that can be used to improve current gene bank practices. Lastly, we demonstrate the importance of wild *Oryza* species to rice improvement.

3.2 Current Status of Germplasm Collections

Gene banks are facilities that ensure the long-term conservation and continued availability of crops and CWRs. The five largest collections of rice genetic resources are maintained by the International Rice Genebank (IRG) at the International Rice Research Institute (IRRI) (132,141 accessions), Dale Bumpers National Rice Research Center (DBNRRC) (36,280 accessions), Africa Rice Center (21,815 accessions), EMBRAPA National Center for Research on Rice and Beans (EMBRAPA/CNFAP) (20,797 accessions), and the National Small Grains Germplasm Research Facility (NSGRF) (18,715 accessions) (Genesys 2021). However, of the five major rice repositories, only IRRI conserves a significantly huge number of wild rice species and accessions (Table 3.2). The China National Rice Research Institute is said to house 70,000 rice accessions (CGRIS 2021), but germplasm data is not easily accessible. Meanwhile, the National Germplasm Guangzhou Wild Rice Nursery (NGWRN Guangzhou) and the National Germplasm Nanning Wild Rice Nursery (NGWRN Nanning) manage the largest field gene bank collections of wild *Oryza* (CGRIS 2021). The National Institute of Genetics (NIG) of Japan also conserves a relatively large number of wild rice accessions (Oryzabase 2021).

The wild *Oryza* collections are compared in Table 3.2. It is highly possible that duplicate accessions exist between and within gene banks. There are some differences in taxonomic treatment. NIG does not recognize *O. nivara* and *O. schweinfurthiana* as distinct species, regarding the former as the annual ecotype of *O. rufipogon* and the latter as the tetraploid form of *O. punctata*. A few accessions were labelled with taxonomically unaccepted species names (e.g., *O. perennis*, *O. breviligulata*). Such entries were lumped with the weedy and hybrid accessions in Table 3.2.

IRRI has the most comprehensive collection in terms of species coverage (23 out of 25 wild *Oryza* species), followed by NIG (21 species) and NGWRN Guangzhou (19 species). NGWRN Nanning maintains the world's most massive collection of *O. rufipogon* followed closely by NGWRN Guangzhou which also holds the largest collection of *O. officinalis* and *O. latifolia*. NIG has the highest number of *O. grandiglumis* and *O. longiglumis* accessions while EMBRAPA/CFAP has the biggest collections of *O. alta* and *O. latifolia*. IRRI conserves the largest number of accessions of the other 14 wild species (Table 3.2).

Overall, *O. rufipogon* occupies the bulk of wild rice ex situ collections, accounting for 61%, assuming that there are no duplicate accessions (Fig. 3.1). *O. nivara* (11%) and *O. officinalis* (7%) are also abundantly present in the collections (Fig. 3.1). *O. nivara* and *O. rufipogon* (Fig. 3.2) are thought to be the wild progenitors of Asian rice (Cheng et al. 2003; Yamanaka et al. 2002; Ohtsubo et al. 2004; Hu et al. 2006). Their populations occupy open habitats along rivers and streams, on marshes, or near cultivated rice fields, making them easily accessible for germplasm collection. *O. officinalis* is one of the most widely distributed wild rice species (Table 3.2) and grows in a variety of habitats (e.g., forest margins, ditches and lakes, agricultural farms). Weedy accessions also make up a considerable portion of the collection (9%). Weedy populations are natural hybrid populations between a cultivated variety and a wild rice population or between two cultivated varieties. They are also recognized as important genetic resources that can be used for rice improvement. While the primary gene pool of rice is relatively well represented in gene banks, the presence of other series is limited (Fig. 3.1; Table 3.2). Most of the species with few accessions have a restricted distribution range. *O. longiglumis* and *O. schlechteri* can only be found in Irian Jaya and Papua New Guinea. *O. malampuzhaensis* and *O. neocaledonica* are endemic to the Western Ghats of India and New Caledonia, respectively (Table 3.2).

Table 3.2 Wild *Oryza* species and their representation in rice gene bank collections

Species	Genome	Distribution	Number of available accessions							
			IRRI	DBNRRC	Africa Rice	EMBRAPA/ CNFAP	NSGRF	NGWRN Guangzhou	NGWRN Nanning	NIG
O. barthii A. Chev.	AA	Africa	188	1	37	3	4	6	0	20
O. glumipatula Steud.	AA	South and Central America	52	1	0	98	19	44	0	26
O. longistaminata Chev. et Roehr	AA	Africa	213	0	1	1	0	11	30	62
O. meridionalis Ng	AA	Tropical Australia	53	0	0	0	0	6	1	28
O. rufipogon Griff.	AA	Tropical and subtropical Asia, Tropical Australia	845	4	0	3	11	3479	4055	363
O. nivara Sharma et Shastry	AA	Tropical and subtropical Asia	1458	2	0	5	13	80	5	0
O. punctata Kotschy ex Steud.	BB	Africa	43	0	1	2	0	20	8	10
O. schweinfurthiana Prod.	BBCC	Africa	25	0	0	0	0	0	0	10[a]
O. malampuzhaensis Krish. et Chandr.	BBCC	India	10	0	0	0	0	3	0	0
O. minuta J. S. Presl. et C. B. Presl.	BBCC	Philippines and Papua New Guinea	59	0	0	0	0	0	0	14
O. eichingeri Peter	CC	South Asia and East Africa	22	0	0	0	0	4	2	5

(continued)

Table 3.2 (continued)

Species	Genome	Distribution	Number of available accessions							
			IRRI	DBNRRC	Africa Rice	EMBRAPA/ CNFAP	NSGRF	NGWRN Guangzhou	NGWRN Nanning	NIG
O. officinalis Wall ext. Watt	CC	Tropical and subtropical Asia, Tropical Australia	274	1	0	2	1	491	230	86
O. rhizomatis Vaughan	CC	Sri Lanka	20	0	0	0	0	3	0	3
O. alta Swallen	CCDD	South and Central America	12	1	0	42	2	6	5	6
O. grandiglumis (Doell) Prod.	CCDD	South and Central America	9	0	0	41	1	4	2	60
O. latifolia Desv.	CCDD	South and Central America	54	1	1	20	3	109	5	25
O. australiensis Domin.	EE	Tropical Australia	36	0	0	0	0	22	0	23
O. brachyantha Chev. et Roehr.	FF	Africa	17	0	0	0	0	5	0	17
O. granulata Nees et Arn ex Wall.	GG	South and Southeast Asia	23	0	0	0	0	2	0	6
O. meyeriana (Zoll. et Mor ex Steud.) Baill.	GG	Southeast Asia, Andaman and Nicobar Islands	8	0	0	0	0	9	3	8
O. neocaledonica Morat	GG	New Caledonia	0	0	0	0	0	0	0	0

O. longiglumis Jansen	HHJJ	Irian Jaya and Papua New Guinea	5	0	0	0	0	0	0	12
O. ridleyi Hook f.	HHJJ	Southeast Asia and Papua New Guinea	12	0	0	0	0	2	0	4
O. schlechteri Pilger	HHKK	Irian Jaya and Papua New Guinea	0	0	0	0	0	0	0	0
Oryza coarctata Roxb.	KKLL	Tropical Asia	1	0	0	0	0	0	0	1
Weedy, natural hybrids, and accessions with uncertain or ambiguous classification			1130	3	0	95	0	26	3	11
		Total	4569	14	40	312	10	4332	4349	800

[a] Recorded as tetraploid *O. punctata* in the database.

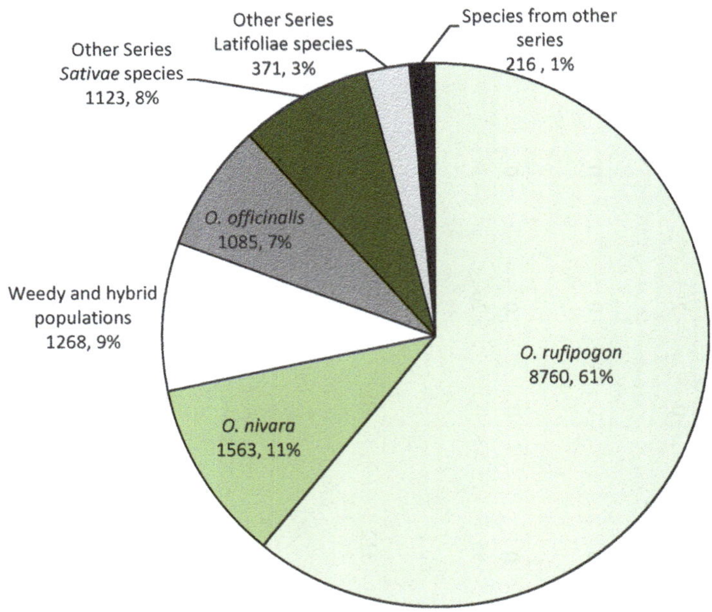

Fig. 3.1 Wild *Oryza* accessions conserved across major rice gene banks. The number of available accessions precedes the percentage values

3.3 Priority Species and Collection Gaps

Red flags should be raised for the endangered species *O. neocaledonica* (Fig. 3.3) and *O. schlechteri* as they remain absent in gene bank collections. Previous attempts to conserve them ex situ have failed. These species should be studied in their natural environments. Information on habitat microclimate and other ecological requirements are critical to the success of their ex situ conservation. Germplasm collection should be conducted before they become extinct in the wild.

O. coarctata and *O. longiglumis* are also poorly represented in gene bank collections. Both species are categorized as data deficient by the International Union for Conservation of Nature (IUCN 2021). Increased ex situ presence is also needed for the near-threatened *O. rhizomatis* and vulnerable *O. malampuzhaensis* (IUCN 2021).

The geographic coverage of the IRRI wild rice collection is displayed in Fig. 3.4. Only eight wild *Oryza* species exhibited complete (100%) country-level representation of their corresponding geographic range. Six species have moderate-to-good geographic coverage (57–87% representation), while 11 species have poor coverage with 0–50% representation. Most of the species with poor geographic coverage are from Africa and South America. Hence, these two regions should be prioritized for germplasm collection. The countries that are most poorly covered in the collection

Fig. 3.2 Plant and seeds of *O. rufipogon,* IIRGC 93216 (**a**), and *O. nivara,* IRGC 86493 (**b**)

(i.e., with at least two uncollected species) are shown in Table 3.3. Seventeen of the 30 countries are from Africa, ten from South America, and three from Southeast Asia. Countries with the most number of uncollected species are Central African Republic, Ivory Coast, Sudan, and Zimbabwe.

Fig. 3.3 Spikelets of the endangered wild rice species *O. neocaledonica*

Several NIG accessions fit into some of the IRRI collection gaps. NIG accessions that correspond to geographic populations that are absent in the IRRI collection are *O. coarctata* from India, *O. officinalis* from Borneo, *O. brachyantha* and *O. schweinfurthiana* from Chad, and *O. latifolia* from Argentina, Brazil, and Paraguay. Geographic data of accessions conserved by EMBRAPA/CNFAP, NGWRN Guangzhou, and NGWRN Nanning are not publicly available online. Perhaps major wild rice gene banks all over the world should coordinate to compare the provenance of collections and determine duplicate and unique accessions. Only then can we have a complete picture of the state of wild rice ex situ conservation.

3.4 In Situ Conservation

CWRs should be allowed to traverse their evolutionary paths by adapting to the biotic and abiotic challenges that they encounter in their natural habitats. Hence, a good combination of in situ protection and ex situ preservation will ensure the effectiveness of CWR conservation. However, the known in situ sites exclusively dedicated to wild rice conserve mainly the species *O. rufipogon*. In China, there are ten conservation sites distributed along the southern provinces (Song et al. 2005). In Thailand, two in situ areas are located inside the compounds of separate rice research centers (Kaewcheenchai et al. 2018). In Laos, a wild rice conservation site, characterized by a complex of *O. rufipogon* populations growing in a large swamp, was established (Fig. 3.5). In Guangxi province, China, two conservation sites for *O. officinalis* were established in 2005 (Zhao et al. 2012).

Genetic analyses revealed vast and complex diversity residing within the in situ populations in Laos (Wang et al. 2012) and Thailand (Kaewcheenchai et al. 2018).

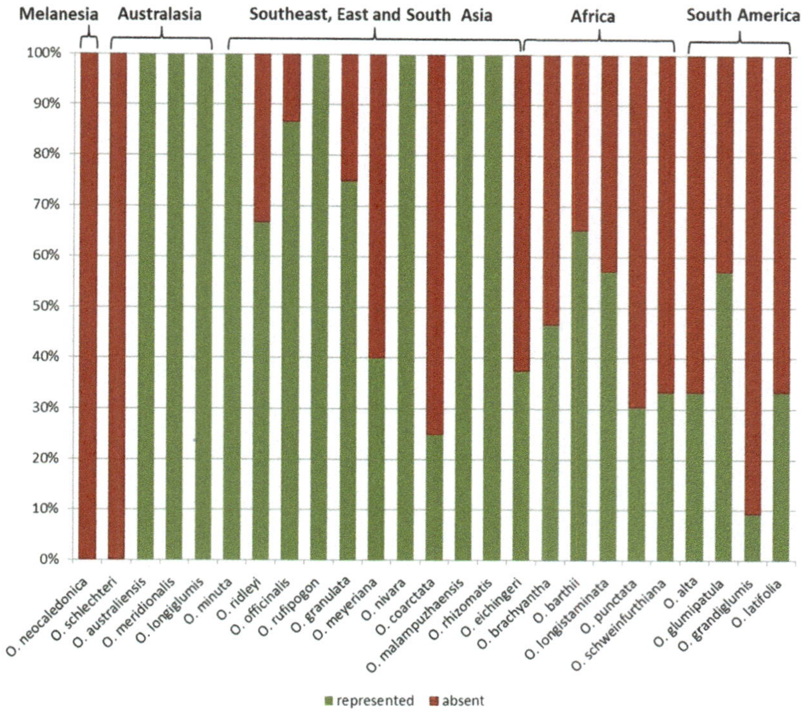

Fig. 3.4 Geographic coverage of the IRG, IRRI wild rice collection. The percentage values pertain to the country-level representation of each species (number of countries where the species was collected/number of countries where the species was recorded). The species are arranged according to their geographic region

Different levels of genetic differentiation between and within populations were observed (Wang et al. 2012; Kaewcheenchai et al. 2018). In Laos, unique polymorphisms were detected in certain areas of the conservation site (Wang et al. 2012). This implies that ex situ conservation may not entirely capture the actual diversity of the species and its populations in nature. In fact, Xie et al. (2010) found that gene bank accessions were not as genetically diverse as their in situ counterparts. Similarly, Liu et al. (2017) detected differences in population structure between ex situ collections and in situ populations of *O. rufipogon* in Dongxiang, China.

Gao et al. (2012) studied numerous natural populations of *O. rufipogon* in China, proposed conservation strategies, and identified priority populations for in situ protection. The same should be done for the other wild species of the genus *Oryza*. In situ conservation is a must for species with recalcitrant seeds (*O. coarctata*); strongly dormant seeds (series *Meyerianae* and *Ridleyanae*); and inability to flower or produce seeds under ex situ conditions (*O. neocaledonica* and *O. schlechteri*).

Table 3.3 Priority countries and species that are not represented in the IRG, IRRI collection. The countries and species are arranged according to geographic region

	Asia				
	O. officinalis	*O. granulata*	*O. ridleyi*	*O. meyeriana*	*O. coarctata*
Cambodia	x	x	x		
India				x	x
Myanmar		x	x		
Angola					
Benin					
Central African Republic					
Chad					
Congo					
Ethiopia					
Ghana					
Ivory Coast					
Liberia					
Malawi					
Rwanda					
South Africa					
Sudan					
Swaziland					
Zaire					
Zambia					
Zimbabwe					
Argentina					
Belize					
Bolivia					
Colombia					
Ecuador					
French Guiana					
Guyana					
Honduras					
Paraguay					
Peru					

	Africa						South America			
	O. eichingeri	*O. brachyantha*	*O. barthii*	*O. longistaminata*	*O. punctata*	*O. schweinfurthiana*	*O. alta*	*O. glumipatula*	*O. grandiglumis*	*O. latifolia*
Cambodia										
India										
Myanmar										
Angola				x	x	x				
Benin			x		x	x				
Central African Republic	x	x	x	x	x					
Chad		x				x				
Congo		x			x	x				
Ethiopia			x		x					
Ghana		x	x		x					
Ivory Coast	x		x		x	x				
Liberia			x	x						
Malawi				x	x					
Rwanda	x			x						
South Africa				x	x	x				
Sudan		x	x	x	x					
Swaziland					x	x				
Zaire	x	x			x					
Zambia				x	x	x				
Zimbabwe			x	x	x	x				
Argentina									x	x
Belize							x			x
Bolivia									x	x
Colombia							x	x	x	
Ecuador									x	x
French Guiana									x	x
Guyana								x		x
Honduras							x			x
Paraguay									x	x
Peru							x		x	x

Fig. 3.5 In situ conservation site for *O. rufipogon* in Thong Mang Village, Vientiane Municipality, Xaythany District, Laos

Ideally, wild rice populations are conserved when they reside in protected areas. Sjahril et al. (2015) discovered that *O. officinalis* populations are well preserved in the conservation sites of South and West Sulawesi, Indonesia. However, the area is said to be threatened by rapid urban expansion. Table 3.4 provides a list of protected areas where wild *Oryza* species were recorded (Vaughan 1990; Maxted and Kell 2009). Additionally, Joshi et al. (2008) identified the Rupandehi and Kapilvastu districts in Nepal as candidate areas for in situ conservation on account of their high wild rice species diversity and extreme soil and climatic conditions. There is a need to assess and monitor the status of wild rice populations in specially designated in situ conservation sites as well as in protected areas to guarantee their continued existence in nature. Moreover, in situ protection sites should be identified and established in South America and Africa.

3.5 Genetic Diversity and Ecogeographic Studies

3.5.1 Population Diversity

A clear understanding of species variation patterns is essential in designing practical conservation strategies. The predominantly outcrossing *O. rufipogon* displayed low population differentiation across its distribution range (Banaticla-Hilario et al. 2013a), and at the country level (Gao and Zhang 2005; Shishido et al. 2006; Kuroda et al. 2007). Likewise, the highly outcrossing species *O. longistaminata*

Table 3.4 Protected areas where wild *Oryza* species were recorded (updated Vaughan (1990) and Maxted and Kell (2009))

Country	Protected area	Species
India	Parambikulam Game Reserve, Kerala	*O. granulata* and *O. officinalis*
	Bhoothathankettu Forest Reserve, Kerala	*O. rufipogon*
	Karulai Range Teak Plantation/Forest Reserve, Kerala	*O. granulata*, *O. malampuzhaensis* and *O. nivara*
	Simlipal National Wildlife Park, Orissa	*O. granulata*, *O. nivara* and *O. officinalis*
Sri Lanka	Polonnaruwa, ruins of ancient capital	*O. nivara*
	Puwarasankulam Forest Reserve	*O. rhizomatis*
	Ritigala Strict Natural Reserve	*O. eichingeri*
	Weerakulicholai-Elavankulam Forest Reserve	*O. rhizomatis*
	Wilpattu National Park	*O. rhizomatis*
	Wilpotha Forest Reserve	*O. rhizomatis*
	Yala Strict Nature Reserve and Yala (Ruhuna) National Park	*O. nivara* and *O. rhizomatis*
Thailand	Khao Sam Lan National Park near Saraburi	*O. officinalis* and *O. ridleyi*
	Mae Sae Valley Forest Reserve and Botanic Garden near Chiang Mae	*O. granulata*
	Pukae Botanic Garden near Saraburi	*O. officinalis*
	Tony Nga Chang Park near Hat Yai	*O. ridleyi*
Indonesia	Gunung Lorentz National Park	*O. schlechteri*
	Ujung Kulon National Park	*O. officinalis*
	Gunung Leuser Nature Reserve, Aceh/N. Sumatra	*O. meyeriana*
Papua New Guinea	Bismarck-Ramu National Park	*O. schlechteri*
	Neiru Wildlife Management Area and Kikori Marine Park/Reserve	*O. schlechteri*
	Tonda Wildlife Management Area	*O. longiglumis* and *O. minuta*

exhibited high intrapopulation and low interpopulation genetic diversity (Kiambi et al. 2005; Melaku et al. 2013; Mangosongo 2020). The same variation pattern was shown by *O. latifolia* and *O. grandiglumis* (Veasey et al. 2011). In contrast, high genetic variability between populations was displayed by the inbreeding *O. nivara* (Kuroda et al. 2007; Banaticla-Hilario et al. 2013a), the predominantly inbreeding *O. glumipatula* (Brondani et al. 2005; Karasawa et al. 2007; Abreu et al. 2015), and *O. granulata* (Wu et al. 2004; Wu and Gao 2013). In China, a high degree of population differentiation was observed in the predominantly outcrossing *O. officinalis* (Gao 2005).

Such information can be used in establishing species-specific protocols for germplasm collection and conservation, and in situ preservation. Species with low intrapopulation variability and high population differentiation can be aptly represented in ex situ collections by few individuals from numerous populations.

In this case, a small number of individuals can be planted for regeneration purpose. Moreover, several distinct populations should be marked for in situ protection. On the other hand, collecting many individuals from a small number of populations, using a large number of individuals for regeneration, and conserving a few, extensive in situ populations may be suitable for species with high intrapopulation and low interpopulation diversity.

3.5.2 Ecogeographic Populations

Genetic diversity analyses have detected distinct, geographic populations within several wild *Oryza* species. The Australian populations of *O. rufipogon* are distinct from the rest of the species as revealed by chloroplast sequences (Waters et al. 2012), microsatellite markers (Banaticla-Hilario et al. 2013a), and morphological data (Banaticla-Hilario et al. 2013b). Asian *O. rufipogon* displayed further genetic differentiation between accessions from 1) South Asia and Myanmar and 2) Southeast Asia (Banaticla-Hilario et al. 2013a). A similar variation pattern was observed in *O. officinalis* where higher latitude (India, China, Northern Thailand, and Myanmar) accessions separated from lower latitude (Malesia, Vietnam and Central Thailand) accessions (Naredo et al. 1998). Divergence of South Asian and Southeast Asian populations was also detected in *O. nivara* (Banaticla-Hilario et al. 2013a and 2013b). Based on single sequence repeat (SSR) data, populations from Nepal appeared to be genetically distinct from the rest of *O. nivara* (Banaticla-Hilario et al. 2013a).

Regional divergence was also observed in African species. Single-nucleotide polymorphism (SNP) analysis showed three regional population groups of *O. barthii* corresponding to (1) Central Africa; (2) Niger River Delta and West Africa; and (3) Chad Lake region and Austral Africa (Orjuela et al. 2014). Labroo (2017) detected differentiation in *O. longistaminata* where genetic subpopulations were recognized as (1) Northwestern African; (2) Pan-African; and (3) Southern African. SSR data revealed divergence of African *O. eichingeri* accessions from disjunct Sri Lankan populations (Quilloy 2008).

Even species with limited distribution range exhibited geographic divergence. RAPD, SSR, and hybridization data recognized three geographic populations of *O. meridionalis* namely (1) Irian Jaya, (2) Queensland, and (3) Northern Territory and Western Australia (Juliano et al. 2005). *O. rhizomatis* displayed differentiation between the northern and southeastern populations in Sri Lanka, as evidenced by RAPD and SSR data (Bautista et al. 2006). However, this was disputed by Rajkumar et al. (2015) as AFLP and morphological markers showed no significant differences among accessions collected throughout the range of *O. rhizomatis*.

For some species, genetic divisions are not just limited to geographic regions, and exist at finer spatial and ecoclimatic scales. Subnational differentiation was observed in *O. rufipogon* in Malaysia (Ngu et al. 2010), Laos, and Northern Thailand (Pratepha 2012); *O. glumipatula* in Brazil (Veasey et al. 2011; Karasawa et al. 2012); *O. granulata* in China (Wu and Gao 2013); and *O. meyeriana* in China

(Ya-tao et al. 2008) and the Philippines (Bon et al. 2011). *O. rufipogon* in China separated into two main ecogeographic groups (a south subtropical population and a middle subtropical population) that were both further divided into geographic populations (Wang et al. 2008). In India, differentiation of *O. rufipogon* into (1) Pro-Indica, (2) Pro-Aus, and (3) Mid-Gangetic groups was also correlated to the ecology and geography of populations (Singh et al. 2017). Likewise, *O. longistaminata* displayed an ecogeographic division between populations from the Amhara and Gambella regions in Ethiopia (Melaku et al. 2019, 2020).

Ecology-driven differentiation has been detected in certain species. SSR markers grouped *O. rufipogon* populations in Hainan, China, according to the type of their habitats (Yibo et al. 2010). Bressan et al. (2018) detected divergence between mountaintop and lowland populations of *O. glumipatula* in Brazil. Wang et al. (2019) observed that ecological factors, rather than spatial distance, played a great role in shaping the genetic structure of *O. rufipogon* in China.

Knowing the ecogeographic variation patterns of species can help in establishing in situ sites and ex situ collections, and in identifying conservation gaps. Gene banks with limited capacity may opt to maintain regional populations rather than represent the entire species range in their collections. On the other hand, some gene banks may want to capture species genetic diversity at a finer scale. This can be done by conserving populations that correspond to intraspecific genetic divisions within the subnational geographic scope (e.g., provincial, town, or local level). Since ecology is a key driver to species evolution, germplasm collection and in situ protection should focus on targeted areas and cover the range of species habitats. For *O. rufipogon* in China, high-diversity priority areas for conservation have been determined (Wang et al. 2019) and local populations for immediate protection have been identified (Yibo et al. 2010). Similar efforts should be done for the rest of the wild rice species. Ecogeographic variation patterns can also be considered in studying the wild species of *Oryza* since different genetic population groups may possess distinct traits and exhibit varied responses to environmental conditions. However, only a few wild rice species have been well analyzed. Ecogeographic studies on *O. alta, O. australiensis, O. brachyantha, O. coarctata, O. longiglumis, O. ridleyi, O. minuta, O. malampuzhaensis, O. punctata, O. schweinfurthiana, O. schlechteri,* and *O. neocaledonica* are lacking.

3.5.3 Gene Flow

Reproductive barriers between species from the same series or genome group are not completely sealed. Introgression with cultivated rice has been detected in *O. rufipogon* in China (Song et al. 2005; Ji et al. 2016), Laos (Kuroda et al. 2007), Malaysia (Ngu et al. 2010), Thailand (Wedger et al. 2019), and globally (Wang et al. 2019). This was also observed in *O. nivara* in Laos (Kuroda et al. 2007) and *O. glumipatula* in Costa Rica (Fuchs et al. 2016). As mentioned by Lu (2012), one of the greatest challenges for CWR in situ conservation is crop-wild introgression since it can change the genetic structure of wild populations, affecting its integrity.

Thus, it is crucial that in situ populations are isolated (geographically or through artificial barriers) from their crop relative. Germplasm-collecting missions should target wild populations that are distant from cultivated fields.

Hybridization and introgression can also occur between closely related species that co-occur within the same locality. A natural hybrid population of *O. minuta* and *O. officinalis* was studied by Sana et al. in 2003. Gene flow has also been reported between *O. nivara* and *O. rufipogon* (Zhou et al. 2008; Banaticla-Hilario et al. 2013a) and between *O. glumipatula* and *O. grandiglumis* (Lentini and Espinosa 2005 as cited by Vaughan et al. 2005). SSR data revealed reinforced reproductive isolation barriers between sympatric populations of *O. nivara* and *O. rufipogon* compared to non-sympatric populations, indicative of ecological sympatric speciation (Banaticla-Hilario et al. 2013a). Regeneration protocols should be reviewed and revised to ensure that inter- and intraspecific gene flow between gene bank accessions is prevented or minimized at the least. Under ex situ conditions, species and populations are no longer protected by ecogeographic isolation, necessitating the use of pollen barriers to preserve the genetic integrity of accessions.

In contrast, in situ protection should not hinder intraspecific gene flow. Yang et al. (2007) observed that enclosing the in situ conservation area (in Dongxiang, China) with brick walls caused genetic differentiation among subpopulations. Hence, protective structures such as solid walls should be installed cautiously to allow gene flow between conspecific populations. In situ structures should also not impede species dispersal by obstructing dispersal agents such as bodies of water and animals.

3.6 Other Conservation-Related Research

3.6.1 Seed Studies

Strong seed dormancy limits the availability and use of an accession. In 1998, Naredo et al. tested the response of wild *Oryza* species to various dormancy-breaking treatments. Of the 16 wild species, only *O. officinalis* and *O. ridleyi* required chemical treatment. For the remaining species, successful dormancy breaking was achieved by a combination of dry heat treatment, dehulling, and specific temperature conditions (Naredo et al. 1998). Similar results were obtained by Waheed et al. (2012) in *O. rufipogon* and *O. longistaminata*. Soaking the seeds in 5% H_2O_2 solution was found to be effective in breaking the dormancy of *O. rufipogon* populations in Dongxiang, China (Jian Kun et al. 2011).

Dehulling prior to germination is an effective way of breaking wild rice seed dormancy. However, it is time consuming and labor intensive, limiting the number of accessions that can be germinated at a given time. Timple et al. (2018) subjected intact seeds of 15 wild *Oryza* species to heat and chemical treatments and recommended the following dormancy-breaking methods: (1) dry heat treatment at 50 °C and sowing directly in KNO3; (2) dry heat treatment followed by pre-soaking in GA3 or H2O2 for 18 h, and then soaking in distilled H2O for another 18 h before sowing, and for thick-hulled species; and (3) dry heat treatment followed by pre-soaking in HNO3 for 18 h, and then soaking in GA3 or H2O2 for 18 h before sowing. However, only a single accession from each species was analyzed and strongly dormant species *O. granulata, O. meyeriana, O. longiglumis*, and *O. ridleyi* were not included in the study. The recommended treatments can be further optimized, and variation of seed dormancy within species should be examined so that more efficient, species-specific dormancy-breaking protocols can be established.

The rate of dormancy loss differs between and within species. Seed dormancy of *O. meridionalis* was broken faster than *O. rufipogon* under natural conditions (Belairs et al. 2015). Complete loss of dormancy was observed in *O. eichingeri, O. grandiglumis, O. latifolia*, and *O. punctata,* 2 months after harvesting, while *O. glumipatula* and *O. rufipogon* displayed prolonged dormancy (Veasey et al. 2011). Varying levels of dormancy were also observed in different ecotypes of *O. glumipatula* (Veasey et al. 2011).

Timple and Hay (2018) determined that initial drying at 45 °C followed by drying at 15% RH and 15 °C enhanced the seed longevity of wild *Oryza* accessions. Other than this, there are no recent published studies on wild rice seed longevity. It is important to know the seed life span of the different *Oryza* species under storage conditions. The schedule of testing for accession viability monitoring can be tailor-made to suit the seed longevity of species. Viability testing can be done at longer intervals for long-lived species while species with short-lived seeds should be monitored at shorter intervals.

3.6.2 Molecular Authentication

Species authentication is essential to gene bank management. The taxonomic name conveys all the known information about the species. Hence, an incorrect classification or species label defeats the purpose of accession. However, for the genus *Oryza*, there are cases when morphological markers are not enough to determine the species identity of accessions. Species from the same series/genome groups would have overlapping character traits, making it difficult to visually delineate between species. Several molecular markers have been developed to aid in identifying wild *Oryza* germplasm. Yamaki et al. (2013) established codominant insertion/deletion (INDEL) markers that can discriminate between genome types and between certain species while Lou et al. (2021) developed an INDEL marker set that can discriminate alleles between the species of series *Sativae* (AA genome group). Brozynska et al. (2014) proved that chloroplast sequencing methods can be used in barcoding cultivated and wild rice species in Australia. Subsequently, Song et al. (2017) developed chloroplast molecular markers that can provide reliable identification of 12 wild species and 2 subspecies of Asian rice. Ndjiondjop et al. (2018) developed SNP markers that can distinguish *O. barthii, O. longistaminata, O. glaberrima*, and *O. sativa* accessions from each other. Finally, Zhang et al. (2021) identified super DNA barcodes as the best markers for discriminating 21 wild *Oryza* species, compared to other conventional and rice-specific barcodes. These molecular markers can be used in verifying the taxonomic identity of incoming accessions and in detecting contaminations and mishandling errors during routine gene bank operations.

3.6.3 Tissue Culture

Wild *Oryza* accessions are mainly conserved in field and seed gene banks. But for some species, in vitro techniques can be an effective method for germplasm preservation. In vitro conservation involves growing tissue cultures in flasks or tubes, under sterile and controlled conditions.

For *O. alta,* protocols for obtaining regeneration plantlets from leaf tissues have been developed (Ni et al. 2007) and the best media for callus formation and for plantlet regeneration have been determined (Liang et al. 2014). Tissue culture protocols have also been optimized for *O. nivara* (Jazeela et al. 2009), *O. eichingeri* (Wang et al. 2019), and *O. rufipogon* (Zeliang et al. 2010; Paramasivam and Harikrishna 2020). Recent studies on the suitability of tissue culture for the rest of the wild rice species are either lacking or unavailable publicly.

In vitro conservation is highly recommended for the recalcitrant species *O. coarctata*, the endangered and hard-to-regenerate species of *O. neocaledonica* and *O. schlechteri*, and other accessions that are difficult to grow ex situ. A duplicate collection of tissue cultures can serve as a safety backup in case the vegetatively maintained accessions in the field gene banks are lost.

3.6.4 Endophytic Microbiome

Optimum plant health is vital in producing good-quality seeds for gene bank safekeeping. Plant-dwelling symbionts (also known as endophytes) play a key role in improving plant fitness and defense by facilitating nutrient uptake, promoting growth, regulating metabolism, and facilitating hormone signaling that enables the plant to adapt to biotic and abiotic stressors (Liu et al. 2017).

The endophytic microbiome of *O. brachyantha* was shown to promote plant growth and induce tolerance against pests and diseases (Acharya et al. 2017). Meanwhile, Peng et al. (2021) hypothesized that the dominant microbiome of *O. longistaminata* may have enabled the species to survive nutrient-poor soils. A diverse array of beneficial endophytes have been found in *O. officinalis* (Yuan et al. 2018), *O. latifolia* (Peng et al. 2009), *O. granulata* (Yuan et al. 2011), *O. glumipatula* (Junior et al. 2013), and *O. australiensis* (Yuan et al. 2014). New species have been discovered from wild rice plants. A new species of endophytic fungi (*Harpophora oryzae* Yuan et al.) was isolated from *Oryza granulata* (Yuan et al. 2011) while novel nitrogen-fixing bacterial species *Enterobacter oryzae* (Peng et al.) and *Rhizobium wuzhouense* (Yuan et al.) were isolated from *O. latifolia* (Peng et al. 2009) and *O. officinalis* (Yuan et al. 2018), respectively. Comparative studies revealed differences in the microbiomes of cultivated and wild rice species, with the latter harboring more diverse and abundant bacterial communities than the former (Shenton et al. 2016; Tian et al. 2016).

The potential use of endophytic microbiomes in improving the health of wild rice plants should be explored further. Incorporating these symbionts in ex situ conservation practices may be a better way of establishing and maintaining healthy regeneration plots and vegetatively maintained accessions, as it may reduce the need for chemical inputs.

3.7 Utilization

Wild rice species are essential resources of genes and traits that can help breeders overcome the weaknesses of many new varieties. The use of wild *Oryza* species in plant breeding will continue to increase because of its available unlimited genetic diversity. However, the association of these species with several undesirable traits like weediness, high seed shattering, excessive biomass, and photosensitivity is a big challenge to exclude, without losing the more useful genes like high yield and high grain quality, insect pest and disease resistance, and abiotic stress tolerance. The introgression of useful genes from the wild accessions to cultivated species will still remain an important strategy in the future. Until now, the biggest bottleneck in the transfer of useful genes into new-generation high-yielding inbreds and commercial hybrids is the incompatibility barrier in conventional crossing methods.

Both cultivated rice, *O. sativa* and *O. glaberrima*, and AA genome (series *Sativae*) wild species are cross-compatible. Many interspecific hybrids have been produced to transfer valuable genes from the AA genome wild species through conventional hybridization and backcrossing. Hybridization between cultivated rice and non-AA genome wild species is very difficult. Interspecific reproductive barriers make it difficult to transfer useful genes from the non-AA genome wild species to new rice varieties. To bypass incompatibility barriers and increase efficiency, tissue culture and embryo rescue techniques are employed. A study done by Shimizu-Sato et al. (2020) demonstrated that a wide range of wild Oryza accessions respond well to agrobacterium-mediated genetic transformation, embryo rescue, and modified regeneration protocols. The success in generated transgenic plants provides insights on the study of wild *Oryza* genetic resources for molecular breeding applications.

With the recent advances in molecular breeding, genomics, and genetic transformation, new successes have been achieved in using wild relatives for rice improvement. Table 3.5 presents a list of useful wild rice traits that can be tapped to enhance modern rice varieties. Nevertheless, a new research to focus on pre-breeding work must be revived in order to fully utilize the use of wild species on the evaluation and assessment in terms of their reaction to the new strains of pest and disease as well as interaction of environmental stresses brought about by climate change is a new area to explore and the end product is the genetic and phenotypic profiles.

Table 3.5. Useful traits derived from wild *Oryza* species

Donor species	Accession or population used	Gene	Trait	Stage of growth	Reference/s
Oryza australiensis		Rca (Rubisco activase)	Heat stress tolerance	Vegetative phase	Scafaro et al. (2018)
		Pi40	Rice blast disease resistance		Jeung et al. (2007)
		Xa32	Bacterial leaf blight resistance		Zheng et al. (2009)
		Bph10	Brown plant hopper (BPH) resistance		Ishii et al. (1994), Zhao et al. (2016)
		Bph18	BPH resistance		Ji et al. (2016)
Oryza barthii	Chromosome segment substitution lines (CSSLs)	Some QTLs and genes	Panicle length, number of primary branches, number of grains per panicle, grain length, grain width, and 100-grain weight	Reproductive stage	Bessho-Uehara et al. (2017)
	Near isogenic lines (NILs)	Dominant dwarf gene, *Dd7*	Dwarf plant architecture	All phases	Zhao et al. (2018)
			Increased stomatal size		Chatterjee et al. (2020)
	IRGC 106238		Iron toxicity tolerance		Bierschenk et al. (2020)
Oryza brachyantha	Monosomic alien addition lines (MAALs)		Yellow stem borer resistance	Vegetative stage	Narain et al. (2016)
Oryza coarctata	Hybrid progenies		Saline tolerance and submergence tolerance	All phases	Islam et al. (2017)
Oryza glumipatula		*Pi68*	Rice blast disease resistance		Devi et al. (2020)
Oryza latifolia		*Bph12*	BPH resistance		Qiu et al. (2012)
Oryza longistaminata		*Xa21*	Bacterial leaf blight resistance		Khush (1990)
		Pi57	Rice blast disease resistance		Xu et al. (2015)

Oryza minuta		*Pil3*	Rice blast disease resistance		Narain et al. (2016)
		qKI307–2 qKI209–3	Rice blast disease resistance		Rahman et al. (2011)
	IRGC 101133	*Xa35*	Bacterial leaf blight resistance		Guo et al. (2010)
	IRGC 101141	*Bph20*	BPH resistance		Rahman et al. (2009)
		Bph16	BPH resistance		Hirabayashi et al. (2004)
		Bph23	BPH resistance		Ram et al. (2010)
	IRGC 101141	*Xa27*	Bacterial leaf blight resistance		Gu et al. (2004)
Oryza meridionalis	IRGC 105290		Iron toxicity tolerance		Bierschenk et al. (2020)
Oryza meyeriana		*xa32*	Bacterial leaf blight resistance		Ruan et al. (2008)
		qBBR5	Bacterial leaf blight resistance		Chen et al. (2012)
Oryza nivara		*Xa33*	Bacterial leaf blight resistance		Kumar et al. (2012)
		Xa38	Bacterial leaf blight resistance		Bhasin et al. (2012)
		qBLAST8, qBLAST8	Rice blast disease resistance		Eizenga et al. (2013)
		qShB6	Sheath blight disease resistance		Eizenga et al. (2013)
	IRGC 100593		Iron toxicity tolerance		Bierschenk et al. (2020)
	IRGC 103837		Iron toxicity tolerance		
		Grh4	Green rice leafhopper		Fujita (2004)
		Grh6	Green rice leafhopper		Phi et al. (2019)
		Bph34	BPH resistance		Kumar et al. (2018)

(continued)

Table 3.5. (continued)

Donor species	Accession or population used	Gene	Trait	Stage of growth	Reference/s
	IRGC 81825	*Xa38*	Bacterial leaf blight resistance		Ellur et al. (2016)
Oryza officinalis		*Xa29*	Bacterial leaf blight resistance		Tan et al. (2004)
		qBph3 and *qBph4*	BPH resistance		Hu, et al. (2015)
		Bph6	BPH resistance		Ali and Chowdury (2014)
		Bph10	BPH resistance		Thi Lang and Chi Buu (2003)
		bph11	BPH resistance		Hirabayashi (1998)
		Bph13	BPH resistance		Renganayaki et al. (2002)
		Bph14	BPH resistance		Du et al. (2009)
		Bph15	BPH resistance		Yang et al. (2004)
	IRGC 100896	*Bph13*	BPH resistance		Ali and Chowdury (2014)
		Wbph7 and Wbph8	White-backed plant hopper (WBPH) resistance		Tan et al. (2004)
		qSbph3d, qSbph7a qSbph12b	Small brown plant hopper resistance		Zhang et al. (2014)
Oryza rufipogon		*Xa23*	Bacterial leaf blight resistance		Yinglun et al. (2006)
		Xa30	Bacterial leaf blight resistance		Xuwei et al. (2007)

		Bph29	BPH resistance		Ali and Chowdury (2014)
	IRGC 105293		Iron toxicity tolerance		Bierschenk et al. (2020)
	IRGC 106431		Iron toxicity tolerance		Bierschenk et al. (2020)
		bph18, bph19	BPH resistance		Li et al. (2010)
		Bph20, Bph21	BPH resistance		Yang et al. (2012)
		bph22, bph23	BPH resistance		Hou et al. (2011)
		bph24	BPH resistance		Deen et al. (2010)
		Bph27	BPH resistance		Huang et al. (2013)
		bph29	BPH resistance		Wang et al. (2015)
		Bph30	BPH resistance		Yang et al. (2012)
		Bph36	BPH resistance		Li, et al. (2019)
		Bph38	BPH resistance		Yang et al. (2020)
		qWbph2, qWbph5, qWbph9	WBPH resistance		Chen et al. (2010)
		Grh5	Green rice leafhopper resistance		Fujita et al. (2006)

References

Abreu AG, Rosa TM, de Oliveira Borba TC et al (2015) SSR characterization of *Oryza glumaepatula* populations from the Brazilian Amazon and Cerrado biomes. Genetica 143: 413–423. https://doi.org/10.1007/s10709-015-9840-7

Acharya S, Padhi S, Bose L, Dangar T (2017) Structure and function of phytonic *Bacillus thuringiensis* of wild rice genotype *Oryza brachyantha* A. Chev and Roehr Oryza 54:431. https://doi.org/10.5958/2249-5266.2017.00059.5

Ali MP, Chowdhurry TR (2014) Tagging and mapping gene and QTLs of *Nilapravata lugens* resistance in rice. Euphytica 195(1):1–30

Allen E, Gaisberger H, Magos Brehm J et al (2019) A crop wild relative inventory for Southern Africa: a first step in linking conservation and use of valuable wild populations for enhancing food security. Plant Genet Resour Char Util 17:128–139. https://doi.org/10.1017/S1479262118000515

Banaticla-Hilario MCN, Sosef MSM, McNally KL et al (2013b) Ecogeographic variation in the morphology of two Asian wild rice species, *Oryza nivara* and *Oryza rufipogon*. Int J Plant Sci 174:896–909. https://doi.org/10.1086/670370

Banaticla-Hilario MCN, van den Berg RG, Hamilton NRS, McNally KL (2013a) Local differentiation amidst extensive allele sharing in *Oryza nivara* and *O. rufipogon*. Ecol Evol 3:9. https://doi.org/10.1002/ece3.689

Bautista NS, Vaughan D, Jayasuriya AHM et al (2006) Genetic diversity in AA and CC genome *Oryza* species in southern South Asia. Genet Resour Crop Evol 53:631–640. https://doi.org/10.1007/s10722-004-3227-4

Bellairs SM, Kernich B (2015) Temperature affects the dormancy and germination of sympatric annual (*Oryza meridionalis*) and perennial (*O. rufipogon*) native Australian rice (Poaceae) and influences their emergence in introduced para grass (*Urochloa mutica*) swards. Aust J Bot 63: 687–695

Bessho-Uehara K, Furuta T, Masuda K et al (2017) Construction of rice chromosome segment substitution lines harboring *Oryza barthii* genome and evaluation of yield-related traits. Breed Sci 67:408–415. https://doi.org/10.1270/jsbbs.17022

Bhasin H, Bhatia D, Raghuvanshi S (2012) New PCR-based sequence-tagged site marker for bacterial blight resistance gene Xa38 of rice. Mol Breed 30:607–611. https://doi.org/10.1007/s11032-011-9646-y

Bierschenk B, Tagele MT, Ali B et al (2020) Evaluation of rice wild relatives as a source of traits for adaptation to iron toxicity and enhanced grain quality. PLoS One 15:1–17. https://doi.org/10.1371/journal.pone.0223086

Bon SG, Borromeo TH, Altoveros NC (2011) Genetic Variability Potential and Morphometric Characteristics of Five Geographic Populations of *Oryza meyeriana* (Zoll.et Mor. ex. Steud.) Baill. Found in the Philippines. Philipp J Crop Sci 36:28–36

Bressan EA, Rolim SG, Sebbenn AM et al (2018) Conservation and genetic diversity of populations of *Oryza glumaepatula* Steud. in ferruginous mountaintop lakes of the Brazilian Amazonia. Flora Morphol Distrib Funct Ecol Plants 238:183–190. https://doi.org/10.1016/j.flora.2017.07.003

Brondani RP, Zucchi MI, Brondani C et al (2005) Genetic structure of wild rice *Oryza glumaepatula* populations in three Brazilian biomes using microsatellite markers. Genetica 125:115–123. https://doi.org/10.1007/s10709-005-4916-4

Brozynska M, Furtado A, James Henry R (2014) Direct chloroplast sequencing: comparison of sequencing platforms and analysis tools for whole chloroplast barcoding. PLoS One 9. https://doi.org/10.1371/journal.pone.0110387

Castañeda-Álvarez NP, Khoury CK, Achicanoy HA et al (2016) Global conservation priorities for crop wild relatives. Nat Plants 2:5. https://doi.org/10.1038/NPLANTS.2016.22

CGRIS (2021) Chinese crop germplasm resource information system. http://cgris.net/cgris_english.html. Accessed 20 Mar 2021

Chatterjee J, Thakur V, Nepomuceno R et al (2020) Natural Diversity in Stomatal Features of Cultivated and Wild *Oryza* species. Rice 13. https://doi.org/10.1186/s12284-020-00417-0
Chen J, Huang DR, Wang L, Liu GJ, Zhuang JY (2010) Identification of quantitative trait loci for resistance to white-backed planthopper, Sogatella furcifera, from an interspecific cross *Oryza sativa*× *O. rufipogon*. Breed Sci 60:153–159. https://doi.org/10.1270/jsbbs.60.153
Chen LN, Yang Y, Yan CQ, Wang XM, Yu CL, Zhou J, Zhang WL, Cheng Y, Cheng XY, Chen JP (2012) Identification of quantitative trait loci for bacterial blight resistance derived from *Oryza meyeriana* and agronomic traits in recombinant inbred lines of *Oryza sativa*. J Phytopathol 160(9):461–468. https://doi.org/10.1111/j.1439-0434.2012.01931.x
Cheng C, Motohashi R, Tsuchimoto S et al (2003) Polyphyletic origin of cultivated rice: Based on the interspersion pattern of SINEs. Mol Biol Evol 20:67–75. https://doi.org/10.1093/molbev/msg004
Contreras-Toledo AR, Cortés-Cruz MA, Costich D et al (2018) A crop wild relative inventory for Mexico. Crop Sci 58:1292–1305. https://doi.org/10.2135/cropsci2017.07.0452
Deen R, Ramesh K, Gautam S, Rao Y, Lakshmi V, Viraktamath B, Brar DS, Ram T (2010) Identification of new gene for BPH resistance introgressed from *O. rufipogon*. Rice Genet Newsl 25:70–71
Dempewolf H, Baute G, Anderson J et al (2017) Past and future use of wild relatives in crop breeding. Crop Sci 57:1070–1082. https://doi.org/10.2135/cropsci2016.10.0885
Dempewolf H, Eastwood RJ, Guarino L et al (2014) Adapting agriculture to climate change: a global initiative to collect, conserve, and use crop wild relatives. Agroecol Sustain Food Syst 38: 369–377. https://doi.org/10.1080/21683565.2013.870629
Devi SJSR, Singh K, Umakanth B, Vishalakshi B, Rao KVS, Suneel B, Sharma SK, Kadambari GKM, Prasad MS, Senguttvel P, Syamaladevi DP, Madhav MS (2020) Identification and characterization of a large effect QTL from *Oryza glumaepatula* revealed Pi68(t) as putative candidate gene for rice blast resistance. Rice 13(1):1–13. https://doi.org/10.1186/s12284-020-00378-4
Diederichsen A, Schellenber MP (2018) Genetic resources of crop wild relatives: a Canadian perspective. In: Greene SL, Williams KA, Khoury CK et al (eds) North American crop wild relatives: conservation strategies. Springer, Cham, pp 1–346
Du B, Zhang W, Liu B, Hu J, Wei Z, Shi Z, He R, Zhu L, Chen R, Han B, He G (2009) Identification and characterization of Bph14, a gene conferring resistance to brown planthopper in rice. Proc Natl Acad Sci 106:163–168. https://doi.org/10.1073/pnas.0912139106
Engels JJ, Thormann I (2020) Main challenges and actions needed to improve. Plants 9:968
Eizenga G, Prasad B, Jackson A, Jia M (2013) Identification of rice sheath blight and blast quantitative trait loci in two different *O. sativa/O. nivara* advanced backcross populations. Molec Breed 31:889–907. https://doi.org/10.1007/s11032-013-9843-y
Ellur RK, Khanna A, Gopala Krishnan S et al (2016) Marker-aided incorporation of Xa38, a novel bacterial blight resistance gene, in pb1121 and comparison of its resistance spectrum with xa13 + Xa21. Sci Rep 6:2–9. https://doi.org/10.1038/srep29188
Fitzgerald H, Palmé A, Asdal Å et al (2019) A regional approach to Nordic crop wild relative in situ conservation planning. Plant Genet Resour Char Util 17:196–207. https://doi.org/10.1017/S147926211800059X
Food and Agriculture Organization of the United Nations (2017) Voluntary guidelines for the conservation and sustainable use of crop wild relatives and wild food plants. FAO, Rome
Fuchs EJ, Martínez AM, Calvo A et al (2016) Genetic diversity in *Oryza glumaepatula* wild rice populations in Costa Rica and possible gene flow from O. sativa. Peer J 2016:1–16. https://doi.org/10.7717/peerj.1875
Fujita D (2004) Introgression of a resistance gene for green rice leafhopper from *Oryza nivara* into cultivated rice, *Oryza sativa* L. Rice Genet Newsl 21:64–66
Fujita D, Doi K, Yoshimura A, Yasui H (2006) Molecular mapping of a novel gene, Grh5, conferring resistance to green rice leafhopper (*Nephotettix cincticeps* Uhler) in rice, Oryza sativa L. Theor Appl Genet 113:567–573. https://doi.org/10.1007/s00122-006-0270-x

Gao LZ (2005) Microsatellite variation within and among populations of *Oryza officinalis* (Poaceae), an endangered wild rice from China. Mol Ecol 14:4287–4297. https://doi.org/10.1111/j.1365-294X.2005.02758.x

Gao L-Z, Li D-Y, Wu X-Q et al (2012) In situ conservation of wild rice populations: a targeted study of common wild rice *Oryza rufipogon* from China. Am J Plant Sci 03:854–868. https://doi.org/10.4236/ajps.2012.37103

Gao LZ, Zhang CH (2005) Comparisons of microsatellite variability and population genetic structure of two endangered wild rice species, *Oryza rufipogon* and *O. officinalis*, and their conservation implications. Biodivers Conserv 14:1663–1679. https://doi.org/10.1007/s10531-004-0537-y

Genesys (2021) The global gateway to genetic resources. Data provided by the International Rice Research Institute, Dale Bumpers National Rice Research Center, Africa Rice Center, EMBRAPA National Center for Research on Rice and Beans, and the National Small Grains Germplasm Research Facility. https://www.genesys-pgr.org/. Accessed 19 Mar 2021.

Gu K, Tian D, Yang F, Wu L, Sreekala C, Wang D, Wang G-L, Yin Z (2004) High-resolution genetic mapping of Xa27 (t), a new bacterial blight resistance gene in rice, *Oryza sativa* L. Theor Appl Genet 108:800–807. https://doi.org/10.1007/s00122-003-1491-x

Guo S, Zhang D, Lin X (2010) Identification and mapping of a novel bacterial blight resistance gene Xa35(t) originated from *Oryza minuta*. Sci Agric Sin 43:2611–2618

Hirabayashi H (1998) Identification of brown planthopper resistance gene derived from *O. officinalis* using molecular markers in rice. Breed Sci 48(1):82

Hirabayashi H, Ideta O, Sato H, Takeuchi Y, Ando I, Nemoto H, Imbe T, Brar DS, Ogawa T (2004) Identification of a resistance gene to brown planthopper derived from *Oryza minuta* in rice. Breed Res 6(1):285

Hou L-Y, Yu P, Xu Q, Yuan X-P, Yu H-Y, Wang Y-P, Wang C-H, Wan G, Tang S-X, Peng S-T, Wei X-H (2011) Genetic analysis and preliminary mapping of two recessive resistance genes to brown planthopper, Nilaparvata lugens Stål in rice. Ric Sci 18:238–242. https://doi.org/10.1016/S1672-6308(11)60033-4

Hu H, Mu J, Zhang HJ et al (2006) Differentiation of a Miniature Inverted Transposable Element (MITE) system in Asian rice cultivars and its inference for a diphyletic origin of two subspecies of Asian cultivated rice. J Integr Plant Biol 48:260–267. https://doi.org/10.1111/j.1744-7909.2006.00221.x

Hu J, Xiao C, Cheng M-X, Gao G-J, Zhang Q-L, He Y-Q (2015) A new finely mapped *Oryza australiensis*-derived QTL in rice confers resistance to brown planthopper. Gene 561(1): 132–137. https://doi.org/10.1016/j.gene.2015.02.026

Huang D, Qiu Y, Zhang Y, Huang F, Meng J, Wei S, Li R, Chen B (2013) Fine mapping and characterization of BPH27, a brown planthopper resistance gene from wild rice (*Oryza rufipogon* Griff.). Theor Appl Genet 126(1):219–229. https://doi.org/10.1007/s00122-012-1975-7

International Union for Conservation of Nature (2021) The IUCN red list of threatened species. https://www.iucnredlist.org/. Accessed 23 Mar 2021

Ishii T, Brar D, Multani D, Khush G (1994) Molecular tagging of genes for brown planthopper resistance and earliness introgressed from *Oryza australiensis* into cultivated rice. *O sativa Genome* 37:217–221. https://doi.org/10.1139/g94-030

Islam T, Biswas S, Mita UH et al (2017) Characterization of progenies from intergeneric hybridization between *Oryza sativa* L. and *Porteresia coarctata* (Roxb.) Tateoka. Plant Tissue Cult Biotechnol 27:63–76. https://doi.org/10.3329/ptcb.v27i1.35013

Jaseela F, Sumitha VR, Nair GM (2009) Somatic embryogenesis and plantlet regeneration in an agronomically important wild rice species *Oryza nivara*. Asian J Biotechnol 1:74–78

Jeung J, Kim B, Cho Y et al (2007) A novel gene, Pi40(t), linked to the DNA markers derived from NBS-LRR motifs confers broad spectrum of blast resistance in rice. Theor Appl Genet 115: 1163–1177

Ji H, Kim S-R, Kim Y-H, Suh J-P, Park H-M, Sreenivasulu N, Misra G, Kim S-M, Hechanova SL, Kim H, Lee G-S, Yoon U-H, Kim T-H, Lim H, Suh S-C, Yang J, An G, Jena KK (2016) Map-based cloning and characterization of the BPH18 gene from wild rice conferring resistance to Brown Planthopper (BPH) insect pest. Sci Rep 6:1–13. https://doi.org/10.1038/srep34376

JianKun WP, Luo XD, Fu XQ et al (2011) Studies on seed dormancy of Dongxiang wild rice (*Oryza rufipogon* Griff.) and its interspecific offspring. Hybrid Rice 26:74–77

Joshi KB, Bahadur HK, Upadhyay PM (2008) Ex situ and in situ management of wild and weedy rice in Nepal using a geographical information system. Plant Genet Resour Newsl 69–74

Juliano AB, Naredo MEB, Lu BR, Jackson MT (2005) Genetic differentiation in *Oryza meridionalis* Ng based on molecular and crossability analyses. Genet Resour Crop Evol 52: 435–445. https://doi.org/10.1007/s10722-005-2259-8

Júnior FPI, Duarte Pereira GM, Perin L et al (2013) Diazotrophic bacteria isolated from wild rice *Oryza glumaepatula* (Poaceae) in the Brazilian Amazon. Rev Biol Trop 61:991–999. https://doi.org/10.15517/rbt.v61i2.11238

Kaewcheenchai R, Promnart U, Soontrajarn K et al (2018) Diverse genetic variation in maternal lineages with high heterogeneity among in situ-conserved wild rice (*Oryza rufipogon* Griff.) developed in Thailand. Breed Sci 68:614–621. https://doi.org/10.1270/jsbbs.16105

Karasawa MMG, Vencovsky R, Silva CM et al (2007) Mating system of Brazilian *Oryza glumaepatula* populations studied with microsatellite markers. Ann Bot 99:245–253. https://doi.org/10.1093/aob/mcl248

Karasawa MMG, Vencovsky R, Silva CM et al (2012) Comparison of microsatellites and isozymes in genetic diversity studies of *Oryza glumaepatula* (Poaceae) populations. Rev Biol Trop 60: 1463–1478. https://doi.org/10.15517/rbt.v60i4.2055

Khoury CK, Carver D, Greene SL et al (2021) Crop wild relatives of the United States require urgent conservation action. Proc Natl Acad Sci U S A 117(33351–33):357. https://doi.org/10.1073/PNAS.2007029117

Khoury CK, Greene SL, Krishnan S et al (2019) A road map for conservation, use, and public engagement around North America's crop wild relatives and wild utilized plants. Crop Sci 59: 2302–2307. https://doi.org/10.2135/cropsci2019.05.0309

Khush GS, Bacalangco E, Ogawa T (1990) A new gene for resistance to bacterial blight from *O. longistaminata*. Rice Genet Newsl 7:121–122

Kiambi DK, Newbury HJ, Ford-Lloyd BV, Dawson I (2005) Contrasting genetic diversity among *Oryza longistaminata* (A. Chev et Roehr) populations from different geographic origins using AFLP. Afr J Biotechnol 4:308–317. https://doi.org/10.4314/ajb.v4i4.15100

Kumar K, Sarao PS, Bhatia D, Neelam K, Kaur A, Mangat GS, Brar DS, Singh K (2018) High-resolution genetic mapping of a novel brown planthopper resistance locus, Bph 34 in *Oryza sativa* L. X *Oryza nivara* (Sharma & Shastry) derived interspecific F2 population. Theor Appl Genet 131:1163–1171. https://doi.org/10.1007/s00122-018-3069-7

Kumar PN, Sujatha K, Laha G, Rao KS, Mishra B, Viraktamath B, Hari Y, Reddy CS, Balachandran SM, Ram T, Madhav MS (2012) Identification and fine-mapping of Xa33, a novel gene for resistance to *Xanthomonas Oryzae* pv. *Oryzae Phytopathol* 102:222–228. https://doi.org/10.1094/PHYTO-03-11-0075

Kuroda Y, Sato YI, Bounphanousay C et al (2007) Genetic structure of three *Oryza* AA genome species (*O. rufipogon, O. nivara* and *O. sativa*) as assessed by SSR analysis on the Vientiane Plain of Laos. Conserv Genet 8:149–158. https://doi.org/10.1007/s10592-006-9156-3

Labroo M (2017) Unlocking the genetic diversityof the undomesticated rice relative *Oryza longistaminata*. University of Illinois at Urbana-Champaign, Urbana

Lentini Z, Espinosa AM (2005) Coexistence of weedy rice and rice in tropical America - gene flow and genetic diversity. In: Gressel J (ed) Crop Ferality and Volunteerism. Taylor & Francis, Boca Raton, pp 305–322

Liang YT, Pan YH, Chen CB, Zhang ZQ (2014) Study on differentiation and dedifferentiation of Oryza alta Swallen. Southwest China J Agric Sci 27(3):905–909

Li R, Li L, Wei S, Wei Y, Chen Y, Bai D, Yang L, Huang F, Lu W, Zhang X, Li X (2010) The evaluation and utilization of new genes for brown planthopper resistance in common wild rice (*Oryza rufipogon* Griff.). Molec Entomol 1:1. https://doi.org/10.5376/me.2010.01.0001

Li Z, Xue Y, Zhou H, Li Y, Usman B, Jiao X, Wang X, Liu F, Qin B, Li R, Qiu Y (2019) High-resolution mapping and breeding application of a novel brown planthopper resistance gene derived from wild rice (*Oryza rufipogon* Griff). Rice 12(1):41. https://doi.org/10.1186/s12284-019-0289-7

Liu H, Carvalhais LC, Crawford M et al (2017) Inner plant values: diversity, colonization and benefits from endophytic bacteria. Front Microbiol 8:1–17. https://doi.org/10.3389/fmicb.2017.02552

Lou HS, Bhattarai K, Simon EV et al (2021) Development of a genome-wide InDel marker set for allele discrimination between rice (Oryza sativa) and the other seven AA-genome Oryza species. Sci Rep 11:8962. https://doi.org/10.1038/s41598-021-88533-9

Lu B-R (2012) The challenge of in situ conservation of crop wild relatives in the biotechnology era-a case study of wild rice species. In: Maxted N, Ehsan Dulloo M, Ford-Lloyd BV, Frese L et al (eds) Agrobiodiversity conservation: securing the diversity of crop wild relatives and landraces. CAB International, Wallingford

Brehm MJ, Kell S, Thormann I, Gaisberger H, Dulloo M, Maxted N (2019) New tools for crop wild relative conservation planning. Plant Genet Resour Charact Util 17(2):208–212. https://doi.org/10.1017/S1479262118000527

Mangosongo HM, Mneney EE, Harvey J, Kyallo M (2020) Assessment of genetic diversity of the wild rice (*Oryza longistaminata*) populations in Tanzania using microsatellite. Markers 46:828–839

Maxted N, Kell SP (2009) Establishment of a global network for the in situ conservation of crop wild relatives: status and needs. FAO Comm Genet Resour Food Agric 5:1–266

Melaku G, Haileselassie T, Feyissa T, Kiboi S (2013) Genetic diversity of the African wild rice (*Oryza longistaminata* Chev. et Roehr) from Ethiopia as revealed by SSR markers. Genet Resour Crop Evol 60:1047–1056. https://doi.org/10.1007/s10722-012-9900-0

Melaku G, Huang L, Zhang S et al (2020) Genetic relatedness among Ethiopian *Oryza longistaminata* populations and other AA genome Oryza species. Plant Growth Regul 91: 175–183. https://doi.org/10.1007/s10725-020-00597-1

Melaku G, Labroo M, Liyu H et al (2019) Genetic diversity and differentiation of the African wild rice (Oryza longistaminata Chev. et Roehr) in Ethiopia. Sci African 2:e00138. https://doi.org/10.1016/j.sciaf.2019.e00138

Mondal TK, Henry RJ (eds) (2018) The Wild Oryza genomes, 1st edn. Springer, Cham

Narain A, Kar MK, Kaliaperumal V, Sen P (2016) Development of monosomic alien addition lines from the wild rice (*Oryza brachyantha* A. Chev. et Roehr.) for introgression of yellow stem borer (*Scirpophaga incertulas* Walker.) resistance into cultivated rice (*Oryza sativa* L.). Euphytica 209:603–613. https://doi.org/10.1007/s10681-016-1633-2

Naredo MEB, Juliano AB, Lu BR et al (1998) Responses to seed dormancy-breaking treatments in rice species (*Oryza* L.). Seed Sci Technol 26:675–689

Ndjiondjop MN, Semagn K, Zhang J et al (2018) Development of species diagnostic SNP markers for quality control genotyping in four rice (*Oryza* L.) species. Mol Breed 38:1. https://doi.org/10.1007/s11032-018-0885-z

Ngu MS, Sabu KK, Lim LS et al (2010) Genetic structure of *Oryza rufipogon* Griff. Natural populations in Malaysia: implications for conservation and genetic introgression of cultivated rice. Trop Plant Biol 3:227–239. https://doi.org/10.1007/s12042-010-9060-3

Ni W, Cheng F, Zhang J (2007) Gaining regeneration plantlets from embryoids in leaf tissue culture of *Oryza alta* L. Shanghai J Agri Sci 23:18–20

Ohtsubo H, Cheng C, Ohsawa I et al (2004) Rice retroposon p-SINE1 and origin of cultivated rice. Breed Sci 54:1–11. https://doi.org/10.1270/jsbbs.54.1

Orjuela J, Sabot F, Chéron S et al (2014) An extensive analysis of the African rice genetic diversity through a global genotyping. Theor Appl Genet 127:2211–2223. https://doi.org/10.1007/s00122-014-2374-z

Oryzabase (2021) Integrated Rice Science Database. https://shigen.nig.ac.jp/rice/oryzabase/. Accessed 20 Mar 2021

Paramasivam S, Ann HJ (2020) Effect of culture media and conditions on callus induction and plant regeneration of Malaysian wild rice *Oryza rufipogon* IRGC 105491. Res J Biotechnol 15:120–127

Peng G, Zhang W, Luo H et al (2009) *Enterobacter oryzae* sp. Nov., a nitrogen-fixing bacterium isolated from the wild rice species *Oryza latifolia*. Int J Syst Evol Microbiol 59:1650–1655. https://doi.org/10.1099/ijs.0.005967-0

Peng X, Xie J, Li W et al (2021) Comparison of wild rice (Oryza longistaminata) tissues identifies rhizome-specific bacterial and archaeal endophytic microbiomes communities and network structures. PLoS One 16:1–16. https://doi.org/10.1371/journal.pone.0246687

Phi CN, Fujita D, Yamagata Y, Yoshimura A, Yasui H (2019) High-resolution mapping of GRH6, a gene from *Oryza nivara* (Sharma et Shastry) conferring resistance to green rice leafhopper (*Nephotettix cincticeps* Uhler). Breed Sci 69:439–446. https://doi.org/10.1270/jsbbs.19029

Phillips J, Magos Brehm J, van Oort B et al (2017) Climate change and national crop wild relative conservation planning. Ambio 46:630–643. https://doi.org/10.1007/s13280-017-0905-y

Prathepha P (2012) Genetic diversity and population structure of wild rice, *Oryza rufipogon* from Northeastern Thailand and Laos. Aust J Crop Sci 6:717–723

Qiu Y, Guo J, Jing S, Zhu L, He G (2012) Development and characterization of japonica rice lines carrying the brown planthopper-resistance genes BPH12 and BPH6. Theor Appl Genet 124: 485–494. https://doi.org/10.1007/s00122-011-1722-5

Quilloy SM (2008) Assessment of genetic diversity in the diploid species belonging to the *Oryza officinalis* complex. University of the Philippines at Los Banos, Los Banos

Rahman L, Khanam S, Roh JH, Koh HJ (2011) Mapping of QTLs involved in resistance to rice blast (*Magnaporthe grisea*) using *Oryza minuta* introgression lines. Czech J GenetPlant Breed 47:85–94. https://doi.org/10.17221/19/2011-CJGPB

Rahman ML, Jiang W, Chu SH, Qiao Y, Ham T-H, Woo M-O, Lee J, Khanam MS, Chin J-H, Jeung J-U, Brar DS, Jena KK, Koh H-J (2009) High-resolution mapping of two rice brown planthopper resistance genes, Bph20 (t) and Bph21 (t), originating from *Oryza minuta*. Theor Appl Genet 119:1237–1246. https://doi.org/10.1007/s00122-009-1125-z

Rahman W, Magos Brehm J, Maxted N (2019) Setting conservation priorities for the wild relatives of food crops in Indonesia. Genet Resour Crop Evol 66:809–824. https://doi.org/10.1007/s10722-019-00761-1

Rajkumar G, Weerasena OVDSJ, Fernando KKS (2015) A study of genetic diversity in Oryza rhizomatis D. A. Vaughan accessions using AFLP markers and morphological traits. Trop Plant Res 2:10–16

Ram T, Deen R, Gautam S, Ramesh K, Rao Y, Brar D (2010) Identification of new genes for brown planthopper resistance in rice introgressed from *O. glaberrima* and *O. minuta*. Rice Genet Newsl 25:67–69

Renganayaki K, Fritz AK, Sadasivam S, Pammi S, Harrington SE, McCouch SR, Kumar SM, Reddy AS (2002) Mapping and progress toward map-based cloning of brown planthopper biotype-4 resistance gene introgressed from Oryza officinalis into cultivated rice, O. sativa. Crop Sci 42:2112–2117. https://doi.org/10.2135/cropsci2002.2112

Ricepedia (2021) The online authority on rice. http://ricepedia.org/. Accessed 9 Apr 2021

Riordan EC, Nabhan GP (2019) Trans situ conservation of crop wild relatives. Crop Sci 59:2387–2403. https://doi.org/10.2135/cropsci2019.06.0356

Ruan HH, Yan CQ, An DR, Liu RH, Chen JP (2008) Identifying and mapping new gene xa32 (t) for resistance to bacterial blight (*Xanthomonas Oryzae* pv. Oryzae, Xoo) from *Oryza meyeriana* L. Northwest Agric J 17:170–174

Scafaro AP, Atwell BJ, Muylaert S et al (2018) A thermotolerant variant of rubisco activase from a wild relative improves growth and seed yield in rice under heat stress. Front Plant Sci 871:1–11. https://doi.org/10.3389/fpls.2018.01663

Shenton M, Iwamoto C, Kurata N, Ikeo K (2016) Effect of wild and cultivated rice genotypes on rhizosphere bacterial community composition. Rice 9:42. https://doi.org/10.1186/s12284-016-0111-8

Shimizu-Sato S, Tsuda K, Nosaka-Takahashi M et al (2020) Agrobacterium-mediated genetic transformation of wild *Oryza* species using immature embryos. Rice 13:33. https://doi.org/10.1186/s12284-020-00394-4

Shishido R, Kikuchi M, Nomura K, Ikehashi H (2006) Evaluation of genetic diversity of wild rice (*Oryza rufipogon* Griff.) in Myanmar using simple sequence repeats (SSRs). Genet Resour Crop Evol 53:179–186. https://doi.org/10.1007/s10722-004-3226-5

Singh J, Kakade DP, Wallalwar MR et al (2017) Evaluation of potential dna barcoding loci from plastid genome: intraspecies discrimination in rice (Oryza species). Int J Curr Microbiol Appl Sci 6:2746–2756. https://doi.org/10.20546/ijcmas.2017.605.308

Sjahril R, Panga NJ, Riadi M et al (2015) Sustainability of populations of wild rice species in natural in situ conservation sites of South and West Sulawesi. Int J Agric Syst 3:205–211. https://doi.org/10.20956/IJAS.V3I2.108

Song Y, Chen Y, Lv J et al (2017) Development of chloroplast genomic resources for *Oryza* species discrimination. Front Plant Sci 8:1–10. https://doi.org/10.3389/fpls.2017.01854

Song Z, Li B, Chen J, Lu BR (2005) Genetic diversity and conservation of common wild rice (*Oryza rufipogon*) in China. Plant Species Biol 20:83–92. https://doi.org/10.1111/j.1442-1984.2005.00128.x

Tan G, Ren X, Weng Q et al (2004) Mapping of a new resistance gene to bacterial blight in rice line introgressed from *Oryza officinalis*. Yi Chuan Xue Bao 31:724–729

Thi Lang N, Chi Buu B (2003) Genetic and physical maps of gene Bph10 controlling brown plant hopper resistance in rice (*Oryza sativa* L.). Omonrice 11:35–41

Tian L, Zhou X, Ma L et al (2016) Root-associated bacterial diversities of *Oryza rufipogon* and *Oryza sativa* and their influencing environmental factors. Arch Microbiol 199:563–571

Timple SE, Hay FR (2018) High-temperature drying of seeds of wild *Oryza* species intended for long-term storage. Seed Sci Technol 46:107–112. https://doi.org/10.15258/sst.2018.46.1.10

Timple SE, Hay FR, Mercado MFO et al (2018) Response of intact seeds of wild rice (*Oryza*) species to dry heat treatment and dormancy-breaking chemicals. Seed Sci Technol 46:157–173. https://doi.org/10.15258/sst.2018.46.1.16

Tyack N, Dempewolf H, Khoury CK (2020) The potential of payment for ecosystem services for crop wild relative conservation. Plan Theory 9:1305. https://doi.org/10.3390/plants9101305

Vaughan DA (1990) In-situ conservation of wild rice in Asia. Rice Genet Newsl 7:90

Vaughan DA, Kadowaki K, Kaga A, Tomooka N (2005) On the phylogeny and biogeography of the genus Oryza. Breed Sci 55:113–122

Vaughan DA, Ge S, Kaga A, Tomooka N (2008) Phylogeny and biogeography of the genus *Oryza*. Biotechnol Agric For 62:219–234. https://doi.org/10.1007/978-3-540-74,250-0_17

Veasey EA, de Bressan EA, Zucchi MI et al (2011) Genetic diversity of American wild rice species. Sci Agric 68:440–446. https://doi.org/10.1590/S0103-90162011000400008

Waheed A, Ahmad H, Abbasi FM (2012) Different treatment of rice seed dormancy breaking, germination of both wild species and cultivated varieties (*Oryza sativa* L.). J Mater Environ Sci 3:551–560

Wang K, Wambugu PW, Zhang B, Wu AC, Henry RJ, Gilbert RG (2015a) The biosynthesis, structure and gelatinization properties of starches from wild and cultivated African rice species (*Oryza barthii* and *Oryza glaberrima*). Carbohydr Polym 129:92–100. https://doi.org/10.1016/j.carbpol.2015.04.035

Wang L, Wang B, Fu J et al (2019) Establishment of induction, differentiation and seedling rooting system for young embryonic calli of *Oryza eichingeri*. Acta Agric Jiangxi 31:10–15

Wang MX, Zhang HL, Zhang DL, Qi YW, Fan ZL, Li DY, Pan DJ, Cao YS, Qiu ZE, Yu P, Yang QW, Wang XK, Li ZC (2008) Genetic structure *of Oryza rufipogon* Griff. in China. Heredity (Edinb) 101:527–535. https://doi.org/10.1038/hdy.2008.61

Wang YP, Bounphanousay C, Kanyavong K et al (2012) Population structural analysis of an in-situ conservation site for wild rice in Laos. Genes Genet Syst 87:311–322. https://doi.org/10.1266/ggs.87.311

Waters DLE, Nock CJ, Ishikawa R et al (2012) Chloroplast genome sequence confirms distinctness of Australian and Asian wild rice. Ecol Evol 2:211–217. https://doi.org/10.1002/ece3.66

Wedger MJ, Pusadee T, Wongtamee A, Olsen KM (2019) Discordant patterns of introgression suggest historical gene flow into Thai weedy rice from domesticated and wild relatives. J Hered 110:601–609. https://doi.org/10.1093/jhered/esz030

Wu C-J, Cheng Z-Q, Huang X-Q et al (2004) Genetic diversity among and within populations of *Oryza granulata* from Yunnan of China revealed by RAPD and ISSR markers: implications for conservation of the endangered species. Plant Sci 167:35–42

Wu CY, Gao L-Z (2013) SNP analysis of six nuclear gene fragments for studying population genetics of *Oryza granulata* from China. Plant Divers Resour 35:537–546

Xie J, Agrama HA, Kong D et al (2010) Genetic diversity associated with conservation of endangered Dongxiang wild rice (*Oryza rufipogon*). Genet Resour Crop Evol 57:597–609. https://doi.org/10.1007/s10722-009-9498-z

Xu P, Dong L, Zhou J, Li J, Zhang YU, Hu F, Liu S, Wang Q, Deng W, Deng X, Tharreau D, Yang Q, Tao D (2015) Identification and mapping of a novel blast resistance gene Pi57(t) in Oryza longistaminata. Euphytica 205(1):95–102. https://doi.org/10.1007/s10681-015-1402-7

Xuwei J, Chunlian W, Qing Y (2007) Breeding of near-isogenic line CBB30 and molecular mapping of Xa30(t), a new resistance gene to bacterial blight in rice. Sci Agric Sin 40:1094–1100

Yamaki S, Ohyanagi H, Yamasaki M et al (2013) Development of INDEL markers to discriminate all genome types rapidly in the genus *Oryza*. Breed Sci 63:246–254. https://doi.org/10.1270/jsbbs.63.246

Yamanaka S, Nakamura I, Nakai H, Sato Y (2002) Dual origin of the cultivated rice based on molecular markers of newly collected annual and perennial strains of wild rice species. Genet Resour Crop Evol 50:1–10

Yang H, You A, Yang Z, Zhang F, He R, Zhu L, He G (2004) High resolution genetic mapping at the Bph15 locus for brown planthopper resistance in rice (Oryza sativa L.). Theor Appl Genet 110(1):182–191. https://doi.org/10.1007/s00122-004-1844-0

Yang L, Li RB, Li YR, Huang FK, Chen YZ, Huang SS, Huang LF, Liu CH, Ma ZF, Huang DH, Jiang JJ (2012) Genetic mapping of bph20(t) and bph21(t) loci conferring brown planthopper resistance to Nilaparvata lugens Stål in rice (*Oryza sativa* L.). Euphytica 183:161–171. https://doi.org/10.1007/s10681-011-0437-7

Yang M, Lin J, Cheng L, Zhou H, Chen S, Liu F, Li R, Qiu Y (2020) Identification of a novel planthopper resistance gene from wild rice (*Oryza rufipogon* Griff.). Crop J 8:1057–1070. https://doi.org/10.1016/j.cj.2020.03.011

Yang Q, Yu L, Zhang W (2007) The genetic differentiation of Dongxiang wild rice (*Oryza rufipogon* Griff.) and its implications for In-Situ conservation. Sci Agric Sin 40:1085–1093

Ya-tao W, Xin-xiang A, Chuan-zhang F et al (2008) ISSR analysis on genetic diversity of the 34 populations of Oryza meyeriana distributed in Yunnan province, China. Rice Sci 15:13–20

Yibo D, Xinwu P, Qianhua Y et al (2010) Ecological, morphological and genetic diversity in *Oryza rufipogon* Griff. (Poaceae) from Hainan Island. China Genet Resour Crop Evol 57:915–926. https://doi.org/10.1007/s10722-009-9523-2

Yinglun F, Xuewei C, Chunlian W (2006) Mapping the rice bacterial blight resistance gene Xa23 with RFLP markers and converting RFLP to STS marker. Acta Agronomica Sinica 32:931–935

Yuan H, Yan H, Yang F et al (2014) Molecular characterization and phylogenetic analysis of endophytic nitrogen-fixing bacteria in Oryza australiensis. Chin J App Environ Biol 20:571–577. https://doi.org/10.3724/SP.J.1145.2013.11040

Yuan T, Liu L, Huang S et al (2018) Rhizobium wuzhouense sp. Nov., isolated from roots of Oryza officinalis. Int J Syst Evol Microbiol 68:2918–2923. https://doi.org/10.1099/ijsem.0.002921
Yuan ZL, Su ZZ, Mao LJ et al (2011) Distinctive endophytic fungal assemblage in stems of Wild Rice (*Oryza granulata)* in China with special reference to two species of *Muscodor* (Xylariaceae). J Microbiol 49:15–23. https://doi.org/10.1007/s12275-011-0213-3
Zeliang PK, Pattanayak A, Iangrai B et al (2010) Fertile plant regeneration from cryopreserved calli of *Oryza rufipogon* Griff. and assessment of variation in the progeny of regenerated plants. Plant Cell Rep 29:1423–1433. https://doi.org/10.1007/s00299-010-0932-7
Zhang W, Dong Y, Yang L, Ma B, Ma R, Huang F, Wang C, Hu H, Li C, Yan C, Chen J (2014) Small brown planthopper resistance loci in wild rice (Oryza officinalis). Mol Genet Genomics 289:373–382. https://doi.org/10.1007/s00438-014-0814-8
Zhang W, Sun Y, Liu J et al (2021) DNA barcoding of *Oryza*: conventional, specific, and super barcodes. Plant Mol Biol 105:215–228. https://doi.org/10.1007/s11103-020-01054-3
Zhao Y, Chen C, Rong J et al (2012) Population clonal diversity and fine-scale genetic structure in *Oryza officinalis* (Poaceae) from China, implications for in situ conservation. Genet Resour Crop Evol 59:113–124. https://doi.org/10.1007/s10722-011-9672-y
Zhao Y, Huang J, Wang Z, Jing S, Wang Y, Ouyang Y, Cai B, Xin XF, Liu X, Zhang C, Pan Y (2016) Allelic diversity in an NLR gene BPH9 enables rice to combat planthopper variation. Proc Natl Acad Sci 11, 12:850–855. https://doi.org/10.1073/pnas.1614862113
Zhao Z, Zhang C, Liu X et al (2018) Genetic analysis and fine mapping of a dominant dwarfness gene from wild rice (Oryza barthii). Plant Breed 137:1–10. https://doi.org/10.1111/pbr.12553
Zheng CK, Chun-Lian W, Yuan-Jie Y, Liang YT, Kai-Jun Z (2009) Identification and molecular mapping of Xa32 (t), a novel resistance gene for bacterial blight (*Xanthomonas Oryzae pv. oryzae*) in rice. Acta Agronomica Sinica 35:1173–1180. https://doi.org/10.1016/S1875-2780(08)60089-9
Zhou HF, Zheng XM, Wei RX et al (2008) Contrasting population genetic structure and gene flow between *Oryza rufipogon* and *Oryza nivara*. Theor Appl Genet 117:1181–1189. https://doi.org/10.1007/s00122-008-0855-7

GIS and Its Role in Plant Genetic Resources Use and Conservation

4

Ricardo T. Bagarinao

Abstract

Providing the biological underpinning for agricultural production, plant genetic resources (PGR) form the essential ingredients for global food security, livelihoods, and development aspirations especially for developing countries. Unfortunately, they are threatened with habitat destruction, overexploitation, pest infestation, and climate change, thereby making PGR conservation imperative worldwide. As the conservation contains a spatial element, the geographic information system (GIS), which has a capacity to store, process, manage, and analyze big spatial data and link several databases, can be utilized to effect conservation initiatives. This chapter aims to elucidate the use of geographic information system (GIS) in plant genetic resource (PGR) conservation and management by reviewing existing literature in GIS as a tool for PGR conservation. The review reveals that GIS performs diverse functions in the conservation and management of PGR such as the identification of conservation sites, development of accession explorations and collection plans, and georeferencing of germplasm in gene banks. Prospectively, the ability of GIS to integrate different kinds of georeferenced data can enable researchers to precisely map the spatial distributions of these species and correlate them with relevant socioeconomic and ecological and physical data to be able to develop a more contextualized conservation strategy. Lastly, it can also assist PGR conservationists, managers, and even farmers develop a pest monitoring map to effectively address the impacts of pest infestation on landraces or in wild species.

R. T. Bagarinao (✉)
University of the Philippines Open University, Los Banos, Laguna, Philippines
e-mail: rtbagarinao@up.edu.ph

S. Ramamoorthy et al. (eds.), *Plant Genetic Resources, Inventory, Collection and Conservation*, https://doi.org/10.1007/978-981-16-7699-4_4

Keywords

Plant genetic resources · Geographic information system · Gene bank · Germplasm · Conservation

4.1 Introduction

The food security of a nation depends largely on its plant genetic resources (PGR). Fowler and Lower (2004) indicated that PGR provide the biological underpinning for agricultural production. They regarded PGR as the essential ingredient for the security of food production, livelihoods, and development aspirations of countries anywhere in the globe. This calls for an effective and efficient conservation and management of PGR because of the expansion of the global population (Ogwu et al. 2014).

The success of conserving PGR depends largely on the completeness and quality of the data stored in gene bank databases or as collected from conservation sites. As the conservation activities require information about the environment where the individual plant species or their population is thriving (Parra-Quijano et al. 2012), the data in gene bank databases should contain geographical locations of the germplasm collection. According to Guarino et al. (2002), these data are critical for the effectiveness of the whole process of conservation since they contain information about the population of the wild plant species, or their cultivated relatives identified to be thriving in certain known areas. Parra-Quijano et al. (2012) also emphasized that geographical locations of the stands can provide information about the adaptability of the individual plant or population to biotic or abiotic factors operating in the place. Geographic locations can be represented by the specific coordinates of the areas where these plants are thriving. However, Hijmans et al. (1999) observed that in most cases, gene bank databases contain either imprecise, erroneous, or missing geographic coordinates, making the data in these databases less useful for the designing of conservation plans.

Meanwhile, van Zonneveld et al. (2012) emphasized the need to optimize the conservation efforts if PGR will continue to be used in sustaining food production that will meet the burgeoning global requirement. These initiatives can be optimized when spatial patterns of the genetic diversity of plant species utilized in the production system can be visualized and better understood. As Jones et al. (2007) had indicated, visualizing the pattern of variations within a certain area can provide insights about the complex nature of the spatial structure of the genetic resources in such area. This leads to the creation of a more detailed prioritization of areas for conservation especially for in situ initiatives (van Zonneveld et al. 2012).

The capacity of GIS in manipulating georeferenced data and overlaying thematic map layers has been receiving significant attention from individuals and institutions engaged in PGR exploration, collection, and conservation in recent decades. These capacities were explored to address issues in geographical information or rectify erroneous coordinates of germplasm collection and visualization of patterns and relationships in geographical data to develop more efficient conservation designs for PGR. For instance, Hijmans et al. (1999) reported that GIS has been an effective tool in checking and correcting the dataset including the coordinates of wild potatoes in Bolivia. They observed that more than 50% of accessions in the database have erroneous geographic locations. Using GIS, they were able to correct the location of accessions previously recorded as residing in oceans, highly separated, or in doubtful altitude (Hijmans et al. 1999). Using GIS, van Zonneveld et al. (2012) had also identified areas with high potential for in situ conservation for cherimoya (*Annona cherimola* Mill.) in Ecuador and Peru. They were able to identify potential in situ conservation and targeted collection sites for cherimoya using the GIS-based analysis of a large microsatellite dataset across the Andean distribution range. Moreover, Fortuna et al. (2009) had discovered well-defined modules or groups of patches of Mediterranean woody plant species with genetically similar populations in Southern Spain using spatial analysis. These modules were thought of as conservation or management units for these woody plant species.

This chapter aims to review and discuss the various applications of GIS with respect to PGR conservation. Specifically, it discusses the system of GIS as a tool for PGR-related analyses and elucidates on how they have been utilized in the efficient conservation of PGR. Prospective uses of the system are also presented and discussed.

4.2 Plant Genetic Resources for Food Security

In 1970, during the International Biological Program conference, the idea of PGR was first presented. Since then, the term was being defined in several ways (Hammer and Teklu 2008). For instance, FAO previously defined PGR as the plant materials involved in the generative and vegetative reproduction with high agricultural values. Later, they considered PGR to refer to the heritable plant materials with societal, economic, and scientific values. According to them, these materials can be widely used in various ecological, biochemical, cytogenetic, pathological, and evolutionary research as well as in production-related studies like agronomic or breeding propensities (FAO 1989). Given these utilities, Brockhaus and Oetmann (1996) considered PGR as essential for food production and thus the establishment of food security across the globe aside from being important in restoring forest resources. However, Maxted et al. (2008) had expanded this definition to include the crop wild relatives, modern and obsolete cultivars, ecotypes and landraces, and genetic stocks and breeding lines.

But regardless of how PGR is being defined, it is clear that they are essential in ensuring global food security. In fact, FAO (2015) emphasized that the lack of genetic diversity will increase the risks to food security given the expanding global population. If unmanaged, it may result in a devastating malnutrition especially among developing countries. For instance, the potato blight in mid-nineteenth century and the banana fungal disease black Sigatoka in the 1990s had resulted in devastating famine in Ireland and significantly reduced production of banana in Central America, respectively.

The role of PGR in food security could not be underestimated. Several studies (e.g., Esquinas-Alcazar 2005; Kantar et al. 2015; Halewood et al. 2018) highlighted the importance of PGR in addressing the world's hunger. Esquinas-Alcazar (2005) emphasized that addressing the food requirement of the world's population of 8.3 billion by 2030 necessitates the wise use and conservation of plant genetic resources and agricultural biodiversity. Thus, putting these resources at risk will compromise the health and life of millions of people due to hunger and malnutrition. Currently, there are more than 800 million people worldwide who are suffering from malnutrition and hunger while 15 million people, most of whom are children, have died each year (Esquinas-Alcazar 2005). The condition may be more acute in developing countries where 90% of the world's population reside. Thus, Ronald (2011) suggested improving PGR use and conservation to keep the pace of food production with the growing population and reduce the impacts of climate change and shrinking environmental resources on food supply. Esquinas-Alcazar (2005) emphasized the same concern to ensure a more reliable, sustainable, and accessible food supply that could feed the expanding global population.

Kantar et al. (2015) suggested the use of crop wild relatives (CWR) for crop improvement. They argued that CWR are rich sources of genetic diversity, which can be utilized to maintain and improve food production. Dempewolf et al. (2017) reported that wild plants related to domesticated crops are important genetic resources that can be utilized for plant breeding. Being a large pool of genetic diversity, CWR can provide materials to develop new allelic variations for breeding programs, which may produce crop varieties that are highly adaptive to changing disease pressures, climatic conditions, farming practices, and market demands (Dempewolf et al. 2017). Gur and Zamir (2004) opined that traits from CWR can be introgressed into domesticates with relative ease because of their close evolutionary relationships. This introgression process may help the new crop variety adapt to the agricultural environment, hence better crop yield. Ladizinsky (1998) indicated that domesticating CWR can lead to increased productivity although it can narrow the genetic basis of crop species. Such increase in yield has been observed in the study of Gur and Zamir (2004) involving tomato (*Solanum lycopersicum*). They reported that the hybrids from the pyramided genotypes yielded more than 50% compared to the leading variety in the market. The experiment was done during the wet and dry conditions of the fields, which received about 10% of irrigation water (Gur and Zamir 2004). Hedden (2003) also cited a similar observation involving the domestication of rice and wheat. N'Danikou and Tchokponhoue (2019) corroborated this observation and reported that there was a drastic increase in wheat and rice productivity when genetic introgression involving the dwarfing genes was done during the green revolution. Both crops, which were important during the green revolution, had remained to be major food resources for smallholders at present (N'Danikou and Tchokponhoue 2019).

The criticality of PGR to ensure global food security stems from the interdependence of countries on their access to plant genetic resources. Such interdependence is a product of long tradition of global dispersal of PGR, adaptation, and crop domestication. For instance, Galluzzi et al. (2016) had indicated that all countries have become highly dependent upon the PGR located beyond their borders because of crop domestication. And many have projected that such interdependence will increase soon because of the changing global food systems and diet patterns (Khoury et al. 2014) and climatic conditions (Lane and Jarvis 2007). Jarvis et al. (2008) projected that greatly afflicted countries by the impacts of the changing climate will be the most food insecure, thereby badly needing germplasm that is adaptive to climate change. Hence, crop varieties adaptive to climate change will most likely be exchanged to these countries. However, Ainsworth et al. (2008) suggested conducting a comprehensive assessment of the adaptation needs and suitable genetic resources for these countries.

In addition, the global recognition of the importance of PGR to ensure a sustainable supply of food across the globe has been epitomized with the inclusion of the multilateral system of access and benefits sharing in Article 11 of the International Treaty on PGR for Food and Agriculture (Galluzzi et al. 2016). The global interdependence on PGR has been added as one criterion so that crops may be included in the multilateral system; the other criterion is their importance in food security (Fowler and Lower 2004). According to Esquinas-Alcazar (2005), the treaty "promotes the development of national integrated approaches to the exploration, collection, characterization, evaluation, conservation, and documentation of their PGR for food and agriculture" (p. 950). It has been considered as an important instrument to ensure that benefits resulting from using PGR are shared fairly and equitably (FAO 2009). Most importantly, it recognizes the role of farmers in the diversification of crops for the sustainability of agricultural production, thereby ensuring food security throughout the globe.

4.3 PGR Conservation: Strategies and Challenges

Given the importance of PGR in achieving global food security, conservation initiatives are imperatives to sustain availability and accessibility of these resources. Currently, two conservation strategies were implemented, namely, ex situ and in situ conservation. These strategies differ in the location by which the initiatives are being implemented. According to FAO (2009), initiatives for ex situ conservation are done outside of the PGR natural habitat. Research institute or gene bank plays a vital role in these initiatives. In contrast, initiatives for in situ conservation are done in PGR ecosystems and natural habitats and are aimed to maintain viable populations of species or help the species recover in their natural habitats (FAO 2009). For domesticated plants, the conservation initiatives are implemented in areas where they acquired their distinguishing properties (FAO 2009).

Initiatives for ex situ conservation are implemented to protect a wider spectrum of plant species that may range from wild and weedy to domesticated species (Jaramillo and Baena 2002). Dulloo et al. (2010) had indicated that such a conservation approach can help maintain the life of the germplasm materials and seeds to avoid genetic diversity loss. For instance, the major cereals and legume crops in the past had their seeds being conserved in gene banks. Ex situ conservation was able to provide an opportunity to study the biology of and understand the threats to these species, which eventually helped in their successful recovery (Dulloo et al. 2010). The success of their recovery was owing to the seed banking techniques where dried seeds are stored at low temperatures to increase their viability. Dickie et al. (1990) had indicated that temperature, seed moisture content, and relative humidity are important factors for seed longevity.

Aside from avoiding genetic erosion, Frankel et al. (1995) pointed out that ex situ conservation initiatives are essential in protecting important materials from the wild species for research and crop improvement. For instance, the conserved materials from the wild species of sugarcane had helped improve the traits of the modern sugarcane as a crop. As Jaramillo and Baena (2002) reported, the pedigree of the modern sugarcane now includes the wild *Saccharum spontaneum* that has increased its vigor and disease resistance, thus increasing its yield.

The use of gene banks as spaces for ex situ conservation should not be underestimated. Scientists, researchers, and even the public especially the farmers have witnessed the long-term benefits of conserving and securing biological diversity in gene banks (Rabara et al. 2018). Diez et al. (2018) emphasized the importance of gene banks in preserving cultivated biodiversity when modern varieties start substituting the landraces. Their creation can essentially safeguard the future of these landraces, and therefore maintain their diversity (Diez et al. 2018). Gene banks usually house biodiversity that helps advance global plant breeding programs (McCouch et al. 2013). Fowler and Hodgkin (2004) had also indicated similar importance. In addition, Tanksley and McCouch (1997) pointed out that the preserved genetic materials in gene banks can be used in breeding programs so that varieties that can potentially resist environmental as well as climatic changes can be created. Diez et al. (2018) reported that the establishment of gene banks allowed the immense evolution of biological technologies that supported the generation of more variability and new data sources for breeding. Such variability and data sources allow plant breeders to explore and develop new and improved cultivars, i.e., cultivars with traits that are desirable to farmers and conservationists. These traits may include large seed, higher yield potential, photosensitivity, and resistance to pests and diseases (Govindaraj et al. 2015). For instance, the study on wild rice, *Oryza nivara*, revealed that the species can provide protection from the impacts of the grassy stunt virus to almost all Asian tropical rice varieties (Plucknett et al. 1987). These varieties have benefited from the conserved accessions of the wild variety in gene banks. Rabara et al. (2018) also reported that at least one accession conserved in the international rice gene bank has been integrated in more than 4300 rice varieties worldwide.

On the other hand, the in situ conservation strategy maintains species' populations in their natural habitat where they occur either as uncultivated plant communities or as part of the existing agroecosystems in farmers' fields (Brush 1995). Though it has previously received skepticism from breeders (Valkoun and Damania 1993), it is now looked at as a more plausible backup to gene bank collection (Brush 2000). It has been considered as a less expensive alternative for maintaining collections (Brush 1995). Engels (2002) considered it as a conservation strategy that permits crop adaptation and evolution to continue. For instance, Valkoun and Damania (1993) reported that the International Center for Agricultural Research in the Dry Areas (ICARDA) is a good backup for the ex situ collections of legumes and cereals. They reported that studies at ICARDA and elsewhere had demonstrated the preservation of an extensive genetic diversity of the wild gene pool in the populations in the natural habitat which may be impossible to attain using the standard ex situ collection procedures. Moreover, they noted that the approach has preserved genetic diversity that is much larger and continuously evolving than the ex situ collections.

Maxted et al. (1997) identified two approaches for in situ conservation of PGR, namely (a) genetic reserves and (b) on-farm management. The former approach involves conserving wild species in their natural habitat (Maxted et al. 1999) while the latter involves conserving landraces within the traditional farming systems (Hammer et al. 2002). Genetic reserves approach is implemented to conserve CWR. The approach was considered to be more practical than the ex situ conservation because of the involvement of a large number of species and the difficulty of collecting and conserving the entire genetic diversity using the ex situ approach (Maxted et al. 2005). According to Maxted et al. (1999), conserving CWR using this approach can manage and monitor the genetic diversity of the wild populations in their natural habitats identified for long-term conservation.

In contrast, on-farm approach to conservation involves the sustainable management of the genetic diversity of the landraces developed locally by farmers in the traditional agri-silvicultural or agricultural systems (Maxted et al. 1997). The conservation process will include their associated weedy and wild species. Sustainably managing this genetic diversity is expected to maximize their direct benefits to the local farmers especially those who are in the marginal environments (Di Falco and Chavas 2006; Ceccarelli 2014). In addition, the approach can also involve those crops that are either grown for sale or used primarily for home consumption (Stolton et al. 2006).

Despite the unparalleled benefits of conserving the PGR, the process is not without any challenges. For instance, Gollin (2020) reported that ex situ conservation has a very high average cost. This could probably be due to high costs of storage, monitoring, and regeneration (Singh et al. 2012) although Koo et al. (2003a) and Kim et al. (2003) indicated a relatively low marginal cost for conserving an additional accession in gene banks if these capital costs are taken out of the equation. Singh et al. (2012) suggested that estimating the costs for ex situ conservation should consider the candidate crops and their breeding behavior and the seed size and composition. They observed that the costs of conservation are highly sensitive to these factors. For instance, they reported that the cost of conservation becomes higher when it involves larger population size and controlled pollination. Koo et al. (2003b) also reported that low multiplication rate, space requirement, and growth habit had contributed to higher costs.

In addition, both ex situ and in situ conservation approaches are faced with issues related to collection and definition of area for conservation, respectively. For instance, Vidal et al. (2020) had indicated the criticality of establishing collection plans so that the representativeness of species under consideration will be ensured. Unfortunately, there is difficulty in determining the best strategy that maximizes the collection of domesticated species that are conserved in situ. As Vidal et al. (2020) had indicated, the magnitude of species diversity in a specific geographical area is practically undefined. This creates a challenge in determining the number of samples and the populations to be included in the sampling activity. This is crucial when financial and human resources are limited. Consequently, only a small number can be considered in the collection activity. However, such number must best represent the diversity of landraces in the in situ conservation areas to make the collection activity more efficient (Vidal et al. 2020).

Another challenging issue in PGR conservation is the incomplete passport information of accessions preserved in gene banks. Anglin et al. (2018) emphasized the importance of a complete passport information in the collection, use, and conservation of accessions. They emphasized that such an information is useful in identifying the accessions that contain the trait the users want. In most cases, accessions conserved in gene banks have no information about the environment from where they were collected and their unique traits making it difficult to adequately evaluate them (Anglin et al. 2018). For instance, the environmental information of accessions can play a crucial role in determining the accession that optimally performs in varying environments. In fact, Anglin et al. (2018) suggested that such information should encompass wider range of seasons and geographic locations so that users can choose the right accession. Without this, they would not be able to predict how the trait of their selected accession responds to environmental variations.

Hijmans et al. (1999) had identified the absence of a complete geographical address of the accessions in gene bank as a pressing issue in ex situ conservation. They observed that in many instances databases in gene banks have either missing, imprecise, or wrong coordinates of collection sites. However, the presence of georeferenced gene bank data can make a significant contribution in plant genetic resource conservation and use. For instance, Jones et al. (1997) pointed out that the analysis of georeferenced gene bank data can assist in investigating the taxonomic structure of collections. On the other hand, Frankel et al. (1995) had indicated that such data can be used in identifying areas of high diversity while Guarino et al. (1998) emphasized their usefulness in targeting PGR for developing core collections and breeding programs as well as in designing and selecting in situ conservation areas. Hijmans et al. (1999) reported that geographic location data in gene bank are essential in germplasm collection management and usage and in planning, designing, and implementing future collection activities. They can also be utilized to correlate climate data with specific traits of plant species, which Galluzzi et al. (2020) recognized as critical factors for intensifying any program for plant breeding. Plant breeding is essential in the creation of climate-adaptive varieties, which may improve food production and security in countries highly vulnerable to the changing climate. Saina et al. (2013) indicated that changes in climatic conditions had a direct link with food insecurity, which is acute in developing countries where impacts of climate change are exacerbated with poverty.

On the other hand, Parra-Quijano et al. (2012) emphasized the importance of the information about the unique traits of PGR in plant breeding. They reported that breeders need this information in selecting the parentals that contain their desired traits. In most cases, breeders need the germplasm that is adaptive to certain environmental conditions and provide the highest yield. Thus, the absence of a well-structured documentation of accession in gene banks will not only devalue these resources but also not help breeders improve their breeding programs (Weise et al. 2020).

For the in situ conservation strategy, the information about the extent, amount, and spatial distribution of genetic variations must be enhanced. Such an enhancement can be useful in understanding the variations between plant species and between their different populations. Hodgkin (1997) indicated that these variations can be expected to be large across the spatial distributions of a particular trait or group of traits. For instance, Hamrick and Godt (1990) reported that great variations among wild plant species can be attributed to the geographic distribution of their traits, which is affected by their historical development across locations and the selection process involving the plant species that people made (Hodgkin 1997). To maximize any conservation efforts, it is important to identify the locations of plant populations with high allelic richness. For instance, Vidigal et al. (2016) reported that strong climatic and altitudinal clines were detected among the wild accessions of *Arabidopsis thaliana* for most of their life history traits. Thus, such information must be used as a basis for planning, designing, and implementing an in situ conservation for this species. It is also important to consider that Hodgkin (1997) also suggested that for allogamous species, sites with different ecogeographic characteristics should be located. This is in response to Schoen and Brown's (1991) report that variations of autogamous species were much greater between their populations across space. Hodgkin (1997) further suggested that ecogeographic and agroecological survey methods should be combined with GIS and landscape analysis so that data on species distribution and the environments in which they occur can be generated and utilized in any conservation actions. Conservation actions are usually limited and thereby must be carefully planned to maximize the amount of useful conserved diversity (Hodgkin 1997).

4.4 GIS in PGR Conservation

Geographic information system or GIS has been instrumental in addressing issues associated with PGR conservation and use. The system usually allows the users to display database information geographically and links two or more previously unrelated databases (Liang et al. 2005). In addition, Mahlein (2016) had indicated that GIS can be used to integrate a range of geographical information into a single analytical model. The model can be used to georeference location data into cartographic projections, which can then be utilized to connect spatial objects such as geographic locations of the accessions with the attribute information such as the traits of the species (Liang et al. 2005). This is an important step to make efficient conservation and collection activities of PGR.

Visualizing and analyzing the relationship between the geographic locations and the trait of accessions can be facilitated greatly by using GIS. The USGS (n.d.) defined GIS as a computer system used in analyzing and displaying geographically referenced information. Dempsey (2019) defined it as a system that can be used in capturing, manipulating, analyzing, managing, and presenting different geographical data types. Meanwhile, Guarino (1995) considered GIS as a tool for inputting, verifying, editing, storing, retrieving, and managing data. The first three processes involve the inputting of data into the GIS environment for spatial processes and analyses. For instance, data from maps such as plant distribution, elevation, and soil can be entered into the GIS environment so that they can be linked into the associated attribute of or information about the accession (Guarino 1995). Accession data on the latitude and longitude in gene bank can also be linked to digitized maps and/or remote sensing (Guarino et al. 2002).

Geographical data stored in GIS can be either vector or raster files. The vector files are presented as points, lines, and/or polygons, which can represent the density, sizes, and/or shapes of the areas where plant species are distributed (Guarino et al. 2002). On the other hand, the raster data are stored as regularly shaped cells such as an interpolated climate data or data on elevations which influence plant species distribution. Both data formats are utilized to generate and analyze enormous quantities of environmental, phenotypic, and genotypic data necessary for PGR conservation (Halewood et al. 2018).

The element of data manipulation and analysis allows the user to integrate various datasets or make datasets compatible so that they can be combined and analyzed (Guarino 1995). In addition, Bailey (1994) indicated that data analysis in a GIS environment can be deterministic such as the use of network analysis, algebraic projection, and three-dimensional modeling. It can also be statistical such as the use of data interpolation, pattern and trend analysis, and spatial correlation (Moghadama et al. 2018). Once the analysis has been done, users can select the data output that will be generated from the system. Results can be presented as overlays or maps in various scales, tables, graphs, and even animations (Guarino et al. 2002). They can also be presented as a three-dimensional visual of the selected attributes or a set of attributes.

With such functionalities, Parra-Quijano et al. (2012) underscored the role of GIS in PGR conservation especially in the analysis of the spatial relationship between the environmental conditions and the plant species to determine the adaptive traits of the latter. Hijmans and Spooner (2001) also emphasized the importance of GIS in analyzing the spatial characteristics of the sites from which germplasms and/or accessions have been collected. These characteristics include the geographical distances of the germplasms and/or their spatial distribution patterns. GIS has been useful in improving the efficiency of some PGR conservation activities including field explorations, identification of under-conserved areas, creation of core collections, and/or selection of the appropriate germplasm for breeding programs

(Hijmans and Spooner 2001). For instance, Zair et al. (2021) reported that GIS has been useful in identifying genetic reserves in the Fertile Crescent of the Middle East for the conservation of crop wild relatives. They were able to identify critical areas for collection of ex situ conservation, specifically in the areas of Turkey, Syria, and Iraq. On the other hand, Fitzgerald et al. (2019) used GIS in planning for in situ conservation of the priority crop wild relatives in the Nordic region in Northern Europe. The GIS was useful in the complementarity as well as ecogeographic analyses, which identified the regional potential genetic reserve networks. They reported that the use of GIS had supported the identification of the minimum number of conservation sites. These sites have been observed to potentially support a maximum number of species targeted for a maximum intraspecific diversity within and outside of the existing conservation areas (Fitzgerald et al. 2019).

GIS has also been instrumental in rectifying the erroneous location of passport data of germplasms in gene banks. Hijmans et al. (1999) reported that more than 50% of the accession data of wild potatoes collected through 18 expeditions in Bolivia had an error in geographic locations. Using GIS, they were able to detect accessions in wrong or impossible places. Using overlay analysis and visual inspection, Hijmans et al. (1999) were able to detect and rectify these errors in the accessions' passport data. Semwal and Ahlawat (n.d.) emphasized the importance of having correct georeferenced data of accessions conserved in gene banks. According to them, these data are critically useful in predicting the accessions' habitats that will change due to some environmental processes such as climate change as well as in identifying future collection sites of germplasms (Semwal and Ahlawat n.d.). Tas et al. (2019) also emphasized that such data can form the basis for creating a nationwide in situ genetic reserve network. GIS can be used to perform complex analyses as well as visualization of results to facilitate authorities' decision-making process and implementation of conservation policies (van Zonneveld et al. 2012). In addition, the increased availability and accessibility of better geographical and environmental data as well as the advances in both software and hardware have increased the capability of GIS technology to link genetic data from different in situ PGR reserve sites. Hyman et al. (2013) suggested combining these improved data with spatial analysis methods so that environmental conditions across sites can be compared and national agroecological maps can be created (Hyman et al. 2013). They further suggested linking genotype to phenotype to determine the vast potential for improving crop adaptation efforts. As Jarvis et al. (2005) had pointed out, spatial analysis can be useful in understanding the variations of the genetic structure of species through careful analysis of genotype-environment interactions. The advent of functional genomics and improved quantification of adaptive traits can greatly facilitate this process.

4.5 Prospects of GIS for PGR Conservation

The various functionalities of GIS have great prospects for the betterment of plant genetic resource conservation, especially in these challenging times. The rapidly advancing information and communication technologies as well as the genetic technologies can be exploited to advance germplasm exploration and collection, select in situ conservation site, and create ecogeographic and genomic information-driven passport data in gene banks. This information once collected can possibly develop methodologies that have adequate capability to accurately predict the interactions between genotypes and environmental factors, and eventually allows for the development of crops that are tailored to a particular set of environmental conditions (Halewood et al. 2018).

In addition, the rapidly growing fields of geoinformatics and data science can make GIS an essential tool with great potential in the analysis and management of such environmental factors as soil fertility, pest population, and soil moisture that affect plant growth and health (Roberts and Kobayashi 2011). Once managed, the process of germplasm improvement can be properly facilitated. In turn, this is expected to enhance tolerance of crops to pathogens and abiotic stresses while they are made to adapt to different systems of cropping and environmental conditions. Advances in the GIS technology in combination with cloud-based, big data-driven applications can speed up plant breeding programs (Roberts and Kobayashi 2011). This is critical especially in the coming years where the impacts of climate change and other disasters are projected to significantly affect the global food production systems. Myers et al. (2017) projected significant reductions in food production under certain extreme weather conditions.

Moreover, Delgado et al. (2019) had pointed out that combining spatial data management and big data analytics can greatly facilitate the development and use of climate-resilient, next-generation, and environment-friendly crop production system. The combination of spatial data and big data analytics will help project agricultural fields as a nonhomogeneous management unit. As such, Delgado et al. (2019) have suggested to treat them as a varied substrate supporting plant growth. In this manner, farmers would be able to maximize their disease control and management schemes. Mahlein (2016) had indicated that this can lead to increased crop production and more efficient use of plant genetic resources.

GIS can also be used to effectively study, locate, and conserve the genetic resources of endangered, threatened, and rare species. The availability and accessibility of digitized data on species diversity, distribution, taxonomy, and characterization of these species can be combined with relevant social, physical, biotic, and abiotic variables within a GIS platform to achieve these goals (Williams 2001). The ability of GIS to integrate different kinds of georeferenced data (Guarino et al. 2002) can be tapped to precisely map the spatial distributions of these species and correlate them with socioeconomic and ecological and physical data to be able to develop a more contextualized conservation strategy. According to Romero and Peña-Claros (2009), this conservation strategy, where all important variables that affect plant growth and health are being integrated, creates more favorable in situ conservation outcomes of plant genetic resources. They reported that combining the ecological, sociocultural, political, market access, and economic settings of forests will more likely increase the potential of sustaining the improvement in the management of these resources as well as the overall forest conservation outcomes. Improving in situ conservation outcomes can greatly complement the ex situ storage of germplasm by allowing the uninterrupted evolutionary dynamic of the conserved populations and maintain a sufficient population size (Williams 2001).

Several studies (e.g., Beckler et al. 2005; Dminic et al. 2010; Mazzi and Dorn 2012) also indicate the potential use of GIS in managing pests that threaten the survival and growth of plant species important in securing global food production. According to Dminic et al. (2010), the GIS can potentially analyze the spatial and temporal behavior of insect populations in in situ conservation sites by overlaying multiple spatial data layers. Overlaying multiple spatial data layers can help investigators analyze and understand the ecological behavior of the pests in situ including the spatial behavior of their natural enemies and crop hosts (Deleon et al. 2017). Dminic et al. (2010) even expanded this potential to a regional scale by linking analyses of the pest's ecology from individual in situ conservation sites to various landscapes. Moral-García (2006) explored the potential use of GIS in combination with geostatistics and global positioning system (GPS) in predicting the spatial distribution and temporal behavior of the *Helicoverpa armigera* Hb. (Lepidoptera: Noctuidae) across the Extremadura region of the southwestern Spain. Using those tools, they were able to create maps that showed the spatial distributions of the pest across the region. The maps were then being utilized in developing effective pest control and management strategies as well as a precise agricultural production system. Nansen and Elliott (2016) also utilized GIS in detecting insect stresses in agricultural fields. Deleon et al. (2017) suggested that the spatial and temporal data of crops and the monitoring data of pests can be geo-analyzed so that pest monitoring maps can be developed. The maps can visually represent the densities of insect populations relative to the economic thresholds of the farms. They can then be used to estimate the time and intensity of the appearances of the targeted pests on the agricultural fields or even in in situ conservation sites. If this is done, farmers or conservation managers can develop appropriate and timely mitigating measures or interventions to reduce the impacts of pest infestations on the crops or on the plant species being conserved (Dminic et al. 2010). In this way, in situ management, protection, and conservation of plant genetic resources are being enhanced.

The timely application of interventions plays a critical role in plant genetic resource management and conservation. Mazzi and Dorn (2012) reported that pests are continually exacting a significant toll on farm production or in situ conservation despite the high intensity of agrochemical inputs. However, using GIS and other ecogeographical tools like GPS, the spatial-temporal abundance and dispersion of pests can potentially be measured and analyzed. The analysis can in turn be used to estimate the extent of damage inflicted on crops or on the conserved plant species (Mazzi and Dorn 2012). This analysis can greatly be facilitated with the capacity of GIS to manipulate multiple geospatial data and perform geostatistical analyses of spatial interactions (Beckler et al. 2005). The created knowledge may be shared with farmers and plant genetic resource managers so that they can develop more sustainable management and conservation strategies.

4.6 Concluding Remarks

The pressing issue of genetic erosion due to overexploitation of our plant resources, habitat destruction, and climate change highlights the need to manage, conserve, and preserve the global plant genetic resources efficiently and effectively. This is necessary because PGR forms a major part of the global food supply chain. Securing the global food production system depends on the availability, accessibility, and viability of PGR worldwide.

PGR conservation, either in situ or ex situ, is data intensive and requires the inclusion of the geographical data in either the passport data in gene banks or places where they are naturally thriving. As these data are georeferenced, they can be subjected to GIS-based analysis so that information on plant-environment interactions can be generated while monitoring maps can be created. These data can then be utilized in creating crops with improved traits, making PGR collection, utilization, and conservation decisions, and formulating and implementing conservation policies. The advances in GIS technology can also provide knowledge on the statistical and spatial relationships of the sociocultural, physical, and biological interactions surrounding the conserved plant genetic resources. Such knowledge can be utilized in predicting the impacts of the changing cultural, social, economic, political, ecological, and climatic conditions on the future plant as well as pest populations so that suitable habitat conditions can appropriately be identified for conservation.

However, though GIS can perform effectively in analyzing spatial based data, it may not be able to address all analytical issues that beset PGR conservation. Data integration in GIS environment may be limited when data are fragmented and dispersed across organizations and international borders, when there is no universal system for recording and tracking PGR-related data, and when policies hamper the sharing of data within and across the public and private sectors (Halewood et al. 2018). Thus, it will be of universal benefit if PGR-related data across the globe will be made open access and recorded and managed following a universally acceptable system.

References

Ainsworth EA, Leakey ABD, Ort DO, Long SP (2008) Letters: FACE-ing the facts: inconsistencies and interdependence among field, chamber and modelling studies of elevated [CO2] impacts on crop yield and food supply. New Phytol 179(1):5–9

Anglin NL, Amri A, Kehel Z, Ellis D (2018) A case of need: linking traits to Genebank accessions. Biopreserv Biobank 16(5):337–349. https://doi.org/10.1089/bio.2018.0033

Bailey TC (1994) A review of statistical spatial analysis in geographical information systems. In: Fotheringham S, Rogerson P (eds) Spatial analysis and GIS. Taylor & Francis, London, pp 13–44

Beckler AA, Chandler LD, French BW (2005) Using GIS in area-wide pest management: a case study in South Dakota. Trans GIS 9(2):109–127

Brockhaus R, Oetmann A (1996) Aspects of the documentation of in situ conservation measures of genetic resources; Plant Genet. Res Newslett 108:1–16. Retrieved from https://bit.ly/3EPwSnb

Brush SB (1995) In situ conservation of landraces in centers of crop diversity. Crop Sci 35(2): 346–354

Brush SB (2000) The issues of *in situ* conservation of crop genetic resources. In: Brush SB (ed) Genes in the field: on-farm conservation of crop diversity. Lewis, Washington

Ceccarelli S (2014) GMO crops, organic agriculture and breeding for sustainability. Sustainability 6:4273–4286. https://doi.org/10.3390/su6074273

Deleon L, Brewer MJ, Esquivel IL, Halcomb J (2017) Use of a geographic information system to produce pest monitoring maps for South Texas cotton and sorghum land managers. Crop Prot 101:50–57

Delgado JA, Short NM Jr, Roberts DP, Vandenberg B (2019) Big data analysis for sustainable agriculture on a geospatial cloud framework. Front Sustain Food Syst 3:54. https://doi.org/10.3389/fsufs.2019.00054

Dempewolf H, Baute G, Anderson J, Kilian B, Smith C, Guarino L (2017) Past and future use of wild relatives in crop breeding. Crop Sci 57:1070–1082

Dempsey C (2019) What is GIS? GIS Lounge. https://www.gislounge.com/what-is-gis/

Di Falco S, Chavas JP (2006) Crop genetic diversity, farm productivity and the management of environmental risk in rainfed agriculture. Eur Rev Agric Econ 33(3):289–314

Dickie JB, Ellis RH, Kraak HL, Ryder K, Tompsett PB (1990) Temperature and seed storage longevity. Ann Bot 65(2):197–204. https://doi.org/10.1093/oxfordjournals.aob.a087924

Diez MJ, De la Rosa L, Martín M, Guasch L, Cartea ME, Mallor C, Casals J, Simó J, Rivera A, Anastasio G, Prohens J, Soler S, Blanca J, Valcárcel JV, Casañas F (2018) Plant Genebanks: present situation and proposals for their improvement. The case of the Spanish network. Front Plant Sci 9(1794):1–13

Dminic I, Kozina A, Bazok R, Barcic JI (2010) Geographic information systems (GIS) and entomological research: a review. J Food Agric Environ 8:1193–1198

Dulloo ME, Hunter D, Borelli T (2010) *Ex situ* and *in situ* conservation of agricultural biodiversity: major advances and research needs. Not Bot Hort Agrobot Cluj 38(2):123–135

Engels J (2002) Home gardens—a genetic resources perspective. In: Watson JW, Eyzaguirre PB (eds) Proceedings of the second international home gardens workshop: contribution of home gardens to *in situ* conservation of plant genetic resources in farming systems. International Plant Genetic Resources Institute, Rome

Esquinas-Alcazar J (2005) Protecting crop genetic diversity for food security: political, ethical, and technical challenges. Sci Soc 6:946–953

FAO (1989) Plant genetic resources: their conservation in situ for human use. FAO, Rome

FAO (2009) International treaty for plant genetic resources for food and agriculture. Food and Agriculture Organization of the United Nations, Rome

FAO (2015) Genetic resources for food security and nutrition. FAO, Rome

Fitzgerald H, Palmé A, Asdal Å, Endresen D, Kiviharju E, Lund B, Rasmussen M, Thorbjörnsson H, Weibull J (2019) A regional approach to Nordic crop wild relative in situ conservation planning. Plant Genet Res Charact Util 17(2):196–207. https://doi.org/10.1017/S147926211800059X

Fortuna MA, Albaladejo RG, Fernández L, Aparicio A, Bascompte J (2009) Networks of spatial genetic variation across species. Proc Natl Acad Sci 106(45):19044–19049. https://doi.org/10.1073/pnas.0907704106

Fowler & Hodgkin (2004) Plant genetic resources for food and agriculture: assessing global availability. Annual Rev 29:143–179

Fowler C, Lower R (2004) The politics of plant breeding. Plant Breeding Rev 25:21–56
Frankel OH, Browny AHD, Burdon JJ (1995) Conservation of plant biodiversity. Cambridge University Press, Reino Unido, p 299
Galluzzi G, Halewood M, Noriega IL, Vernooy R (2016) Twenty-five years of international exchanges of plant genetic resources facilitated by the CGIAR genebanks: a case study on global interdependence. Biodivers Conserv 25:1421–1446. https://doi.org/10.1007/s10531-016-1109-7
Galluzzi G, Seyoum A, Halewood M, Noriega IL, Welch EW (2020) The role of genetic resources in breeding for climate change: the case of public breeding programmes in eighteen developing countries. Plan Theory 9(1129):1–19. https://doi.org/10.3390/plants9091129
Gollin D (2020) Conserving genetic resources for agriculture: economic implications of emerging science. Food Sec 12:919–927
Govindaraj M, Vetriventhan M, Srinivasan M (2015) Importance of genetic diversity assessment in crop plants and its recent advances: an overview of its analytical perspectives. Genet Res Int 2015:431487. https://doi.org/10.1155/2015/431487
Guarino L (1995) Geographic information systems and remote sensing for the plant germplasm collector. In: Guarino L, Ramanatha Rao V, Reid R (eds) Collecting plant genetic diversity. CAB International, Wallingford, pp 315–328
Guarino L, Jarvis A, Hijmans RJ, Maxted N (2002) Geographic information systems (GIS) and the conservation and use of plant genetic resources. In: Engels JMM, Ramanatha Rao V, Brown AHD, Jackson MT (eds) Managing plant genetic diversity. CABI Publishing, Wallingford
Guarino L, Maxted N, Sawkins M (1998) Analysis of georeferenced data and the conservation and use of plant genetic resources. Linking genetics and geography: emerging strategies for managing crop biodiversity. ASA/CSSA/SSSA Annual Meetings, 26–31 October 1997. Anaheim
Gur A, Zamir D (2004) Unused natural variation can lift yield barriers in plant breeding. PLoS Biol 2(10):e245. https://doi.org/10.1371/journal.pbio.0020245
Halewood M, Chiurugwi T, Hamilton RS, Kurtz B, Marden E, Welch E, Michiels F et al (2018) Plant genetic resources for food and agriculture: opportunities and challenges emerging from the science and information technology revolution. New Phytol 217(4):1407–1419. https://doi.org/10.1111/nph.14993
Hammer K, Gladis TH, Diederichsen A (2002) *In situ* and on-farm management of plant genetic resources. Eur J Agron 19:509–517
Hammer K, Teklu Y (2008) Plant genetic resources: selected issues from genetic erosion to genetic engineering. J Agric Rural Dev Trop Subtrop 109(1):15–50
Hamrick JL, Godt MJW (1990) Allozyme diversity in plant species. In: Brown AHD, Clegg MT, Kahler AL, Weir BS (eds) Plant population genetics, breeding, and genetic resources. Sinauer, Sunderland
Hedden P (2003) The genes of the green revolution. Trends Genet 19(1):5–9
Hijmans RJ, Spooner DM (2001) Geographic distribution of wild potato species. Am J Bot 88 (11):2101–2112. Retrieved from https://bsapubs.onlinelibrary.wiley.com/doi/epdf/10.2307/3558435
Hijmans RJ, Schreuder M, De la Cruz J, Guarino L (1999) Using GIS to check co-ordinates of genebank accessions. Genet Resour Crop Evol 46:291–296
Hodgkin T (1997) Some current issues in the conservation and use of plant genetic resources. In: Ayad WG, Hodgkin T, Jaradat A, Rao VR (eds) Molecular genetic techniques for plant genetic resources. Report of an IPGRI workshop. International Plant Genetic Resources Institute, Rome
Hyman G, Hodson D, Jones P (2013) Spatial analysis to support geographic targeting of genotypes to environments. Front Physiol 18:4–40. https://doi.org/10.3389/fphys.2013.00040
Jaramillo S, Baena M (2002) *Ex situ* conservation of plant genetic resources: training module. International Plant Genetic Resources Institute, Colombia
Jarvis A, Upadhyaya H, Gowda CCL, Aggarwal PK, Fujisaka S, Anderson B (2008) Climate change and its effect on conservation and use of plant genetic resources for food and agriculture

and associated biodiversity for food security. Thematic background study. http://www.fao.org/3/i1500e/i1500e16.pdf

Jarvis A, Yeaman S, Guarino L, Tohme J (2005) The role of geographic analysis in locating, understanding, and using plant genetic diversity. Methods Enzymol 395:279–298. https://doi.org/10.1016/S0076-6879(05)95017-9

Jones P, Jarvis A, Hyman G, Beebe SE, Pachico DH (2007) Climate proofing agricultural research investments. SAT eJ 4:1–29

Jones PG, Beebe SE, Tohme J, Galwey NW (1997) The use of geographical information systems in biodiversity exploration and conservation. Biodivers Conserv 6:947–958. https://doi.org/10.1023/A:1018311513176

Kantar MB, Sosa CC, Khoury CK, Castañeda-Álvarez NP, Achicanoy HA, Bernau V, Kane NC, Marek L, Seiler G, Rieseberg LH (2015) Ecogeography and utility to plant breeding of the crop wild relatives of sunflower (*Helianthus annuus* L.). Front Plant Sci 6(841):1–11. https://doi.org/10.3389/fpls.2015.00841

Khoury CK, Bjorkman AD, Dempewolf H, Ramirez-Villegas J, Guarino L, Jarvis A, Rieseberg LH, Struik PC (2014) Increasing homogeneity in global food supplies and the implications for food security. Proc Nat Acad Sci USA 111(11):4001–4006. https://doi.org/10.1073/pnas.1313490111

Kim B, Pardey PG, Wright B (2003) Saving seeds: the economics of conserving crop genetic resources ex situ in the future harvest centers of the CGIAR. CABI, Wallingford

Koo B, Pardey PG, Wright B (2003a) The price of conserving agricultural biodiversity. Nat Biotechnol 21(2):126–128

Koo B, Pardey PG, Wright B (2003b) Conserving genetic resources for agriculture: counting the cost. In: Pardey PG, Koo B (eds) Research at glance, biotechnology and genetic resources policies. Brief 6. International Food Policy Research Institute, Washington

Ladizinsky G (1998) Plant evolution under domestication. Kluwer Academic Press, Dordrecht, p 262

Lane A, Jarvis A (2007) Changes in climate will modify the geography of crop suitability: agricultural biodiversity can help with adaptation. SAT eJ 4(1):1–12

Liang LY, Hua LT, Ma'some, D.M. (2005) Traffic accident application using geographic Information system. J East Asia Soc Transp Stud 6:3574–3589. Retrieved from https://citeseerx.ist.psu.edu/viewdoc/download?doi=10.1.1.515.5359&rep=rep1&type=pdf

Mahlein AK (2016) Plant disease detection by imaging sensors—parallels and specific demands for precision agriculture and plant phenotyping. Plant Dis 100:241–251

Maxted N, Dullo M, Eastwood A (1999) A model for genetic reserve conservation: a case study for Coffea in the Mascarene Islands. Botanica Lithuanica 2:61–78

Maxted N, Ford-Lloyd BV, Jury SL, Kell SP, Scholten MA (2005) Towards a definition of a crop wild relative. Biodivers Conserv 14:1–13

Maxted N, Hawkes JG, Ford-Lloyd BV, Williams JT (1997) A practical model for in situ genetic conservation. In: Maxted N, Ford-Lloyd BV, Hawkes JG (eds) Plant genetic conservation: the *in situ* approach. Chapman and Hall, London

Maxted N, Iriondo J, Dulloo E, Lane A (2008) Introduction: the integration of PGR conservation with protected area management. In: Iriondo JM, Maxted N, Dulloo E (eds) Plant genetic population management. CAB International, Wallingford, pp 1–22

Mazzi D, Dorn S (2012) Movement of insect pests in agricultural landscapes. Ann Appl Biol 160(2):97–113. https://doi.org/10.1111/j.1744-7348.2012.00533.x

McCouch S, Baute GJ, Bradeen J, Bramel P, Bretting PK, Buckler E, Burke JM, Charest D, Cloutier S et al (2013) Agriculture: feeding the future. Nature 499(7456):23–24

Moghadama ST, Toniolo J, Mutani G, Lombardi P (2018) A GIS-statistical approach for assessing built environment energy use at urban scale. Sustain Cities Soc 37:70–84

Moral-García FJ (2006) Analysis of the spatiotemporal distribution of *Helicoverpa armigera* Hb. In a tomato field using a stochastic approach. Biosyst Eng 93:65–259. https://doi.org/10.1016/j.biosystemseng.2005.12.011

Myers SS, Smith MR, Guth S, Golden CD, Vaitla B, Mueller ND, Dangour AD, Huybers P (2017) Climate changes and global food systems: potential impacts on food security and undernutrition. Annu Rev Public Health 38(1):259–277

N'Danikou S, Tchokponhoue DA (2019) Plant domestication for enhanced food security. Springer, Cham. https://doi.org/10.1007/978-3-319-69626-3_96-1

Nansen C, Elliott N (2016) Remote sensing and reflectance profiling in entomology. Annu Rev Entomol 1:139–158. https://doi.org/10.1146/annurev-ento-010715-023834

Ogwu MC, Osawaru ME, Ahana CM (2014) Challenges in conserving and utilizing plant genetic resources (PGR). Int J Genet Mol Biol 6(2):16–22. https://doi.org/10.5897/IJGMB2013.0083

Parra-Quijano M, Iriondo JM, Torres E (2012) Applications of ecogeography and geographic information systems in conservation and utilization of plant genetic resources. Span J Agric Res 10(2):419–429

Plucknett DL, Smith NJH, Williams JT, Anishetty NM (1987) Gene banks and the world's food. Princeton University Press, Princeton

Rabara RC, Ferrer MC, Jara-Rabara J, Sotto RC (2018) Securing diversity for food security: the case of conservation and use of rice genetic resources, new visions in plant science. IntechOpen, Özge Çelik

Roberts DP, Kobayashi DY (2011) Impact of spatial heterogeneity within spermosphere and rhizosphere environments on performance of bacterial biological control agents. In: Maheshwari DK (ed) Bacteria in agrobiology: crop ecosystems. Springer, Berlin

Romero C, Peña-Claros M (2009) Beyond tropical forests adoption: contextualizing conservation strategies. Biotropica 41:653–655. https://doi.org/10.1111/j.1744-7429.2009.00582.x

Ronald P (2011) Plant genetics, sustainable agriculture, and global food security. Genetics 188(1): 11–20. https://doi.org/10.1534/genetics.111.128553

Saina CK, Murgor DK, Murgor FAC (2013) Climate change and food security. In: Silvern S, Young S (eds) Climate change and food security, environmental change and sustainability. IntechOpen, Rijeka. https://doi.org/10.5772/55206

Schoen DJ, Brown AHD (1991) Intraspecific variation in population gene diversity and effective population size correlates with the mating system. Proc Nat Acad of Sci USA 88:4494–4497

Semwal DP, Ahlawat SP (n.d.) Applications of geoinformatics in plant genetic resources studies. E-Publication (NBP-16-03). http://www.nbpgr.ernet.in/Downloadfile.aspx?EntryId=7271

Singh AK, Varaprasad KS, Venkateswaran K (2012) Conservation costs of plant genetic resources for food and agriculture: seed genebanks. Agric Res 1(3):223–239. https://doi.org/10.1007/s40003-012-0029-3

Stolton S, Maxted N, Ford-Lloyd B, Kell S, Dudley N (2006) Food stores: using protected areas to secure crop genetic diversity. In: The Argument for Protection Series. World Wide Fund for Nature, London

Tanksley SD, McCouch SR (1997) Seed banks and molecular maps: unlocking genetic potential from the wild. Science 277:1063–1066. https://doi.org/10.1126/science.277.5329.1063

Tas N, West G, Kircalioglu G, Topaloglu SB, Phillips J, Kell S, Maxted N (2019) Conservation gap analysis of crop wild relatives in Turkey. Plant Genet Resour Characteriz Utiliz 17(2):164–173. https://doi.org/10.1017/S1479262118000564

USGS (n.d.) What is a geographic information system (GIS)? https://on.doi.gov/3uIZr0U

Valkoun J, Damania AB (1993) Methodologies for in situ conservation of plant genetic resources. JIRCAS international symposium series 2, pp 168–173. https://www.jircas.go.jp/sites/default/files/publication/intlsymp/intlsymp-2_168-173.pdf

van Zonneveld M, Scheldeman X, Escribano P, Viruel MA, Van Damme P, Garcia W, Tapia C, Romero J, Sigueñas M, Hormaza JI (2012) Mapping genetic diversity of cherimoya (*Annona cherimola* Mill.): application of spatial analysis for conservation and use of plant genetic resources. PLoS One 7(1):e29845. https://doi.org/10.1371/journal.pone.0029845

Vidal R, de Almeida Silva NC, Ogliari JB (2020) Old tools as new support for on farm conservation of different types of maize. Sci Agric 77(1):1–9. https://doi.org/10.1590/1678-992X-2018-0091

Vidigal DS, Marques ACSS, Willems LAJ, Buijs G, Méndez-Vigo B, Hilhorst HWM, Bentsink L, Picó FX, Alonso-Blanc C (2016) Altitudinal and climatic associations of seed dormancy and flowering traits evidence adaptation of annual life cycle timing in *Arabidopsis thaliana*. Plant Cell Environ 39(8):1737–1748

Weise S, Lohwasser U, Oppermann M (2020) Document or lose it—on the importance of information management for genetic resources conservation in Genebanks. Plan Theory 9(1050):1–13. https://doi.org/10.3390/plants9081050

Williams DE (2001) New directions for collecting and conserving peanut genetic diversity. Peanut Sci 28:135–140

Zair W, Maxted N, Brehm JM, Amri A (2021) Ex situ and in situ conservation gap analysis of crop wild relative diversity in the Fertile Crescent of the Middle East. Genet Resour Crop Evol 68:693–709

Plant Diversity of Mount Makiling Forest Reserve: Implications to Management and Conservation

5

Damasa B. Magcale-Macandog, Vachel Gay V. Paller, Nerissa K. Torreta, Ivy Amor F. Lambio, Annalee S. Hadsall, Inocencio Jr. Buot, Marjorie D. delos Angeles, Cleofas R. Cervancia, Sarena Grace L. Quiñones, and Jeoffrey M. Laruya

Abstract

Mt. Makiling is a low mountain straddling between Laguna and Batangas provinces harboring diverse and unique forest ecosystems from lowland evergreen rainforests to mossy montane forests at its peak of 1130 masl. It was proclaimed a Forest Reserve with an area of 4224.37 hectares from 100 to 200 masl to the summit. Mt. Makiling is a crucial forest reserve as it has diverse and valuable products which are economically important. It is also one of the 170 priority areas for biodiversity conservation instituted by the government. Mt. Makiling has an environment that is conducive for the growth of various vegetation types based on the altitude, namely upper montane rain forest at higher than 1000 masl, lower montane forest at beyond 750 masl, lowland evergreen forest at around 100–500 masl, and grassland vegetation at below 100 masl. Researchers from the Institute of Biological Sciences conducted a vegetation survey at the various elevational gradients every 100 m along the northeastern slopes of Mt. Makiling. Twenty-four species of mosses from 19 families were collected in various altitudinal gradients of Mt. Makiling. Forest floor census reveals that there are 44 fern species identified from 22 genera and 17 families. Generally, there was a higher number of understory species in the low and middle elevation levels up to 600 masl. At higher elevation levels, the number of understory species decreased greatly. Loganiaceae is the most represented family of understory species in the low elevation areas (<100–400 masl) followed by Meliaceae, Araceae, Dipterocarpaceae, and Urticaceae. The vascular epiphytes recorded in the survey belong to six fern families (Aspleniaceae, Davalliaceae,

D. B. Magcale-Macandog (✉) · V. G. V. Paller · N. K. Torreta · I. A. F. Lambio · A. S. Hadsall · I. J. Buot · M. D. delos Angeles · C. R. Cervancia · S. G. L. Quiñones · J. M. Laruya
Institute of Biological Sciences, University of the Philippines Los Baños College, Los Baños, Laguna, Philippines
e-mail: dmmacandog@up.edu.ph

S. Ramamoorthy et al. (eds.), *Plant Genetic Resources, Inventory, Collection and Conservation*, https://doi.org/10.1007/978-981-16-7699-4_5

Grammitidaceae, Lomariopsidaceae, Polypodiaceae, and Vittariaceae). Pollination and floral visitation of bees are also important ecosystem activities observed in Mt. Makiling. In a number of both new and established Philippine Giant Honey Bees, Apis breviligula Maa, colonies were located at elevations of 30–60 m above sea level within Mt. Makiling. Mt. Makiling is currently vulnerable to the illegal conversion of forest areas into agricultural areas and informal housing. The various human activities in Mt. Makiling produced changes in the richness and distribution of the species in the forest. The baseline data contained in the Makiling Biodiversity Information System (MaKIBIS) is important to support the conservation plans and safeguard the future of the species in Mt. Makiling.

Keywords

Mt. Makiling · Tropical forest reserve · MaKIBIS · Understory · ferns · Philippine giant honey bees · Vascular epiphytes

5.1 Introduction

Mount Makiling is an inactive volcanic cone in the Philippines. It is situated on the south-central portion of Luzon Island (14°08" north longitude and 12°12" east latitude) within the provinces of Laguna and Batangas, along the southern shore of Laguna Lake. This comparatively low mountain has a peak of only 1130 m, but harbors diverse and unique forest ecosystems from lowland evergreen rain forests to mossy montane forests (Figs. 5.1 and 5.2).

Although Mount Makiling was established as one of the first national parks in the Philippines in 1933 thru Proclamation No. 552, it was reclassified as a Forest Reserve, with an area of 4224.37 ha from 100 to 200 masl to the summit, under Republic Act 6967 and placed under the jurisdiction of the University of the Philippines. It is also one of the 170 biodiversity conservation priority areas established by the government (Mallari et al. 2001).

The Mt. Makiling Forest Reserve (MFR) has a rocky and generally steep topography (Fabunan et al. 2008). Its climate belongs to the Type I class by which it has two pronounced seasons, namely wet and dry. The reserve is predominantly wet during the year. It is only dry from January to April and wet from May to December (Caldecott 1993). It is relatively cold in the area, with an average temperature of 27 °C. At night, it becomes colder with a fall in temperature to below 15 °C. Mt. Makiling has an annual mean temperature ranging from 25.9 to 29.3 °C, with the hottest temperature during the months of April to July while the coldest during December to February. The climate is determined primarily by the east and west monsoon and the movement of intertropical convergence zone. The average annual rainfall recorded at the nearest national weather station (800 m) ranged from 1645 to 2299 mm for the past 5 years (2015–2020).

Fig. 5.1 Map of the Philippines and inset topographic map showing the location of Mt. Makiling Forest Reserve (adapted from Lambio and Buot Jr. 2007)

The reserve covers the north-eastern slope of the mountain. The forest area below 750 masl is covered mainly of hardwood species and several forest plantations. Though these areas have rich vegetation, it can be considered as secondary growth forest since human activities were observed and households were found in some sites. On the other hand, the higher elevations (>750 masl) are mossy forest with rocky and steep topography predominantly covered with epiphytes, narra, and mahogany species.

The soil in the reserve is loam-clay derived from volcanic tuff and ash, except on the southwestern slopes facing Sto. Tomas. The soil in this area is classified as white-clay from which the town near it derived its name, Puting Lupa. The pH of the soil ranges from 4.3 to 6.5 (Pancho 1973, 1983).

Only a certain percentage of MFR is considered as closed canopy forest. Some areas became sites for establishments such as the National Arts Center, College of Forestry and Natural Resources of the University of the Philippine Los Baños (UPLB), and the Boy Scouts of the Philippines. It also features recreational sites and ecotourism spots such as the Makiling Botanic Gardens, Mudspring, Flat Rocks, Pook ni Maria Makiling, and panoramic views from the peaks.

Mt. Makiling is one of the most famous mountains in Luzon for its mystic affiliation to the goddess Maria Makiling and its appeal to visitors like hikers and campers since it is near urban areas in Southern Tagalog and Metro Manila. Visitors usually leave marks and garbage; thus pollution and vandalism are evident throughout the reserve. Also, different illegal activities such as hunting and bird trading are said to be practiced by some settlers.

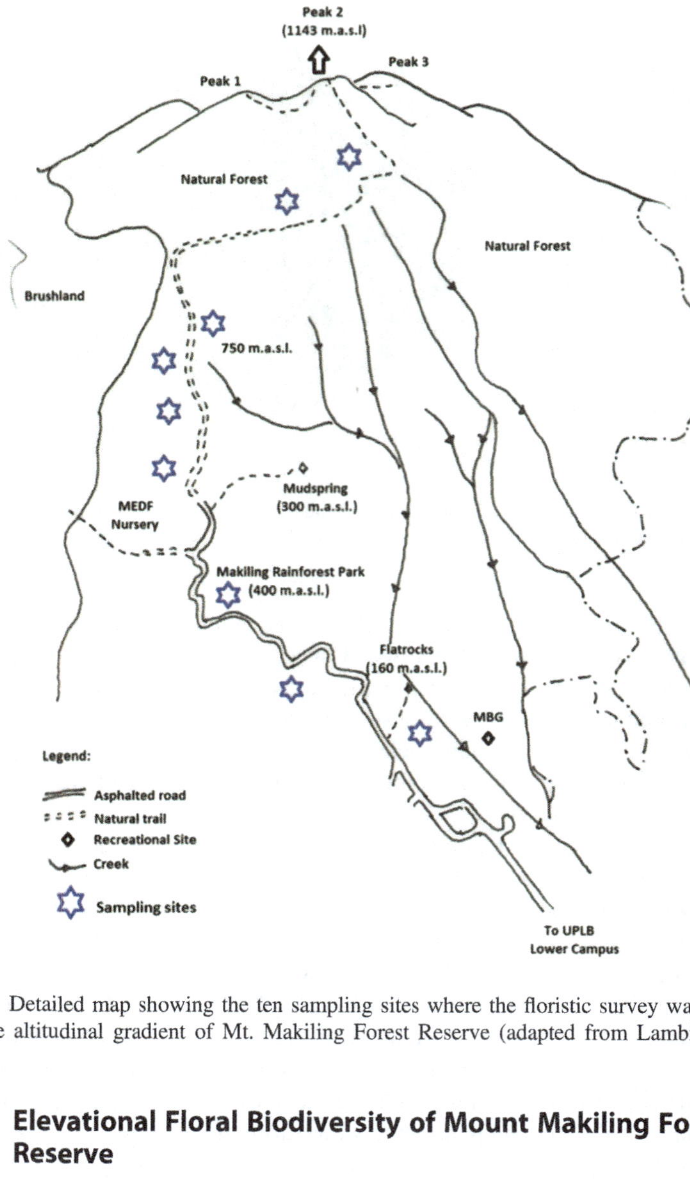

Fig. 5.2 Detailed map showing the ten sampling sites where the floristic survey was conducted along the altitudinal gradient of Mt. Makiling Forest Reserve (adapted from Lambio and Buot Jr. 2007)

5.1.1 Elevational Floral Biodiversity of Mount Makiling Forest Reserve

Diverse ecosystems can be found in tropical rain forest such as Mt. Makiling. In fact, it is one of the most famous biological areas in the Philippines (Pancho 1973). Mt. Makiling has distinct and diverse ecosystems, which function as habitats for various organisms in the mountain.

Moreover, Mt. Makiling has an environment that is conducive for the growth of various vegetation types based on the altitude. As recognized by Fernando et al.

(2004), the four types of vegetation in Mt. Makiling are upper montane rain forest which is above 1000 masl, lower montane forest which is above 750 masl, lowland evergreen forest which is around 100–500 masl, and grassland vegetation which is below 100 masl. Based on the habitat information analysis in Makiling Biodiversity Information System (MakiBIS) 2.0, an online biodiversity database of Mt. Makiling, there are about 434 plant species which can be found in the upper montane rain forest, 110 species in lower montane forest, 474 species in lowland evergreen forest, and around 28 species in grassland vegetation (Magcale-Macandog et al. 2021; Pancho and Gruezo 2006, 2009, 2012). Floral biodiversity surveys conducted in Mt. Makiling reported 148 tree species with 87 indigenous or native species and 44 endemic species (Malabrigo Jr. et al. 2016; Castillo et al. 2018; Abraham et al. 2010). The endemic tree species make up 32% of the country's endemism profile.

There are numerous forest plantations and hardwood species in the forest area under 750 masl (Combalicer et al. 2011). On the other hand, the mossy forests are in higher elevation above 750 masl. The species of plants is helpful in identifying altitudinal zonation because specific species can be found abundantly under certain climatic factors such as temperature and humidity which change with the altitude.

A rapid biodiversity assessment has been conducted by Abraham et al. (2010) at the crater area of Mt. Makiling between the months of April and August for 2 years since 2014. The mountain's crater area as well as its forest cover and biodiversity are the least explored by scientists. This study, which was conducted as the first systematic assessment of such areas, revealed that the crater area has mainly undisturbed mid-mountain and mossy forests where endemic, rare, and endangered plants and animals are nurtured. There are at least 416 ferns and flowering plant species, 118 arthropod species, 117 vertebrate species, and no gymnosperms recorded by the study. The study also revealed that about 28% of the plants in the mountain are considered endemic to the country, having at least nine rare and endangered species. Today, several species are found and recorded to be unique or abundant to the mountain.

5.2 Bryophytes

Researchers from the Institute of Biological Sciences conducted a vegetation survey of bryophytes at the various elevational gradients every 100 m along the northeastern slopes of Mt. Makiling (Paller et al. 2012). Nine sampling sites, 20–50 m from the trail, were established at alternating sides of the trail except at altitudes 750–950 masl due to the steepness of the region. Within each sampling site, moss patches were observed on soil banks, tree trunks, and rock surfaces. Bryophyte species were identified by physical examination and microscopic observation of the sample.

A total of 24 species of mosses belonging to 19 families were collected in various altitudinal gradients of Mt. Makiling. Appendix A shows a list of moss species recorded along different elevations of the northeastern slope of Mt. Makiling.

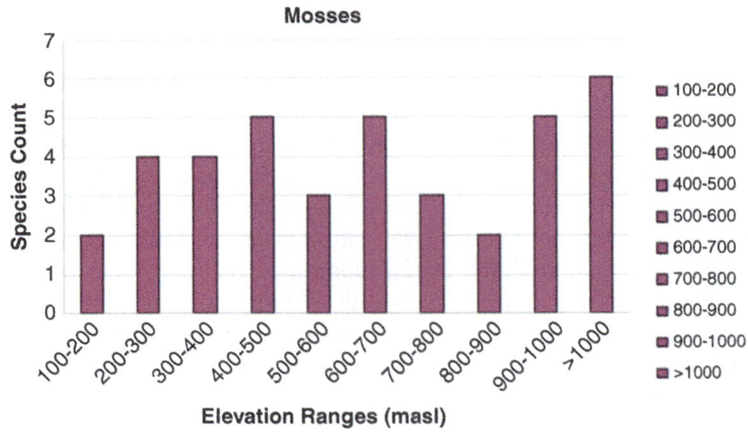

Fig. 5.3 Number of moss species recorded at various elevational levels along the northeastern slopes of Mt. Makiling

The highest number of moss species was recorded at high elevations greater than 1000 masl with six species (Fig. 5.3). This was followed by elevations at 900–1000, 40–500, and 600–700 masl with five species each.

In most cases, there were one to two species which were found to be common in some of the sampling sites. Indices of similarity showed that zones 6 and 7, 8 and 6, and 10 and 8 exhibited greater similarity indices (0.5–0.67) compared to the other zones. These zones were found in higher altitudes where moisture is most abundant to support the growth and development of mosses.

Clustering results show that species composition in high elevation areas (900→1000 masl) is distantly related to the other elevational levels (Fig. 5.4). This cluster is composed of *Camptochaete arbuscula, Ctenidium pubescens, Dicranella heteromalla, Isopterygiopsis pulchellum, Leucobryum candidum, L. glaucoma, Mnium* sp., *Pseudohypnella verrucosa,* and *Sphagnum cristatum.* The second cluster is composed of elevation levels from 200 to 500 masl. Species associated in this cluster include *Cratoneuropsis relaxa, Hylocomium sp., Hypnum chrysogaster, Mnium* sp., *Papillidiopsis* sp., *Rhadocarpus purpurascens, Rhizogonium distichum,* and *Thuidium furfurosum.* The third cluster is composed of elevation levels at 100–200, 500–600, and 800–900 masl. Species associations found in this cluster include *Calyptochaeta sp., Ctenidium pubescens, Hennediella heimii, Himantocladium sp., Hypnum triquetrum, Leucobryum glaucum, Mnium* sp., *Orthorrhynchium elegans, Papillidiopsis* sp., *Polytrichum formosum, Rhizogonium distichum,* and *Taxithelium novaeguineae.*

The principal component analysis (PCA) plot shows the location of elevations along the PC1 and PC2 axes. Vectors represent each individual moss species. Vectors that are close to each other and have small angles represent closely the associated moss species while vectors located in opposite quadrants represent the

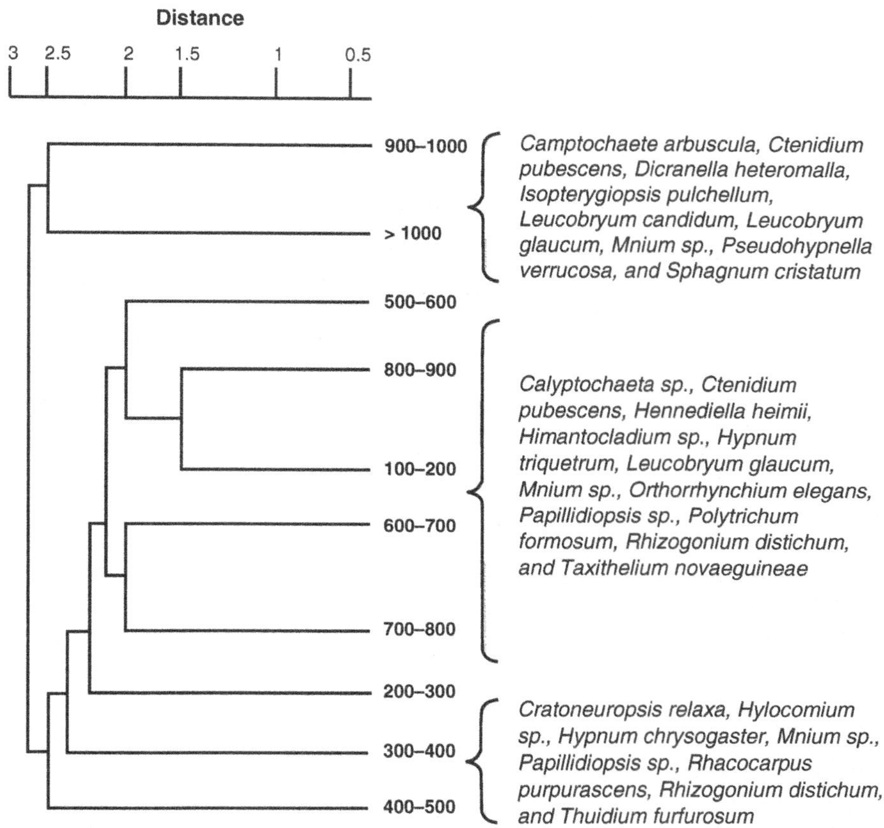

Fig. 5.4 Clustering of moss species recorded at various elevational levels along the northeastern slopes of Mt. Makiling

moss species that are negatively associated with each other. On the other hand, vectors at right angles represent species that are not associated with each other.

Results of PCA revealed that the high elevation areas greater than 800 masl are positively correlated to PC1 (Fig. 5.5). Moss species closely associated with high elevation areas are *Ctenidium pubescens*, *Camptochaete arbuscula*, *Sphagnum cristatum*, *Dicranella heteromalla*, and *Leucobryum glaucum*. Elevational areas at 100–300 and 600–800 masl are located in the upper left quadrant of the PCA plot while elevational areas from 300–500 masl are located in the lower left quadrant of the plot. *Papillidopsis sp.* is closely associated to 300–400 masl. At low elevation areas (100–300 masl), the closely associated species is *Polytrichum formosum*. *Hypnum triquetrum* is closely associated to areas with 700–800 masl elevation.

In 2019, a survey was conducted by Alcala et al. (2019) to compare the moss diversity along the altitudinal gradients in Sipit and Makiling trails of the MFR. Mosses are considered as one of the critical parts of ecosystem diversity. In fact, a

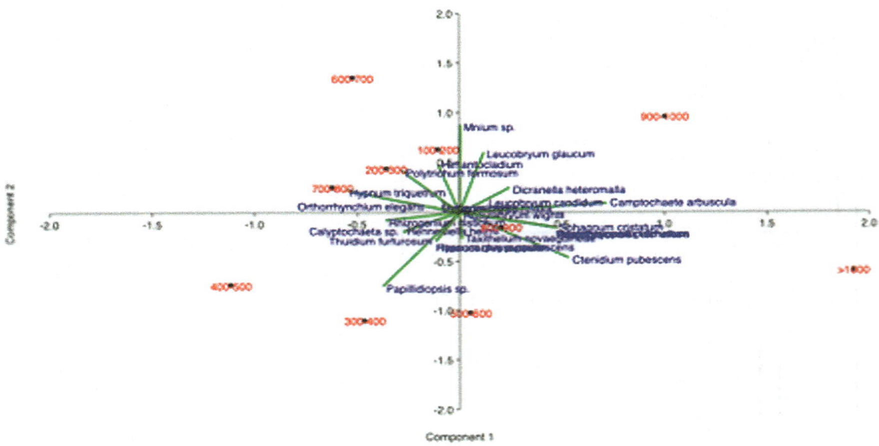

Fig. 5.5 PCA plot of mosses recorded at various elevational levels along the northeastern slopes of Mt. Makiling

total of 77 moss species were discovered in all of the plots established. From the overall number of species, 48 and 43 species were identified from Sipit and Makiling trails, respectively. Compared to higher elevation areas, there are more disturbances in lower elevation areas, thus having low species richness. Indeed, as the elevation increases, the species richness, abundance, and diversity increase as well. Meanwhile, the species richness, diversity, and abundance of mosses decrease as the temperature increases. Generally, the increasing elevation in both trails positively affects the diversity of mosses. The results of the correlation analysis made on moss species composition for the two sites supported this generalization on elevation and species diversity. On the other hand, the abrupt decrease in species diversity on sampling plots located at 600 masl and 900 masl (Sipit Trail) and 800 masl and 1000 masl (Makiling Trail) is perhaps because of the disturbances or presence of canopy gaps in the sites.

5.3 Ferns and Fern Allies

Ferns and fern allies found across the elevational gradient of Mt. Makiling were documented by Delos Angeles and Buot Jr. (2015, 2018). Forest floor census reveals that a total of 44 fern species were identified belonging to 22 genera and 17 families. Thirty-nine species of ferns were found within the subquadrants while the remaining five fern species were found outside of the established subquadrants but were within the sampling area. These five fern species were *Dicranopteris linearis (Burn), Dipteris conjugata Reinw*, *Microsorum membranifolium,* and *Oleandra neriiformis*. These species of ferns were collected for documentation.

The Philippines has a total of 1100 species of lycophytes and ferns distributed among 154 genera and 34 families (Pelser et al. 2011; Amoroso et al. 2016;

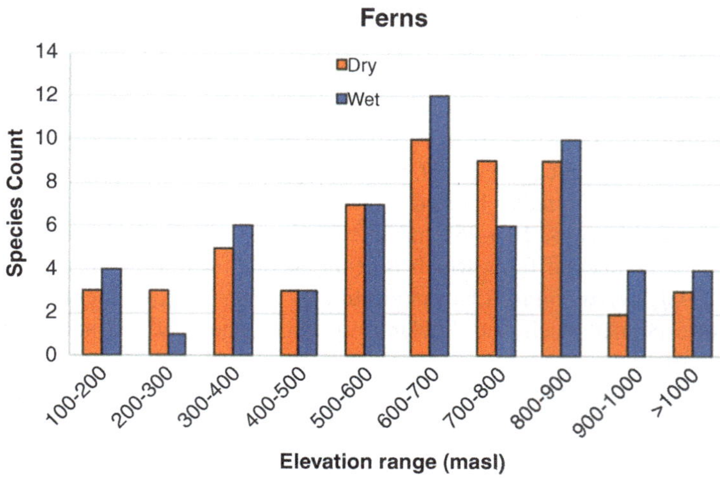

Fig. 5.6 Number of fern species recorded at different elevation levels in the northeastern slopes of Mt. Makiling Forest Reserve

Barcelona 2002). Copeland (1958) accounted a total of 20 families, 155 genera, and 950 species and pioneered Philippine fern studies by assembling his three-volume treatment of Philippine ferns entitled "Fern flora of the Philippines." In 1972, Price attempted to collate and update all important fern literature, including that of Copeland through his unpublished manuscript entitled "The Pteridophytes of Mt. Makiling and Vicinity." The manuscript included keys, synonymies, original descriptions, distribution, ecology, common names, species examined, comments, and an index or list of ferns found at Mt. Makiling and its vicinity.

Diversity of ferns along the altitudinal gradient on Mount Makiling was higher during the wet season than during the dry season (Fig. 5.6). There is an increasing trend of fern diversity across the altitudinal gradient; however, more fern species were recorded at 600–700 masl during both the wet and the dry seasons. This is partly due to the existence of a body of water and a rocky substrate providing a favorable habitat for ferns. Report showed that elevations starting from 500 masl to 900 masl have higher number of recorded fern species. On the other hand, very low (100–300 masl) and high elevation areas (900 to >1000 masl) have low number of fern species.

During the dry season, site 6, at an elevation of 650 masl, had the highest value for the Shannon index of diversity (1.818288). During the wet season, site 6 also had the highest diversity index of 1.789949572. The established subquadrants were located within the wilderness zone, characterized by a rocky substrate enclosed by canopies of trees. In addition, a stream was present in the site which provided sufficient moisture for ferns to grow. Among the sampling sites, the presence of rocks in the area provides a suitable substrate for ferns to attach to. Since the area had a closed canopy, moisture in the soil had a lower rate of evaporation.

5.3.1 Fern Zonation Along the Altitudinal Gradient

Using fern density values along the altitudinal gradient, the fern vegetation was classified with the aid of a hierarchical cluster analysis (average linkage). The dendrogram (Fig. 5.7) showed three distinct clusters during the dry season at a square Euclidean distance of eight. The resulting clusters were designated as zones. Zone 1: *Blechum-Selaginella* located at the lowest portion of the dendrogram was situated at an altitude of 100–450 masl. Zone 2: *Selaginella-Bolbitis-Microsorum* had the most number of ferns which is at an elevation of 500–700 masl. Zone 3: *Selaginella-Sphaerostephanos* has an altitude of 800–900 masl. Incidentally, these zones corresponded with the lowland, mid-montane, and montane zones of Brown (1919).

Data collected during the wet season were also subjected to hierarchical cluster analysis. The dendrogram (Fig. 5.8) illustrated three zones: Zone 1: *Bolbitis-Ferns sp. 1* located at the lowest portion of the dendrogram was situated at an altitude of 150–250 masl; Zone 2: *Lygodium-Christella-Sphaerostephanos* at an elevation of 350–650 masl, with the most number of ferns in the dry season; and Zone 3: *Sphaerostephanos-Selaginella* at an elevation of 750–950 masl.

At both seasons, Zone 2 had the highest number of fern species since sites 5 and 6 have a number of ravines. Ravines are characterized by crevices as well as ditches. A ravine had been an excellent habitat for ferns because its topography keeps the anthropogenic activities to a minimum or even nil and has good moisture content. Moisture is essential, especially during sexual reproduction in ferns, carrying the male gamete to its female counterpart. Generally, the zonation pattern of pteridophytes on Mt. Makiling follows the usual zonation pattern for woody vegetation. It had been noted that the identified zones of ferns in this study coincided with the identified zones of Brown (1919).

Appendix A shows a list of pteridophyte species recorded during the dry and wet seasons along different elevations of the northeastern slope of Mt. Makiling.

5.4 Magnoliophytes

The rocky and steep landscape in Mt. Makiling forest is mostly covered with mahogany, narra, and epiphytes. The 148 indigenous tree species are found within the permanent biodiversity monitoring area within the MFR and 47 are endemic. The most abundant species is magabuyo (*Celtis luzonica*) with 2287 individuals, followed by balobo (*Diplodiscus paniculatus*), tinaang pantai (*Drypetes maquilingensis*), and apanang (*Mallotus cumingii*). Over 90% of the trees are poles and saplings while the large trees account for only 0.6%. Around 29 species are either in the Philippine Red List or the International Union for Conservation of Nature (IUCN) Red List of Threatened Species. Also in the list are the six critically endangered premium timber species such as makaasim (*Syzygium nitidum*), kamagong (*Diospyros discolor*), narra (*Pterocarpus indicus*), bagtikan (*Parashorea

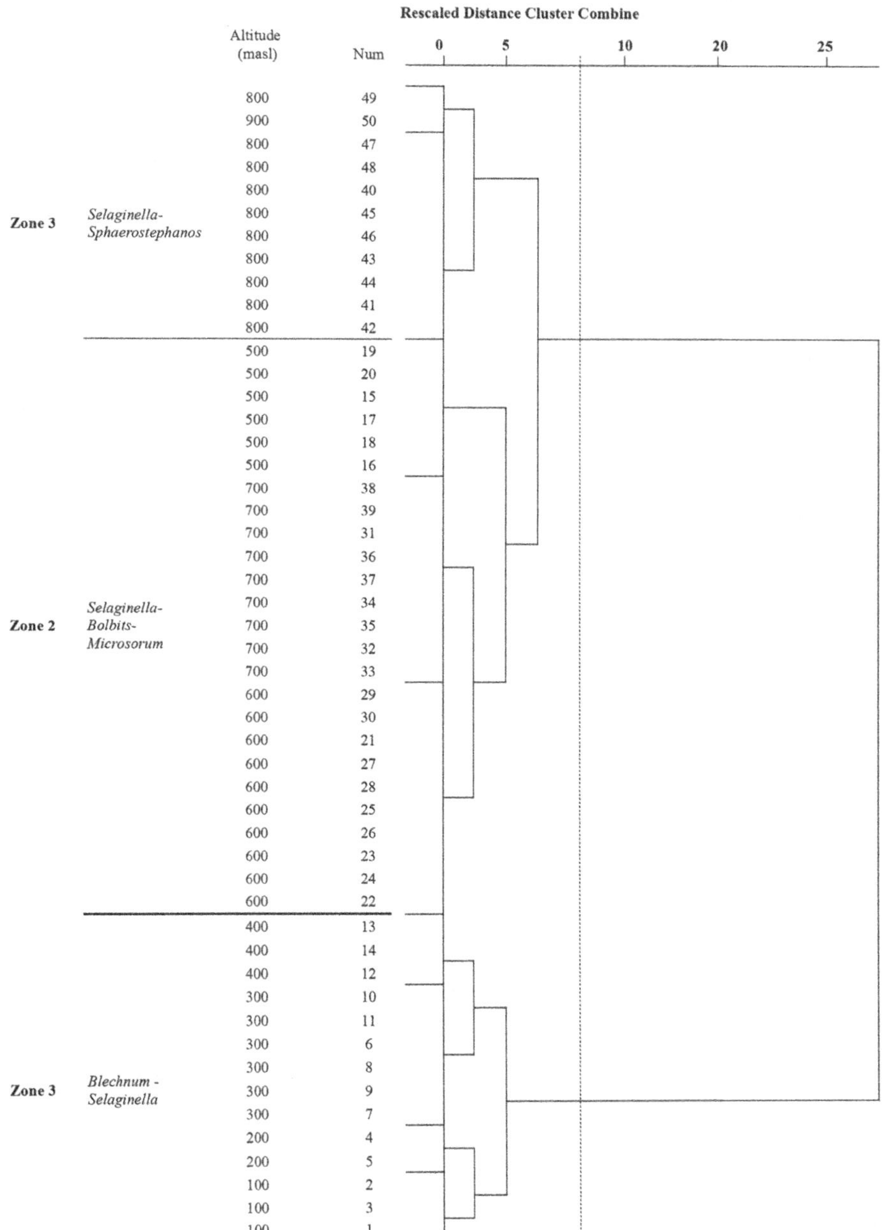

Fig. 5.7 Dendrogram of the 25 subquadrants from the 10 sampling sites during the dry season by average linking clustering using the SPSS software. Three zones were identified: Zone 1 at 150–450 masl; Zone 2 at 550–750 masl; Zone 3 at 850–1050 masl (adopted from Delos Angeles and Buot Jr. 2015)

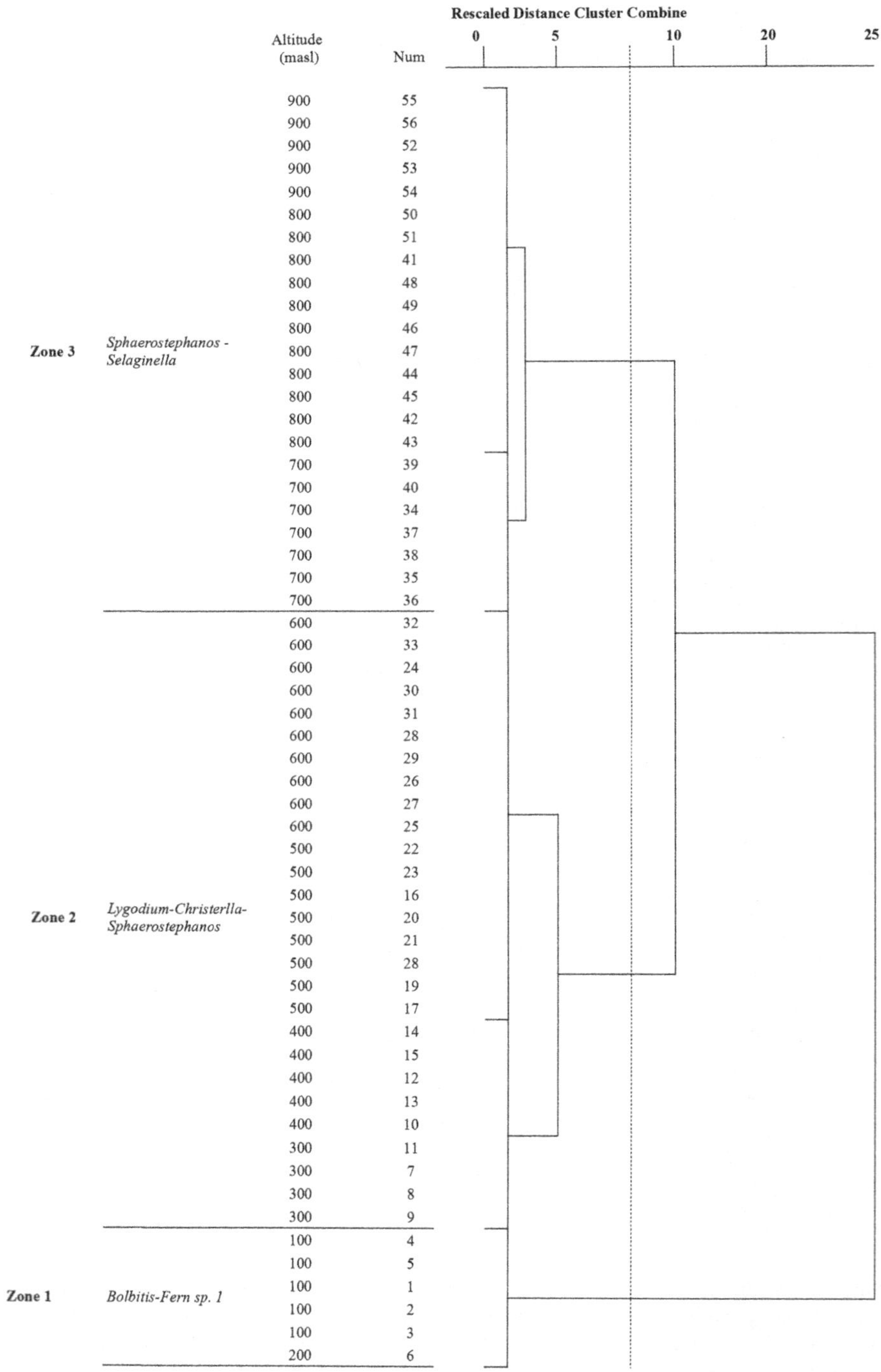

Fig. 5.8 Dendrogram of the 25 subquadrants from the nine sampling sites during the wet season by average linking clustering using the SPSS software. Three zones were identified: Zone 1 at 150–250 masl.; Zone 2 at 350–650 masl.; Zone 3 at 750–950 masl (adopted from Delos Angeles and Buot Jr. 2018)

malaanonan), guijo (*Shorea guiso*), and white lauan (*Shorea contorta*) (Malabrigo Jr. et al. 2016).

In another vegetation survey, a total of 1266 tree individuals of 135 species were recorded in the 2-ha long-term ecological research plot (LTERP) in Mt. Makiling through the recent botanical survey conducted by Castillo et al. (2018). The 135 tree species are distributed among 95 genera and 58 families. Out of this number, four species are considered exotic or invasive, 87 species are indigenous or native, and 44 species are endemic. For the entire study area, such endemic species have 32% of the country's endemism profile. Meanwhile, the number of threatened species recorded in the area is 29. The study showed that those threatened species dominate in the LTERP area. Of the ten dominant species found, three species are classified as critically endangered (*Parashorea malaanonan* Blanco, Merr, and *Diospyros blancoi* A. DC.) and seven species are threatened. As per the Species Importance Value, the three species that dominate the study area the most are *Diplodiscus paniculatus* Turcz, *Celtis luzonica* Warb, and *Pinanga insignis* Becc. The MFR is shown to be more diverse, intact, and preserved, as indicated by the results of the tree diversity analysis.

5.5 Understory Species

During the 2011 vegetation survey conducted by the Institute of Biological Sciences (Paller et al. 2012) at various elevation levels along the northeastern slopes of Mt. Makiling, the highest number of species was recorded at 200–300 masl elevation, followed by 500–600 masl elevation (Fig. 5.9). Generally, there was a higher number of species in the low and middle elevation levels up to 600 masl. At higher elevation levels, the number of understory species decreased greatly.

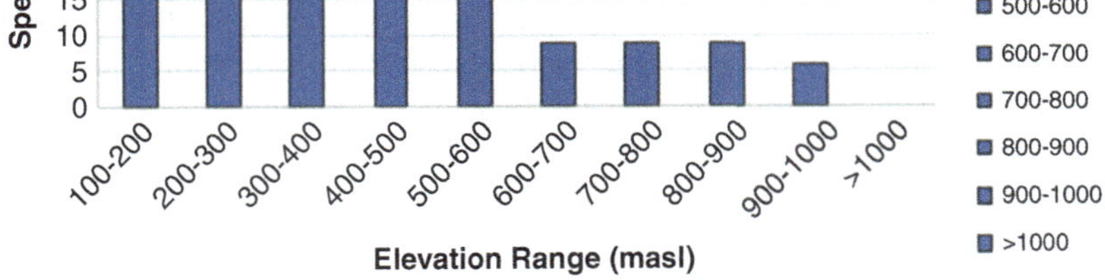

Fig. 5.9 Number of species recorded at various elevational levels in the northeastern slopes of Mt. Makiling

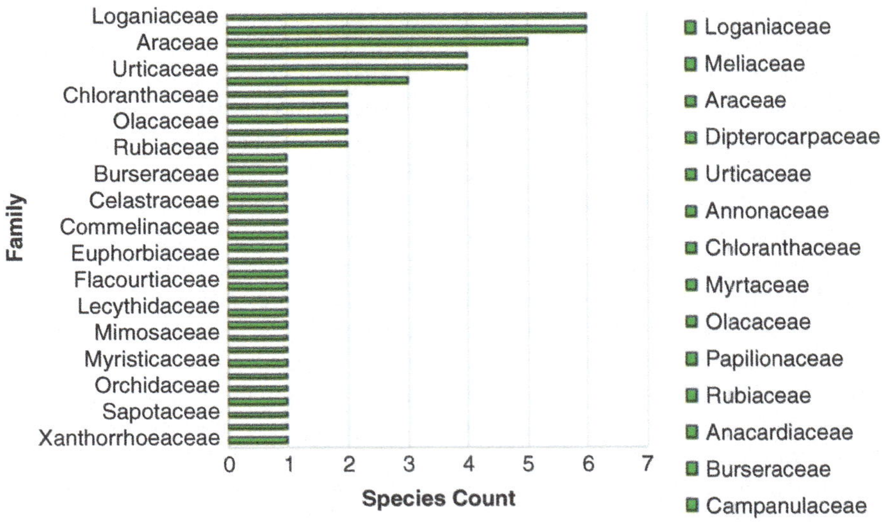

Fig. 5.10 Species count of families of understory vegetation recorded in the low elevation levels (<100–400 masl) in the northeastern slopes of Mt. Makiling

Loganiaceae is the most represented family of understory species in the low elevation areas (<100–400 masl) with six species (Fig. 5.10). This was followed by Meliaceae (6 spp.), Araceae (5 spp.), Dipterocarpaceae (4 spp.), Urticaceae (4 spp.), and Annonaceae, Chloranthaceae, Myrtaceae, Olacaceae, Papilionaceae, and Rubiaceae (2 spp. each). All the other families have one species each.

In the middle elevation from 400 to 600 masl, Araceae is the most represented family with nine species, followed by Meliaceae (7 spp.), Moraceae (4 spp.), Commelinaceae (3 spp.), and Sapindaceae (3 spp.) (Fig. 5.11). Annonaceae, Marantaceae, Orchidaceae, Rubiaceae, and Urticaceae have two species each. The rest of the other families have one species each.

In the high elevation areas (>600 masl), Araceae is the most represented family with six species, followed by Marantaceae, Meliaceae, and Moraceae with three species each (Fig. 5.12). Lauraceae and Acanthaceae have two species each. All the other families are represented by one species each.

Results of cluster analysis showed that the species compositions at each elevational gradient from 100 to 600 masl are distinct from each other. On the other hand, the species compositions at high elevation levels from 700 to more than 1000 masl are closely associated (Fig. 5.13). The species associations found in this cluster include *S. pluriformis, H. lanceolata, P. flava, Aglaonema sp., A. zebrina, A. paeoniifolius, H. philippinensis,* and others. The cluster of species composition recorded at 200–300 masl elevation is distantly related to the other elevation levels. The species associations found at 200–300 masl include *Aglaonema* sp., *A. konjac, A. paeoniifolius, E. aureum, L. longiflora, S. chinensis, S. glabra, T. macrocarpa, P. malaanonan, D. pilosanthera, A. ghaesembilla, B. insulana, H. alcalae,*

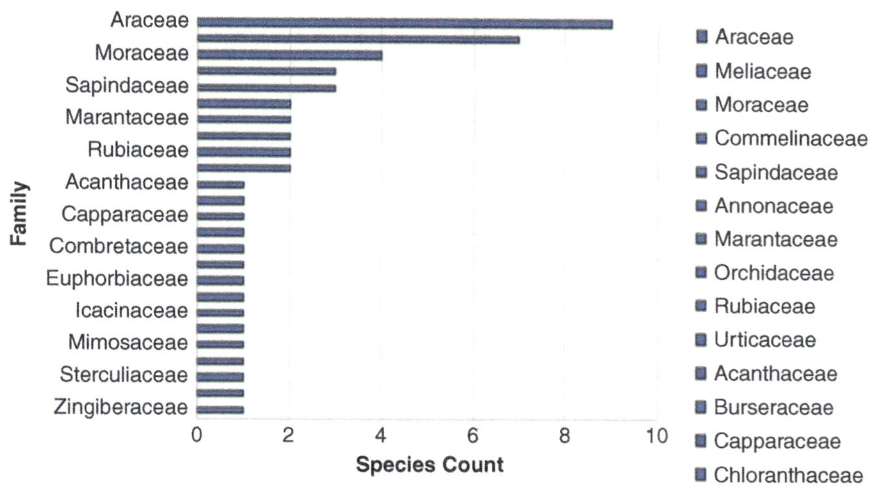

Fig. 5.11 Species count of families of understory vegetation recorded in the mid elevation levels (400–600 masl) in the northeastern slopes of Mt. Makiling

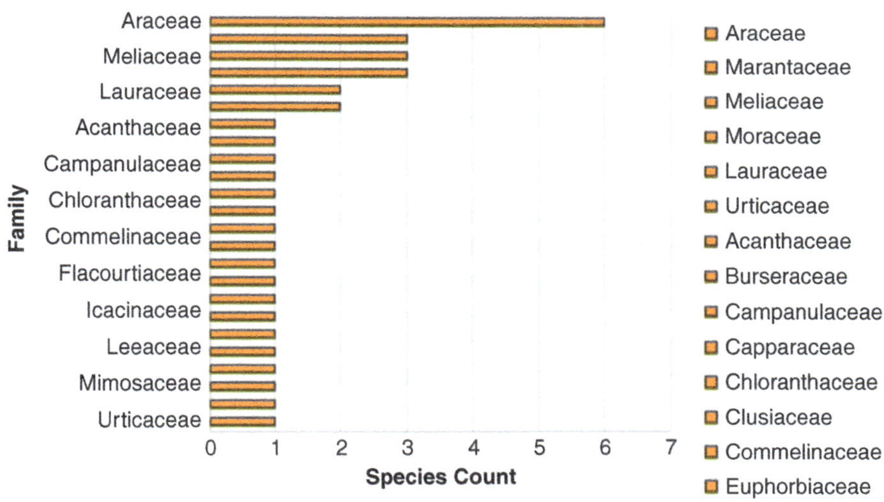

Fig. 5.12 Species count of families of understory vegetation recorded in the high elevation levels (>600 masl) in the northeastern slopes of Mt. Makiling

C. capitulate, and *A. argentea*. This was followed by the species composition at 100–200 masl including *B. nitida, A. luzonensis, P. arborea, P. flava, A. paeoniifolius, Caryota* sp., *S. chinensis, C. elatior, T. nitens, A. hispida, N. cumingii, L. ovalis, H. alcalae, P. spectabilis, D. cannaeformis, A. argentea,* and *A. edulis*.

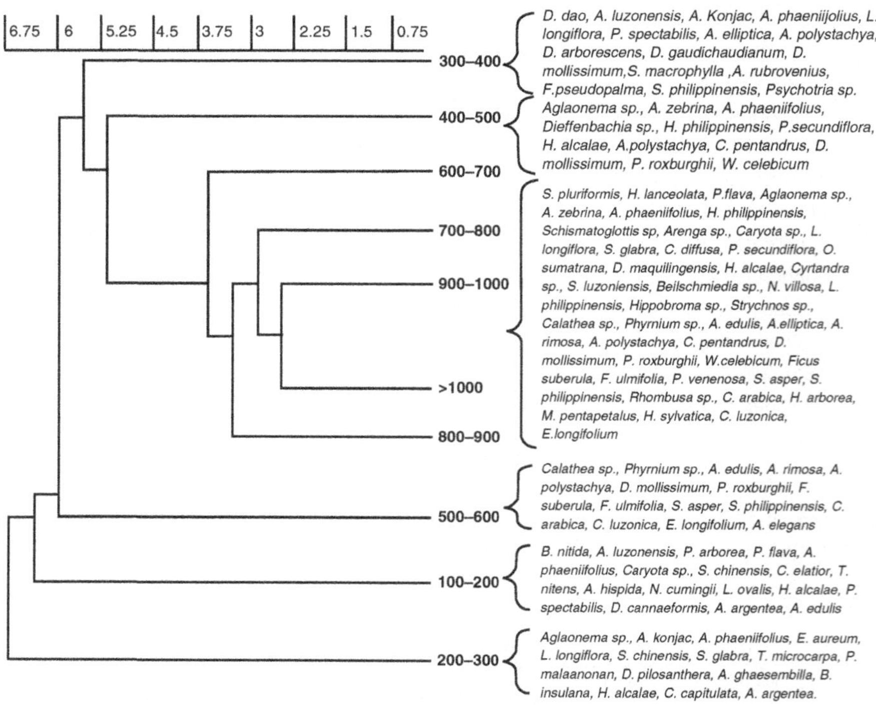

Fig. 5.13 Clustering of understory vegetation species recorded at various elevational levels along the northeastern slopes of Mt. Makiling

Generally, it would be observed that species similarity between adjacent stands would be high since propagules would be easier dispersed over short distances. However, in the case of the observed understory species, very low similarity indices were noted between gradients. This could be attributed to the type of forest canopy in the area. Forest overstory vegetation dictates the type of understory vegetation. Also, there would be very high competition between species for the limited resources of the forest floor; thus competitive exclusion might be working for the success of some species in the forest understory. Most abundant understory plants belong to the aroid family, with their dark-colored and wide-bladed leaves. It was also noted that similarities in species composition between seasons per site were also low, indicating the importance of water as a resource and condition for the dominance of few species per season.

Elevation level of 200–300 masl is located in the far right along PC1. Other elevation levels at 100–200 masl and 300–400 masl are also located in the positive side of PC1 (Fig. 5.14). On the left side of PC 1 are elevation levels above 600 masl. Along PC2, the elevation class at 500–600 masl is highly positively correlated.

PCA biplot shows that majority of the understory species are positively correlated to low elevation areas from 100 to 300 masl, i.e., right side of the PCA biplot

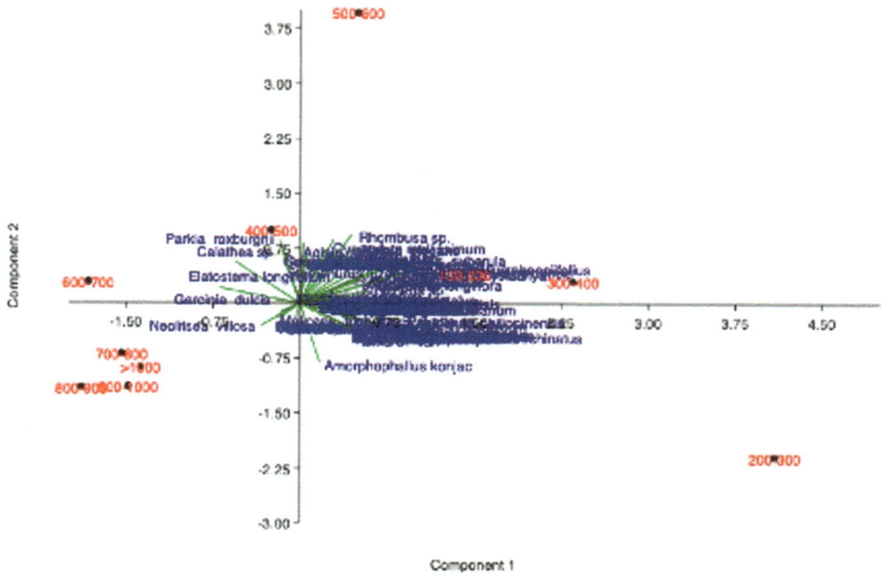

Fig. 5.14 PCA biplot for understory vegetation recorded at various elevation levels along the northeastern slopes of Mt. Makiling

(Fig. 5.14). *Parkia roxburghii* and Calathea sp. are closely related to 400–500 masl. *Garcinia dulcis* is closely correlated to 600–700 masl while *Neolitsea villosa* is closely correlated to high elevation areas at 700 to >1000 masl.

5.6 Vascular Epiphytes

Epiphytes can be defined as plants that live on the surface of another plant, not necessarily obtaining food from that plant. Primarily, epiphytes derive physical support.

The vascular epiphytes recorded in the survey conducted by the Institute of Biological Sciences (Paller et al. 2012) belong to six fern families (Aspleniaceae, Davalliaceae, Grammitidaceae, Lomariopsidaceae, Polypodiaceae, and Vittariaceae). For the angiosperms, three monocot families (Araceae, Orchidaceae, and Pandanaceae) and six dicot families (Apocynaceae, Araliaceae, Gesneriaceae, Melastomataceae, Nepenthaceae, and Piperaceae) with representative epiphytic species were likewise recorded. Vascular epiphyte species recorded during the dry and wet seasons along different elevations of the northeastern slope of Mt. Makiling are included in Appendices F, G, and H.

At an altitude of 100 masl up to the peak of Mt. Makiling, different fern species were encountered. The ferns can be said to make the most successful epiphytes since they possess characteristics that enable them to survive stressful conditions like low light intensity and ability to withstand drying. *Drynaria quercifolia*, a common fern epiphyte under the family Polypodiaceae, makes use of its cabbage-like sterile leaf to collect humus. This humus absorbs water and allows the species to survive water-stress conditions. *Asplenium nidus* (family Aspleniaceae) has whorled bowl-like fronds that similarly collect and retain the humus. Like in Drynaria, this humus acts as compost pile that later on releases the nutrients needed by the plants. Epiphytic angiosperms like members of the family Orchidaceae (Dendrochilum sp., Bulbophyllum sp., Dendrobium sp.) at higher altitudes (from 900 masl up to the peak) have thick fleshy stems that can store water.

5.6.1 Economic Use

Mt. Makiling is a crucial forest reserve. It has diverse and valuable products which are economically important. Some of the forest products are not sold but are consumed by the local community. The highest percentage of the total economic value came from fodder (34%), followed by ornamental plants (27%), wildlife (22%), and planting and seed materials (9%). The lowest percentage values came from bamboo, fuelwood, vines, and rattan. The total amount generated from these products is just about 3% of the total yearly household income of the farmers in Mt. Makiling. Despite this low percentage, these products still contribute significantly to the livelihood of the farmers since there are no alternative livelihood activities in their area (Diamante-Camacho et al. 2009).

There are many plant species in the forest that have economic value. In MakiBIS 2.0, the different plant uses stored in its database denote their identified economic use and their potential. The economic uses of the floral species identified in Mt. Makiling are characterized as either medicinal, ornamental, edible, construction, or source of raw materials. Below are brief descriptions of these categories.

Medicinal: These plants are significant for traditional and modern medicine. The database of MakiBIS 2.0 contains information for around 393 species of plants with medicinal features.

Ornamentals: There are many plants in the forest which have decorative features. About 67 species of plants listed in the database are harvested for garden plants, landscaping, and other horticultural uses.

Edible plants: These plants have edible parts. There are 249 edible plant species listed in the database, which are both locally consumed and highly cultivated.

Construction/building materials: These plants are harvested because of its wood quality which is best for building materials, light construction utilization, furniture making, and carpentry. Mt. Makiling used to provide timber for commercial purposes; however, the regulations imposed preserved around 140 species of timber.

Raw materials: These are non-wood materials from the plants. The database has listed about 115 plant species which are used as the source of dye, fiber, pulp, resin, tannins, gums, and paper materials.

Others: These are the plants used as green manure, soil erosion measures, and reforestation species, and for other cultural purposes. The database has 58 plant species which are used for these purposes.

5.6.2 Threats

MakiBIS 2.0 contains information on the conservation status of the flora species in Mt. Makiling (Magcale-Macandog et al. 2021). These were gathered from the country's Department of Environment and Natural Resources (DENR 1997) and the International Union for Conservation of Nature (IUCN 2015). About 70 floral species in Mt. Makiling are threatened based on the category of DENR Administrative Order 2007–01 and 72 species based on the IUCN threatened category. About 40 plant families have threatened species. The highest number of threatened species was recorded for Dipterocarpaceae (16 species). These are species of hardwood trees which are cut down for construction utilization (Magcale-Macandog et al. 2021).

Mt. Makiling is currently vulnerable to the illegal conversion of forest areas into agricultural areas and informal housing (Fernando et al. 2008). These housings, which are usually in the not steep slopes, are constantly expanding. There were only 280 households in Mt. Makiling in 2003 but the recent record showed that there were already around 1225 households (Paller et al. 2012). Moreover, the plantations, leased areas, agroforestry, grasslands, and agricultural farms are within this forest.

There are only certain areas in Mt. Makiling which are closed canopy. Other areas are used for the Boy Scouts of the Philippines, UPLB-College of Forestry, and National Arts Center. Also, there are areas for ecotourism and recreational activities such as Flat Rocks, Makiling Botanic Garden, Mudspring, and Pook ni Maria Makiling. The forest is also popular for its clean spring water and green forest. Many campers and hikers, usually from Metro Manila and Southern Tagalog, go to Mt. Makiling because of these tourist attractions. However, these tourists commonly leave tracks, and thus there are many pollution, vandals, and damages in the forest.

Despite the numerous expeditions and comprehensive flora and faunal surveys done on Mt. Makiling, much of the ecology and biology of its rich species remain poorly known. Valuable baseline information regarding the biology and ecology of organisms on the mountain is essential to support conservation initiatives and safeguard the future of organisms on Mt. Makiling. A detailed account of species diversity is fundamental in establishing the relationships among different groups of organisms present in the area and determining how these interrelationships affect the structure and functioning of the ecosystem as a whole (Staley and Reysenbach 2001). In devising participatory strategies for the rehabilitation and protection of Mt. Makiling biodiversity, it is imperative to have a baseline data on its species composition and their identification using morphological and molecular characters. Species with potential medicinal and industrial uses can be conserved and managed for sustained utilization.

The various human activities in Mt. Makiling produced the changes in the richness and distribution of the species in the forest. The UPLB administration, with the help of the Institute of Forest Management, is developing and implementing plans and measures in regulating the current condition towards biodiversity conservation of Mt. Makiling since it is under its management. Several consultations and awareness campaigns on biodiversity conservation have been conducted with the local communities in the forest. Also, UPLB is strictly monitoring and regulating the number of tourists visiting the forest. The illegal collection of samples and specimens from the forest is strictly forbidden by UPLB.

5.6.3 Way Forward

With all this information, it is evident that many expeditions and comprehensive surveys in Mt. Makiling were already done. This detailed information on the species diversity is important for establishing relationships among various groups of organisms living in the forest and determining the effects of their relationship to their function and structure in the ecosystem (Staley and Reysenbach 2001). With this vast amount of ecological and biological information in Mt. Makiling, the use of information and communication technologies will greatly help in its management.

There has been a great development in information dissemination and communication technologies since the late 1990s. This advancement in technological capacity has made information and education readily available to anyone. In the current condition where there are increasing rates of ecological imbalance and depletion of forest covers, crucial studies in conserving the resources should not only include paper maps and policies. To implement the results and recommendations of different studies, it is very crucial to maximize the use of the present information and communication technologies (Roy et al. 2002).

Actual information is crucial to enhance management efficiency. To lower the cost and enhance the efficiency and speed of information use and dissemination, the development of information system is needed for data sorting, processing, transmission, and application. The kind of information to be developed is directly associated with the effects of information use. The management of biodiversity information is a crucial part of the works in biodiversity as the decision-making on sustainable development and biodiversity conservation must be based on accurate and comprehensive information and data (Xiao-Ting 2012).

The growth in information technology has caused a progressive change in managing biodiversity and scientific explorations. In the past two decades, the biodiversity researchers around the world have tried to establish databases of various sizes and management systems to maintain biodiversity databases (Xiao-Ting 2012).

In the case of MFR, a geodatabase was already created to support the management decision system. A study showed the initial results in developing a geodatabase and the utilization of geographic information system (GIS) in planning, designing, and implementing biodiversity conservation and forest renewal programs (Bantayan et al. 2008). The geodatabase was created by mapping and characterizing a total area of 11,000 ha at 100-ha grids, 25-ha quadrants, 1-ha cells, 0.04-ha^2, and 0.01-ha granules. It was designed after the GAME model made by Bantayan (2006). GAME is an acronym for GIS-based assessment, monitoring, and evaluation. It utilizes regular-shaped polygons to give description of watersheds. Using the GAME model, the environmental parameters were generated at 10 m × 10 m granule including biodiversity indicators, elevation, ground cover, basal area, number of trees and tree species, mean tree height, and stem density. This enables the determination of "hotspots" for forest renewal and biodiversity conservation (Bantayan et al. 2008).

For Biodiversity Information System (BIS), its development is a crucial basis for biodiversity conservation and development. In 1980, the biodiversity concept was first proposed and after many years, BIS was introduced. In the 1990s, many countries conducted researches in this field. After two decades, the joint work of the biodiversity scientists, biodiversity information experts, and computer specialists has resulted in much progress in the BIS and many different kinds of systems have been developed (Xiao-Ting 2012).

A BIS is an information system for the storage, management, analysis, search and retrieval, presentation, and publishing of data about the organisms, their interaction, and relationship with the ecosystem (Bisby 2000). It helps the users to determine the gaps and habitat/species relationship and helps in the development of biodiversity conservation plans by setting priority sites. This database combined with other detailed inventories helps in determining areas for bioprospecting. Also, its development is for visualization and faster retrieval of information (Roy et al. 2002).

In the case of Mt. Makiling, the MakiBIS has been developed to serve as a repository for these data (Lapitan et al. 2013). Species information included in the MakiBIS are taxonomy, botanical description, habitat, mode of reproduction, endemicity, economic use and conservation status, habitat, general description, and photo. The baseline data on its biology and ecology is important to support the conservation plans and safeguard the future of the species in Mt. Makiling (Magcale-Macandog et al. 2021).

References

Abraham ERG, Gonzalez JCT, Castillo ML, Lit IL Jr, Fernando ES (2010) Forest cover and biodiversity profile of the crater area of Mt. Makiling, Luzon, Philippines. Asia Life Sci 4:48–82

Alcala AA, Angeles MDD, Buot IE Jr (2019) Fern species diversity across various land use types of Mt. Makiling, Luzon Island, Philippines. Biodivers J Biol Divers 20:9

Amoroso VB, Coritico FP, Fritsch PW (2016) Species richness and conservation status of ferns and lycophytes in Mt. Hamiguitan range wildlife sanctuary, Davao Oriental, Philippines. Philippine J Sci 145(2):17–23

Bantayan NC (2006) GIS in the Philippines: principles and applications in forestry and natural resources. Philippine Agriculture and Resources Research Foundation, Manila

Bantayan NC, Abraham ERG, Fernando ES (2008) Geodatabase development for forest restoration and biodiversity conservation in the Mt. Makiling Forest reserve, Philippines. Philippine Agric Sci 9:14

Barcelona JF (2002) Philippine pteridophyte collections as a resource for conservation planning. Fern Gazette 16:307–312

Bisby FA (2000) The quiet revolution: biodiversity informatics and the internet. Science 2895488: 2309–2312

Brown WH (1919) Vegetation of Philippine Mountains: the relation between the environment and physical types at different altitudes. Bureau of Science, Manila, p 434

Caldecott JO (1993) Opportunities for biodiversity management in the Makiling Forest reserve. ERMP, Manila

Castillo ML, Castillo LA, Canceran MS, Gonzalvo KJP, Barua LD, Alegre AC, ... , Bantayan NC. 2018. Distribution, diversity and spatial analysis of tree species in a long-term ecological research plot in Molawin-Dampalit watershed, mount Makiling Forest reserve Asian J Biodiv, 91 5

Combalicer MS, Kim D, Lee DK, Combalicer EA, Cruz RVO, Im S (2011) Changes in the forest landscape of Mt. Makiling forest reserve, Philippines. Forest Sci Technol 72:60–67

Copeland EB (1958) Fern flora of the Philippines. Manila Bureau of Printing, Manila

Delos Angeles MD, Buot IE Jr (2015) Diversity and distribution of Pteridophytes along the altitudinal gradient of the northeastern slope of a secondary Forest in Mt. Makiling, Philippines. Int J Ecol Conserv 16:25–46

Delos Angeles MD, Buot IE Jr (2018) Altitudinal zonation of ferns along the northeastern slope of Mt. Makiling. Philippines SYLVATROP Technl J Philippine Ecosyst Nat Res 28(2):47–68

Department of Environment and Natural Resources/United Nations Environment Program. DENR/UNEP (1997) Philippine biodiversity: an assessment and action plan

Diamante-Camacho L, Sofronio C, Camacho, Yeo-Chang Y (2009) Values of forest products in the Makiling forest reserve (MFR), Philippines. Forest Sci Technol 5(2):35–44. https://doi.org/10.1080/21580103.2009.9656346

Fabunan M, Gerardo M, Maguslog J, Salumbre R, Surquia J, Diesmos M (2008). Comparative study of lizard populations in Mt. Makiling, Laguna, Dissertation, University of Santo Tomas

Fernando ES, Co LC, Lagunzad DA, WSM G, Barcelona JF, Madulid DA, Zamora PM (2008) Threatened plants of the Philippines: a preliminary. Asia Life Sci 3:1–52

Fernando ES, Sun BY, Suh MH, Kong HY, Koh KS (2004) Flowering plants and ferns of Mt, Makiling. ASEAN-Korea Environmental Cooperation Unit, Seoul National University, Korea, Seoul, p 368

IUCN (2015) The IUCN red list of threatened species version 2015-4. www.iucnredlist.org. Accessed 17 Oct 2020

Lambio IA, Buot IE Jr (2007) Floristic composition of Woody species along the altitudinal gradient on Mt. Makiling. Institute of Biological Sciences. University of the Philippines Los Baños, Laguna, College, Laguna

Lapitan FRG, Magcale-Macandog DB, Raymundo MDS, dela Cruz AG. (2013) Makiling biodiversity information system MakiBIS: development of an online species information system for mount Makiling. J Nat Stud 12(1):1–11

Magcale-Macandog DB, Lapitan FRG, Laruya JM, Valerio JIF, Aguila JCD, Mesina CAL, Santos TMF, Cuevas ANT, Baylon KD, Silapan IM, Dimalibot R, Edrial JD, Larida NJA, Natuel FA, Perez MGLD, Quinones SGL (2021) Updating of Makiling biodiversity information system (MakiBIS) and analysis of biodiversity data. Biodiversitas 22(1):211–226

Malabrigo PL Jr, Umali AGA, Tiburan CL Jr, Pampolina NM, Balatibat JB, Abasolo WP, Luna AC, Boncodin JC (2016) Tree diversity and stand structure or permanent biodiversity monitoring area in Mt. Makiling. Asian J Biodiv 7:1. https://doi.org/10.7828/ajob.v7i1.885

Mallari NA, Tabaranza B, Crossby MJ (2001) Key conservation sites in the Philippines. Bookmark, Inc., Makati City, p 485

Paller VGV, Cervancia CR, Macandog DM, Raymundo AK, De Guia APO, Pagaduan DC, Dimalibot JC, Lacaste AV, de Chavez ERC, Lauan MCB, Hadsall AS, Torreta NK, Nacorda OS, Buot IE Jr, Lambio IAF, Mendoza BC, Lit II Jr, dale Cruz CPP, DAP F (2012) Biodiversity along elevational gradients of the northeastern slope of Mt. Makiling Forest Reserve. UPLB College, Laguna

Pancho JV (1973) Mt. Makiling as a station for biological research. Philipp Agric 57:21–33

Pancho JV (1983) Vascular Flora of Mount Makiling and vicinity Luzon: Philippines, part 1 Kalikasan, the Philippine journal of biology. Kalikasan Press, Manila

Pancho JV, Gruezo WSm. (2006) Vascular Flora of mount Makiling and vicinity Luzon: Philippines, part 2. National Academy of science and technology NAST Philippines. Department of Science and Technology, Bicutan, Taguig City and Institute of Biological Sciences, University of the Philippines Los Baños, College, Laguna, p 626

Pancho JV, Gruezo WSm. (2009) Vascular Flora of mount Makiling and vicinity Luzon: Philippines, part 3. National Academy of science and technology NAST Philippines. Department of Science and Technology, Bicutan, Taguig City and Institute of Biological Sciences, University of the Philippines Los Baños, Laguna, p 496

Pancho JV, Gruezo WSm. (2012) Vascular Flora of mount Makiling and vicinity Luzon: Philippines, part 4. National Academy of science and technology NAST Philippines. Department of Science and Technology, Bicutan, Taguig City and Institute of Biological Sciences, University of the Philippines Los Baños, Laguna, p 405

Pelser PB, Barcelona JF, Nickrent DL (2011) Co's digital flora of the Philippines. http://www.philippineplants.org. Accessed 14 Jan 2021

Roy PS, Saran S, Ghosh S, Prasad N, Karnatak H, Talukdar G (2002) Development of biodiversity information system for north East India using internet GIS. Symposium on geospatial theory, processing and applications, ISPRS, Ottawa. https://citeseerx.ist.psu.edu/viewdoc/download?doi=10.1.1.97.5320&rep=rep1&type=pdf#:~:text=The%20Biodiversity%20Information%20System%20allows,in%20identifying%20areas%20for%20bioprospecting. Accessed 18 Oct 2020

Staley JT, Reysenbach AL (2001) Biodiversity of microbial life: foundation of Earth's biosphere. Wiley-Liss, New York, p 15

Xiao-Ting D (2012) Research Progress on biodiversity information system. Int Conf Comput Sci Serv Syst 2012:296–299. https://doi.org/10.1109/CSSS.2012.81

Native/Heirloom Rice in the Cordilleras: Status, Conservation, and Utilization

6

Romeo Jr. A. Gomez, Belinda A. Tad-awan, Alexander W. Fagyan, Joyce N. Paing, and Esther Josephine D. Sagalla

Abstract

The chapter discusses the characteristic nature of native or heirloom rice varieties in the Cordillera Region, in the northern part of the Philippines. Such varieties are cultivated in so-called traditional rice-based ecosystems (TRBEs), which are intrinsically highly organic production systems. These are situated in indigenous cultural communities where community life is centered around the cultivation of such traditional rice varieties (TRVs). Presented herein are typical observations and similar empirical research data generated across select provinces within the region which include Benguet, Mountain Province and Ifugao, among others. The pedon in such TRBEs is generally characterized to be acidic to slightly acidic and has relatively high to very high organic matter content. Soils of these nature have been associated with the cultivation of TRVs, which are closely intertwined with organic farming. Germplasm and phenotypic diversity of TRVs are generally

R. J. A. Gomez (✉)
Biology Department, Benguet State University, La Trinidad, Benguet, Philippines

Office of the Vice President for Research and Extension, Benguet State University, La Trinidad, Benguet, Philippines
e-mail: r.gomezjr@bsu.edu.ph

B. A. Tad-awan
Agronomy Department, Benguet State University, La Trinidad, Benguet, Philippines

Office for Research Services, Benguet State University, La Trinidad, Benguet, Philippines

A. W. Fagyan
Soil Science Department, Benguet State University, La Trinidad, Benguet, Philippines

J. N. Paing
Biology Department, Benguet State University, La Trinidad, Benguet, Philippines

E. J. D. Sagalla
Agronomy Department, Benguet State University, La Trinidad, Benguet, Philippines

S. Ramamoorthy et al. (eds.), *Plant Genetic Resources, Inventory, Collection and Conservation*, https://doi.org/10.1007/978-981-16-7699-4_6

high, as these have been the varieties cultivated for a long period of time, across several generations. Generally, these varieties may be categorized as glutinous or non-glutinous types. Conservation of TRVs is strongly attributed to the indigenous or traditional ecological knowledge systems (TEKS) customarily practiced in their cultivation. These farming practices are seen to be ecologically sound and have contributed to their continued existence in the TRBEs. The relevance of the TRVs in the indigenous communities lies in terms of food security and cultural expression. Such continued utilization has all the more heightened their conservation as precious genetic and indigenous community resources.

Keywords

Native · Heirloom rice · Cordillera · Status · Conservation · Utilization

6.1 Introduction

Many traditional societies in the Southeast Asian region revolved around the culture of native varieties of rice, here in this book oftentimes referred to as "heirloom" rice. Being the staple food on the table, it has consequently defined many of the indigenous cultures, traditions, and practices that are associated with the everyday life of many indigenous people. In many parts of the region such as Japan, the cultivation of native rice varieties has been associated with the popular and "prized" *sake*, many forms of rice cakes, and other products and derivatives, which form a part of many cultural celebrations. In the Philippines, and in the Cordilleras in general, cultivation of such varieties represents a part and parcel of the indigenous culture. They are basically preferred to some extent, as they represent the precious genetic resources in their localities. Characteristically, having distinctive taste and aroma, and other notable features, they have been given cultural significance owing to the fact that these have been handed down by their ancestors for many generations. The Ifugao Rice Terraces (IRT), for one, are known as a "Living Cultural Landscape" under the UNESCO Heritage site. Quite recently, it has also been added to the list of Food and Agricultural Organization (FAO)-led recognition as "Globally Important Agricultural Heritage Systems (GIAHS)" (Gomez Jr 2014, 2020). This is a strong recognition of the universal significance of the IRT, which is centered on native rice varieties. It is worthy to note that the Japan International Cooperation Agency (JICA) has been pouring international advocacy and support into the Province of Ifugao. Indeed, it is a well-known fact that the traditional rice varieties are culturally significant as they have an integral role in the "rice rites" of the traditional Ifugao culture. Similarly, in other parts of the Cordilleras, there abound several native or "heirloom" rice varieties that have been documented by a number of researchers, including the PhilRice and the International Rice Research Institute. In this chapter, a number of databases on heirloom rice in other areas, such as Benguet and Mountain Province, are discussed in some detail. As mentioned earlier, people from these areas demonstrate a very close affinity to native or heirloom rice, as much of their everyday living and culture are associated with the production of these varieties. Moreover, the expression of culture and tradition lies, to a certain extent, on how people value and evaluate their native resources—be it the forests, rivers, genetic resources, etc., and in this particular case, heirloom rice.

6.2 Framework of the Study

The foregoing discussion essentially revolves around traditional rice varieties—their relevance in many indigenous cultural communities in the Cordilleras. Indigenous peoples in this part of the region value these resources because of their role in addressing food sufficiency and security (Gomez Jr 2014) as well as the central role in defining and giving sense to many indigenous knowledge and practices important in their everyday lives. Thus, the "status" referred to herein first exemplifies the cultural and economic importance and relevance they afford toward these precious resources. Secondly, this status is also described in terms of the germplasm and its variety as observed in many places in the region. Key sets of empirical data generated from the field, across a good number of sites, are offered in order to have a good sense of the diversity and distribution of such genetic resources. Accompanying such information would be the nature of the soil that to some extent characteristically and generally supports the kind of organic production system appropriate for the heirloom varieties of rice (read: native or traditional). The chapter continues on with the discussion on aspects of conservation and utilization of such indigenous rice varieties as precious genetic and biological resources. These practices are very important with the understanding and value in mind that in order to have these varieties maintained and sustained over long periods of time across generations, community people must have a good sense and experience on the usefulness, appropriateness, and functionality of such biological resources vis-à-vis survival, daily living, and benefits derived. An enumeration and review of key indigenous practices are offered to allow the reader to peek into how they primarily relate with traditional or heirloom rice and their continuous upkeep—whether in the households, in the farms, or within the community.

6.2.1 Soil/Pedon

The cultivation of heirloom rice varieties has been part and parcel of the everyday life of people in the Cordilleras, since time immemorial. As such, this staple food of the indigenous cultural communities (ICCs) has been at the forefront in their life and culture. Concomitant with the cultivation of these traditional rice varieties, there are fairly common characteristics of the pedon that support this kind of indigenous agroecological production systems. Field surveys across similar agroecosystems in the last two decades or so in the Cordilleras generally reveal acidic to slightly acidic pH and relatively high to very high organic matter content. The latter feature is understandable as these indigenous or traditional agroecosystems are excellent exemplification of organic agricultural production systems, albeit with some indications of the entry of modern agricultural technologies of late.

In some surveys in Kiangan and Banaue areas, in the province of Ifugao for instance, the pH range is shown to be from 4.7 to 5.3, with organic matter ranging from 3.6 to 5.0% (Gomez Jr and Pacardo 2005). In some other areas in Benguet, rice-cultivated areas showed pH ranges from 4.4 to 4.8 (acidic) and organic matter content of 1.75–2.08% in Palina, Kibungan (Gomez Jr and Baniaga 2018). In other areas, such as in Kapangan (Benguet), pH, organic matter, and estimated nitrogen content had values of 5.3–6.6%, 13.3–17.3%, and 0.925%, respectively (Ingosan et al. 2008). In the case of Sadanga, Mountain Province, characteristically, the clay textural class of the soil has a high water-holding capacity ranging from 53.43 to 60.65% (Fagyan 2012). Moreover, the soil of the rice terraces in that area had organic matter content ranging from 4.0 to 4.9%, moderately acidic with a pH of 5.3–5.9.

6.2.2 Soil Nutrient Management Practices

To preserve the fertility status of the rice paddies in the growing of traditional rice, farmers practice green manuring. Farmers use available organic materials present in the farms and its surroundings to include weeds and grasses intermixed into the rice paddies or sunflower, weeds, and grasses. Others also mix sunflower, weeds, and grass legume pods (Faroden 2006). This green manure is applied at 1.32–1.82 tons/ha. Farmers have been growing the traditional rice for almost 41 years while others have been grown for less than 40 years. Other strategies which many farmers still adhere to include the incorporation of rice straw and weeds into the soil to maintain the availability of organic fertilizer application in the devoted areas for traditional rice production.

According to CECAP and PhilRice (2000), rice straw when completely decomposed would release nitrogen which is needed by the young plants. The removal of the rice straw from the field lowers the soil nutrient contents. The removal of one ton of straw from a hectare of field loses about 5–8 kg N, 1.05–2.7 kg P, 12–17 kg K, 0.5–1.0 kg S, and 40–70 kg Si. On the other hand, an integrated management of nitrogen that involves application of inorganic and organic sources of plant nutrients should be considered as an essential component of management practices for rice production provided that agronomic data and socioeconomic conditions are well defined. There are some organic sources of nitrogen such as rice straw, weeds, compost, green manures (*Azolla sp., Sesbania sp., Crotalaria sp.,* ipilipil), and others. According to the respondents in the two sites, kuliplip (Azolla) is dominant in the rice fields. In addition, they affirmed that kuliplip adds nutrients to the soil based on their observations and experiences. *Azolla* is a tiny, aquatic fern where the blue-green alga Anabaena lives and traps atmospheric nitrogen. Azolla reproduces rapidly that enough herbage can be produced for plowing under to supply half or all the nitrogen needed by the rice crop.

6.2.3 Germplasm of Traditional Rice Varieties

The collection and conservation of traditional rice germplasm are critical especially in the improvement and utilization of rice cultivars in the Philippines. As of 2014, about 307 traditional rice varieties have been conserved in the gene bank of the Philippine Rice Research Institute (PhilRice), which is mandated to conduct research and development projects for the Philippine rice. The PhilRice germplasm for traditional rice consists of 215 indica, 89 javanica, and 3 japonica varieties. These traditional varieties were found to have moderate to high phenotypic diversity based on Shannon-Weaver diversity index for both qualitative and quantitative traits. Moreover, culm number, panicle number, and caryopsis ratio were some of the traits that explained 48% of the diversity in the germplasm (Rabara et al. 2014).

In the Cordillera Administrative Region or CAR (17.3513^0 N, 121.1719^0 E), which is located in the northern part of the Philippine highlands, rice (*Oryza sativa*) is grown in about 49,431 ha of agricultural land. Out of the 7.2 million metric tons (MT) of rice produced in the Philippines, an estimated amount of 133,372 MT is produced in CAR, i.e., a 1.86% share in production (PSA 2019). Notwithstanding, the most genetically diverse pigmented rice landraces or traditional varieties are found in CAR in addition to two other regions located in the southeastern part of the country. Specifically, a mean allele richness of 1.86 $\pm$ 0.22 was found among the pigmented rice landraces of CAR (Mbanjo et al. 2019).

Within the six provinces of CAR are several rice farmers who continue to cultivate traditional rice varieties or landraces. Complementary to the continuous cultivation of traditional rice varieties is the collection and conservation by several research institutions in the country and State Universities and Colleges (SUCs). One such institution is the International Rice Research Institute or IRRI which released a Varietal Catalogue showcasing the traditional rice varieties present in three provinces of CAR, namely Ifugao (16.8331^0 N, 121.1710^0 E), Kalinga (17.4740^0 N, 121.3542^0 E), and Mountain Province (17.0663^0 N, 121.0335^0 E). The unique traits of 62 out of 152 traditional rice varieties, dubbed as heirloom rice varieties, were featured. These heirloom rice varieties were collected by PhilRice and the Regional Field Office of the Department of Agriculture in the country. In addition to the vegetative, early reproductive, several quantitative, and postharvest traits of each of the varieties, resistance to rice blast and bacterial blight diseases was also studied. Twenty-three percent of the traditional rice varieties were found to be resistant to one strain of the rice blast disease and 43% were resistant and moderately resistant to one strain of bacterial blight. *Intan*, *Kintan*, and *Kinyas* were among the resistant traditional varieties evaluated (Cope et al. 2020).

In Benguet (16.5577^0 N, 120.8029^0 E), another province of CAR, a total of 157 traditional rice varieties or landraces were collected and studied (Tad-awan and Sagalla 2011). Several of these rice landraces are continuously cultivated by rice farmers whereas some are conserved at Benguet State University, a SUC in the province. The different rice landraces were collected from 12 out of 13 municipalities or towns of Benguet with varying elevations (Table 6.1). The highest frequencies ($\geq$23) of collected rice landraces came from three rice-growing municipalities or

Table 6.1 Place of collection and number of traditional rice varieties collected in Benguet, Philippines (2011)

Place of collection	Elevation (masl)	Number of traditional rice varieties
Atok	540–2660	11
Bakun	180–2660	23
Bokod	580–2680	8
Buguias	1240-2780	5
Itogon	160–2080	5
Kabayan	840–2920	19
Kapangan	240–1740	27
Kibungan	140–2580	29
Mankayan	540–2340	7
Sablan	100–1500	8
Tuba	40–2240	6
Tublay	480–1820	9
Total		157

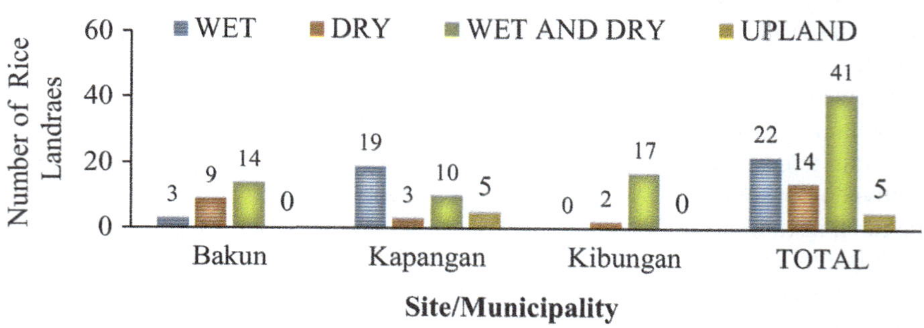

Fig. 6.1 Number of rice landraces documented/collected per municipality/site classified according to season when cultivated

towns with elevations ranging from 140 to 2660 masl. In one municipality, namely Kapangan (16.6816^0 N, 120.6890^0 E), on-farm conservation of rice landraces is one of their priority programs.

In a more recent study in three major heirloom rice-producing municipalities of Benguet (Tad-awan et al. 2020), a total of 82 rice landraces were documented and collected: 37 in Kapangan, 26 in Bakun, and 19 in Kibungan (Fig. 6.1). Majority of the documented rice landraces are planted for both wet and dry seasons (41): 22 for wet season, 14 for wet season, and 5 for uplands. In Bakun and Kibungan, majority of the rice landraces are planted in both seasons, while in Kapangan, majority of the rice landraces are planted for wet season only. There were also five rice landraces for upland planting during the wet season in Kapangan.

Among the 26 rice landraces planted by farmers in Bakun, the most cultivated variety is *Balisanga* (Table 6.2). Other landraces cultivated include *Bongkitan*,

Table 6.2 Most widely cultivated rice landraces (>10%) by farmers in Bakun, Kapangan, and Kibungan

Municipality/rice landrace	% farmers	Season	
		Wet	Dry
Bakun			
Balatinaw	19.7	/	/
Balisanga	28.0	/	/
Bongkitan	21.0	/	/
Kandiling	21.1	/	/
Ngimol	19.7	/	/
Kapangan			
Balatinaw	11.9	/	x
Balatingi	16.5	/	/
Bingawan	10.1	/	/
Bongkitan	19.3	/	/
Kulot	11.9	/	x
Ngalabngab	11.9	/	x
Saba	11.9	x	/
Topel	12.8	/	x
Wagwag	16.5	/	x
Kibungan			
Balatinaw	16.0	/	/
Balisanga	22.7	/	/
Bongkitan	24.0	/	/
Camporo	16.0	/	/
Diket red	14.7	/	/
Kabal	20.0	/	/
Lasbakan	14.7	x	/
Oklan	20.0	/	/

Kandiling, Balatinaw, and *Ngimol* which are cultivated in both wet and dry seasons. In Kapangan, *Bongkitan* is the most cultivated variety (Table 6.4). *Balatingi* and *Wagwag* are cultivated by 16.5% of the farmers. *Bingawan, Balatinaw, Kulot, Ngalabngab, Saba,* and *Topel* are cultivated by 10.1–12.8%. *Bongkitan* is the most planted variety by the farmers in Kibungan (Table 6.2). Other landraces commonly planted are *Balisanga, Kabal, Oklan*, and Talokitok. These landraces are grown by the respondents during the wet and dry seasons.

It was observed that the number of collected/documented rice landraces was high in Benguet. This coincides with the report of the Department of Agriculture Cordillera Administrative Region (DA-CAR) citing that Benguet province had the highest number of rice landraces cultivated though it is not known as a rice-producing province. The report indicates that 6–18 traditional landraces each in 12 barangays of Atok and Kibungan, Benguet, are still being grown by the farmers (Domoguen 2011). The report of DA-CAR and the result of this present study are in agreement with the study of Tad-awan and Sagalla (2015) which documented, collected, and characterized 90 rice landraces in Benguet.

6.3 Phenotypic Diversity of Traditional Rice Varieties

In the study of Tad-awan and Sagalla (2011) the phenotypic characterization of rice landraces revealed a low mean diversity of both vegetative and grain phenotypes. Few individual phenotypic traits such as sterile lemma color of grain showed a high diversity index (0.8). On the other hand, cluster analysis from phenotypic characterization of the grains of 157 rice landraces revealed 7 clusters with 2 clusters containing only 1 rice landrace each (*Lablabi* and *Lasbakan*). These landraces are unique in their sterile lemma shape and color and lemma color of apiculus. Another cluster classified 7 rice landraces as having potential for high grain yield due to high frequencies of grain and caryopsis width and 100 grain weight (Table 6.3). Some of these rice landraces with potential for high yield are *Gal-ong*, *Bongkitan*, and *Gulitan*. Other rice landraces, namely *Sabul* and *Tudoy*, showed potential not only for high grain yield but also for pest resistance (Table 6.4).

The low diversity of grain and vegetative phenotypes indicates homogeneity among the 157 rice landraces conserved at Benguet province. Molecular characterization of selected landraces may however reveal a higher diversity and identify desirable traits in the rice germplasm. It is also noteworthy to confirm the potential for high yield and pest resistance of selected rice landraces through growth and yield trials.

Table 6.3 Mean diversity indices for grain characters of 157 rice landraces

Grain character	Diversity index	Description	Most diverse phenotype
Qualitative	0.22	Low diversity	Sterile lemma color (0.80)
Quantitative	0.16	Low diversity	Awn length (0.34)

Table 6.4 Total number of rice landraces per cluster and their grain phenotypes

Cluster	Number of rice landraces	Grain phenotypes
1	83	Awn length (0–7 cm) Awn thickness (0–2 mm) Awn color (gold to black)
2	58	Lemma and palea color (straw to purple furrows) Lemma and anthocyanin coloration of keel (absent to medium) Lemma anthocyanin coloration of area below apiculus (absent to medium)
3	4	Grain length (5–10 mm) Caryopsis length (4–8 mm)
B 4	7	Grain width (3–5 mm) 100 grain weight (3–5 g) Caryopsis width (2–4 mm)
5	1	Sterile lemma shape (absent to triangular) Sterile lemma color (straw to purple)
6	2	Sterile lemma length (2–4 mm) Lemma shape of apiculus (pointed to curved)
7	1	Lemma color of apiculus (straw to black)

Table 6.5 Maturity and computed grain yield of glutinous rice landraces selected by farmers and researchers

	Maturity (days from transplanting to ripening)		Computed grain yield ($t\ ha^{-1}$)		
Rice landrace	Year 1	Year 2	Year 1	Year 2	Reason for selection
Glutinous					
Balatingi	144	149	1.36	0.62	High yield
Bayabas	148	155	1.72	0.85	Early maturing and high yield
Lablabi	144	173	1.40	0.42	Early maturing and high yield
Makabsog	138	165	1.24	0.65	High yield
Muri	142	162	1.08	0.69	Early maturing
Non-glutinous					
Ba-ay	140	141	1.42	0.98	Early maturing and high yield
Balisanga	142	121	1.66	1.08	Early maturing
Kayamsing	146	118	1.40	1.08	Early maturing and high yield
Maltiha	142	120	1.24	1.12	High yield
Sapaw	143	131	1.20	0.86	Early maturing and high yield

Growth and yield trials of glutinous and non-glutinous rice landraces during the wet season of two cropping years (2010–2011 and 2011–2012) were conducted in Kapangan (Tad-awan and Sagalla 2015). Kapangan (16.6816^0 N, 120.6890^0 E; 1000 masl) is one of the rice-growing municipalities or towns in Benguet where a high frequency of rice landraces was collected (Table 6.5). Among the glutinous and non-glutinous rice landraces, *Bayabas* and *Balisanga* produced the highest yield, respectively, in both cropping years (Table 6.5). It is noteworthy that both varieties also matured early especially during the second year of cropping. Computed grain yield of more than 1 $t\ ha^{-1}$ was produced by glutinous and non-glutinous rice landraces alike during the first cropping year. As expected, however, computed grain yield notably decreased during the second cropping year. The common reasons for selecting a landrace among farmers and researchers were high yield and early maturity. High tillering ability as indicated by the number of productive tillers was also one of the reasons for selection.

Growth and yield trials are critical in identifying rice landraces with desirable traits, e.g., high yield and early maturity. The participation of farmers in such trials is also important especially in the selection and recommendation of productive rice landraces.

6.4 Glutinous and Non-glutinous Traditional Rice Varieties

The traditional rice varieties grown in the Cordillera region are presented in Table 6.6. The non-glutinous rice is the primary food for consumption daily. A total of 44 white-grain-colored rice varieties which include Benggawan, Kalanyag, Kuyugyo, and Oklan are among others, while 15 red-grain-color rice varieties would

Table 6.6 Traditional rice varieties grown

Non-glutinous rice		Glutinous rice		
White-grain color	Red-grain color	White-grain color	Red, light orange, brown-grain color	Violet, black violet-grain color
Benggawan	Akangan	Bongkitan/Waray	Yangaw/ilekwet	Ominio
Kalanyag	Chorchor-os	Diket	Talangkay	Balatinaw/Manmansa
Kuyogyo	Kastila	Dikalot	Diket	Sungduan
Oklan	Lasbakan	Sayapot	Bollikaw	Balatingi
Lad-ocan	Toling	Longgot	Makabsug	Kotinao
Pokpoklo	Saluyaw	Lambasi	Bangaw	
Raminad	Linob-ong		Damba	
Taywan	Kurot		Bontayog	
Binaay	Chapochapor			
Monay	Intan			

Table 6.7 Total traditional rice production in Mountain Province

Municipality	No. of farmers	Average yield (MT/Ha)	Average production/year (MT)
1. Barlig	1030	3.27	2624.10
2. Bauko	2580	3.23	1990.50
3. Besao	980	3.24	769.6
4. Bontoc	2120	3.35	1585.50
5. Natonin	1356	3.25	4120.90
6. Sabangan	1026	3.26	688
7. Sadanga	1396	3.48	3634.20
8. Sagada	867	3.22	645.3
9. Tadian	2268	3.57	2332.20
Total	13,623	28.97	18,390.30

include Akangan, Chorchor-os, Kastila, Lasbakan, etc. The glutinous rice varieties like Ominio, Waray, Balatinaw, Kotinao, Kinogong, and Kitneban are all originated from the community (Fagyan 2012; Solimen and Dumaslan 2009). Most of the glutinous rice varieties are prepared as *kankanen* or delicacies and *tapey* or rice wine during special occasions and ritual observance. Moreover, some of the traditional rice varieties cultivated are drought resistant and are dependent on rainwater irrigation also called aerobic rice. These include Balisanga, Bayag, Bongkitan, Gupong, Waray, Oklan, and Benggawan (Paing et al. 2019).

The production of traditional rice varieties is substantially supporting the family's daily consumption of rice. Table 6.7 reveals the nine municipalities of Mountain Province that still adhere to rice farming. Notice the average production per year of more than 18 metric tons.

6.5 Conservation

6.5.1 Traditional Rice Conservation Practices

Women folks are repositories of indigenous knowledge (IK) or traditional ecological knowledge systems (TEKS) in crop production. In Sadanga, women select good seed varieties during harvest. This is because they are familiar than men with the unique characteristics of each rice variety and they carry it as a lifetime objective in seed selection for rice production that eventually conserves their rice plant genetic resources for the generations to come. Furthermore, some TEKS are still observed as conservation approach in the local community such as madmad, pudong, alluyon, pa-ad, and inayan (Paing et al. 2019; Ingosan et al. 2008). Men have known more TEKS than women, probably because during the interview the head of the family represents the whole household. Both genders have common beliefs and are knowledgeable of TEKS which they have inherited from their foreparents. In farming livelihood practices, they need to perform some rituals by way of butchering chicken or pig before and after seed sowing and during harvest (Table 6.8). People learn to

Table 6.8 Some traditional ecological knowledge systems as conservation strategies

Selected TEKS practices	Description
Madmad/e ayag/eserbi/ ibuni/liyaw/denet	Before and after planting or harvest of rice, rituals are performed by butchering chicken and *tapey* (rice wine) as an offering As thanksgiving for bountiful harvest and food supply, good yield, more blessings for the next cropping, to wash "malas" or bad luck and that their crops will not be destroyed by calamity and pest/ diseases. Accordingly, it serves as an offering in exchange for a good harvest
Pudong	A knot which is tied on a corner of the paddies (using grass) or by using a chicken feather tied on a stick By placing this on the rice paddies after rice planting and during harvest. It signifies that nobody should enter and pass-through during harvest so that the production will not decrease; otherwise *makawas* or the harvested rice will not last long; protection from unseen spirit
Alluyon/ubbu	The spirit of communal unity, work cooperation among the local communities to achieve a particular goal. In rice cultivation, a reciprocal work arrangement on each farmer's farm through *binnadang* where the owner of the farm provides meals for the labor Spread of blessing for the entire community so that their crops will grow healthy and for the harvest to last longer. Enhance camaraderie, socialization, cooperation, and good working relationships
Pa-ad/ngilin	Rest day for 1 day or more after farm-related activities (*tunod/ esek*) Practice of the community as their rest day and to avoid accident. No one should go to the rice field to work or collect vegetables
Inayan	Belief in supernatural being and respect of living things

care for and manage their lands generally by ways of living with nature involving various rituals, customs, ceremonies, and traditions (Sangha et al. 2018). Similarly, Ingosan et al. (2008) cited different TEKS and rituals being performed during the earlier times and that are still being practiced because they believed that these are sources of blessings to enhance rice production.

6.6 Utilization

The traditional rice in bundled panicles and processed food products are used for offerings in family and community rituals like wakes, wedding, and house construction (Fagyan 2012). The glutinous rice is processed by women into wine, fermented rice, tupig, saliket, patopat, and sinab-ang as offerings in the performance of community festivals like "fegnas." Nursing mothers prepare *segget* or rice milk for their young. Sinab-ang/champorado is prepared as food and snack during land preparation, transplanting, and harvesting; saliket (glutinous rice cooked plane without sugar or coconut milk) for hospitality during wakes and weddings; and tapey (rice wine and fermented rice) for packed lunch and a drink to provide a refreshing feeling during harvest. During harvest, no one starts to go to the field without rice wine because *tapey* invites good spirit to drive away bad lucks. In addition, men and women drink and eat rice wine after hauling the palay to the rice granary. Staple rice though is sold by women if there is extra after keeping for the family. After harvest, non-matured rice is gathered and processed as iluban/du-um/chu-um which is a delicacy made from burning the un-matured rice grains, pounding when cooled, and winnowing.

6.7 Conclusion and Implications

It is noteworthy to state again at this point that in a broad sense, the traditional or heirloom rice varieties have been cultivated in rice pond fields, or paddies, whose soils are generally rich in organic matter. This is attributed to the fact that nutrient cycling as an ecological function has been continuously observed in such organic production system. A vast array of studies have shown commonly that material and nutrient import to the system is very low; thus, it is characterized as a low-input agroecosystem.

The enormous germplasm diversity is also very evident, confirming the rich genetic resources on rice landraces in this northern region of the Philippine main island of Luzon. These heirloom rice varieties have relatively high yield and drought and pest resistance, among other notable characteristics. Molecular characterization of selected landraces may, all the more, reveal a higher diversity and identify desirable traits in the rice germplasm. These include the glutinous and non-glutinous rice landraces in the various study areas.

Due to the influence of the agroecological and sociocultural factors, the traditional or indigenous cultivation of such varieties has contributed much to its existence for long periods of time. Furthermore, the utilization of such primary agroecosystem products as daily subsistence for the family, processed food products, delicacies during special cultural occasions, and rice-wine (locally called *tapuy*) as offering during community rituals has provided various significant meanings to the lives of the locals. Such importance has contributed to maintaining and conserving this material resource through the ages.

An apparent implication of this observation is on the context of local policy-making on the conservation of these genetic resources and traditional rice-based ecosystems (TRBEs) in general. A concerted effort from the academe, the local government units, as well as the local people in conducting information campaign on the environmental, cultural, social, agricultural, and nutritional significance and value of these precious genetic resources is thus essential.

References

[CECAP and PhilRice] Central Cordillera Agricultural Programme and Philippine Rice Research Institute (2000) Highland rice production in the cordillera. CECAP/PhilRice, Banaue, Ifugao/ Munoz, Nueva Ecija, Philippines

Cope AE, Reano RA, Luis JS, Santelices AP, Padilla JJ, Acuin CS (eds) (2020) Philippine cordillera heirloom rice. International Rice Research Institute (IRRI), Los Banos

Domoguen, RL (2011) Cordillera losing its heirloom rice varieties. The Sun Star Baguio Newspaper

Fagyan HS (2012) Women's participation in the production and product development of heirloom rice in Sadanga, Mountain Province. Master's thesis, Open University, Benguet State University

Faroden JC (2006) Physico-chemical properties of selected rice terraces in Sadanga, Mountain Province. Master's thesis. Graduate School, Benguet State University, La Trinidad

Gomez RA Jr (2014) Food sufficiency in the terraces: the case of globally-important agricultural heritage systems in Ifugao, Central Cordillera, Philippines. Int J Ecol Conserv 7:1. https://doi.org/10.7718/ijec.v7i1.723

Gomez RA Jr (2020) Spatial and productivity measurements in traditional rice-based ecosystems (TRBEs) in Highland cordillera. In: Buot IE (ed) Methodologies supportive of sustainable development in agriculture and natural resources management. SEARCA, Los Banos

Gomez RA Jr, Baniaga AV (2018) Preliminary report of biophysical assessment in Palina, Kibungan: assessment field, validation and documentation of evidence-based ecosystem adaptation (EbA) measures with potential for scaling-up in indigenous cultural communities. GIZ Project. Benguet State University, La Trinidad, Benguet

Gomez RA Jr, Pacardo EP (2005) Survey of the damage in the Ifugao rice terraces and its environmental dynamics. Philippine Agric Scient 88(1):133–137

Ingosan JB, Labaro JS, Sally JL, Gomez RA Jr (2008) Agro-ecological appraisal of traditional rice-based ecosystem. Kapangan Exp BSU Res J 60:& 61

Mbanjo EG, Jones H, Caguiat XG, Carandang S, Ignacio JC, Ferrer MC, Boyd LA, Kretzschamar T (2019) Exploring the genetic diversity within traditional Philippine pigmented rice. Rice 15: 1–18. https://doi.org/10.1186/s12284-019-0281-2

Paing JN, Gomez RA Jr, Pladio LP, Pontino RV, Bulcio JM (2019) Ecosystem services and the role of traditional ecological knowledge system in conservation of biodiversity and food security, terminal report. Benguet State University, La Trinidad

Philippine Statistics Authority (2019) Rice and corn situation outlook, January 2019 round. psa.gov.ph

Rabara R, Ferrer M, Diaz C, Newingham M, Romero G (2014) Phenotypic diversity of farmers' traditional rice varieties in the Philippines. Agronomy 2014:217–241. https://doi.org/10.3390/agronomy4020217

Sangha KK, Preece L, Rosas J, Kegamba JJ, Paudyal K, Warmenhoven T, Ramakrishnan PS (2018) An ecosystem services framework to evaluate indigenous and local peoples' connection with nature. Ecosystem Services. https://www.iucn.org/sites/dev/files/content/documents/2018/indigenous_es_framework_es_journal_2018.pdf

Solimen JA, Dumaslan MR (2009) Baseline survey on rice terraces and indigenous rice production, terminal report. Benguet State University, La Trinidad

Tad-awan BA, Gayomba HC, Masangcay JM, Mauting WS (2020) Germplasm collection and varietal evaluation of heirloom rice landraces in Benguet mountain. J Sci Inter-discip Res

Tad-awan BA, Sagalla EJD (2011) Morphological diversity analysis of rice landraces in Benguet. BSU Res J 66:1–14

Tad-awan BA, Sagalla EJD (2015) Heirloom rice varieties for wet season cropping in the highlands of northern Philippines. J Org Agric Org Agric Soc Philipp 1(1):1–26

Identification and Conservation of Culturally Important Trees Used for Traditional Houses in Ifugao, Philippines

7

Consuelo DI Habito, Rosalie C. Mendoza, Raymundo Dong-e, and Pastor L. Malabrigo Jr

Abstract

The Ifugao Rice Terraces of the Philippine Cordillera is a UNESCO World Heritage Site and a proud heritage of the indigenous Ifugao people. The wooden traditional houses dot the expanse of the rice terraces and forms part of Ifugao's tangible cultural heritage. The passage of time and modern influences have seen a decline in the number and use of these traditional houses. Moreover, little is known about wood species used in its construction. This study sought to identify different wood species from a sample of traditional houses in Kiangan, Ifugao. Thirty-two species of mostly indigenous and endemic wood species were identified. The choice of wood species and their use reflected indigenous knowledge on house construction, functionality, their physical attributes, and availability. Premium species were generally preferred but their dwindling availability has influenced a shift to the use of lesser known wood species and exotics. Better knowledge on wood used for traditional houses can help the younger generation

C. D. Habito (✉)
Faculty of Management and Development Studies, University of the Philippines Open University, Los Baños, Laguna, Philippines
e-mail: cdhabito@up.edu.ph

R. C. Mendoza
Department of Forest Products and Paper Science, College of Forestry and Natural Resources, University of the Philippines Los Baños, Los Baños, Laguna, Philippines
e-mail: rmcalapis1@up.edu.ph

R. Dong-e
Poblation, Hungduan, Philippines

P. L. Malabrigo Jr
Department of Forest Biological Sciences, College of Forestry and Natural Resources, University of Philippines Los Baños, Los Baños, Laguna, Philippines
e-mail: plmalabrigo@up.edu.ph

S. Ramamoorthy et al. (eds.), *Plant Genetic Resources, Inventory, Collection and Conservation*, https://doi.org/10.1007/978-981-16-7699-4_7

of indigenous craftsmen. As such, the study seeks to contribute to cultural preservation by enhancing support for protection and cultivation of important wood species as well as promoting preservation of traditional houses, a tangible emblem of the Ifugao heritage and landscape.

Keywords

Tree identification · Traditional house · Ifugao · Philippines

7.1 Introduction

The Rice Terraces of the Philippine Cordillera mountain range are listed among UNESCO's World Heritage Sites and have been described as "an outstanding example of an evolved, living cultural landscape" (UNESCO 2019). The rice terraces were carved out by the Ifugao, a group of indigenous people, who have lived in these remote mountains for hundreds of years. Using age-old farming practices and planting traditional rice varieties (FAO 2014), the hardy Ifugao people to this day have tilled and sustained the rice terraces as part of their cultural heritage and primary source of livelihood.

The unique design and construction of wooden traditional houses have been part of the rich tangible cultural heritage of the Ifugao people. These structures are completely made of wood and plant materials and are most often found throughout the rice terraces landscape. They typically serve as homes, resting places, or rice granaries for the families who have cultivated the terraced rice paddies. In recent times, however, the growth of villages and modern influences have led to changes in the way many Ifugao people have built their houses and the way they look. Today, the villages where most Ifugao people live are made up of houses that are built using concrete, commercial timber, nails, galvanized iron roofs, and metal windows. Consequently, the number of traditionally constructed Ifugao houses is now far outnumbered by "modern" houses typically seen in other parts of the country. Sadly, the unique traditional house design features of the Ifugao seem to have gradually been relegated to the background. Recently, however, there have been efforts to revive the awareness and appreciation for traditional Ifugao house architecture as part of the overall effort to preserve the rich indigenous culture and to instill ethnic knowledge and pride especially among Ifugao youth. To help in the conservation of the culturally important trees, this research study identified the wood species used in the construction of traditional houses in Kiangan, Ifugao.

7.2 Review of Related Literature

7.2.1 Origin and History of the Ifugao Traditional House

Lambrecht (1929) describes "Where there are many rice fields there are also many houses: either a few big villages having each more than forty houses or a lot of small villages having each five to ten or more houses." As a Catholic priest who dedicated his life to the Christianization of the Ifugaos, Lambrecht was the first outsider to have written detailed accounts of the Ifugao traditional house. Thus, the location of the house, granary, or village was typically within or in close proximity to where the rice fields were located. As Conklin (1980) aptly described, the farmers and their families usually cannot be found far from their rice fields they tilled, as he pictured the life of the Ifugao to be one that revolves around rice culture in the terraced paddy fields. In describing the social organization of Ifugaos, Conklin (1980) further stated that a nuclear family (meaning husband and wife) lived in a "separate house (*bale*)—ideally a raised four-posted house." He continued to describe the house as a "multi-purpose pile dwelling, essentially an expanded and hearth-equipped granary (*alang*), also serves as a shrine in which the family reliquary (*punamhan*) is kept."

The Ifugao culture is also steeped in supernatural beliefs. The traditional house was also briefly mentioned in Ifugao mythology, where Lambrecht (1962) documented that "a higher deity who intended to make a shortcut on its journey to earth can do so by diving down into the top of an Ifugao pyramidal roof." Interestingly, this was one of the rare instances that the traditional house structure was mentioned as part of the Ifugao mythological chronicles.

7.2.2 The Ifugao Traditional House

Lambrecht (1929) describes the unique features of the traditional house as follows: The houses are strongly made, and if the materials used are good, they can last for generations. I have seen many houses constructed five to six generations ago which are still in good condition. The parts are wonderfully fitted together, although no nails are used. They can be taken apart, carried away and put up again in another place, all in a course of one day (Fig. 7.1).

By modern standards, the house is small and windowless and cooking can, over time, blacken the interior and cover it with smoke. However, Lambrecht (1929) observes that "the houses maybe uncomfortable, but an Ifugao has little to store away: just a few baskets, some firewood, and some rice."

7.2.2.1 Types of Ifugao Traditional House

There are basically two general types of the Ifugao traditional house: the hut (*abong*) and the house (*bale*) Lambrecht (1929). There are two variations: the *agamang*, which is a type of hut that serves as a residence for unmarried boys and girls, while the *alang* serves as a granary, and is classified as a type of house. According to Lambrecht (1929), the granary houses found in Kiangan are relatively numerous

Fig. 7.1 Diagrammatic illustration of traditional house

because there is usually much rice to be stored and because the low pitch of the roof does not leave much room for the rice bundles. In the following section, the discussion will describe the different wood species that were identified from the different parts of the typical traditional house. The general process of construction that involves the wooden components of the house will also be described, but without going into the intricate step-by-step details on how the entire house was built, as these were already documented by Lambrecht.

7.2.2.2 Parts of the Traditional House

The main parts of an Ifugao traditional house have been described to include (1) the foundation or support of the house, (2) house cage, (3) roof, and (4) accessories (Lambrecht 1929). Lambrecht described the parts of the house and how the different wooden components were joined together without the use of nails in the process of house construction. Instead of nails, wooden beams or posts were attached using a mortise-tenon joint. For example, the foundation of the house consisted of four sturdy posts (*tukud*) whose lower ends rested on flat stones buried in the ground. The posts' upper ends were then shaped to form a tenon that fit into a squared mortise of the transverse girders (*kuling*). Wooden anti-rat disks (*halipan*) were then inserted into the posts before the transverse girders were jointed to the posts.

The next section of the house is the house cage. Two-side floor beams (*mundilig*) and one center floor beam (*gawaan*) were placed on the two transverse girders (*kuling*). Two front and rear floor joists (*hagpo*) were then mortised to the side floor beams (*mundilig*) and supported by the center floor beam (*gawaan*). The bottom section of the wall boards (*gaob*) was fitted on the grooves on the side floor beams (*mundilig*) which were slightly inclined to the exterior. Another set of wall header/upper wall beams (*huklub*) were positioned on top of the upper end of the wallboards (gaob). The floorboards (*dotal*) rested on the side floor beams (*mundilig*) and the center floor beam (*gawaan*). Two front and two rear special wallboards (*pamadingan*) were fitted into the grooves of the front and rear floor joists (*hagpo*) and penetrated the front and rear purlin/lintel beam (*wanan*) (Lambrecht 1929).

The door space (*panto*) between the beam support (*pamadingan*) was then covered by the door shutter (*onob*). The movable door shutter (*onob*) was hung and slid by strong rattan vines (*uwe*) and locked from the inside by a bar (*dakig*).

Lambrecht (1929) detailed the roof structure as "four *bagad* or roof posts were passed (*nitakatak*) through the two *mundilig* and the two *kuling* and came out at the lower surface of the two *kuling*. Thus, they served to fasten the *mundilig* to the *kuling* and at the same time were kept erect."

Four pieces of purlin/lintel beam (*wanan*) were mortised together and attached to the upper end of the roof posts (*bagad*). A central attic beam (*pumpitolan*) was inserted onto the middle of two opposite-sided purlin/lintel beam (*wanan*). Two king posts/upright timber (*taknang*) were jointed to the middle of the central attic beam (*pumpitolan*). A square piece of wood called the apex frame (*ambubulan*) provides support for the eight long rafters (*bughol*) that form the apex of the roof. The other eight shorter rafters are located only above the level of the apex frame (*ambubulan*) in between the first eight rafters (Lambrecht 1929).

Lambrecht (1929) further described the differences between a house (*bale*) and a granary (*alang*): "In Kiangan, the pitch of the roof is not as steep. The whole house case is then visible.... Sometimes, there is no *pumpitolan* or *taknang* and only an *ambubulan* called *bobong*. The rafters are not so strong. They are only twelve in number and are attached to the *bobong*, which is a flat round board in which grooves are cut for the rafters to fit into. Granaries usually have no *pumpitolan* or *taknang*, while houses have the *taknang* and a *pumpitolan* supported by a board similar to the *pamadingan*, the latter also forming a part of the side wall."

The interior of the traditional house typically has some accessories. In one corner of the house is a fireplace (*pundapulan*). On top of the fireplace are two sideboards or shelves (*huguhug*). The first shelf is for firewood and the second shelf (*batuung*) is for rice (Lambrecht 1929). In addition, there is a horizontal storage shelf (*patie*) that runs from the wall header (*huklob*) to the rafters (*bughol*). If rice is stored in the traditional house (*bale*), there will be a shelf or flooring (*palan*) on the central attic beam (*pumpitolan*). However, most homeowners in Kiangan often have a granary (*alang*). To prevent birds from getting into the granary house, the wallboards (*gaob*) were built to extend to the purlin/lintel beam (*wanan*).

7.2.3 Wood Species Used in Ifugao Traditional Houses

Indigenous knowledge on the characteristics of different wood species suitable for house construction is thought to be a major influence on the choice of wood used by traditional builders. This review sought to gain an overview of wood species that have been used for traditional house building. Lambrecht (1929) made one of the earliest references to the use of *Vitex parviflora* (*amugawon*) tree for posts (*tukud*)—

a variety of wood noted to last for up to four generations. Alberto et al. (2011) also listed some hardwood indigenous species that have been used for building foundations, beams, and posts of the traditional house. These wood species include *Pterocarpus indicus* (*udyo* or *narra*), *V. parviflora (amugawon* or *molave), Pinus kesiya* ssp. *langbianensis* (*halong* or mature Benguet pine), *Weinmannia luzoniensis (tabangawon), Shorea astylosa (bannutan) or yakal, Toona calantas (bagtinon), Shorea contorta (pole, lauan), Streblus glaber (galiwgiwon),* and *Artocarpus ovatus (tobak).* Among these species, *galiwgiwon* and *tobak* were cited as the preferred wood by builders for traditional house parts. For other parts of the house, the following wood species have also been listed as part of the wood varieties used by Ifugao builders: *Shorea astylosa (bannutan)* (yakal, *pfanutan*), *S. negrosensis (pfuchu'an), S. contorta (pfo're/pole), Dracontomelon dao (chal-ligan or daligan, camarag), Agathis philippinensis (almasiga), Lithocarpus woodii (palayon, palajun), Syzygium* sp. *(burtic, futi), Antidesma bunius* (*bunne*), and *Vaccinium cumingianum* var. *cumingianum (gutmo).* There are other historical studies that presented a general list of wood species used in traditional house construction (Madulid 2011; Rondolo 2010; Taguiling 2013; FAO 2014). Most of the historical information described above was obtained from local indigenous knowledge passed on orally through generations and gathered through key informant interviews by the different authors cited.

There have been missed opportunities to gain better knowledge about the wood species used for the Ifugao traditional house. Recently, a community-based project that aimed to restore Ifugao traditional houses for the purpose of promoting ecotourism was completed in the Municipality of Batad in Ifugao province (Macapagal and Bermejo 2017). Through this project, dilapidated traditional houses were repaired and restored by applying traditional carpentry techniques and house-building practices. In another project in the nearby town of Mayoyao, Gawongna (2016) conducted a mapping survey of Ifugao traditional houses. In both these projects, however, there was no deliberate effort to identify the wood from the remaining intact or damaged components of the houses prior to restoration.

A study done in northern Laos looked at traditional housing in northern Laos and how wood preferences might have had an impact on forest biodiversity (Kretz 2009). The study found large differences in the kind of wood species that Lao villagers were using today compared to 50 years ago. The author strongly implied the loss of timber species and biodiversity to explain the observed results. Here, identification of the wood species used in the construction was done by interviews with knowledgeable local villagers. In a similar study, Anania et al. (2013) sought to document and preserve traditional building culture and materials knowledge in one district in Papua New Guinea. They documented several species being used for various wooden components by collecting data from local villages through interviews. In both studies mentioned above, no actual species identification of the intact or damaged wood from the houses was done.

7.2.4 Sustainable Forest Management in Woodlots and Communal Forests

Planting, cultivating, and harvesting cycles proceed with regularity in the paddy rice (*payoh*), swidden fields (*habal*), and woodlots (*muyong*). Interestingly, Conklin (1980) describes "inter-locking cycles for three major agricultural zones (pond fields, swiddens and woodlots) have different durations and exhibit varying intensities with respect to demands for natural resources and labor." Furthermore, Conklin (1980) notes, "how to plan, balance, manage and control these transformations is the concern of every Ifugao household." In the seasonal calendar, *muyong* planting activities start in the last week of July to November, followed by a period of nonpeak activity (dry season) from December to late March. The woodlot harvesting is from late March to the third week of June. The peak activity of woodlot harvesting is in preparation for the increased demands for firewood during the rice harvest season in late June and July. From three agricultural zones described by Conklin (1980), Butic and Ngidlo (2003) instead recognized five components of the agroecological zones to include the woodlot (*muyong*), swidden fields (*habal*), terraced paddies (*payo*), settlement districts (*boble*), and riverbeds (*wangwang*). They described the assisted natural regeneration strategy as embedded in the traditional forest management of the *muyong* and called it the *muyong* system. Invariably, indigenous knowledge system and practices in forest management, soil and water management, agroforestry, biodiversity, and silviculture are efficiently and sustainably played out. Camacho et al. (2012) also used the *muyong* system as a traditional forest conservation knowledge of the Ifugaos. Added to this was the ala-a system pertaining to communal forest management. While it is recognized that both *muyong* and *ala-a* systems are integral to the Ifugao way of life, their management practices are quite different. Whether the forest is private or community owned, what is more important is that resources are sustainably managed and bounded by customary laws and practices. Practices related to observance of customary forest laws, stand management, biodiversity, and ecosystem service maintenance were elucidated by Camacho et al. (2016).

Agnoletti and Rotherham (2015) have pushed strongly for the concept of biocultural diversity in recognition of the "dynamic interplay between biological and cultural diversity at the landscape level and its implications for livelihoods and well-being." Moreover, Ellis et al. (2021), in an analysis of global land-use history, have recently emphasized the role of indigenous peoples in sustainable land management and biodiversity conservation. This concept of biocultural diversity is best exemplified by the indigenous Ifugao community and their practice of sustainable land management. While this indigenous land management has transformed the Ifugao mountainous landscape into paddy fields, swidden areas, and managed forests, the interlocking activities of the indigenous peoples and low-intensity land use have contributed to biodiversity conservation. Ethnobotanical surveys of these interlocked ecosystems in Ifugao have shown diverse floral resources (Madulid 2011; Rondolo 2010; Taguiling 2013; FAO 2014; Rabena et al. 2015). The indigenous Ifugao's traditional management and environmental stewardship of the landscape as incorporated in their indigenous knowledge systems should be recognized as viable, sustainable, and effective.

7.3 Materials and Methods

7.3.1 Study Area

Traditional houses in the Municipality of Kiangan, Ifugao province, located in the northern part of Luzon island in the Philippines were visited and included in the study. Kiangan was chosen primarily because the town is one of the five inscribed World Heritage Sites collectively known as the Ifugao Rice Terraces of the Philippine Cordilleras. A preliminary inventory and mapping of Ifugao traditional houses done in 2018 established community relations and facilitated the selection and access to the houses that were included in the study.

Kiangan is a fourth-class municipality[1] and is described as the oldest town in the province of Ifugao (Fig. 7.2). Its name is derived from *Kiyyangan*, an old village located near the Ibulao River. Based on the 2015 census, the town has a population of 7048 and is divided into 14 barangays[2] in an area of about 200 square kilometers.

7.3.2 Identification of Wood Species

Wood species identification was done using the method described by Meniado et al. (1975). Briefly, wood species were identified in situ by examining cross-section specimens of wood. Each wood specimen was photographed using a portable Dino-Lite Edge AF4915ZT USB digital microscope (30× magnification) connected to a laptop computer and a digital camera. Each photograph was compared with known samples in wood species databases. Consultation visits to the Forest Products Research and Development Institute (FPRDI) Xylarium of the Department of Science and Technology (DOST) were also done.

Some of the wood species had no available wood samples in the Xylarium and even in available literature; hence, plant specimens (bark, leaves, flowers, fruits) were photographed and identified in situ with the help of the indigenous traditional house builders and plant taxonomists who identified and verified the scientific names of all the wood species samples from the traditional houses. The morphological identification of the plant specimens also resulted in the verification of the identity of all the wood samples.

[1]In the Philippines, local government units are classified according to income categories; municipalities with the highest income are classified as first-class municipalities while those with the lowest income are those that belong to sixth-class municipalities.

[2]The barangay or village is the lowest administrative division; it is headed by an elected official, the Barangay Captain, who has a tenure of 3 years.

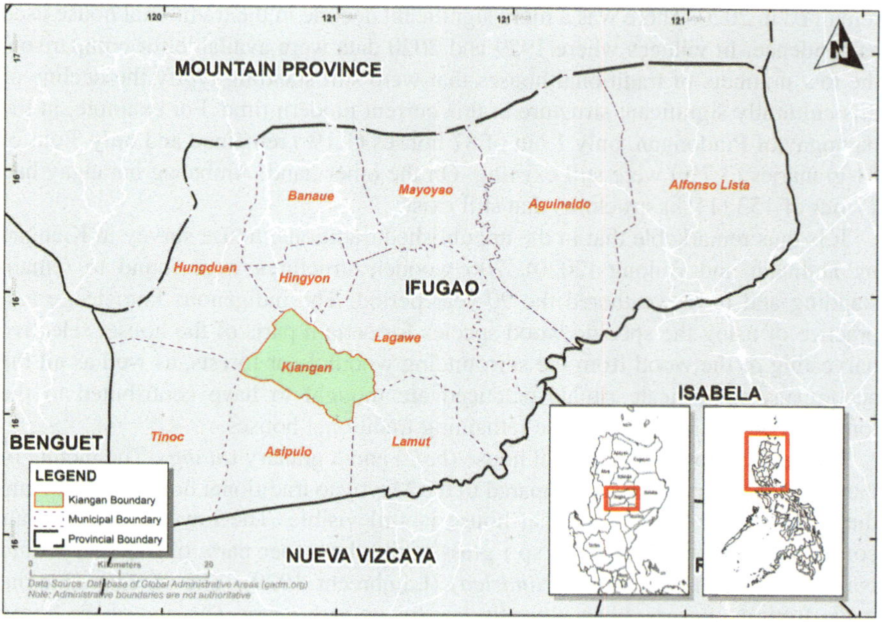

Fig. 7.2 Administrative boundaries of towns of Ifugao, Philippines, highlighting the municipality of Kiangan

7.4 Results and Discussion

7.4.1 Status of Traditional Houses in Ifugao

The traditional house has long been considered to be an integral part of the Ifugao rice terrace landscape. Lambrecht (1929) aptly described the relevance of the traditional house in the broader Ifugao landscape by writing "neither groups of houses nor even single houses are to be found far from the rice terraces." Despite the seemingly small space it provides, Lambrecht (1929) enumerates the four uses of the traditional house, "A shelter against the rain, a sleeping place at night, a place to cook his food and to store his few belongings—this is all the Ifugao needs."

A comparison of the inventory done by Lambrecht (1929) with an unpublished inventory on traditional houses by Dulnuan and Bulong (2020) showed a loss of 115 traditional Ifugao houses, a decline of roughly 65% over a 90-year period. Of the 143 traditional houses used as granary as reported by Lambrecht (1929), granaries still remain in use. This estimate was calculated by comparing the counts of traditional houses in villages included in his report. Since the total number of traditional houses was not reported in all the villages in Lambrecht's (1929) report, it is likely that the true overall rate of decline would be much higher, considering that in barangays such as Bolog, Julongan, and Poblacion, not even one traditional house remained in 2020. There was a more significant decline in the traditional house used as residence. In villages where 1929 and 2020 data were available for comparison, the low numbers of traditional houses that were still standing typify the decline of this culturally significant structure in this current modern time. For example, in the barangay of Pindongan, only 1 out of 87 houses (1.1%) remained and only 3 out of 81 granaries (3.7%) were still existing. On the other hand, Ambabag barangay had 17 out of 153 (11%) structures that still exist.

It is thus remarkable that in the unpublished traditional house survey in Kiangan by Dulnuan and Bulong (2020), 206 wooden structures were found to remain standing and have weathered the 90-year period. The indigenous knowledge and practice of using the specific wood species for certain parts of the house, selective harvesting of the wood from the surrounding woodlots or forests, as well as all the accompanying intricate rituals practiced are thought to have contributed to the longevity and durability of these remaining traditional houses.

Figure 7.3 shows a traditional house (*bale*) and a granary (*alang*). The incline of the roof is not very steep as compared to the Mayoyao traditional house. In Kiangan, the house cage of the traditional house is still visible. The lower roof materials consisted of runo (*Miscanthus* sp.) grass while the upper parts of the roof would consist of cogon (*Imperata cylindrica*) (Lambrecht 1929). The roofing material made from cogon grass can typically last for up to 6 years. The rest of the house parts except for the rattan vines used to fasten the grass to the rafters are made of wood.

Modifications of the traditional house are a sign of its changing function and relevance in the indigenous community. As shown in Fig. 7.4, the cogon grass pyramidal roof has been replaced with a more durable galvanized iron two-sided pitched or peak roof. Moreover, the traditional wooden wallboards are replaced by interwoven bamboo strips (*sawali*).

Fig. 7.3 Grass-roofed traditional house (*bale*) (left) and a granary (*alang*) (right) in Kiangan, Ifugao

Fig. 7.4 Modified traditional house (*bale*) (left) and a granary (*alang*) (right) in Kiangan, Ifugao

7.5 Identification of Wood Species Used for Ifugao Traditional Houses

To be able to preserve and maintain the traditional houses in Ifugao, it is imperative that the wood species traditionally used in building *bale* and *alang* be conserved as well. As presented in Table 7.1, a total of 32 wood species belonging to 22 genera in 17 families were identified from different house parts of the traditional houses located in Kiangan, Ifugao. Of the 17 families of trees, Dipterocarpaceae, Moraceae, and Lauraceae (with four species each), followed by Sapotaceae with three species, were the dominant tree families found. Except for *Gmelina arborea* (gmelina) and *Swietenia macrophylla* (mahogany) which were exotics, the rest (30) of the wood species commonly used in the traditional houses were native to the country. Of the 27 native trees identified to species level, 14 (52%) were found to be Philippine endemics (found nowhere else on the planet). Out of 32 species, 30 (94%) species were hardwood (angiosperms) species while the remaining two (6%) were softwoods (gymnosperms).

It is important to note that five of the native species identified were included in the updated national list of threatened Philippine plants as categorized by the Department of Environment and Natural Resources issued as the Department Administrative Order (DAO) 2017-11 (DAO 11 2017). These included the three dipterocarp species, i.e., *S. astylosa* (critically endangered), *S. contorta* (vulnerable), and *S. negrosensis* (vulnerable); *P. indicus* (vulnerable); and *V. parviflora* (endangered). *S. astylosa* and *P. indicus* were both categorized as endangered species in the International Union for Conservation of Nature (IUCN) Red List (2021).

Some of the species identified in this study were also listed in ethnobotanical and biodiversity studies of Madulid (2011), Rondolo (2010), Alberto et al. (2011), and

Table 7.1 Taxonomic list of all the wood species commonly used for traditional houses in Ifugao including their endemism, and conservation status as categorized in the DAO 2017-11 and the IUCN 2021 red list of threatened species

Family name	Scientific name	Local name	Endemism	DAO 2017-11	IUCN 2021
Cannabaceae	*Trema orientalis* Blume	*Analdong*	NE		
Cunoniaceae	*Weinmannia luzoniensis* S. Vidal	*Tabangawon*	PE		
Dipterocarpaceae	*Shorea astylosa* Foxw.	*Bannutan*	PE	CR	EN
Dipterocarpaceae	*Shorea contorta* Vidal	*Pole*	PE	VU	
Dipterocarpaceae	*Shorea negrosensis* Foxw.	*Pfuchu'an*	PE	VU	
Dipterocarpaceae	*Shorea* sp.[a]	*Bannutan*			
Ericaceae	*Vaccinium cumingianum* var. *cumingianum* S. Vidal	*Gutmo*	PE		
Fabaceae	*Pterocarpus indicus* Willd.	*Udyo*	NE	VU	EN
Fagaceae	*Lithocarpus woodii* (Hance) A. Camus	*Palayon*	PE		
Hypericaceae	*Cratoxylum formosum* (Jack) Dyer	*Aliguyon*	NE		
Lamiaceae	*Gmelina arborea* Roxb.	*Gmelina*	EX		
Lamiaceae	*Vitex parviflora* Juss.	*Amuwagon*	NE	EN	
Lauraceae	*Cryptocarya* sp.[a]	*Kawahi*			
Lauraceae	*Litsea cordata* Hook. f.	*Bakkan*	NE		
Lauraceae	*Litsea luzonica* (Blume) Fern. Vill.	*Bini*	PE		
Lauraceae	*Machilus philippinensis* Merr.	*Dulawon*	NE		
Meliaceae	*Sandoricum vidalii* Merr.	*Baccuwog*	PE		
Meliaceae	*Swietenia macrophylla* King	*Mahogany*	EX		
Moraceae	*Artocarpus ovatus* Blanco	*Tobak*	PE		
Moraceae	*Ficus benguetensis* Merr.	*Latbang*	NE		
Moraceae	*Ficus variegata* Blume	*Ludping*	NE		
Moraceae	*Streblus glaber* Merr. Corner.	*Galiwgiwon*	PE		
Myrtaceae	*Syzygium polycephaloides* (C. B. Rob.) Merr.	*Bahag*	PE		
Myrtaceae	*Syzygium sp.*[a]	*Bultik*			
Phyllanthaceae		*Tuwol*	NE		

(continued)

Table 7.1 (continued)

Family name	Scientific name	Local name	Endemism	DAO 2017-11	IUCN 2021
	Bischofia javanica Blume				
Pinaceae	*Pinus kesiya* ssp. *langbianensis* Royle ex Gordon	*Halong*	NE		
Podocarpaceae	*Dacrydium pectinatum* de Laub.	*Dumliing*	NE		
Rubiaceae	*Neonauclea media* Elmer	*Tikom*	NE		
Rubiaceae	*Neonauclea reticulata* (Havil.) Merr.	*Analtap*	NE		
Sapotaceae	*Palaquium globosum* H.J. Lam.	*Dallakan-babae*	PE		
Sapotaceae	*Palaquium merrillii* Dub.	*Dallakan*	PE		
Sapotaceae	*Palaquium montanum* Elmer	*Dallakan-lalake*	PE		

[a] Native species but without species name, difficult to determine if endemic or not.
NE nonendemic (it can also be found in other parts of the world), *PE* Philippine endemic; *EX* exotic, *CR* critically endangered, *VU* vulnerable, *EN* endangered.

Taguiling (2013) as the wood species used in the construction of the traditional Ifugao houses. Lambrecht (1929) mentioned *V. parviflora (amugawon)* as a strong hardwood used as a house foundation. Madulid (2011) mentioned 11 species used for house construction. Only *A. ovatus (tobak), V. parviflora (amugawon),* and *G. arborea* (gmelina) were common in Madulid's report and this study as the rest of the species were not encountered. Rondolo (2010) listed 113 species used for house construction and only 11 species were identified in this study. Alberto et al. (2011) listed 15 wood species which were also recorded in this study while the list of Taguiling (2013) included 17 wood species also identified in this study's field samples.

In the past, wood identification has been reported and presented as an independent study especially on commercially important timber and its extensive use in culturally important tangible heritage cannot be dismissed. In this study, the conservation of these culturally important trees used in the construction of the Ifugao traditional house takes equal importance to the identification of the wood. This is the first time that the appreciation and visualization of the wood and its actual tree specimens are presented together (Table 7.2).

Table 7.2 Species used for Ifugao traditional houses with corresponding photos of the transverse section of the wood (30× magnification) and fresh plant specimens

Name of species	Transverse section of the wood	Fresh plant specimen
Cannabaceae *Trema orientalis* Blume Analdong[a]	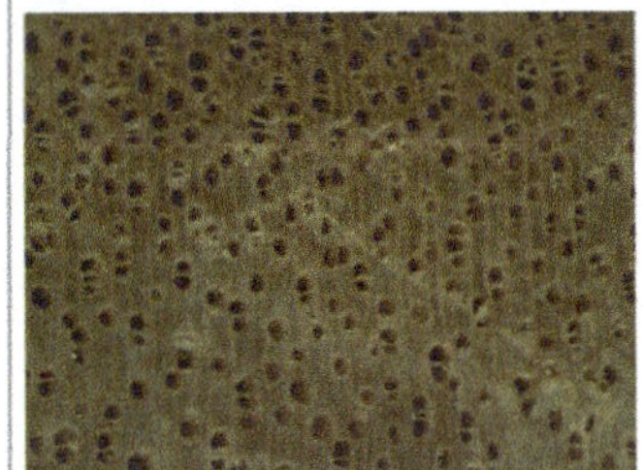	
Cunoniaceae *Weinmannia luzoniensis* S. Vidal Tabangawon[a]		
Dipterocarpaceae *Shorea* sp. Bannutan[a]		

Dipterocarpaceae *Shorea astylosa* Foxw. Bannutan[a]	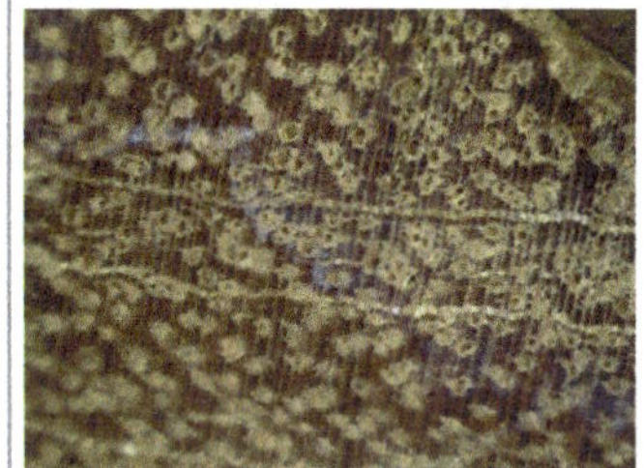	
Dipterocarpaceae *Shorea contorta* Vidal Pole[a]	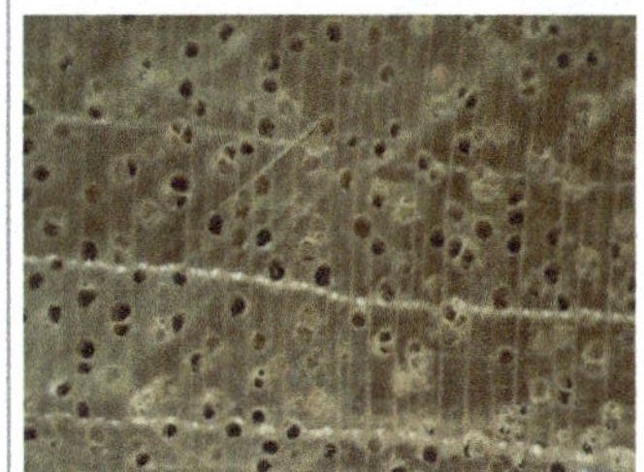	
Dipterocarpaceae *Shorea negrosensis* Foxw. Pfuchu'an[a]		

(continued)

Table 7.2 (continued)

Name of species	Transverse section of the wood	Fresh plant specimen
Ericaceae *Vaccinium cumingianum* var. *cumingianum* S. Vidal Gutmo[a]	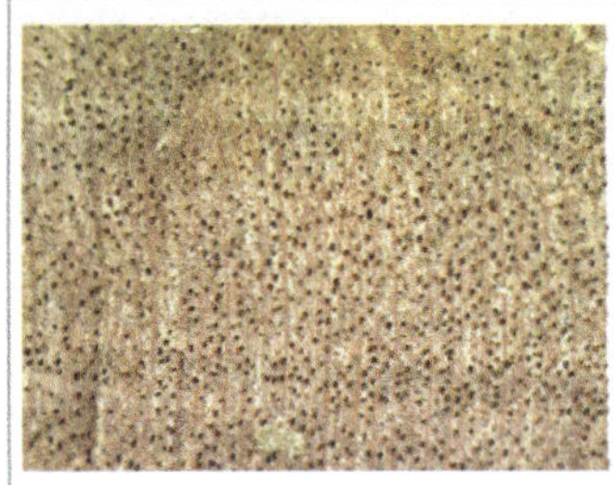	
Fabaceae *Pterocarpus indicus* Willd. Udyo[a]	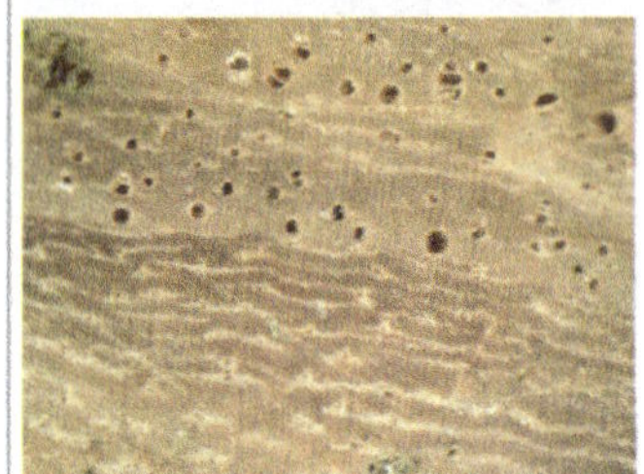	
Fagaceae *Lithocarpus woodii* (Hance) A. Camus Palayon[a]		

Hypericaceae *Cratoxylum formosum* (Jack) Dyer Aliguyon[a]	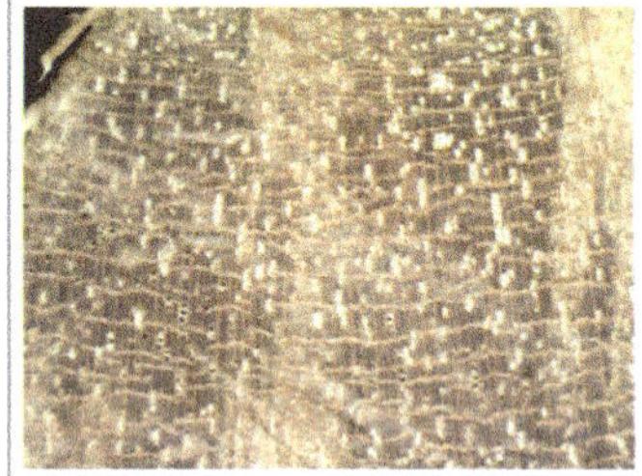	
Lamiaceae *Gmelina arborea* Roxb. Gmelina[a]	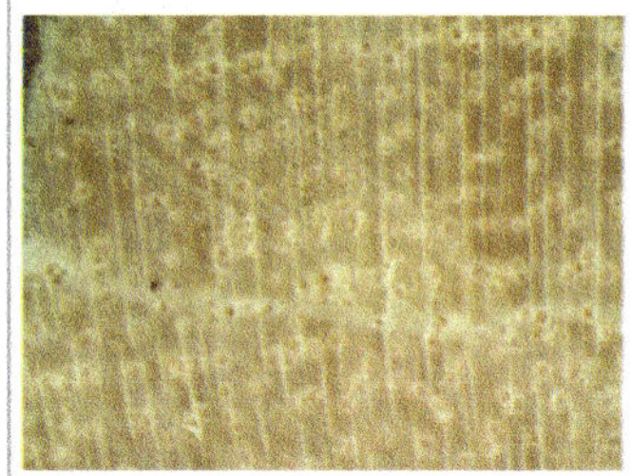	
Lamiaceae *Vitex parviflora* Juss. Amugawon[a]	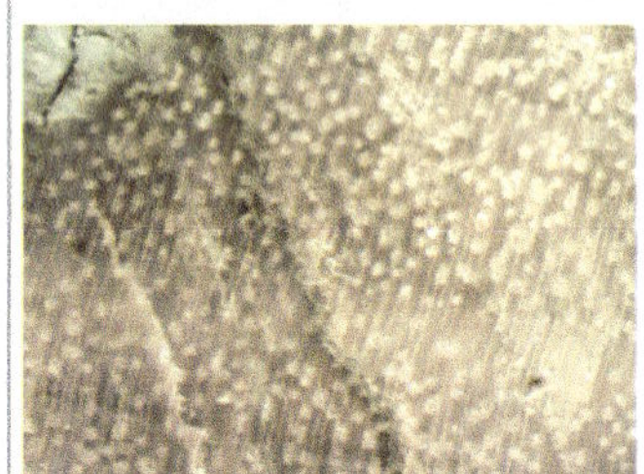	

(continued)

Table 7.2 (continued)

Name of species	Transverse section of the wood	Fresh plant specimen
Lauraceae *Cryptocarya* sp. Kawahi[a]		
Lauraceae *Litsea cordata* Hook. f. Bakkan[a]	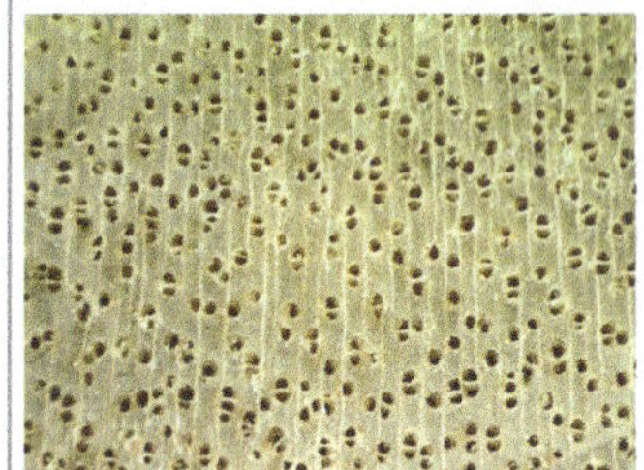	
Lauraceae *Litsea luzonica* (Blume) Fern. Vill Bini[a]	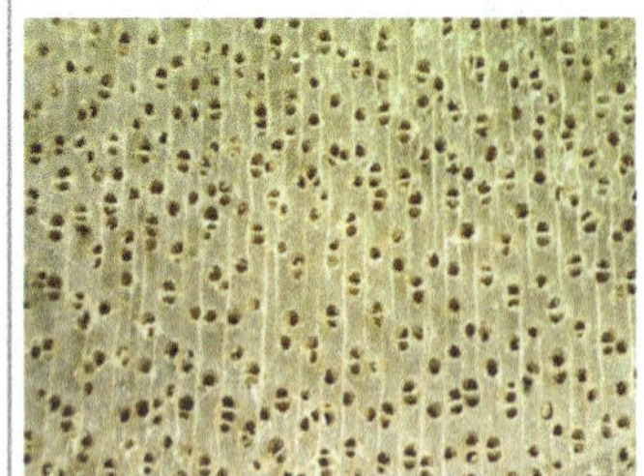	

Lauraceae *Machilus philippinensis* Merr. Dulawon[a]	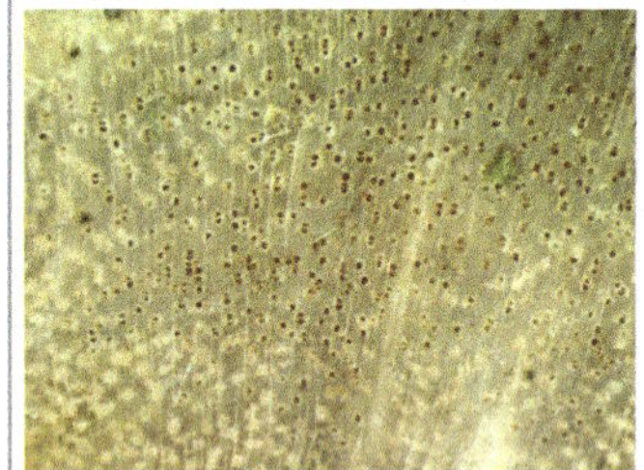	
Meliaceae *Sandoricum vidalii* Merr. Baccuwog[a]	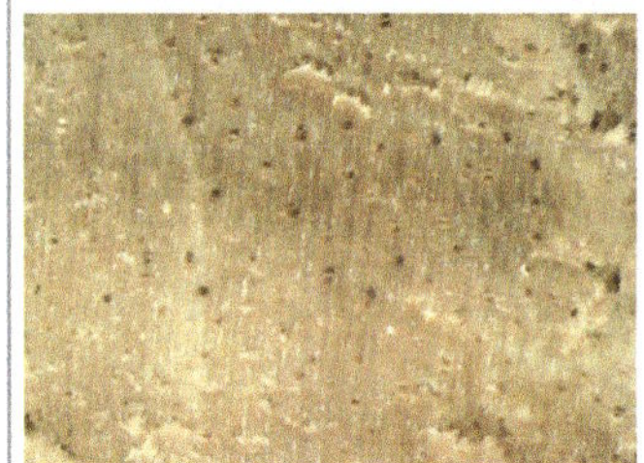	
Meliaceae *Swietenia macrophylla* King Mahogany[a]	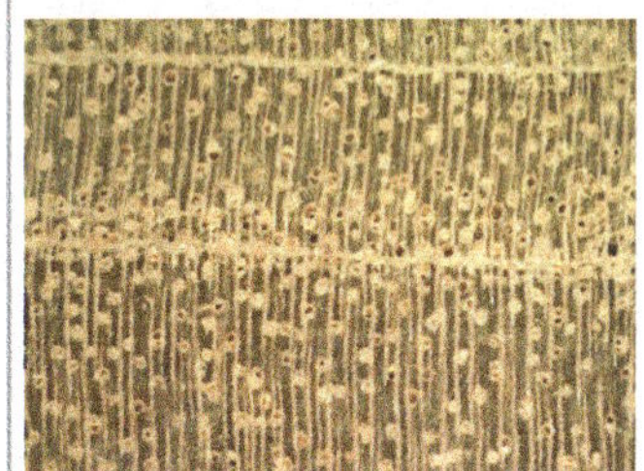	

(continued)

Table 7.2 (continued)

Name of species	Transverse section of the wood	Fresh plant specimen
Moraceae *Artocarpus ovatus* Blanco Tobak[a]	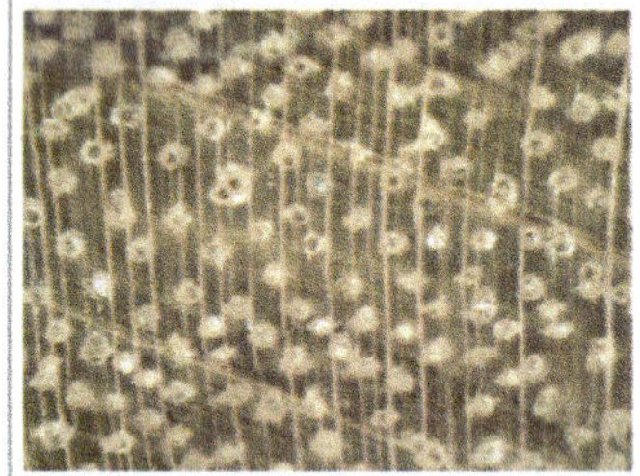	
Moraceae *Ficus benguetensis* Merr. Latbang[a]	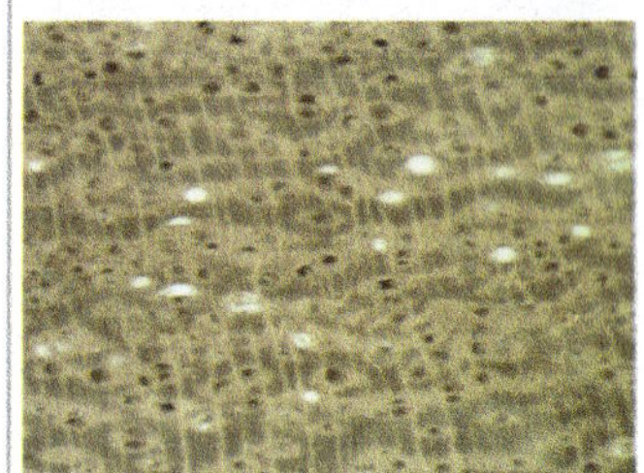	
Moraceae *Ficus variegata* Blume Ludping[a]		

Moraceae *Streblus glaber* (Merr.) Corner Galiwgiwon[a]		
Myrtaceae *Syzygium* sp. Bultik[a]		
Myrtaceae *Syzygium polycephaloides* (C. B. Rob.) Merr. Bahag[a]		

(continued)

Table 7.2 (continued)

Name of species	Transverse section of the wood	Fresh plant specimen
Phyllanthaceae *Bischofia javanica* Blume Tuwol[a]		
Pinaceae *Pinus kesiya* ssp. *langbianensis* Royle ex Gordon Halong[a]		
Podocarpaceae *Dacrydium pectinatum* de Laub. Dumliing[a]		

Rubiaceae *Neonauclea media* Elmer Tikom[a]	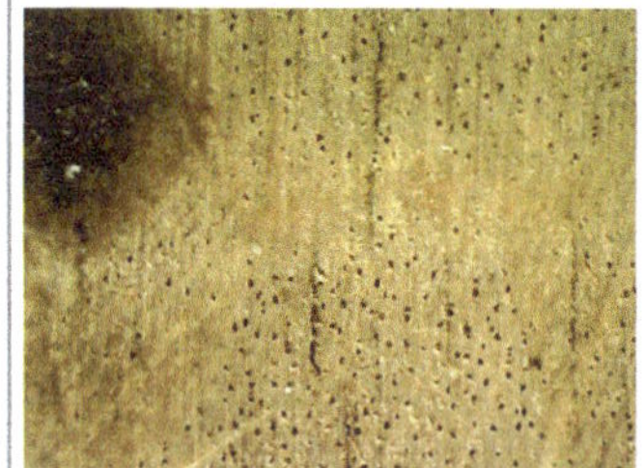	
Rubiaceae *Neonauclea reticulata* (Havil.) Merr. Analtap[a]	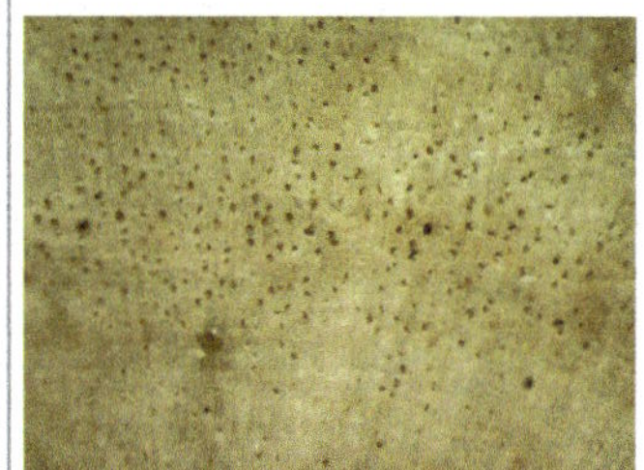	
Sapotaceae *Palaquium globosum* H.J. Lam. Dallakan-babae[a]		

(continued)

Table 7.2 (continued)

Name of species	Transverse section of the wood	Fresh plant specimen
Sapotaceae *Palaquium merrillii* Dub. Dallakan[a]	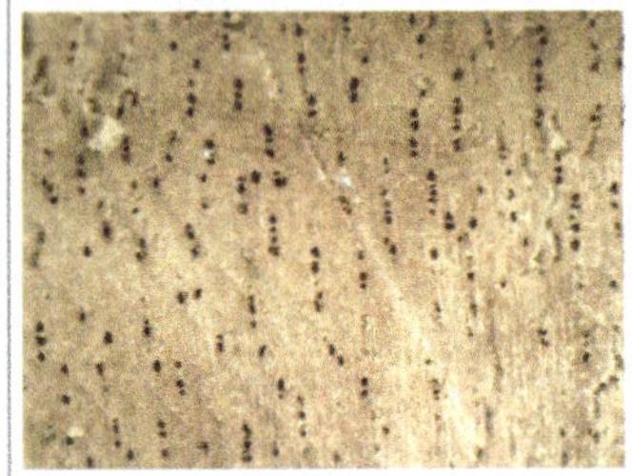	
Sapotaceae *Palaquium montanum* Elmer Dallakan-lalake[a]	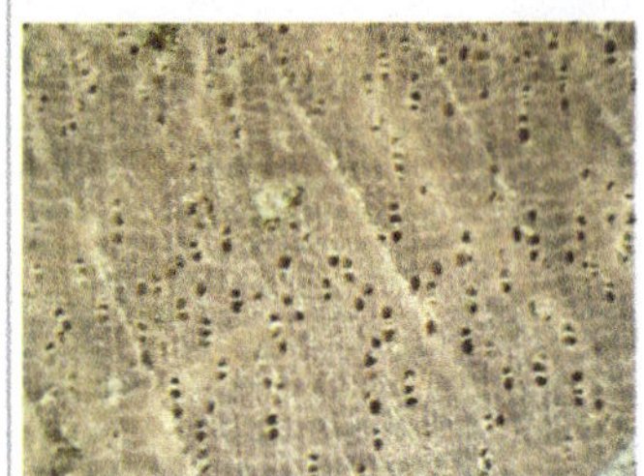	

[a] Local name of the species

7.6 Wood Species Identification of House Components of the Traditional House

Various wood species were traditionally used by the Ifugaos for a specific house part. Indigenous knowledge passed on through generations about wood species with important properties (e.g., hardness, resistance to termite attack) that are suitable for a particular house part played an important role in its choice.

Table 7.3 presents the list of wood species used for various parts of the traditional house according to their taxonomic classification. This is followed by the scientific and local names and the specific house part where it was used, its relative density, strength grouping, and uses. Lambrecht (1929) pointed out that the foundation of an Ifugao traditional house was usually composed of a strong hardwood species, particularly *Vitex parviflora*. Alberto et al. (2011) concurred with Lambrecht (1929) and added that other hardwood species such as *A. ovatus (tobak), L. woodii (palayon), P. globosum (dallakan-babae), P. merrillii (dallakan-lalake), P. montanum (dallakan-lalake), P. kesiya* ssp. *langbianensis (halong), P. indicus (udyo), Shorea* sp., *S. astylosa (bannutan), S. contorta (pole), S. negrosensis (pfuchu'an), S. glaber (galiwgiwon), Syzygium* sp. *(bultik), V. cumingianum* var. *cumingianum (gutmo), and W. luzoniensis (tabangawon)* were also used by Ifugaos for the foundation and skeletal frames of the house. This study validates this listing of 16 wood species, except for *A. philippinensis (almaciga), A. bunius (bunne), D. dao (challigan, dalligan)*, and *T. calantas (bagtinon)*, through an actual visual examination of house parts from dismantled and intact standing traditional houses. Apart from these wood species, 16 additional wood species were reported for the first time in this study suitable for Ifugao traditional house construction as previous studies only mentioned these wood species collectively as being used for traditional house construction.

As described by Lambrecht (1929), the Ifugao traditional house can be divided into three groupings: (a) the foundation, (b) the house cage, and (c) the roof structure. Indigenous knowledge about certain wood species to use for specific house parts is very important. Despite the lack of scientific knowledge on the relative density and strength grouping of wood species used for the construction of the traditional house by native house builders, their indigenous knowledge on certain wood species appears to be appropriate.

The foundation of the house requires certain wood characteristics that would ensure that the foundation posts (*tukud*) and transverse girders (*kuling*) were of high strength and relative density, resistant to termite infestation, and long lasting. Relative density or specific gravity of wood is a measure of the amount of cell wall material in wood and is a useful index of the suitability of wood for various uses. It is a reliable indicator of the wood's strength and working properties. Ten wood species belonging to eight families were recorded to be used as posts (*tukud*). Their relative densities ranged from 0.49 to 0.85 and strength grouping ranged from medium low to high. Of importance are *S. glaber* (*galiwgiwon*) with a 0.82 relative density followed by *S. astylosa* (*bannutan*) and *Shorea* sp. (*bannutan*) with 0.76–0.85 relative density and high strength, *V. parviflora* (*amugawon*) with 0.67

Table 7.3 Family name, scientific name, local name, relative density, strength grouping, and common uses of the woods that comprised the Ifugao traditional houses in Kiangan, Ifugao

Family name	Scientific name	Local name	Component			Relative density (g/cm^3)	Strength grouping	Uses
			Foundation	House cage	Roof structure			
Cannabaceae	*Trema orientalis* Blume	*Analdong*		Wallboards Floorboards		0.33[a]	Moderately low	Wood used as firewood and charcoal, fiber material for pulp and paper, panel product manufacturing such as MDF and particle boards (PROSEA 2020)
Clusiaceae	*Cratoxylum formosum* (Jack) Dyer	*Aliguyon*		Floorboard	Rafter	0.59[b]	Moderately low	Posts, beams, joists, and agricultural implements. Firewood, charcoal
Cunoniaceae	*Weinmannia luzoniensis* S. Vidal[e]	*Tabangawon*	Post transverse girders	Side floor beam Center floor beam	Roof post Lintel beam Central attic beam	0.49–0.54[a]	Medium	Wood for house construction, such as beams and poles. For fine furniture and paneling (Conklin 1967)

Dipterocarpaceae	*Shorea* sp.[e]	*Bannutan*	Post Transverse girders	Upper wall beam	Roof post Lintel beam Central attic beam	0.76–0.85[c] (range values of RD for Yakal group)	High	Wood as useful as *S. astylosa*
	Shorea astylosa Foxw[e]	*Bannutan*	Post Transverse girders	Side floor beam Center floor beam Wallboards Upper wall beam	Roof post Central attic beam	0.76[a]	High	Wood for high-grade construction, bridges and wharves, mine timber, and other installations requiring durability and strength[b]
	Shorea contorta Vidal[e]	*Pole*		Center floor beam Floorboards	Central attic beam	0.45[a]	Medium	Wood for furniture, cabinets, interior finishing, veneer and plywood, flooring, boat planking and framing, and pulp and paper making[b]
	Shorea negrosensis Foxw[e]	*Pfuchu'an*		Upper wall beam	Lintel beam Upright timber Apex frame Rafter	0.50[a]	Medium	Wood for furniture, cabinets, veneer and plywood, sash and millwork, boat planking and decking, for medium construction, door and window frames and flooring[b]

(continued)

Table 7.3 (continued)

Family name	Scientific name	Local name	Component			Relative density (g/cm^3)	Strength grouping	Uses
			Foundation	House cage	Roof structure			
Ericaceae	*Vaccinium cumingianum* var. *cumingianum* S. Vidal[e]	*Gutmo*	Post	Transverse girders Side floor beam Upper wall beam	Roof post Central attic beam	No data	No data	The fruits and leaves are eaten raw. A small tree with a very hard fine wood from which spoons and images are carved (Conklin 1967)
Fabaceae	*Pterocarpus indicus* Willd[e]	*Udyo*	Anti-rat disks	Side floor beam Center floor beam Wallboards Floorboards		0.53[a]	Moderately high	Used for high-class furniture and cabinets, decorative sliced veneer, interior wall paneling, feature flooring (including strip and parquet)[b]
Fagaceae	*Lithocarpus woodii* (Hance) A. Camus[e]	*Palayon*	Post Transverse girders			0.57–0.66[a]	Medium to moderately high	Wood for general house construction, furniture, and interior works

Lamiaceae	*Gmelina arborea* Roxb.	*Gmelina*		Beam support	Shelf	0.43[a]	Medium	The wood is suitable for general utility purposes, especially light construction and structural work, general carpentry, packaging, carvings, utility furniture and decorative veneers, with excellent woodworking properties[b]
	Vitex parviflora Juss.[d]	*Amugawon*	Post Transverse girders	Side floor beam Upper wall beam	Roof post Central attic beam Upright timber Apex frame	0.67[a]	Moderately high	The timber is used for house building (both outside and inside), ship building, and carving[1] (Orwa et al. 2009)
Lauraceae	*Cryptocarya* sp.	*Kawahi*	Anti-rat disks			0.48–0.76[b] (range values of RD of *Cryptocarya* sp.)	Medium to moderately high	Timber is used for light construction, furniture, joinery, interior finishing, paneling, light framing, flooring, moldings, boxes and cases, veneer, plywood, turnery, and brushware

(continued)

Table 7.3 (continued)

Family name	Scientific name	Local name	Component			Relative density (g/cm³)	Strength grouping	Uses
			Foundation	House cage	Roof structure			
	Litsea cordata Hook. f.	*Bakkan*		Upper wall beam	Roof post Central attic beam	0.35[b]	Moderately low	Wood for carvings, sculptures, and pattern making[b]
	Litsea luzonica (Blume) Fern. Vill	*Bini*		Center floor beam Wallboards Beam support	Roof post Upright timber Apex frame	0.34–0.67[b] (range values of RD for *Litsea* sp.)	Moderately low to medium	The wood is soft to moderately hard. It is suitable for decorative work such as interior finishing, paneling, furniture, and cabinet making. It is also suitable for plywood manufacture, while the heavier species are suitable for medium construction under cover[2]
	Machilus philippinensis Merr.	*Dulawon*		Center floor beam Wallboard Floorboard Upper wall beam Beam support	Roof post Central attic beam King post Apex frame	0.46	Medium	Carving, furniture, light construction, joinery, veneer and plywood, musical instruments (Wood Technical Fact Sheet, n.d.)

Meliaceae	*Sandoricum vidalii* Merr.	*Baccuwog*	Anti-rat disks			0.45[b]	Medium	For wood carvings, household implements, paneling, planking, and interior construction[f]
	Swietenia macrophylla king	Mahogany		Door shutter	Rafter	0.54[a]	Medium	For high-class furniture and cabinet work, paneling, doors, and decorative borders (PROSEA 2017)
Moraceae	*Artocarpus ovatus* Blanco[e]	*Tobak*	Post Transverse girders	Side floor beam		0.53[a]	Moderately low	House posts, telegraph, poles, piles, railroad ties, bridge timber[b]
	Ficus benguetensis Merr.	*Ludping*		Center floor beam Beam support		0.38[b] (RD of its similar wood *F. fistulosa*)	Moderately low	The wood is of low quality, but is used for items such as temporary constructions, moldings, interior work, cladding, drawers, laundry tubs, fruit crates, and floats; source of fuel

(continued)

Table 7.3 (continued)

Family name	Scientific name	Local name	Component			Relative density (g/cm^3)	Strength grouping	Uses
			Foundation	House cage	Roof structure			
	Ficus variegata Blume	*Latbang*		Wallboard Door shutter		0.29[a]	Low	The wood is of low quality, but is used for temporary construction, moldings, interior work, cladding, drawers, fruit crates, floats, knife sheaths, and fuel (Fern 2014a, b, c, d)
	Streblus glaber Merr. Corner[e]	*Galiwgiwon*	Post			0.82[b]	High	The fruits are edible. The wood is the hardest and heaviest known to the Ifugaos. Wood is used for pipes and carving spoons (Conklin 1967)
Myrtaceae	*Syzygium* sp.[e]	*Bultik*		Side floor beam	Roof post Lintel beam Rafter	0.74[b]	High	Wood for house construction, such as beams, posts, flooring, and windowsills. For cabinet work and furniture, tool handles,

								washbowls, ties, pilings, and implements (PROSEA 2016)
	Syzygium polycephaloides (C. B. Rob.) Merr.	*Bahag*			Roof post Lintel beam Rafter	0.51–0.75[b]	Medium to moderately high	The wood is used for heavy and light construction, posts and poles, windowsills, furniture, flooring, telegraph poles, ships, bridges, railway sleepers, bottom boards of railway carriages, implements, fiber board, and veneer and plywood
Phyllanthaceae	*Bischofia javanica* Blume	*Tuwol*	Post Transverse girders	Center floor beam Upper wall beam Beam support	Lintel beam Central attic beam Rafter	0.56[a]	Moderately low	Wood for charcoal making, general construction, posts of light material houses, agricultural such as bolo and knife handles; veneer and plywood, pulp and paper[b]

(continued)

Table 7.3 (continued)

Family name	Scientific name	Local name	Component			Relative density (g/cm^3)	Strength grouping	Uses
			Foundation	House cage	Roof structure			
Pinaceae	*Pinus kesiya* ssp. *langbianensis* Royle ex Gordon[e]	*Halong*		Side floor beam		0.53[a]	Medium	Timber for general purpose, particle board and pulp (Orwa et al. 2009)
Podocarpaceae	*Dacrydium pectinatum* de Laub.	*Dumliing*	Post Transverse girders			0.50[a]	Medium	Highly valued for construction and furniture[2]
Rubiaceae	*Neonauclea media* Elmer	*Tikom*	Anti-rat disks	Center beam		0.67[b] (relative density of its similar wood *N. glabra*)	Moderately high	The wood is resistant to termites. It is used for inlay work. The wood is used for fuel. Wood also used for flooring, windowsills, furniture and cabinet work, structural timber intended for high-grade construction, poles, posts, inlaying, turner[b]
	Neonauclea reticulata (Havil.) Merr.	*Analtap*	Transverse girders	Side floor beam		0.67[b] (relative density of its similar wood *N. glabra*)	Moderately high	The wood is used for general construction, bridge building, windowsills,

								flooring, boat building, joinery, furniture and cabinet work, posts, piles, poles, implements and packaging, for the production of core veneer, pulp and paper, and charcoal[b]
Sapotaceae	*Palaquium globosum* H.J. Lam[e]	*Dallakan-babae*		Wallboards Beam support		0.57[b] (RD of its synonym *P. gutta* (Hook.) Burck)	Moderately low to medium	Wood is used for general construction, walling, window and door frames, flooring[b]
	Palaquium merrillii Dub.[e]	*Dallakan*		Wallboards Floorboards Upper wall beam	Central attic beam Upright timber Apex frame	0.42[a]	Moderately low	Wood for house building not in contact with the ground, columns, beams, rafters, canoes, ship planking, paneling, flooring, cigar boxes[b]

(continued)

Table 7.3 (continued)

Family name	Scientific name	Local name	Component			Relative density (g/cm^3)	Strength grouping	Uses
			Foundation	House cage	Roof structure			
	Palaquium montanum Elmer[e]	*Dallakan-lalake*		Center floor beam Wallboards Floorboards	Lintel beam Rafter	0.56–0.61[b] (range values of RD of *Palaquium* sp.)	Moderately low to medium	Wood is in general suitable for house construction, columns, beams, rafters, decorative doors and veneers, and paneling[g]

[a] Alipon and Bondad (2008).
[b] Zanne et al. (2009).
[c] Lambrecht (1929).
[d] Alberto et al. (2011).
[e] Escobin et al. (2015).
[f] Fern (2014a, b, c, d).
[g] Meniado and Francisco (1975)

relative density and moderately high strength, *Lithocarpus woodii (palayon)* with a 0.57–0.66 relative density and medium-moderately high strength, *Bischofia javanica (tuwol)* with a 0.56 relative density and moderately low strength, *A. ovatus* (*tobak*) with a 0.53 relative density and moderately low strength, *Weinmannia luzoniensis (tabangawon)* with a relative density of 0.49–0.54 and medium strength, and *D. pectinatum (dumliing)* with a 0.5 relative density and medium strength. Only *Vaccinium cumingianum* var. *cumingianum (gutmo)* lacks relative density and strength grouping information. Of these, only S. *glaber* (*galiwgiwon*) was used for posts while the other species were used for other parts of the traditional house. For the transverse girders (*kuling*), ten wood species belonging to nine families were also used with the inclusion of *N. reticulata* (*analtap*) of the family Rubiaceae. In addition, *D. pectinatum (dumliing)* was the only gymnosperm used for foundation parts of the house.

Another unique feature of the house foundation is the anti-rat disks (*halipan*). In as much as the traditional house can be used as residence or granary, it is important that the harvested rice grains which are usually stored on the third level of the house be protected from rat infestation. As such, the anti-rat disks are an important addition to the house foundation. Four wood species belonging to four families are used as anti-rat disks. These are *P. indicus* (*udyo*), *Cryptocarya* sp. (*kawahi*), *S. vidalii* (*baccuwog*), and *N. media* (*tikom*). *Cryptocarya* sp. (*kawahi*) and *S. vidalii* (*baccuwog*) were used only as anti-rat disks. These were not recorded to have been used in other parts of the traditional house.

The house cage would consist of the side floor beams (*mundilig*) and the center floor beam (*gawaan*), the floor (*dotal*) and wall (*gaob*) boards, upper wall beams (*huklob*), the beam support (*pamadingan*), and the door shutter (*onob*). Nine species belonging to nine families were used as side floor beams and ten species belonging to eight families were used as center beams. *W. luzoniensis (tabangawon)*, *S. astylosa (bannutan), V. cumingianum var cumingianum (gutmo), P. indicus (udyo), V. parviflora (amugawon), A. ovatus (tobak), Syzygium* sp. *(bultik), P. kesiya (halong),* and *N. reticulata (analtap)* were used for side floor beams and *W. luzoniensis (tabangawon), S. astylosa (bannutan), S. contorta (pole), P. indicus (udyo), L. luzonica (bini), M. philippinensis (dulawon), F. benguetensis (latbang), B. javanica (tuwol), N. media (tikom),* and *P. montanum (dallakan-lalake)* were used as center floor beam. Of these, *P. kesiya* ssp. *langbianensis (halong)* of the family Pinaceae was the only gymnosperm used as side floor beams. Two exotics, namely *Gmelina arborea* (gmelina) and *S. macrophylla* (mahogany), were used for beam support and door shutter, respectively, while native trees were used for the rest of the parts of the house cage.

The wallboards and floorboards rest on the side and center floor beams with nine species belonging to six families used for wallboards and seven species belonging to six families used as floorboards. *T. orientalis (analdong), P. indicus (udyo), M. philippinensis (dulawon), P. merrillii (dallakan),* and *P. montanum (dallakan-lalake)* were used for both wall- and floorboards. The upper wall beam rests on top of the wallboards and nine wood species belonging to nine families were used. Nine wood species from seven families were used for the beam support. *S. negrosensis*

(Pfuchu'an) of the family Dipterocarpaceae was used for both upper wall beam and beam support. Moreover, the exotic species *G. arborea (gmelina)* was used for beam support.

The roof structure would consist of the roof post (*bagad*), lintel beam (*wanan*), central attic beam (*pumpitolan*), upright timber (*taknang*), apex frame (ambubulan), rafter (*bughol*), and shelf (*patie*). Overall, ten wood species belonging to six families were used for the roof post, seven species belonging to five families for the lintel beam, and ten species from seven families for the central attic beam. *W. luzoniensis (tabangawon)* and *Shorea* sp. *(bannutan)* were the two wood species commonly used for these roof structures. A total of five wood species from four families were used for the upright timber, five species from four families for the apex frame, and seven species from six families for the rafters. *S. negrosensis (Pfuchu'an)* was used for these roof structures while *S. macrophylla* (mahogany) was the only exotic species used for the rafters.

7.7 Conservation of Trees Used for the Traditional Houses in Ifugao

Conservation of wood species traditionally used by the house builders in Ifugao is crucial in preserving and sustaining their culture of traditional house building. When it comes to tree conservation, the people of Ifugao have developed a unique indigenous forest management system of their private woodlots (*muyong*) and communal forests (*alah*). The *muyong* system is a combination of different indigenous management practices that is deeply embedded in the culture of the Ifugao people. It includes traditional knowledge and practices on forest conservation, watershed rehabilitation, assisted natural regeneration, soil and water conservation, and even agricultural and agroforestry farming systems (Butic and Ngidlo 2003). But just like the cultural degradation of the traditional house building in Ifugao, the *muyong* system has suffered significant deprivation and modifications. For instance, in their enrichment planting (*tanoman*) and forest rehabilitation practices, *muyong* owners are now heavily dependent on the use of widely available exotic species such as *S. macrophylla* (mahogany) and *G. arborea* (gmelina) (Camacho et al. 2016). The proliferation of these exotic species resulted in significant decline in the abundance of native trees in Ifugao, including those wood species used in the traditional house construction.

Aside from stand modifications, the forests in Ifugao also experienced decline primarily due to conversion to agriculture or agroforestry farm (GFW 2021). Data of the global forest watch showed that Ifugao province lost 629,000 ha of tree cover from 2001 to 2020. The primary forest declined by 806 ha, making up 13% of its total tree cover loss in the same time period. As observed by traditional house builders in Hungduan, there has been a noticeable reduction in the forest areas and available mature trees where they can source their wood materials. Most of the time, they need to go to other municipalities and request permission from the muyong owners to gather the target wood species.

With the continuous decline of the forests in Ifugao and the serious endangerment of trees used in traditional houses, implementation of a conservation program for these wood species is extremely urgent and necessary. The identification of the correct name of these wood species from this study is a vital step towards this conservation program. The traditional house builders can serve as indigenous knowledge holders as they can recognize and identify the mostly native wood species suitable for traditional house construction. Native tree nursery establishment and mass propagation of the species have been initiated and must be sustained at the government, school, and community levels. The local government and other stakeholders must find ways to bring these trees back to abundance to provide a sustainable source of wood materials to preserve and maintain the culture of traditional house building. These wood species should be the province's flagship species for conservation, and should be mainstreamed in different planting activities be it in reforestation activities or urban landscaping. More importantly, people of Ifugao should be made aware of the important contributions of these wood species to the lives and culture of the Ifugao's early ancestors.

7.8 Summary and Conclusion

The traditional wood houses of the indigenous peoples of Ifugao are part of the Ifugao Rice Terraces of the Philippine Cordilleras, a UNESCO World Heritage Site. Dismantled and intact traditional houses in Kiangan were examined in situ and wood species used in building the house were identified. Thirty-two wood species belonging to 22 genera in 17 families were identified. Dipterocarpaceae, Moraceae, and Lauraceae (with four species each), followed by Sapotaceae with three species, were the dominant tree families found. Except for *Gmelina arborea* (gmelina) and *Swietenia macrophylla* (mahogany) which were exotics, the rest (30) of the wood species used in the traditional houses were native to the country. Most of the wood species identified were hardwoods (94%) and were preferred by Ifugao traditional house builders. Some wood species were traditionally used by the Ifugao for a specific house part.

Indigenous knowledge regarding wood species that have important properties (e.g., relative density, strength grouping, resistance to termite infestation) and their suitability for particular house parts appears to have been passed on through generations of traditional house builders through oral tradition. Ten wood species belonging to eight families were recorded to be used as posts (*tukud*). These were *S. glaber* (*galiwgiwon*), *S. astylosa* (*bannutan*), *Shorea* sp. (*bannutan*), *V. parviflora* (*amugawon*), *L. woodii (palayon), B. javanica (tuwol), A. ovatus* (*tobak*), *W. luzoniensis (tabangawon), V. cumingianum* var. *cumingianum (gutmo),* and *D. pectinatum (dumliing).* Of these, only S. *glaber* (*galiwgiwon*) was used for posts while the other species were used for other parts of the traditional house. For the transverse girders (*kuling*), ten wood species belonging to nine families were also used with the inclusion of *N. reticulata* (*analtap*) of the family Rubiaceae. In addition, *D. pectinatum (dumliing)* was the only gymnosperm used for foundation parts of the house.

Another unique feature of the house foundation was the anti-rat disks (*halipan*). Four wood species, namely *P. indicus* (*udyo*), *Cryptocarya* (*kawahi*), *S. vidalii* (*baccuwog*), and *N. media* (*tikom*), were used as anti-rat disks. *Cryptocarya* sp. (*kawahi*) and *S. vidalii* (*baccuwog*) were exclusively used only as anti-rat disks.

The house cage used 26 wood species belonging to 15 families. Except for the two exotics, namely *G. arborea* (gmelina) and *S. macrophylla* (mahogany), used for beam support and door shutter, respectively, native trees were used for the rest of the house cage parts of the traditional house. Moreover, *P. kesiya* var. *langbianensis (halong)* of the family Pinaceae was the only gymnosperm used as side floor beams.

Eighteen wood species belonging to ten families formed the roof structure. *W. luzoniensis (tabangawon)* and *Shorea* sp. *(bannutan)* were the two wood species commonly used for these roof posts and beams. As for the upright timber, apex frame, and rafters, *S. negrosensis (Pfuchu'an)* was used while *S. macrophylla* (mahogany) was the only exotic species used for the rafters.

This study is groundbreaking in its in situ examination of actual transverse sections of house parts from dismantled and standing Ifugao traditional houses located in Kiangan. Transverse wood sections were examined, photographed, and identified using appropriate wood identification references and the FPRDI DOST Xylarium. A team of traditional house builders, plant taxonomists, ecologists, and wood specialists examined trees in the surrounding forests in Ifugao to confirm scientific names and local names of the wood specimens examined. Hence, outputs of this project strongly complement indigenous knowledge on traditional house construction by linking the wood species with important attributes such as relative density, resistance to termites, strength, and durability. Better knowledge about wood species and their uses for traditional houses can provide useful information for the younger generation of indigenous builders and craftsmen in the restoration and preservation of the still standing traditional houses in their respective communities.

With the decline in the availability of the premium wood species from the surrounding woodlots and communal forests, traditional house builders can serve as indigenous knowledge holders of wood species important in traditional house construction. If combined with the practice of muyong and ala-a system of forest management by muyong private and community owners, civic organizations, national, provincial, and local government units, and local communities in native tree establishment and mass propagation all can work together towards environmental stewardship of the rice terrace landscape and conservation of biodiversity. Assisted natural regeneration, protection, and cultivation of vital wood species and lesser known species, including softwood species and exotics, can also provide options for traditional house construction. The findings of this study thus hope to make a modest contribution to enhancing local indigenous knowledge in support of new construction and restoration efforts of the traditional house, which is an enduring feature of the Ifugao heritage and landscape.

Acknowledgements The authors wish to thank the World Wood Day Foundation (WWDF) and the International Wood Culture Society (IWCS) for providing financial support for the conduct of the study, Prof. Eulalie D. Dulnuan and Dr. Marissa P. Bulong for the assistance and collaboration, and Mr. Patrick Polpog for his invaluable indigenous knowledge on traditional house construction. In addition, the authors would also like to thank the Forest Products Research and Development Institute-Department of Science and Technology ((FPRDI-DOST) for providing access to their xylarium and assistance in wood species identification. The authors are also grateful to Forester Gerald T. Eduarte for providing the map of the study site.

References

Agnoletti M, Rotherham ID (2015) Landscape and biocultural diversity. Biodivers Conserv 24: 3155–3165

Alberto EG, Alcayna EA, Babang MB et al (2011) Ifugao indigenous knowledge (IK) workbook. Ifugao State University, Ifugao

Alipon MA, Bondad EO (2008) Strength grouping of Philippines timbers for various uses. Forest Products Research and Development Institute, Department of Science and Technology, Los Banos

Anania M et al (2013) Identification of plant materials used in the general construction of traditional houses in Serongko Village, Finschhafen District, Morobe Province, Papua New Guinea. In Paper presented at 7th Huon seminar, Papua New Guinea University of Technology, Lae, Papua New Guinea, 13–14 Nov 2013

Butic M, Ngidlo R (2003) Muyong forest of Ifugao: assisted natural regeneration in traditional forest management. In: Dugan PC et al (eds) Advancing assisted natural regeneration (ANR) in Asia and the Pacific. RAP, Bangkok

Camacho LD, Combalicer MS, Yeo-chang Y et al (2012) Traditional forest conservation knowledge/technologies in the Cordillera, Northern Philippines. Forest Policy Econ 22:3–8. https://doi.org/10.1016/j.forpol.2010.06.001

Camacho LD, Gevaña DT, Carandang AP et al (2016) Indigenous knowledge and practices for the sustainable management of Ifugao forests in Cordillera, Philippines. Int J Biodivers Sci Ecosyst Serv Managr 12(1-2):5–13. https://doi.org/10.1080/21513732.2015.1124453

Conklin HC (1967) Ifugao ethnobotany 1905-1965: the 1911 Beyer-Merrill report in perspective. Econ Bot 21(3):243–272

Conklin HC (1980) Ethnographic atlas of Ifugao: a study of environment, culture, and society in northern Luzon. Yale University Press, New Haven and London

DENR Administrative Order (DAO) 11 (2017) Updated National List of threatened Philippine plants and their categories. Department of Environment and Natural Resources, Republic of the Philippines, Quezon City, Manila. http://www.bmb.gov.ph/index.php/e-library/laws-and-policies/denr-administrative-orders/dao-2017-2019?download=197:denradministrative-order-2017-11. Accessed 14 Jan 2021

Dulnuan E, Bulong M (2020) Documentation and mapping of traditional native houses in the Municipality of Kiangan, Ifugao Terminal Report. National Commission for Culture and the Arts, Philippines. Unpublished report.

Ellis EC, Gauthier G, Goldewijk KK et al (2021) People have shaped most of terrestrial nature for at least 12,000 years. PNAS 118:17. https://doi.org/10.1073/pnas.2023483118

Escobin RP, America WM, Pitargue FC Jr et al (2015) Revised wood identification handbook for Philippine timbers. DOST-FPRDI, Laguna

Fern K (2014a) Useful tropical plants: *Dacrydium novoguineense*. http://tropical.theferns.info/viewtropical.php?id=Dacrydium+novoguineense. Accessed 4 Apr 2020

Fern K (2014b) Useful tropical plants: *Ficus variegata*. http://tropical.theferns.info/viewtropical.php?id=Ficus+variegata. Accessed 4 Apr 2020

Fern K (2014c) Useful tropical plants: *Litsea luzonica*. http://tropical.theferns.info/viewtropical.php?id=Litsea+luzonica. Accessed 4 Apr 2020

Fern K (2014d) Useful tropical plants: *Pterocarpus indicus*. http://tropical.theferns.info/viewtropical.php?id=Pterocarpus+indicus. Accessed 4 Apr 2020

Food and Agriculture Organization (2014) In: Borelli S, Conigliaro M (eds) Assessing and promoting trees outside forests (TOF) in Asian rice production landscapes. FAO, Rome

Gawongna JM (2016) A survey of the state of Mayoyao native houses. Upland Farm 24(1):20–28

Global Forest Watch (2021). https://www.globalforestwatch.org/map/country/PHL/33/?mainMap. Accessed 17 Apr 17 2021

IUCN (2021) The IUCN Red List of threatened species. https://www.iucnredlist.org, 2021–3.

Kretz E (2009) Traditional housing in northern Laos: Wood preferences and impact on the forest biodiversity. https://stud.epsilon.slu.se/12333/1/kretz_e_171030.pdf. Accessed 17 Feb 2021

Lambrecht F (1929) Ifugao villages and houses. Publications of the Catholic Anthropological Conference. Vol. I, No. 3, Washington, DC

Lambrecht F (1962) The religion of the Ifugao. Proceedings of the fifth annual Baguio religious acculturation conference (January-April). Philipp Sociol Rev 10:33–40

Macapagal RA, Raoul AB III (2017) The Batad Kadangyan ethnic lodges project: community-based indigenous tourism in a UNESCO World Heritage rice terrace cultural landscape. In Proceedings of the II international conference on best practices in world heritage: people and communities

Madulid DA (2011) Philippine native plants: *Leptospermum flavescens*. Philstar Global. https://www.philstar.com/othersections/gardening/2011/11/05/744338/leptospermum-flavescens. Accessed 4 Apr 4 2020

Meniado JA, Francisco NT (1975) Wood identification handbook for Philippine timbers, vol 1. Department of Science and Technology-Forest Products Research and Development Institute/CARAMBAS Government Printing Office, Manila

Orwa C, Mutua A, Kindt R et al (2009) Agroforestree Database: a tree reference and selection guide version 4.0. World Agroforestry Centre, Kenya. http://worldagroforestry.org/output/agroforestree-database, 4.0, http://worldagroforestry.org/output/agroforestree-database

PROSEA (2016) Plant resources of South-East Asia: *Syzygium*. https://uses.plantnetproject.org/en/Syzygium_(PROSEA_Timbers). Accessed 4 Apr 2020

PROSEA (2017) Plant resources of South-East Asia: *Swietenia*. https://uses.plantnet-project.org/en/Swietenia_(PROSEA). Accessed 4 Apr 2020

PROSEA (2020) Plant resources of South-East Asia. https://uses.plantnet-project.org. Accessed 10 Mar 2020

Rabena MA, Macandog DM, Cuevas VC et al (2015) A vegetation inventory of a traditional secondary forest (muyong) in Kinakin, Banaue, Ifugao, northern Luzon, Philippines. Philippine J Syst Biol 9:10–32

Rondolo MT (2010) The changing Ifugao woodlots; Implications for indigenous Ifugao plant knowledge and biodiversity. PhD dissertation, Australian National University

Taguiling NK (2013) Macrofloral biodiversity conservation in Ifugao. Eur Sci J 4:1

UNESCO (2019) The Rice Terraces of the Philippine Cordilleras. https://whc.unesco.org/en/list/722/. Accessed 19 Feb 2019

Wood Technical Fact Sheet (n.d.). https://www.fpl.fs.fed.us/documnts/TechSheets/Chudnoff/SEAsian_Oceanic/htmlDocs_SEAsian/Machilusspp.html. Accessed 13 Mar 2020

Zanne AE, Lopez-Gonzales G, Coomes DA et al (2009) Global wood density database. Dryad Identifier: http://hdl.handle.net/10255/dryad.235

8 Pteridophyte Diversity in the Puerto Princesa Subterranean River National Park (PPSRNP), Palawan Island, Philippines

Marjorie D. Delos Angeles and Augusto A. Asis, Jr

Abstract

Checklists are vital information towards identifying conservation priorities over an area. This study provides an enumeration of fern species found in the Puerto Princesa Subterranean River National Park (PPSRNP), Palawan Island, Philippines. A transect walk was conducted to document all fern species encountered during the study. A total of 39 fern species belonging to 27 genera and from 14 families were documented. This study also presents two new local records of *Tectaria aspidioides* Copel and *Tectaria taccifolia* (Fée) M.G.Price. The fern species *Adiantum stenochlamys* Baker is listed as Vulnerable whereas *Christensenia aesculifolia* (Blume) Maxon and *Tectaria psomiocarpa* S.Y. Dong are listed as Endangered based on DAO 2017-11. Continuous monitoring and inventory of floral resources in the PPSRNP will serve as reference in providing future strategic plans such as establishing areas for conservation as well as delineating ecotourism sites for the enhancement of plant and ecosystem value.

Keywords

Endemic species · Fern flora · Sabang

M. D. Delos Angeles (✉)
University of the Philippines Los Baños, Los Baños, Laguna, Philippines
e-mail: mddelosangeles1@up.edu; mddelosangeles1@up.edu.ph

A. A. Asis, Jr
Puerto Princesa Subterranean River National Park Management Office, Puerto Princesa City, Palawan, Philippines

S. Ramamoorthy et al. (eds.), *Plant Genetic Resources, Inventory, Collection and Conservation*, https://doi.org/10.1007/978-981-16-7699-4_8

8.1 Introduction

Ferns are one of the organisms which are least studied, least understood, and least utilized. Studies have found that these spore-bearing plants are usually utilized as ornamentals (Zamora and Co 1986; Buot Jr 1999; Banaticla and Buot 2006) and in handicraft manufacture (Zamora and Co 1986), though many of them have been found to have medicinal properties (Amoroso 1987; Zamora and Co 1986) and some are even edible (Amoroso 1990; Zamora and Co 1986). The study conducted by Barcelona et al. (2006) provided a list of ferns and fern allies of karst forests of Bohol Island, Philippines, and it is the only recorded study conducted on ferns found in forest over limestone. No other study has been recorded to document ferns in stressful environments such as forest over limestone ecosystems. Therefore, there is a need to survey ferns especially in karst landscapes.

In the Philippines, the Puerto Princesa Subterranean River National Park (PPSRNP) is one of the world's most impressive cave systems resulting in a spectacular karst landscape. It is home to an outstanding range of unique flora and fauna. There is a need for biological exploration in the Palawan islands of the Philippines. The assessment of floral communities in a forest over limestone ecosystem here in the Philippines can generate useful information such as a floral inventory, plant community structure, as well as the understanding of plant adaptations. Ecologically, the understory plant assemblages have a fundamental role in diversity, structure, and functional aspects of tropical forests. Varying responses to light level, nutrient availability, and temperature result in different patterns of diversity than tree species (Lasco and Pulhin 2000). Studies on forest over limestone ecosystems here in the Philippines have been conducted; however, there is dearth of data on studies regarding ecological diversity of Pteridophytes in the PPSNRP.

The study aims to document and create a checklist of ferns in the Puerto Princesa Subterranean River National Park, Palawan, Philippines. This inventory will serve as a baseline data for ecotourism. Generally, this study aims to document fern species found in the PPSRNP. The chapter also determines the distribution and conservation status of the identified species in the study site. The specific objectives are:

1. To identify fern species across varying ecological conditions found in the Puerto Princesa Subterranean River National Park, Palawan, Philippines
2. To evaluate/assess fern taxa from varying ecosystems found in the Puerto Princesa Subterranean River National Park, Palawan, Philippines

There is an urgency in generating floristic inventories of islands due to their biological importance and conservation priorities (Campbell and Hammond, 1989). According to Francisco-Ortega et al. (2010), the Philippines is one of the major insular systems in the tropics which do not have a comprehensive flora or even checklists of endemic species. The generation of checklists will be instrumental in the development of more extensive floristic studies in a given area. This study provided a checklist with substantial information regarding ferns present in the Puerto Princesa Subterranean River National Park, Palawan. Hence, the monitoring and inventory of floral resources in the PPSRNP may serve as a reference in providing future strategic plans such as identification of conservation actions and priorities, establishing areas for ecotourism, for the promotion of plant and ecosystem conservation. Furthermore, this checklist will increase conservation awareness, provide educational and research materials for researchers and educators alike, and may aid policy makers towards crafting sustainable use plans and conservation management of the biological resources.

8.2 Methodology

8.2.1 Study Area

The study was conducted in the Puerto Princesa Subterranean River National Park, Palawan Island, Philippines (Fig. 8.1).

8.2.2 Fern Flora Assessment

In documenting fern floral species in the area, a transect walk was conducted. Photographs were taken and fern samples were collected. The collected samples were labeled with field tags before pressed using newspapers. The samples were enclosed in polyethylene plastic bags and soaked with 50% denatured alcohol before drying. Each dried plant sample was brought to the Plant Systematics Laboratory (PSL), Institute of Biological Sciences, University of the Philippines Los Baños, for mounting and identification.

Specimens were identified utilizing original descriptions, taxonomic revisions, and botanical keys Copeland (1958–1960). Specimens were also compared with the collections of Plant Biology Division Herbarium (PBDH), Institute of Biological Sciences, University of the Philippines Los Baños, and from available online resources such as "Co's Digital Flora of the Philippines" (Pelser et al. 2011). Plant samples were also identified with the assistance of experts. Voucher specimens along with their pertinent information were then processed and deposited in the PBDH. The conservation status of each plant species was determined using the Updated National List of Threatened Philippine Plants and Their Categories (DAO No. 2017-11) and the International Union for Conservation of Nature and Natural Resources (IUCN) Red List of Threatened Species (2020).

8.3 Results and Discussion

A total of 39 fern species belonging to 27 genera and from 14 families were recorded from the Puerto Princesa Subterranean River National Park, Palawan Island, Philippines. The family with the most fern species representation was *Pteridaceae* (8 spp.) followed by *Tectariaceae* (5 spp.), then *Hymenophyllaceae* (4 spp.) and

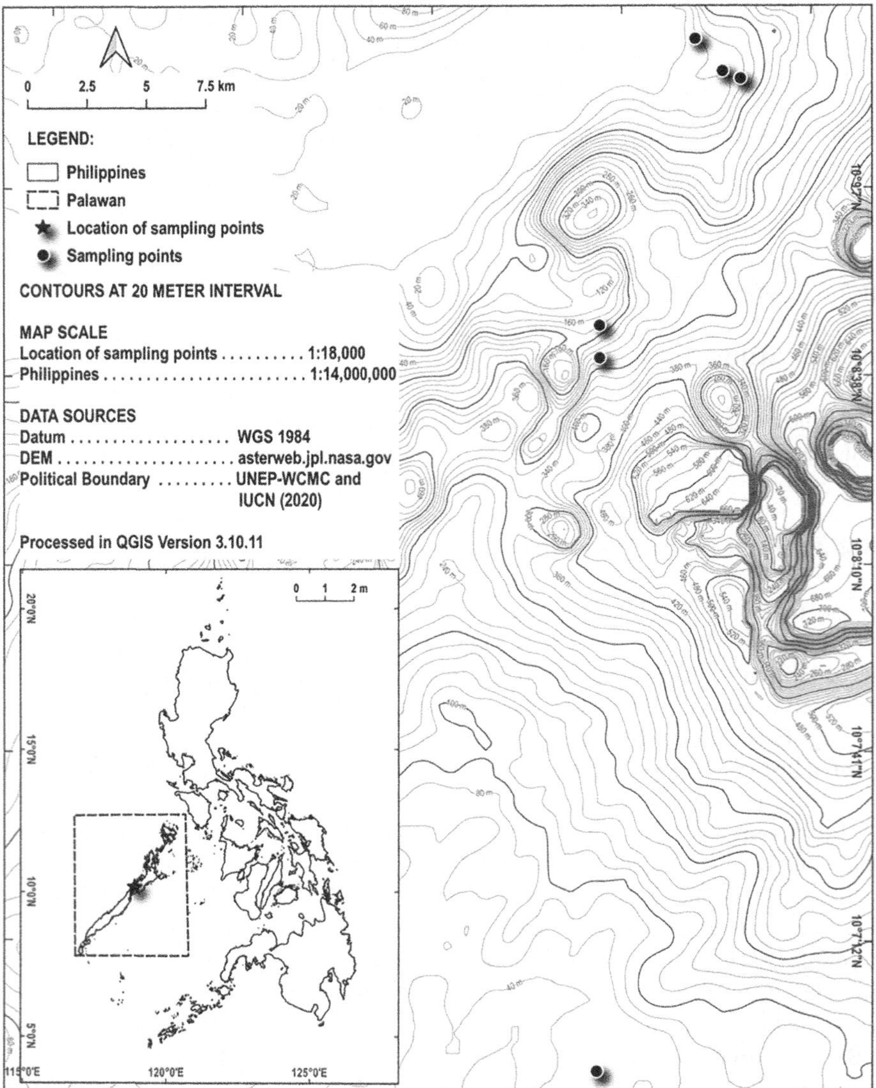

Fig. 8.1 Study site for the fern inventory conducted at the Puerto Princesa Subterranean River National Park (PPSRNP)

Thelypteridaceae (4 spp.), whereas *Aspleniaceae* and *Hymenophyllaceae* were each represented by 3 species. Three fern families were represented by two species, *Athyriaceae*, *Davalliaceae*, and *Lygodiaceae*. The families *Blechnaceae*, *Dryopteridaceae*, *Lindsaeaceae*, and *Marattiaceae* were each represented by a single species. Seven of the 39 fern species were recorded to be epiphytic and the rest are terrestrial in life form.

Fig. 8.2 *Pteris armata* C. Presl, a Philippine endemic fern, showing its (**a**) lamina and (**b**) spores

8.3.1 Notable Fern Flora

Among the recorded fern species, *Pteris armata* C. Presl (Fig. 8.2), *Antrophyum sessilifolium* (Cav.) Spreng, *Tectaria taccifolia* (Fée) M.G.Price (Fig. 8.3), *Tectaria psomiocarpa* S.Y.Dong, *Pneumatopteris nitidula* (C. Presl) Holttum (Fig. 8.4), and *Pronephrium merrillii* (Christ) Holtt. were recorded to be endemic to the Philippines (Table 8.1). This study also reports two new local records of fern species. *Tectaria aspidioides* Copel. is a new local record for the island. This reports the extension of this fern species' local range which is previously recorded to occur from Bohol, Samar, and Mindanao Island. The fern species *Tectaria taccifolia* (Fée) M.G.Price is also reported herewith as a new local record. It is recorded to occur in the island of Mindoro.

The fern species *Adiantum stenochlamys* Baker (Fig. 8.5) and *Drynaria heraclea* (Kunze) T. Moore (Fig. 8.6) are listed as Vulnerable. This species is vulnerable due to the risks of population decline as a result of natural threats such as changes in climate and/or anthropogenic threats whereas *Christensenia aesculifolia* (Blume) Maxon (Fig. 8.7) and *Tectaria psomiocarpa* S.Y.Dong (Fig. 8.8) are listed as Endangered based on DAO 2017-11. These species are considered endangered due to their very small population size in the wild.

8.4 Summary and Conclusions

A total of 39 fern species belonging to 27 genera and from 14 families were recorded from the Puerto Princesa Subterranean. *Pteris armata* C. Presl, *Antrophyum sessilifolium* (Cav.) Spreng, *Tectaria taccifolia* (Fée) M.G.Price, *Tectaria psomiocarpa* S.Y.Dong, *Pneumatopteris nitidula* (C. Presl) Holttum, and

Fig. 8.3 *Tectaria taccifolia* (Fée) M.G. Price, a Philippine endemic fern and new local record for Palawan. (**a**) Lamina; (**b**) fertile leaf

Fig. 8.4 *Pneumatopteris glabra* (Copel.) Holttum, a Philippine endemic fern

Pronephrium merrillii (Christ) Holtt. were found to be endemic to the Philippines. Studies on inventory and documentation of species diversity recorded from poorly studied areas will help in advancing our knowledge regarding biodiversity. Continuous monitoring and inventory of the floral resources in the PPSRNP will serve as reference in providing future strategic plans such as establishing areas for protection and for the promotion of plant and ecosystem conservation.

Table 8.1 Distribution of pteridophyte species documented from the Puerto Princesa Subterranean River National Park (PPSNRP), Palawan Island, Philippines, based on Pelser et al. (2011 onwards)

Family	Scientific name	Life form	Endemicity	IUCN	DAO 2017-11	Coll. no.	Accession number
Aspleniaceae	*Asplenium* cf *normale* D.Don	Terrestrial	–	–	–	MDDLA 154	CAHUP 074077
	Asplenium tenerum G.Forst.	Terrestrial	–	–	–	–	–
	Hymenasplenium subnormale (Copel.) Nakaike	Terrestrial	–	–	–	MDDLA 195	–
Athyriaceae	*Diplazium fraxinifolium* C. Presl	Terrestrial	–	–	–	MDDLA 156	CAHUP 074083
	Diplazium forbesii (Baker) C.Chr.	Terrestrial	–	–	–	–	–
Blechnaceae	*Blechnum orientalis* (L.) C.Presl	Terrestrial	–	–	–	–	–
Davalliaceae	*Davallia denticulata* (Burm.f.) Mett. Ex Kuhn	Epiphytic	–	–	–	MDDLA 188,190	CAHUP 074080
	Davallia divaricata (Blume)	Epiphytic	–	–	–	MDDLA 151	CAHUP 074082
Dryopteridaceae	*Bolbitis heteroclita* (C.Presl) Ching	Terrestrial	–	–	–	–	–
Gleicheniaceae	*Dicranopteris linearis* (Burm.f.) Underw.	Terrestrial	–	–	–	–	–
Hymenophyllaceae	*Cephalomanes atrovirens* C.Presl	Terrestrial	–	–	–	–	–
	Cephalomanes javanicum var. *asplenioides*	Terrestrial	–	–	–	MDDLA 219	CAHUP 074078
	Crepidomanes brevipes (C.Presl) Copel.	Terrestrial	–	–	–	MDDLA 224	–
	Vandenboschia maxima (Blume) Copel.	Terrestrial	–	–	–	MDDLA 175	CAHUP 074106
Lindsaeaceae	*Lindsaea cultrata* (Willd.) Sw.	Terrestrial	–	–	–	MDDLA 211	CAHUP 074086
Lygodiaceae	*Lygodium circinnatum* (Burm.f.) Sw.	Terrestrial/ Climbing	–	–	–	MDDLA 202, 205	CAHUP 074087

(continued)

Table 8.1 (continued)

Family	Scientific name	Life form	Endemicity	IUCN	DAO 2017-11	Coll. no.	Accession number
	Lygodium flexuosum (L.) Sw.	Terrestrial/ Climbing	–	–	–	MDDLA 217	CAHUP 074089
Marattiaceae	*Christensenia aesculifolia* (Blume) Maxon	Terrestrial	–	NE	EN	MDDLA 147	CAHUP 074079
Polypodiaceae	*Drynaria heraclea* (Kunze) T.Moore	Lithophyte	–	–	VU	MDDLA 139, 142	CAHUP 074073
	Leptochilus pedunculatus (Hook. & Grev.) Fraser-Jenk	Epiphytic	–	–	–	MDDLA 153	CAHUP 074085
	Phymatosorus membranifolium (R.Br.) S.G.Lu	Terrestrial	–	–	–	MDDLA 194	CAHUP 074091
	Pyrrosia lanceolata (L.) Farw.	Epiphytic	–	–	–	MDDLA 207,208	CAHUP 074096
Pteridaceae	*Acrostichum speciosum* Willd.	Terrestrial	–	–	–	–	CAHUP 074068
	Adiantum stenochlamys Baker	Terrestrial	–	–	VU	MDDLA 177,180	CAHUP 074071
	Antrophyum reticulatum (G.Forst.) Kaulf.	Epiphytic	–	–	–	MDDLA 182	CAHUP 074075
	Antrophyum sessilifolium (Cav.) Spreng	Epiphytic	Endemic	–	–	MDDLA 218	CAHUP 074076
	Haplopteris elongata (Sw.) E.H.Crane	Epiphytic	–	–	–	MDDLA 213	CAHUP 074084
	Pteris armata C.Presl	Terrestrial	Endemic	–	–	MDDLA 212	CAHUP 074094
	Pteris vittata L. ex Hieron	Terrestrial	–	–	–	MDDLA 193	CAHUP 074095
	Taenitis blechnoides (Willd.) Sw.	Terrestrial	–	–	–		

						MDDLA 143,145	CAHUP 074098
Tectariaceae	*Tectaria aspidioides* Copel.	Terrestrial	–	–	–	MDDLA 172	CAHUP 074100
	Tectaria aurita (Sw.) S.Chandra	Terrestrial	–	–	–	MDDLA 199	CAHUP 074101
	Tectaria decurrens (C.Presl) Copel.	Terrestrial	–	–	–	MDDLA 171	CAHUP 074102
	Tectaria psomiocarpa S.Y.Dong	Terrestrial	Endemic	–	EN	MDDLA 186,187	CAHUP 074103
	Tectaria taccifolia (Fée) M.G.Price	Terrestrial	Endemic	–	–	–	–
Thelypteridaceae	*Cyclosorus interruptus* (Willd.) H.Itô	Terrestrial	–	–	–	–	–
	Pneumatopteris nitidula (C. Presl) Holttum	Terrestrial	Endemic	–	–	–	–
	Pronephrium granulosum (C.Presl) Holttum	Terrestrial	–	–	–	MDDLA 215	CAHUP 074092
	Pronephrium merrillii (Christ) Holtt.	Terrestrial	Endemic	–	–	MDDLA 222	CAHUP 074093

Fig. 8.5 *Pneumatopteris nitidula* (C. Presl) Holttum, a Philippine endemic fern

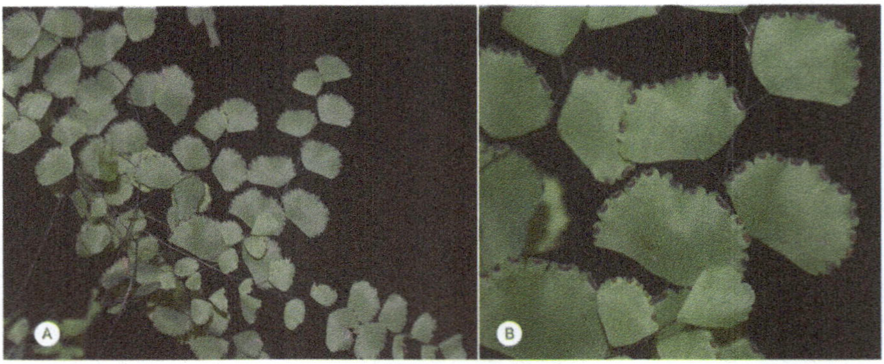

Fig. 8.6 *Aglaomorpha heraclea* (Kunze) Copel., a fern species which is categorized as vulnerable based on the DAO 2017-11. (**a**) Habitat; (**b**) frond (abaxial view)

Fig. 8.7 *Adiantum stenochlamys* Baker, a fern species which is categorized as vulnerable based on the DAO 2017-11

Fig. 8.8 *Christensenia aesculifolia* (Blume) Maxon, a fern species which is categorized as endangered based on the DAO 2017-11

References

Amoroso VB (1987) Medicinal ferns and fern allies of Mindanao. CMU Printing Press, Bukidnon

Amoroso VB (1990) Ten edible economic ferns of Mindanao. Philipp J Sci 199:295–313

Banaticla MCN, Buot IE (2006) Endemic ferns on the northeastern slopes of Mt. Banahaw de Lucban, Sierra Madre mountain range, southern Luzon, Philippines: potentials for domestication and landscaping. J Nat Stud 5:13–22

Barcelona JF, Dolotina NE, Madronero GS, Granert WG, Sopot DD (2006) The ferns and ferns allies of the Karst Forests of Bohol Island. Philippines American Fern J 96(1):1–20

Buot IE Jr (1999) Pteridophytes frequently sold at carbon market, Cebu: implications to urban horticulture and nature conservation. Philipp Scient 36:148–153

Campbell DG, Hammond HD (eds) (1989) Floristic inventory of tropical countries: the status of plant systematics, collections, and vegetation, plus recommendations for the future. The New York Botanical Garden, New York

Copeland EB (1958–1960) Fern flora of the Philippines. Vol. 1-3. Philippine *Schizaea*. Asia Life Sci 5(1):27–34

Francisco-Ortega J, Fa-Guo W, Zhong-Sheng W, Fu-Wu X, Hong L, Han X, Wei-Xiang X, Yo-Bo L, Xi-Qiang S, Gale S, Boufford DE, Maunder M, Shu-Qing A (2010) Endemic seed plants species from Hainan Island: a checklist. Bot Rev 76:295–345. https://doi.org/10.1007/s12229-010-9054-8

Lasco RD, Pulhin FB (2000) Forest land use change in the Philippines and climate change mitigation. Mitig Adapt Strateg Glob Chang 5:81–97

Pelser PB, Barcelona JF, Nickrent DL (2011 onwards) Co's digital flora of the Philippines. http://www.philippineplants.org. Accessed 10 Oct 2021

Zamora PM, Co L (1986) Guide to Philippine flora and fauna. Natural Resources Management Center, Ministry of Natural Resources and University of the Philippines, Quezon City

Situating Plant Genetic Resource in the K-12 Curriculum: A Critical Review

9

Joane V. Serrano, Sherry B. Marasigan, and Janele Ann C. Belegal

Abstract

The Philippines, both a megadiverse country and biodiversity hotspot, requires multi-sectoral conservation efforts to prevent further environmental and resource degradation. Education is a relevant sector for the promotion of biodiversity conservation, which also includes plant genetic resources (PGRs). The Philippines is rich in PGRs, which are considered one of the most essential resources. Along with other ecosystem services and biodiversity resources, PGRs should be properly managed to ensure its availability for present and future use. This chapter argues that education has an important role in PGR, and that its concepts and management strategies should be included in the K-12 curriculum. Therefore, this chapter presents how PGR is situated in the K-12 curriculum, specifically in secondary education, through an archival analysis and a critical perspective. Based on the analysis, PGR was situated in the following parts of the K-12 curriculum: (1) learning competencies, (2) content standard and performance standard, (3) content, (4) core area learning standard and spiraling of concepts, and (5) course description. It is also proposed that PGR can be further situated in the cultural, biodiversity, and agriculture/food security themes of the curriculum. This chapter concludes with recommendations based on the analysis for the improvement of the PGR integration in the K-12 curriculum.

J. V. Serrano (✉) · J. A. C. Belegal
University of the Philippines Open University, Los Baños, Philippines
e-mail: jvserrano@up.edu.ph; janele.belegal@upou.edu.ph

S. B. Marasigan
University of the Philippines Los Baños, Los Baños, Philippines
e-mail: sbmarasigan1@up.edu.ph

S. Ramamoorthy et al. (eds.), *Plant Genetic Resources, Inventory, Collection and Conservation*, https://doi.org/10.1007/978-981-16-7699-4_9

Keywords

Plant genetic resource · PGR · K-12 curriculum · Biodiversity · Conservation · Curriculum

9.1 Plant Genetic Resource in the Philippine K-12 Curriculum

The Philippines is a recognized megadiverse country, along with 16 others. Its rich biodiversity is attributed to its unique geography, and diverse ecosystems and habitats. According to the Convention on Biological Diversity (n.d.), the Philippines "contains between 70 to 80% of the world's plant and animal species, . . . ranks fifth in the number of plant species, . . . maintains 5% of the world's flora, . . . [and covers] at least 25 genera of plants and 49% of terrestrial wildlife" (para. 1). As rich as it is, the Philippine biodiversity provides various services and resources, across ecosystems, species, and animal and plant genetics. It is through this abundant provision that the country's biodiversity is continually threatened by human activities (e.g., overexploitation of natural resources), development-related activities (e.g., conversion of forests into industrial areas), and human population pressure. This environmental degradation has contributed to making the Philippines a biodiversity hotspot (Posa et al. 2008; Liebig et al. 2002; Convention on Biological Diversity n.d.). Hence, timely, immediate, and multi-sectoral conservation responses and efforts are required. Ambal et al. (2012) further implied that "without timely intervention, further degradation of resources will continue and may eventually result in species extinction" (p. 2789).

While biodiversity conservation efforts can often stem from organizations and scientific communities, efforts could also come from the education sector. Ramirez and Santana (2019) suggested that education is an important avenue for promoting biodiversity conservation. This chapter was contextualized in this context, and it focuses on one biodiversity component: plant genetic resources. Plant genetic resources (PGRs) are defined as "plant material with a current or potential value for food, agriculture and forestry" (Brockhaus and Oetmann 1996, as cited in Allem 2000, p. 336). Additionally, according to Moore and Tymowski (2005), the International Treaty on Plant Genetic Resources for Food and Agriculture (ITPGRFA) also provides a similar definition for PGR but puts emphasis on its value for food and agriculture. Rao and Hodgkin (2002) stated that PGR is "among the most essential of the world's natural resources, [and there have been] major advances have been made in conserving them" (p. 2). PGRs are managed through exploration and collection, characterization and evaluation, conservation, exchange and utilization, and documentation or inventory (Sapra 1991, para. 2). This chapter highlights PGR conservation (PGRC) and builds on the narrative that the Philippines "is comparatively rich in [PGR] for food and agriculture" (Liebig et al. 2002, p. 59), which puts grave emphasis on the need for a wider scope of PGRC efforts.

PGRC is the process of collecting and maintaining plant genetic diversity to ensure its availability to meet present and future needs (Khanna and Singh 1991; Rao

2004; Ogwu et al. 2014). There are two approaches to PGRC, namely in situ conservation and ex situ conservation. Literature has thoroughly described the differences of these two PGRC approaches which mainly define the means (inside or outside the natural habitat) where genetic resources are conserved (Khanna and Singh 1991; Rao 2004). There has also been literature on effective methods or techniques under each approach (in situ: Rijal et al. 2000; Satheesh 2000; Aragón-Cuevas et al. 2005; ex situ: Dulloo et al. 2010; Khoury et al. 2010).

Rijal et al. (2000) investigated diversity fairs as a community-based approach and found that it can be an effective strategy for PGRC. Participatory approaches such as community gene fund, biodiversity festival, and grassroots media were also reported as effective strategies for in situ PGRC by Satheesh (2000). Lastly, an investigation on a community seed bank for local maize races in a state in Mexico suggested it as a positive approach to enhance farmer livelihood and further in situ conservation strategies (Aragón-Cuevas et al. 2005). It is noteworthy that these selected studies highlight a participatory approach to in situ conservation.

On the other hand, the selected ex situ conservation studies reflect trends on collection and storage strategies. A 2010 review of ex situ conservation research suggested broadening focus to "[understand] the responses of a wide diversity of species ... to both single and different storage conditions and methods" (Dulloo et al. 2010, p. 126). Similarly, a review of 26 ex situ PGRC strategies revealed "major trends and needs in *ex situ* conservation" which could result in "more efficient and effective conservation and utilization of crop diversity *ex situ*" (Khoury et al. 2010, p. 636).

Given these studies and evidence on effective PGRC techniques, especially through its two established approaches, there is limited research on PGRC through education. PGRC concepts and strategies can be integrated in an education program or curriculum. Sthapit and Friis-Hansen (2000) emphasized that education has important roles in biodiversity conservation, which includes PGR. They stated that "public education ... programmes encourage strong and sustained support for local and national PGR management activities by promoting the role that genetic resources play in development" (Sthapit and Friis-Hansen 2000, p. 164). Hence, education provides an avenue for PGR introduction and discussion for individuals beyond the scientific community.

In this chapter, the critical role of education in PGR, specifically the Philippine K-12 program, is highlighted. The K-12 program is the Philippines' enhanced basic education program, consisting of 10–13 years of education, as implemented by the Republic Act (RA) 10533 (Official Gazette 2013). Sarmiento and Orale (2016) stated that the K-12 program is a "more potent educational model [where] students can prepare themselves for a more rigorous training in higher education ..." (p. 20). PGR, including its concepts and strategies, as well as other biodiversity topics are introduced and thoroughly discussed in higher education, but only in select degree programs or courses (Labonite 2013; Borromeo 2012; Borromeo and Altoveros 2010; Rudebjer et al. 2010). In order for students to understand the importance of PGR and how it is integrated in various subject areas, it is important that basic concepts are introduced in the early grade level and reinforced as the students move up the K-12 program. By including PGR in the K-12 level curriculum, students are introduced to the terminologies, concepts, and how it relates with other areas early on.

Research interest on PGR in the Philippine K-12 program is yet to be established. However, there has been evidence of integrating PGR in the Philippine higher education level. Labonite (2013) reported on the integration of PGRC and management in the curriculum of two bachelor programs. Her study indicated that PGRC and management were integrated only as an elective subject, where only a few students enrolled in (Labonite 2013, p. 194). Borromeo (2012) also reported the institution of an undergraduate course for PGR, and the offering of PGR courses in four Philippine higher education institutions. In Borromeo and Altoveros' study (2010), PGRC principles and strategies were integrated in a postgraduate program; however, enrollment to the said program experienced a decline since 2005 due to lack of scholarships. Despite the enrollment decrease, "students from other degree programs . . . have taken PGR as their minor field of specialization or included PGR courses in their program of course work" (Borromeo and Altoveros 2010, p. 51). Additional evidence of PGR integration in higher education institutions, although in East and Southeast Asia, was documented by Rudebjer et al. (2010).

In addition, there are studies which emphasized the importance of education in PGR strategies. Volk et al. (2019) studied a training program framework for PGR management which could benefit multiple audiences including the educational sector. This framework could be used as reference in developing curricula for PGR integration. Borromeo (2012, p. 25) stressed out "the importance of education . . . to achieve sustainable improvements in [PGRC]." Her study also recommended strengthening "post-graduate education and short-term training on all aspects of [PGRC] to increase the critical mass of PGR workers who are aware of the most recent developments and opportunities in this field" (Borromeo 2012, p. 31). Similarly, Gauchan et al. (2000, p. 191) highlighted the lack of "policies to incorporate curricula . . . in agrobiodiversity aspects in the university." Evidently, these studies gave emphasis on PGR in the higher education level. Unlike the higher education curriculum where PGR courses or programs are only available to select degrees or courses, the K-12 curriculum can have PGR as a mandatory subject.

Through RA 9512, or the National Environmental Awareness and Education Act of 2008, the integration of PGR to the K-12 curriculum is feasible. According to the Official Gazette (2008), this Act covers the integration of environmental education in the curricula of all school levels. Through this chapter, it will be determined if PGR, as a topic encompassed in environmental education, is included in the K-12 curriculum. Hence, this chapter presents how PGR is situated in the K-12 curriculum, specifically in secondary education. Through an archival analysis, this chapter looks into the following documents: (1) Department of Education (DepEd) Order 021, Series of 2019 - Policy Guidelines on the K-12 Basic Education Program (DO 021), and (2) seven (7) Curriculum Guides of Junior High School and Senior High School Subjects relevant to PGR. Then a critical perspective was applied to expound on the analysis of situating PGR in the K-12 curriculum.

9.2 The Need for a Critical Perspective in the K-12 Curriculum

The analysis carried out for this chapter drew upon this suggestion from Ramirez and Santana (2019): "Educators were urged to include environmental education in their curricula, not only in biology and natural science classes, but across the spectrum, as a topic to be discussed in social sciences, and other school subjects" (p. 8). Hence, seven curriculum guides of PGR-relevant subjects from the Philippine K-12 secondary level were analyzed, namely Science K-12, Science 11–12, Araling Panlipunan K-10, Agricultural Crops Production NC I, Agricultural Crops Production NC II, Agricultural Crops Production NC III, and Horticulture NC III. These curriculum guides were then reviewed through a critical perspective.

Horkheimer's critical theory, which he coined in 1937 (Bronner 2009), is applied in this chapter. Specifically, this chapter presents its findings and discussion using the following criteria of Horkheimer's critical theory: it must (1) explain what is wrong with current social reality; (2) identify the actors to change it; and (3) provide both clear norms for criticism and achievable practical goals for social transformation (Critical Theory 2008). Aside from assisting in determining whether PGR, PGR- or biodiversity-related concepts, or PGR-relevant laws and policies are included or applied in the K-12 curriculum, the criteria were utilized for identifying themes wherein PGR was situated or categorized, and for recommending action points for the improvement of the K-12 curriculum in the context of PGR.

9.3 Situating Plant Genetic Resource in the K-12 Secondary-Level Curriculum

9.3.1 The K-12 Secondary Education

The Philippine Secondary Education is mainly composed of two key stages: (1) junior high school which comprises grades 7–10 and (2) senior high school which comprises grades 11–12. The DepEd Order No. 21, series of 2012, or the Policy Guidelines on the Implementation of Grades 1 to 10 of the K to 12 Basic Education Curriculum (BEC) Effective School Year 2012–2013 (Department of Education 2012) states that at the end of junior high school learners must be able to:

> demonstrate communicative competence; think intelligently, critically, and creatively in life situations; make informed and values-based decisions; perform their civic duties; use resources sustainably; and participate actively in artistic and cultural activities and in the promotion of wellness and lifelong fitness. (p. 1)

On the other hand, senior high school learners are expected to prepare and equip graduates for tertiary education, middle skill development, employment, and entrepreneurship. There are four tracks that senior high school students can choose from; these are the academic, technical-vocational livelihood (TVL), sports, and arts and design. The academic track will help students prepare for the subject specialization of the college course they will take in the future. It has four strands, science, technology, engineering, and mathematics (STEM); general academic strand (GAS); humanities and social sciences strand (HUMSS); and accountancy, business, and management (ABM). On the other hand, the TVL track is designed for hands-on learning and application of skills. It also has four strands: the agri-fishery, home economics, industrial arts, and information and communication technology. Technology and livelihood education (TLE) and TVL track specializations may be taken between grades 9 and 12, while exploratory subjects with 40 h per quarter are taken during grades 7–8. The sports track aims to give students an understanding of the basic principles and techniques in relation to physical education and recreation, and lastly the arts and design track prepares students as they wander around the design, performative, and creative industry.

In the analysis, four curriculum guides where PGR was situated were in the TLE subjects and TVL track. The TLE subject follows the guidelines set by the Technical Education and Skills Development Authority or TESDA to allow the students to earn a national certification (NC) required by the industry. The TLE subjects in grades 7–8 are exploratory which means that students are allowed to explore four main areas of TLE: agri-fishery arts, home economics, information and communication technology (ICT), and industrial arts. Students have the liberty to choose a maximum of four TLE mini subjects in grade 7 and another four in grade 8 that the school offers. In grades 7 and 8, the students are not yet able to earn a Certificate of Competency or COC. In grade 9, the student will now choose one course to focus on from among the exploratory courses he or she took in grades 7 and 8. In grade 9, the student now can earn a COC. In grade 10, the student now pursues the TLE specialization course which he or she took in grade 9, which will now allow him or her to get at least a NC Level I or II based on the TLE course he or she chose. If the student chooses the TVL track as a specialty in senior high school, he or she will continue the TLE course he or she studied in grades 9 and 10. Based on this study, most of the subjects where PGRC are integrated are part of the TLE and TVL track.

9.3.2 Parts of the K-12 Curriculum Where PGR Is Situated

This part of the chapter attempts to situate PGR in the various parts of the K-12 curriculum guide. The notion of situating PGR stems from the social constructionist and constructivist view of education wherein students become more actively engaged in learning by situating or anchoring instruction around related and interesting topics (UCD Teaching and Learning, n.d.). By examining PGR as it is situated in the K-12 curriculum, this chapter puts forward the premise that curriculum is viewed as a text or discourse. The discursive nature of the curriculum claims that the world is discursively constructed and that people access the world through discursive means. In a discursive approach, people access the nondiscursive through language. Hence, by examining the K-12 curriculum as a text or discourse, the authors are supporting the notion that PGR can be found in and through the curriculum guides. Foucault (1982) argued that, as cited by Ezzamel and Willmotto (2010), language should not be treated merely as a group of "signs (signifying elements referring to context and representation) but as practices that systematically form the objects of which they speak." Foucault (1982) added that to treat discourses as a sign is to disregard or neutralize its constitutive force (Ezzamel and Willmotto 2010).

Based on the analysis, PGR was situated in the following subject areas and parts of the K-12 curriculum:

9.3.2.1 As Part of Learning Competencies

Grade level/subject/location	Learning competencies
Taken between grades 9 and 12/agricultural crops production (NC I, II) introductory module/page 5 Page 11	1. Explain basic concepts in agricultural crop production
	1. Explain the cultural methods and practices in crop production 2. Formulate adaptive cultural practices in the care and maintenance of a school nursery
-Agricultural crops production (NC 1—agronomic cropping work module/page 20	1. Explain the cultural methods and practices in agronomic crop production
Agricultural crops production (NC II)/lesson 6—performing nursery operations/page 14	LO 3. Handle seeds/planting materials 3.1 Determine planting materials according to kinds and varieties 3.2 Select quality seeds according to prescribed characteristics 3.3 Conduct seed testing to determine the percentage germination of the seedstock in accordance with the standard procedures 3.4 Treat planting materials following standard protocol 3.5 Perform seed scarification for germination purposes based on the type of crop
	LO 5. Conduct propagation activities 5.1. Set up nursery shed according to plant requirement 5.2. Select quality seedlings based on prescribed characteristics 5.3. Select propagation materials according to propagation activity 5.4. Perform plant propagation techniques based on recommended practices 5.5. Maintain germinated seedlings until fully established 5.6. Prick and thin seedlings based on recommended practices
Agricultural crops production (NC II/lesson 9—carrying out harvest and postharvest operations/page 19–20	LO 2. Perform harvesting activity 2.1 Implement appropriate harvesting methods based on best cultural practices for each crop using suitable tools 2.2 Properly handle crops according to postharvest treatment

(continued)

Grade level/subject/location	Learning competencies
	2.3 Harvest crops based on maturity indices and characteristics affecting harvest quality 2.4 Harvest crops while observing Phil GAP principles
	LO 3. Perform postharvest operation. 3.1 Employ postharvest operations based on standard procedures. 3.2 Monitor postharvest equipment to avoid damage to crops in line with manufacturer/ enterprise procedures 3.3 Handle and package according to variety and destination 3.4 Store and stack crops in cool dry place prior to distribution in line with enterprise procedures
	LO 4. Monitor storage pest and diseases 4.1 Identify pests and diseases based on references for storage 4.2 Record and report identified storage pests and diseases to immediate authority
Grades 11–12/horticulture NC III/lesson 12 (implementing a plant nutrition program)/ page 21–22	LO 1. Prepare for implementation of the plant nutrition program 1.2 Identify soils, plant species, and varieties according to farmwork procedures 1.3 Follow soil sampling techniques/ procedures 1.4 Analyze soil composition/nutrients of the area and adjacent lands following the soil sampling procedures 1.5 Identify materials for soil and plant treatments
	LO 3. Determine nutritional problems in plants 3.1 Identify common nutrient deficiency and toxicity problems in plants using visual inspection in accordance with crop production manuals 3.2 Consult supervisor and/or nutritional specialist to determine causes of nutritional or toxicity problems 3.3 Select soil ameliorants to improve soil fertility according to farmwork procedures 3.4 Source soil ameliorants to improve soil fertility according to farmwork procedures
Grades 11–12/horticulture NC III/lesson 15 (establishing horticultural crops)/page 31–32	LO 2. Prepare for horticultural crop establishment 2.1 Monitor soil and weather conditions for optimal seeding conditions according to farmwork procedures 2.2 Recognize soil conservation and sustainable land management 2.3 Confirm soil conservation and

(continued)

Grade level/subject/location	Learning competencies
	sustainable land management practices in accordance with farm requirements and environmental concerns 2.4 Confirm seeding, fertilizer, and pest and weed control against the work plan 2.5 Prepare seeding, fertilizer, and pest and weed control requirements based on manufacturers' specifications using safe handling procedures 2.6 Calculate pesticide requirements and application rates 2.7 Prepare contingency plans for unusual seasonal conditions and pest/disease outbreaks according to farmwork procedures
Grades 11–12/horticulture NC III/lesson 16 (coordinating a horticultural crop maintenance program)/page 34	LO 1. Prepare for the crop maintenance program 1.1 Identify the requirements of the crop maintenance program according to the site plan and farmwork
Grades 11–12/horticulture NC III/lesson 17 (coordinating horticultural crop harvesting)/page 36	LO 1. Prepare for crop harvesting 1.1 Identify crop for harvest according to farmwork plan and procedures 1.2 Identify requirements and procedures to assure crop quality, including suitable weather conditions, farmwork plan, and procedures 1.3 Determine crop maturity based on experience, specification charts, aids, or devices and according to client preferences, farm, and industry quality assurance specifications
Grades 11–12/horticulture NC III/lesson 18 (undertaking propagation activities)/page 39–40	LO 2. Propagate plants 2.1 Apply pretreatment procedures according to standard propagation method and species' requirements 2.2 Carry out propagation techniques according to species' requirements and OHS requirements 2.3 Handle propagation material properly according to species' requirements 2.4 Apply labels according to farmwork procedures 2.5 Apply water and nutrients according to media conditions, plant requirements, and propagation techniques 2.6 Monitor plant health according to species' requirements and farmwork procedures 2.7 Take remedial action to ensure plant health according to species' requirements and farmwork procedures

(continued)

Grade level/subject/location	Learning competencies
	LO 3. Complete propagation activities 3.1 Complete records accurately and at the required time according to the standard recording system
Grades 11–12/horticulture NC III/lesson 19 (undertaking field budding and grafting)/page 40–42	LO 1. Prepare parent material for budding and grafting 1.1 Prepare parent plant according to the species' requirements and organizational procedures 1.2 Employ the method of taking bud or graft material according to the species' requirements and standard grafting/budding procedures
Grades 11–12/horticulture NC III/lesson 19 (undertaking field budding and grafting)/page 40–42	LO 2. Prepare budding and grafting materials 2.1 Select budding/grafting materials for propagation according to species' requirements, crop manual, and farmwork procedures 2.2 Prepare stock in accordance with species' requirements and standard stock preparation procedures 2.3 Handle budding/grafting materials and stock properly according to species and standard grafting procedures 2.4 Maintain viability of materials by using appropriate storage procedures in accordance with the proper storage requirements of the species
Grades 11–12/horticulture NC III/lesson 19 (undertaking field budding and grafting)/page 40–42	LO 3. Bud/graft scion work 3.1 Select budding/grafting method in accordance with species' requirements 3.2 Seal bud/graft according to species and standard budding/grafting procedures 3.3 Apply appropriate labels and identification in accordance with organizational guidelines 3.4 Cite the different factors affecting the success of operations 3.5 Identify problems in budding/grafting operations 3.6 Take corrective actions according to propagation principles 3.7 Complete appropriate records at the required time in accordance with organizational recording guidelines
Grades 11–12/horticulture NC III/lesson 20B (collecting samples for a rural production or horticulture monitoring program)/page 45–46	LO 1. Plan for collection of samples 1.1 Confirm the purpose and scope of sample collection activity from discussion with supervisor or work instructions 1.2 Read/hear sample collection schedule with supervisor

(continued)

Grade level/subject/location	Learning competencies
	1.3 Confirm sample collection schedule with supervisor 1.4 Confirm sampling site location following enterprise guidelines 1.5 Obtain approval for site access following enterprise guidelines 1.6 Identify samples collected and preserved in conjunction with supervisor or by reference to enterprise guidelines 1.7 Assess the range of likely operating conditions, hazards, and difficult/sensitive environments for impact on sampling and testing
Grades 11–12/horticulture NC III/lesson 20B (collecting samples for a rural production or horticulture monitoring program)/page 45–46	LO 3. Carry out sampling and preserving procedures 3.1 Collect samples in accordance with sampling plan and enterprise procedures and industry protocols/guidelines 3.2 Preserve samples in accordance with sampling standards and guidelines 3.3 Prepare samples for external analysis for laboratory procedure in accordance with sampling schedule and laboratory standards 3.4 Pack samples for external analysis for laboratory procedures in accordance with sampling schedule and laboratory standards 3.5 Send samples for external analysis to the laboratory in accordance with sampling schedule and laboratory standards 3.6 Pack hazardous materials in accordance with legislative requirements. 3.7 Transport hazardous materials in accordance with legislative requirements 3.8 Make observations including information on the surrounding area and environmental conditions in accordance with the monitoring schedule 3.9 Follow equipment operation and work practices according to OHS requirements 3.10 Report collection outcomes including presentation of samples in accordance with enterprise guidelines 3.11 Deliver collection outcomes including presentation of samples in accordance with enterprise guidelines
Grades 11–12/horticulture NC III/lesson 20F (maintaining and monitoring environmental work practices)/page 55	LO 4. Maintain environmental records 4.1 Maintain environmental records accurately and legibly in a form accessible for reporting purposes 4.2 Store environmental records securely and accessible for reporting purposes

(continued)

Grade level/subject/location	Learning competencies
	4.3 Maintain internal and external reporting procedures
Grade 11/earth/life sciences/page 7	1. Describe the process of genetic engineering 2. Conduct a survey of the current uses of genetically modified organisms 3. Evaluate the benefits and risks of using GMOs
Grade 10/Araling Panlipunan (social studies)/ second quarter/page 218	*Naipaliliwanag ang kaugnayan ng mga gawain at desisyon ng tao sa pagbabagong pangkapaligiran* (to explain the relationship of the activities and decisions of people in relation to the changes that are happening in the environment) *Nasusuri ang mga kasalukuyang hamon sa pagtamo ng sustainable development* (hal.: consumerism, energy sustainability, poverty, at health inequalities) (to examine the challenges being confronted by sustainable development, e.g.: consumerism, energy sustainability, poverty, at health inequalities) *Napaghahambing ang iba't ibang istratehiya at polisiya na may kaugnayan sa pagtamo ng sustainable development na ipinatutupad sa loob at labas ng bansa* (to compare the different strategies and policies implemented within and outside of the country in relation to the attainment of sustainable development)

9.3.2.2 As Part of Content Standard and Performance Standard

Grade level/subject/location	Content standard	Performance standard
Taken between grades 9 and 12/agricultural crops production (NC 1)–introductory module/age 5 Page 11	The learner demonstrates an understanding of basic concepts and underlying theories in agricultural crop production	The learner independently demonstrates common competencies in agricultural crop production as prescribed by TESDA training regulations
	The learner demonstrates an understanding of concepts, principles, underlying theories, and actual application in crop production and nursery work	The learner independently demonstrates common competencies in crop production and nursery work as prescribed by TESDA training regulations
-Agricultural crops production (NC 1—agronomic cropping work module/page 20	The learner demonstrates an understanding of concepts and principles, underlying theories, and actual application in agronomic crop production and irrigation work	The learner independently demonstrates common competencies in agronomic crop production and irrigation work as prescribed by TESDA training regulations

(continued)

Grade level/subject/location	Content standard	Performance standard
Agricultural crops production (NC II/ lesson 6—performing nursery operations/page 13	The learners demonstrate an understanding of the underlying concepts and principles in performing nursery operations	
Agricultural crops production (NC II/lesson 7—planting crops/page 15	The learners demonstrate an understanding of the underlying concepts and principles in performing planting crops	The learners shall be able to plant crops
Agricultural crops production (NC II/lesson 8—caring and maintaining crops/page 17	The learners demonstrate an understanding of the underlying concepts and principles in caring and maintaining crops	The learners shall be able to care for and maintain crops
Agricultural crops production (NC II/lesson 9—carrying out harvest and postharvest operations/page 18	The learners demonstrate an understanding of the underlying concepts and principles in carrying out harvest and postharvest operations	The learners shall be able to carry out harvest and postharvest operations based on standard procedures
Grades 11–12/horticulture NC III/lesson 10 (preparing land for horticultural crop production)	The learner demonstrates an understanding of concepts, underlying theories, and principles in preparing land for horticultural crop production	The learner independently prepares land for horticultural crop production based on required tasks
Grades 11–12/horticulture NC III/lesson 17 (coordinating horticultural crop harvesting)/page 36	The learner demonstrates an understanding of concepts, underlying theories, and principles in performing horticultural crop harvesting activities	The learner independently performs horticultural crop harvesting activities based on TESDA training regulation and required tasks
Grade 7/science/second quarter (living things and their environment)/page 116	The learners demonstrate an understanding of organisms interacting with each other and with their environment to survive	The learners should be able to conduct a collaborative action to preserve the ecosystem in the locality
Grade 11/earth/life science/ page 8	The learners demonstrate an understanding of: 1. The principles of the ecosystem 2. Biotic potential and environmental resistance 3. How human activities affect the natural ecosystem	The learners shall be able to prepare an action plan containing mitigating measures to address current environmental concerns and challenges in the community

9.3.2.3 As Part of Content

Grade level/subject/location	Topic
Grade 8/science/fourth quarter (living things and their environment)/page 151	Biodiversity – Protection and conservation of endangered and economically important species
Grade 8/science/fourth quarter (living things and their environment)/page 153	Ecosystem – Impact of human activities in an ecosystem
Grades 11–12/horticulture NC III/lesson 10 (preparing land for horticultural crop production)/page 15	Planting plan and its interpretation for a range of crops – Different planting plans (based on crops to be planted) Identification of methods and order of cultivation – Methods of cultivation – Order of cultivation Environmental implications of site cultivation and legislation – Different environmental implications – Environmental rules and regulations/legislations – Assessing environmental implications – Necessary actions to be taken based on assessment
Grades 11–12/horticulture NC III/lesson 12 (implementing a plant nutrition program)/page 21–22	Types of plant species and varieties Analyze nutrients/composition of soil – Soil sampling techniques – Soil analysis Application of materials for soil and plant treatments – Different materials for soil and plant treatments – Sourcing of materials
Grades 11–12/horticulture NC III/lesson 15 (establishing horticultural crops)/page 31–32	Soil and weather requirements for optimal seeding conditions Soil conservation and suitable land management Preparation of the requirements for seeding, fertilizing, and pest and weed control activities: – Fertilizer types, rates of application, and crop nutrient requirements/integrated nutrient management technique – Types of chemical pesticides and alternative pest control methods (nonchemical)/integrated pest management – Calculation of pesticide requirements and application rates
Grades 11–12/horticulture NC III/lesson 16 (coordinating a horticultural crop maintenance program)/page 34	Identification of site and operational requirements – Crop maintenance program – Site plan – Types and application of crop maintenance activities

(continued)

Grade level/subject/location	Topic
Grades 11–12/horticulture NC III/lesson 17 (coordinating horticultural crop harvesting)/ page 36	Identification of crops to be harvested – Maturity parameters for a range of crops – Farm quality procedures and crop characteristics relative to varying market requirements Identification of requirements and procedures – Crop quality – Suitable weather conditions – Environmental factors affecting crop harvesting activities Determination of maturity index – Experience – Specification charts – Age – Devices – Client requirements
Grades 11–12/horticulture NC III/lesson 18 (undertaking propagation activities)/page 39–40	Pretreatment procedures for a range of propagation methods and species Plant propagation processes – Propagation techniques according to species' requirements – Proper handling of propagation materials – Labeling of propagation materials Water and nutrient applications based on: – Media conditions – Plant requirements – Propagation techniques Plant health monitoring and appropriate corrective actions – Monitoring procedures – Common problems in plant propagation – Remedial actions taken
	Record keeping related to propagation activities Records: – Date of propagation – Type of propagation – Number of plants – Source of propagation – Material – Treatments carried out – Spray records
Grades 11–12/horticulture NC III/lesson 19 (undertaking field budding and grafting)/ page 40–42	Preparation of parent plant Methods of budding/grafting – Selection and use of rootstock and scion stock – Cuttings, buds, dormant or green material, nursery rootstock, rootling, and mature plants
	Selection of budding/grafting materials Stock preparation techniques of different horticultural crops

(continued)

Grade level/subject/location	Topic
	Handling requirements for budding/grafting materials according to crop species Maintain viability of materials – Storage requirements for budding/grafting materials – Factors affecting the viability of materials – Maximum viability Proper collection and disposal of wastes and materials related to budding/grafting Environmental codes and regulations
	Selection of budding/grafting methods Procedures/methods for budding/grafting according to crop species Budding/grafting of scion – Bud/graft sealing requirements – Proper labeling and identification of budded materials – Factors affecting the success of budding/grafting operations Record keeping of budding/grafting activities
Grades 11–12/horticulture NC III/lesson 20B (collecting samples for a rural production or horticulture monitoring program)/page 45–46	Horticultural sites/location – Paddocks – Farm building – Work site – Nurseries – Playing fields – Dams Purpose and scope of biological sample collection activities – Sample site – Sample collection schedule Field procedures for sampling and preservation
	Protocol/guideline procedures in sampling activities Procedures in preserving samples – Collect samples – Preserve samples – Prepare and package samples for transport to laboratory according to standards
	Sampling plans and procedures Collecting techniques (record, forward, test, and analyze) Proper handling of collecting equipment according to enterprise procedures Environment act and legislation
Grades 11–12/horticulture NC III/lesson 20F (maintaining and monitoring environmental work practices)/page 55	Environmental issues and concerns – Sustainability reduction and disposal of waste – Water quality – Energy efficiency

(continued)

Grade level/subject/location	Topic
	– Biodiversity – Habitat protection, conservation of natural resources – Air quality – Land contamination – Noise – Soil and salinity management – Fire management. Workplace approaches to environmental practices Prevent and minimize production of pollution Improve workplace maintenance practices

9.3.2.4 As Part of Core Area Learning Standard and Spiraling of Concepts

Grade level/subject/location	Concept
Learning standard (science for the entire K-12)	The learners demonstrate understanding of basic science concepts and application of science-inquiry skills. They exhibit scientific attitudes and values to solve problems critically, innovate beneficial products, protect the environment and conserve resources, enhance the integrity and wellness of people, make informed decisions, and engage in discussions of relevant issues that involve science, technology, and environment
Grade 7–10/science/page 12	Biodiversity and evolution

9.3.2.5 As Part of Course Description

Grade level/subject/location	Course description
Grades 11–12/horticulture NC III/page 5	Under the core competencies in horticulture, it includes (15) establishing horticultural crops, (16) coordinating horticultural maintenance program, (17) coordinating horticultural harvesting, (18) undertaking field budding and grafting, and (19) undertaking propagation activities Elective competencies (which the school can choose 3 lessons among the given 9 lessons): 1. Following site quarantine procedures. 2. Collecting samples for a rural production or horticultural monitoring program. 3. Handling bulk materials in storage area. 4. Preparing grain storage. 5. Complying with industry quality assurance requirements. 6. Maintaining and monitoring environmental work practices. 7. Keeping records for farm business. 8. Performing specialized machinery maintenance. 9. Installing irrigation systems

9.3.3 Themes Wherein PGR Can Be Further Situated in the K-12 Curriculum

Supporting Ramirez and Santana's (2019) affirmation that environmental education should be included in the curriculum of other school subjects aside from biology and natural sciences, PGR in the K-12 curriculum can be further situated under the following themes:

9.3.3.1 Cultural

Culture within the context of PGR in the Philippine K-12 curriculum was somehow integrated or embedded in the agricultural crops production subjects in grades 9–11. This was observed on the learning competencies expected of the students, but the topics were only limited to cultural methods and practices in relation to agricultural crop production. The inclusion of the concept of culture cannot be categorically classified as taking into account the traditions and practices of communities vis-à-vis their indigenous PGRs. According to Phillips (2003), PGRs have varying meanings to people. Some look at it in terms of the practical economic application in agriculture while others are more interested in their symbolic or cultural value to society. Phillips (2003) highlighted the major role that plants play in the economic, social, cultural, and historical life of traditional communities, and that people are also concerned with how the plants that they have developed and taken care of are utilized and valued by others. This very important context of culture was not manifested in the K-12 curriculum that was reviewed. The relationship of biotechnology and its impacts to traditional knowledge about plants and the ecosystem could be a good topic to be added in the curriculum. This could be discussed and integrated in the social studies curriculum which tackles greatly on the value of culture in communities. As such, social studies is seen as one subject area that could actually encompass these varying issues such as ethics, biotechnology, conservation of PGRs, and its relationship to preservation of agricultural traditions and practices.

9.3.3.2 Biodiversity

Biodiversity is one very important concept that cannot be missed in PGR discourses. The concept of biodiversity has repeatedly been included in the science subjects of the K-12 curriculum. Since the approach to lessons in the Philippine K-12 program is spiraling in nature, this has been presented in the different grade levels, particularly in grades 7, 8, and 11. In the grade 7 science curriculum, part of the content standard is the demonstration of learners with an understanding of the interaction of organisms with their environment and their ability to conduct a collaborative action to preserve the ecosystem in a locality. Among the grade 8 level, a particular topic was devoted to biodiversity, with subheading of protection and conservation of endangered and economically important species; likewise, a related topic on the ecosystems is included, focusing on the impact of human activities that affect the ecosystem. This was further reinforced in the grade 11 curriculum, wherein the students are already expected to prepare an action plan containing mitigating measures to address current environmental concerns and challenges in the community. It is notable that there is a progression of how the concept of biodiversity was treated in the science curriculum, from mere understanding of the concept to development of an action plan. There was no direct discussion of the PGR in biodiversity as a concept as stated in the curriculum; however, it could be presumed that embedded in the issue of preservation and conservation of endangered and economically species are the PGRs. This was reinforced in one of the topics among grades 11–12 in the horticulture subject. There was a section which included environmental issues and concerns, with topics on problems in waste disposal, land contamination, and water quality. As mentioned by Gepts (2004), one of the means in achieving a sustainable future for humanity is through proper biodiversity and ecosystem management, particularly on the development of improved cultivars that could enhance agricultural productivity. Furthermore, Ebert and Engels (2020) emphasized that because of extreme changes in the land use, abuse of resources, and human-induced and natural disasters, the systems of production and genetic resources which include crop wild relatives that are found in protected and cultivated land especially in the forests are severely threatened. These changes in the habitat are believed as the primary reasons for the extinction of plant communities. From a critical perspective, these topics are hoped to be included in the discourses of biodiversity and ecosystems in relation to PGRs. However, these topics are only seen in earth/life sciences topics and in biology subject areas, and towards the latter part of the agriculture subject particularly in horticulture. This was not evidently seen in the agricultural crops production subject. Issues like agricultural biodiversity should be heavily emphasized among the students for their further appreciation of the environment and resources. These concerns will have a greater impact on society if discussed in an interactive manner across subject areas of the students. It would better be appreciated if there is a reinforcement and continuity of these topics among the different subjects taken by the students.

9.3.3.3 Agriculture/Food Security

Based on the review of the K-12 curriculum, PGRs are heavily situated in agriculture subject areas of grades 9–12, particularly in agricultural crop production and horticulture. However, it is further observed that most content standards, performance standards, and learning competencies revolve on the technical skills that students must learn in the process of taking these agriculture subjects. The main task of agriculture is to address the issue of freeing people from hunger. Food security is one of the most important elements in the discussions on agriculture. By taking a closer look at the K-12 curriculum, there was no topic that would directly address the concept of food security; however, this could be embedded in the several topics that are covered in grades 9–11 areas of agriculture. Essentially, lessons in agriculture could highlight the importance of PGRs in food security, since erosion of these resources could lead to severe threat to food security of the world. By looking at the learning competencies and topics covered in grades 9–11 in agriculture, there are a lot of lessons that could be discussed in connection with food security. One of the possible key topics that could further be elaborated in the K-12 courses in agriculture is the promotion of sustainable agriculture which could be possible through diversification of crop production; another is the promotion of the development and commercialization of underutilized crops and species (Dulloo 2019). A review of the whole agriculture curriculum could be done to recalibrate the lessons and verify if they are able to cover relevant issues such as food security. However, the issue of food security, as this is a global concern, could also be discussed in other subject areas such as social studies or even in values education. Valuing food, as well as the traditional practices of farmers in order to preserve traditional varieties, could be emphasized in the values education subject. As mentioned above, the concerns of PGRs in relation to food security can transcend several subject areas.

9.4 Conclusion and Recommendations

This chapter presented the critical role of education in PGR, specifically the Philippine K-12 program in the secondary level. The authors reviewed through an archival analysis seven curriculum guides with the premise that curriculum is viewed as a text or discourse. In this chapter, the curriculum guides were critically examined in order to situate how the PGR occurred in and through text.

The following are the main observations and recommendations in this chapter as guided by a critical analysis or perspective:

1. The silence or absence of themes critical to the discussion and demonstration of the concept of PGR was very apparent. Although there were attempts on the inclusion of topics related to PGR in the reviewed curriculum guides, it was clear that there is a need to strengthen and reinforce how the topics are interwoven and integrated in order for the students to have a clear understanding of the critical role PGR plays in the various aspects of human lives and the environment they live in. The authors propose three key themes wherein PGR can be integrated into:
   - Cultural, through the discussion of PGR values, and human relationships with preservation of agricultural traditions and practices
   - Biodiversity, through elaborate and interactive discourses of environmental issues and agricultural biodiversity
   - Agriculture/food security, through the discussions of PGRs in food security, sustainable agriculture, and food values
2. The domains mainly focus on the technical skills (i.e., farming skills, entrepreneurial skills) and not on the conceptual understanding and practical knowledge of PGR. The competencies and outcomes also mainly put emphasis on the memorization of concepts and repeated practice. It is therefore recommended that for PGR to be learned in K-12, there is a need to redesign the curriculum with learning competencies focusing on active learning and inquiry-based approach instead of rote memorization. The Committee on Programs for Advanced Study of Mathematics and Science in American High Schools (2002) proposed a curriculum for understanding that is intentionally designed to provide opportunities for in-depth exploration in a variety of contexts. Their proposed

design is supported by the authors in order to make learning about PGR more meaningful and practical. The curriculum for understanding suggests the following:

- Knowledge is structured around major concepts and principles of the discipline. For example, topics on food security, sustainable development, and climate change must include content and competency related to PGR.
- Links new knowledge to what is already known by presenting concepts in a conceptually and logically sequenced order that builds upon previous learning within and across grade levels: In the curriculum, it was observed that the prior learning and experience of students were not considered. Students usually bring in their own cultural context and the curriculum must be designed to accommodate this prior knowledge.
- The focus must be on the depth of understanding rather than the breadth of content coverage by providing students with multiple opportunities to practice and demonstrate what they learn in a variety of contexts.
- Include structured learning activities that will allow students to experience problem-solving and inquiry in situations that are drawn from their personal experiences and real-world applications.
- Develop students' abilities to make meaningful applications and generalization to new problems and contexts.
- Incorporate language, procedures, and models of inquiry and truth verification that are consistent with the accepted practice of experts in the domain.
- Emphasize interdisciplinary connections and integration and help students connect learning in school with the issues, problems, and experiences that figure prominently in their lives outside of the classroom. Lessons should cover all aspects of agricultural biodiversity, ranging from conservation biology of genetic resources through social sciences to policy and legal aspects and see the interplay of these issues in various subject areas.

3. There is a need to capacitate teachers who will be able to integrate PGR in various subject areas. According to Borromeo (2012), there is a need to increase the critical mass of PGR workers who are aware of the most recent developments and opportunities in this field—this could actually be a call to the teachers who could rise up to the challenge of integrating PGR in lessons. They should also continually educate themselves with the current trends of their discipline and subject areas they are teaching.
4. Amendment of laws and policies related to environmental education: It is recommended by the authors to include topics on climate change, biodiversity, food security, and indigenous knowledge in the scope of environmental education.

References

Allem AC (2000) The terms genetic resource, biological resource, and biodiversity examined. Environmentalist 20(4):335–341

Ambal RGR, Duya MV, Cruz MA, Coroza OG, Vergara SG, de Silva N, Molinyawe N, Tabaranza B (2012) Key biodiversity areas in the Philippines: priorities for conservation. J Threat Taxa 4(8):2788–2796. https://doi.org/10.11609/JoTT.o2995.2788-96

Aragón-Cuevas F, Taba S, Castro-Garcia FH, Hernández-Casillas JM, Cabrera-Toledo JM, Alcalá LO, Ramirez ND (2005) In situ conservation and use of local maize races in Oaxaca, Mexico: a participatory and decentralized approach. In: Taba S (ed) Latin America maize germplasm conservation: regeneration, in situ conservation, core subsets, and prebreeding. International Maize and Wheat Improvement Center, Mexico, pp 26–38. https://repository.cimmyt.org/xmlui/bitstream/handle/10883/592/82524.pdf?sequence=4#page=33

Borromeo TH (2012) Importance of plant genetic resources in sustainable development: global challenges, and solutions being developed in the Philippines. J Develop Sustan Agric 7:23–32. https://www.jstage.jst.go.jp/article/jdsa/7/1/7_1_23/_pdf/-char/ja

Borromeo TH, Altoveros NC (2010) Experiences from University of the Philippines Los Baños. In P. Rudebjer et al. (eds.) Reviewing plant genetic resources education in east and Southeast Asia. In: Proceedings of an international workshop, University of Tsukuba, Japan, 17–-19 November 2009, pp. 49–52. Bioversity International

Brockhaus R, Oetmann A (1996) Aspects of the documentation of in situ conservation measures of genetic resources. Plant Genetics Resources Newsletter No. 108

Bronner SE (2009) Critical theory. In: The international encyclopedia of revolution and protest. Wiley, London, pp 1–6

Committee on Programs for Advanced Study of Mathematics and Science in American High Schools (2002) Learning with understanding: seven principles. In: Gollub JP et al (eds) Learning and understanding: improving advanced study of mathematics and science in U.S. high schools. National Academy Press, Washington, pp 117–133

Convention of Biological Diversity (n.d.) Philippines—main details. https://www.cbd.int/countries/profile/?country=ph. Accessed 13 Apr 2021

Critical Theory (2008) Stanford encyclopedia of philosophy. https://plato.stanford.edu/entries/critical-theory/. Accessed 29 Apr 2021

Department of Education (2012) Policy guidelines on the implementation of grades 1 to 10 of the K to 12 basic education curriculum (BEC) effective school year 2012–2013. https://www.deped.gov.ph/wp-content/uploads/2012/04/DO_s2012_31.pdf

Dulloo ME (2019) Maintaining diversity of plant genetic resources as a basis for food security. In: Farranti P et al (eds) Encyclopedia of food security and sustainability, pp 54–63. https://doi.org/10.1016/B978-0-08-100596-5.22208-7

Dulloo ME, Hunter D, Borelli T (2010) *Ex situ* and *in situ* conservation of agricultural biodiversity: major advances and research needs. Notulae Botanicae Horti Agrobotanici Cluj-Napoca 38(2): 123–135. https://doi.org/10.15835/nbha3824878

Ebert AW, Engels JMM (2020) Plant biodiversity and genetic resources matter! Plan Theory 9(12). https://doi.org/10.3390/plants9121706

Ezzamel M, Willmotto H (2010) Rethinking strategy: contemporary perspectives and debates. Eur Manag Rev 1(1):43–48. https://doi.org/10.1057/palgrave.emr.1500009

Foucault M (1982) The subject and power. In: Dreyfus HL, Rabinow P (eds) Michel Foucault: beyond structuralism and hermeneutics. Harvester Wheatsheaf, pp 202–226

Gauchan D, Subedi A, Shrestha P (2000) Identifying and analyzing policy issues in plant genetic resources management: experiences using participatory approaches in Nepal. In: Friis-Hansen E, Sthapit B (eds) Participatory approaches to the conservation and use of plant genetic resources. International Plant Genetic Resources Institute, Rome, pp 188–193

Gepts PL (2004) Who owns biodiversity, and how should the owners be compensated? Plant Physiol 134(4):1295–1307

Khanna PP, Singh N (1991) Conservation of plant genetic resources. In: Paroda RS, Arora RK (eds) Plant genetic resources conservation and management concepts and approaches. International Board for Plant Genetic Resources, New Delhi

Khoury C, Laliberté B, Guarino L (2010) Trends in *ex situ* conservation of plant genetic resources: a review of global crop and regional conservation strategies. Genet Resour Crop Evol 57:625–639. https://doi.org/10.1007/s10722-010-9534-z

Labonite M (2013) Integrating plant genetic resources conservation and management into the state university's mandates: an initiative for sustainable development. Int J Environ Rural Develop 4(1):190–195

Liebig K, Alker D, Chih KO, Horn D, Illi H, Wolf J (2002) Governing biodiversity: access to genetic resources and approaches to obtaining benefits from their use: the case of the Philippines. German Development Institute, Bonn

Moore G, Tymowski W (2005) Explanatory guide to the international treaty on plant genetic resources for food and agriculture. IUCN, Rome

Official Gazette (2008) Republic Act No. 9512. https://www.officialgazette.gov.ph/2008/12/12/republic-act-no-9512/

Official Gazette (2013) Republic Act No. 10533. https://www.officialgazette.gov.ph/2013/05/15/republic-act-no-10533/. Accessed 23 Apr 2021

Ogwu MC, Osawaru ME, Ahana CM (2014) Challenges in conserving and utilizing plant genetic resources (PGR). Int J Genet Mol Biol 6(2):16–22. https://doi.org/10.5897/IJGMB2013.0083

Phillips PWB (2003) Ethics and biosafety/ownership of plant genetic resources. In: Thomas B (ed) Encyclopedia of plant sciences, pp 289–295. https://doi.org/10.1016/B0-12-227050-9/00232-5

Posa MRC, Diesmos AC, Sodhi NS, Brooks TM (2008) Hope for threatened tropical biodiversity: lessons from the Philippines. Bioscience 58(3):231–240. https://doi.org/10.1641/B580309

Ramirez F, Santana J (2019) Environmental education and biodiversity conservation. In: Environmental education and ecotourism. Springer, Cham, pp 7–11. https://doi.org/10.1007/978-3-030-01968-6_2

Rao NK (2004) Plant genetic resources: advancing conservation and use through biotechnology. Afr J Biotechnol 3(2):136–145

Rao VR, Hodgkin T (2002) Genetic diversity and conservation and utilization of plant genetic resources. Plant Cell Tiss Org Cult 68:1–19

Rijal D, Rana R, Subedi A, Sthapit B (2000) Adding value to landraces: community-based approaches for *in situ* conservation of plant genetic resources in Nepal. In: Friis-Hansen E, Sthapit B (eds) Participatory approaches to the conservation and use of plant genetic resources. International Plant Genetic Resources Institute, Rome, pp 166–172

Rudebjer P, Watanabe K, Sebastian L (2010). Reviewing plant genetic resources education in East and Southeast Asia. In: Proceedings of an international workshop, University of Tsukuba, Japan, 17–19 November 2009, Bioversity International

Sapra RL (1991) Documentation of plant genetic resources. In: Paroda RS, Arora RK (eds) Plant genetic resources conservation and management concepts and approaches. International Plant Genetic Resources Institute, Rome

Sarmiento DH, Orale RL (2016) Senior high school curriculum in the Philippines, USA, and Japan. J Acad Res 1(3):12–23

Satheesh P (2000) Linking to community development: using participatory approaches to in situ conservation. In: Friis-Hansen E, Sthapit B (eds) Participatory approaches to the conservation and use of plant genetic resources. International Plant Genetic Resources Institute, Rome, pp 181–187

Sthapit B, Friis-Hansen E (2000) Overview: increasing public and policy awareness of conservation and use of plant genetic resources. In: Friis-Hansen E, Sthapit B (eds) Participatory approaches to the conservation and use of plant genetic resources. International Plant Genetic Resources Institute, Rome, pp 164–165

UCD Teaching and Learning (n.d.) Constructivism and social constructivism in the classroom. In Education, society, & the k-12 learner. https://courses.lumenlearning.com/teachereducationx92x1/chapter/constructivism-and-social-constructivism-in-the-classroom/. Accessed 1 May 2021

Volk GM, Namuth-Covert D, Byrne PF (2019) Training in plant genetic resources management: a way forward. Crop Sci 59(3):853–857

An Inventory of Commonly Cultivated Plant Species in Permaculture Farms in the Philippines

10

Jabez Joshua M. Flores and Inocencio Jr. Buot

Abstract

Permaculture has been known to intentionally integrate diversity into the design of farms and mimic natural landscapes. This approach is in contrast to the mainstream monocropping system in conventional agriculture. The objectives of the study were to identify what plant species are commonly cultivated in permaculture farms and determine its uses as narrated by farmers and practitioners. The researchers conducted a crop inventory in 12 permaculture sites in the Philippines from August to November in 2018. To survey a 1 ha sampling area, a modified belt transect method with alternating 20 m^2 plots was employed for full enumeration of plant species in each plot. To determine uses, key informant interviews and focus group discussions were conducted among farm staff. A total of 215 plant species were identified with an average species richness of 46 per farm. A comparison of crop inventories revealed that *Colocasia esculenta* and *Capsicum frutescens* were the most commonly cultivated crops found in ten sites (83%). It is followed by *Annona muricata* (nine sites) and *Bambusoideae* (eight). Results revealed that the majority of crops found were cultivated for household consumption.

J. J. M. Flores (✉)
School of Environmental Science and Management, University of the Philippines Los Banos, Los Banos, Laguna, Philippines
e-mail: jmflores4@up.edu.ph

I. J. Buot
Institute of Biological Sciences, College of Arts and Sciences, University of the Philippines Los Baños, Los Banos, Laguna, Philippines
e-mail: iebuot@up.edu.ph

S. Ramamoorthy et al. (eds.), *Plant Genetic Resources, Inventory, Collection and Conservation*, https://doi.org/10.1007/978-981-16-7699-4_10

Keywords

Diversified farming systems · Regenerative agriculture · Biodiversity · Agroecology · Agroforestry · Agroecosystems

10.1 Introduction

Permaculture is a design philosophy for sustainable and regenerative agriculture that promotes biodiversity in the design of farms. Practiced by farmers all over the world, it was originally conceptualized in Australia in the 1970s during the global oil crisis. Permaculture was meant to offer an ecologically conscious alternative to industrial agriculture. Its proponents, Mollison and Holmgren (2002), defined permaculture as "consciously designed landscapes which mimic the patterns and relationships found in nature, while yielding an abundance of food, fiber, and energy for provision of local needs." This design concept encouraged grassroots communities to take action and start their own initiatives and local resource networks. Permaculture came from the combination of the words "permanent" and "agriculture." And later on, it evolved to mean "permanent" and "culture" since the practice transcended agriculture and was applied to different aspects of holistic living among farmers and non-farmers alike.

Industrial (or conventional) farms (Crossley et al. 2020) are known to use a plantation-type monocropped system of production for higher yield and efficient farm management. This is especially evident in the production of major cereal crops such as corn, wheat, rice, and soy. Though economically viable, its long-term effects include soil degradation, water pollution, and consequently loss of wildlife habitats and biodiversity. In contrast, a farm management system based on diversity (Radeschall et al. 2021), like permaculture, mimics natural landscapes and organizes a large variety of crops into zones allocated for both agricultural and natural biodiversity. In addition, crop diversity is also a strategy for food security and climate change adaptation (Medlrum et al. 2018). One permaculture design principle commonly used by practitioners is "use and value diversity" (Principle 10, Holmgren 2002). The word "diversity" in this context may refer to individual skills, farm services offered, or farm system components in general. But in this chapter, we focus on crop diversity.

In a study of permaculture farms in Indonesia, Putro and Miyauro (2020) reported that one distinguishing feature of a permaculture landscape is its biodiversity which mimics that of a forest ecosystem. Unlike the homogenous landscape of conventional farms that consist of only 1–2 main crops in a monoculture design, a permaculture farm has significantly more biodiversity (30–54 plant species) as well as a diversity of farm services and functions being offered (Putro and Miyaura 2020). Dominated mostly by perennial crops, plant and tree species diversity is beneficial for food, forage, fuelwood, windbreaks, and wildlife habitats (Zhao et al. 2021). Other literature reported that permaculture farms are characterized by polyculture cropping (Roberts 2017) using organic farming techniques (Hirschfeld and Van Acker 2019) and integrated crop-livestock systems (Lovell et al. 2010). In other studies, crop diversity has been linked to more employment in the agriculture sector (Garibaldi and Perez-Mendez 2019) and stability in food production (Renard and Tilman 2019).

The objectives of this chapter were to identify what plant species are commonly cultivated in permaculture farms in the Philippines and determine its uses as narrated by farmers. A crop inventory of existing permaculture farms will help decision-making and farm management of practitioners and farmers alike.

10.2 Materials and Methods

10.2.1 Study Areas

A total of 12 self-identified permaculture sites, or farms, from the major island groups of Luzon, Visayas, and Mindanao (Fig. 10.1) were selected for the study using maximum variation sampling (Cohen and Crabtree 2006). Variables considered for selection were climate type, geographic location, land area, and availability of respondents to participate. Fieldwork was conducted from August 31 to November 6, 2018.

10.2.2 Data Collection

A modified belt transect method (Grant et al. 2004) was employed to perform an inventory of plant species in a 1 ha sampling area. The method involved a setup of two intersecting 100 m transect lines with alternating 20 m^2 plots along the lines. Full enumeration of plant species was performed in each of the 38 plots with corresponding photographic evidence. Aerial photos of sampling areas were also taken using a drone. Key informant interviews and focus group discussions with farmers were then conducted to help identify plant species and its uses.

10.3 Results and Discussion

A total of 215 plant species were identified (Table 10.1) with an average of species richness (SR) of 46 per farm. In terms of crop life cycle, perennial crops comprise the majority (75–95%) of plant species on the landscape in all sites. The landscapes in which these plant species were found are arranged in a "zone" pattern based on the frequency of use and energy spent that is unique to permaculture design. Majority of the crops in the inventory were located in the "garden" (Zone 1) and "food forest" (Zone 4) zones. By design, permaculture practitioners plant annual vegetable crops in Zone 1 close to the house structure (Zone 0) for accessibility and better crop management. The "grazing" area (Zone 2) is dedicated to poultry, livestock, and forage crops interfacing with the area for cash crops (Zone 3) such as rice and corn.

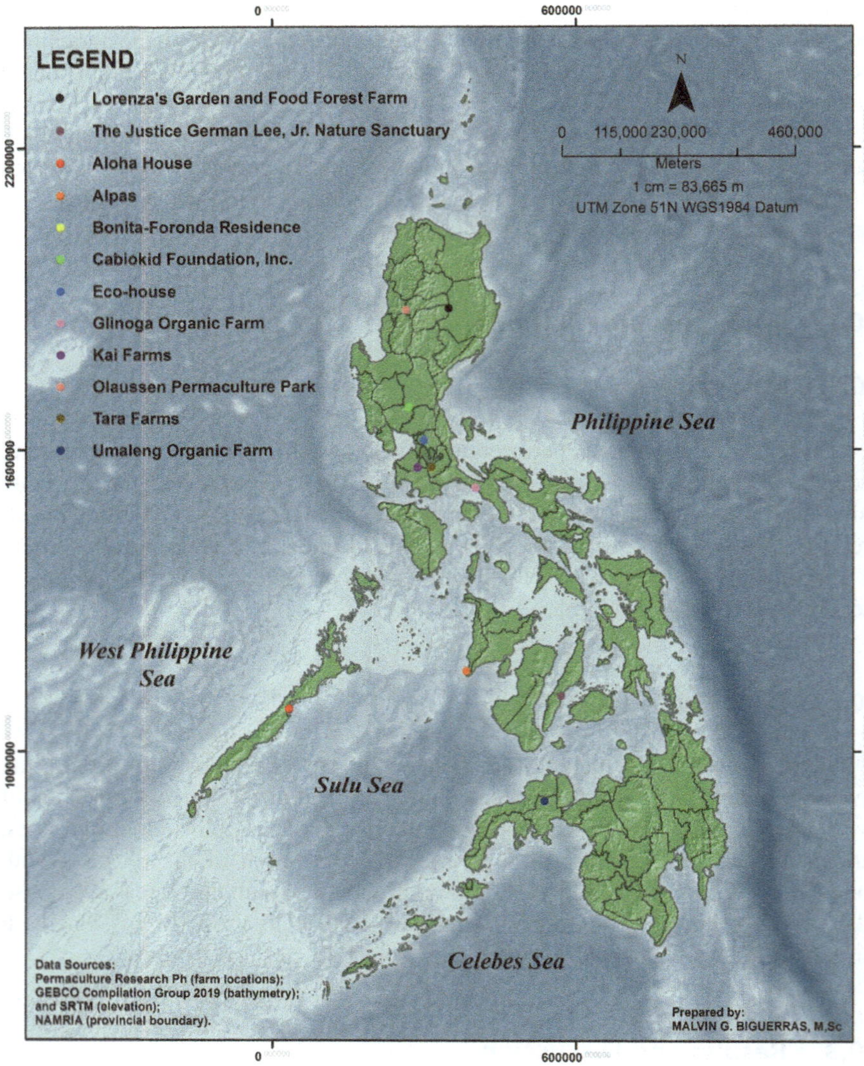

Fig. 10.1 Twelve permaculture sites across the Philippines were visited to conduct a crop inventory

The food forest contains the majority of perennial species for food, fuel, and lumber. Zone 4 also serves as a transition area to the "wilderness" (Zone 5) where plant species beyond the farm property are not disturbed. A prevailing "ecological" perspective of permaculture among practitioners in the Philippines allowed a diversity of plant species to be included in the design of farms which is in direct contrast to the monoculture approach of conventional or industrial farms. With this ecological perspective, plants were not exclusively viewed as food sources for humans but

Table 10.1 A total of 215 plant species were documented and grouped into six use categories: vegetable/cereal crops, tree/fruit-bearing crops, ornamentals, herbs, feed crops, and weeds

Inventory of plant species by use category			
Use category	Scientific name	Common name	Zone 0—House 1—Garden 2—Grazing 3—Cash Crops 4—Food Forest 5—Wilderness (Flores and Buot 2021)
Vegetable/ cereal crops	*Abelmoschus esculentus*	Okra	1
	Alocasia macrorrhizos	Giant taro	1
	Alpinia galanga	Langkawas	1
	Amaranthus	Amaranth	1, 2
	Amorphophallus campanulatus	Elephant foot yam	1
	Arachis hypogaea	Peanut	1
	Brassica rapa L.	Bok choy	1
	Brassica rapa subsp. Chinensis	Chinese cabbage	1
	Brassica juncea	Oriental mustard	1
	Brassica oleracea var. sabellica	Curly kale	1
	Brassica oleracea Alboglabra Group	Chinese kale	1
	Basella alba	Malabar spinach	1
	Pachyrhizus erosus	Jicama	1
	C. lacryma-jobi	Job's tears	1, 3
	Cajanus cajan	Pigeon pea	1
	Capsicum annuum	Siling haba	1, 3
	Capsicum annuum Group	Green bell pepper	1
	Capsicum chinense Habanero Group	Habanero	1
	Capsicum frutescens	Chili pepper	1, 3
	Colocasia esculenta	Taro	1
	Cucumis sativus	Cucumber	1
	Cucurbita	Squash	1
	Curcuma longa	Turmeric	1
	Daucus carota subsp. *Sativus*	Carrot	1
	Dioscorea alata	Purple yam	1
	Diplazium esculentum	Pako fern	1,2

(continued)

Table 10.1 (continued)

Inventory of plant species by use category			
Use category	Scientific name	Common name	Zone 0—House 1—Garden 2—Grazing 3—Cash Crops 4—Food Forest 5—Wilderness (Flores and Buot 2021)
	Eruca vesicaria ssp. *sativa*	Arugula	1
	Ipomoea aquatica	Water spinach	1, 2, 3
	Ipomoea batatas	Sweet potato	1
	Lactuca sativa	Green ice lettuce	1
	Lactuca sativa L. var. *longifolia*	Romaine lettuce	1
	Manihot esculenta	Cassava	1, 3
	Melothria scabra	Cucamelon	1, 2, 3
	Momordica charantia	Ampalaya	1
	Oryza sativa	Rice	3
	Raphanus sativus	Radish	1
	Saccharum officinarum	Sugarcane	3
	Solanum lycopersicum	Tomato	1
	Solanum lycopersicum var. cerasiforme	Cherry tomato	1
	Solanum melongena	Eggplant	1
	Vigna radiata	Mung bean	1, 2, 3
	Vigna unguiculata ssp. *sesquipedalis*	Yardlong bean	1
	Zea mays	Corn	3
	Zingiber officinale	Ginger	1
Tree/fruit-bearing crops	*Acacia sp.*	Acacia	4
	Adonidia merrillii	Manila palm	4
	Ananas comosus	Pineapple	1
	Annona muricata	Soursop	4
	Annona squamosa	Sugar apple	4
	Artocarpus altilis	Breadfruit	4
	Artocarpus cumingiana Trec.	Anubing	4
	Artocarpus heterophyllus	Jackfruit	4
	Averrhoa bilimbi	Bilimbi	4
	Azadirachta indica	Neem	4

(continued)

Table 10.1 (continued)

Inventory of plant species by use category			
Use category	Scientific name	Common name	Zone 0—House 1—Garden 2—Grazing 3—Cash Crops 4—Food Forest 5—Wilderness (Flores and Buot 2021)
	Bambusa vulgaris	Common bamboo	4
	Bixa orellana	Achiote	1, 4
	Caesalpinia pulcherrima	Peacock flower	4
	Carica papaya	Papaya	1, 4
	Carissa carandas	Bengal currant	1, 4
	Castanea	Chestnut	1
	Casuarina equisetifolia	Agoho	4
	Ceiba pentandra	Kapok	4
	Citrofortunella macrocarpa	Calamansi	1, 4
	Citrus × limon	Lemon	1, 4
	Cocos nucifera	Coconut	1, 4
	Corypha elata	Buri palm	4
	Cycas revoluta	Sago palm	4
	Ficus nota	Tibig	4
	Ficus odorata	Pakiling	4
	Fragaria x ananassa	Strawberry	1
	Garuga floribunda	Bogo	4
	Gliricidia sepium	Madre de cacao	1, 2, 4
	Gmelina arborea	Gmelina	4
	Hylocereus sp.	Dragon fruit	1, 2, 3
	Lagenaria siceraria	Calabash	4
	Leucaena leucocephala	Ipilipil	2, 4
	Macaranga tanarius	Parasol leaf tree	4
	Mangifera indica	Mango	4
	Manihot glaziovii	Tree cassava	4
	Manilkara zapota	Chico	4
	Moringa oleifera	Moringa	1, 4
	Morus rubra	Red mulberry	4
	Murraya koenigii	Curry tree	4
	Musa sp.	Banana	1, 4
	Nypa fruticans	Nipa palm	4
	Passiflora edulis	Passion fruit	1

(continued)

Table 10.1 (continued)

Inventory of plant species by use category

Use category	Scientific name	Common name	Zone 0—House 1—Garden 2—Grazing 3—Cash Crops 4—Food Forest 5—Wilderness (Flores and Buot 2021)
	Persea americana	Avocado	4
	Phyllostachys nigra	Black bamboo	4
	Phoenix dactylifera	Date palm	4
	Plumeria alba	Frangipani	1
	Polyalthia longifolia	Indian mast tree	4
	Psidium guajava	Guava	4
	Pterocarpus	Narra	4
	Ribes uva-crispa	Gooseberry	1
	Sambucus nigra	Elderberry	1
	Sandoricum koetjape	Santol	4
	Saribus rotundifolius	Anahaw	4
	Schefflera sp.	Umbrella tree	4
	Spondias purpurea	Siniguelas	4
	Terminalia catappa	Indian almond	4
	Theobroma cacao	Cacao	4
	Thevetia peruviana	Campanilla	4
	Vitex parviflora	Molave	4
Ornamentals	*Acrostichum danaeifolium*	Leather fern	1
	Adiantum formosum	Black stem maidenhair fern	1
	Adiantum trapeziforme	Diamond maidenhair fern	1
	Alpinia purpurata	Red ginger	1
	Alternanthera bettzickiana	Calico plant	1
	Alternanthera ficoidea	Joseph's coat	1
	Aloe vera	Aloe vera	1
	Arachis pintoi	Pinto peanut	1
	Ardisia crenata	Christmas berry	1
	Asplenium musifolium	Pakong babae	1

(continued)

Table 10.1 (continued)

Inventory of plant species by use category			
Use category	Scientific name	Common name	Zone 0—House 1—Garden 2—Grazing 3—Cash Crops 4—Food Forest 5—Wilderness (Flores and Buot 2021)
	Asteraceae	Aster	1
	Bromeliaceae	Bromeliads	1
	Bougainvillea glabra	Bougainvillea	1
	Chrysopogon zizanioides	Vetiver	1
	Coleus blumei	Mayana	1
	Combretum indicum	Chinese honeysuckle	1
	Cordyline fruticosa	Ti	1
	Cosmos bipinnatus	Cosmos	1
	Crassula ovata	Jade plant	1
	Cyclopeltis crenata	Fern	1
	Dieffenbachia amoena	Bakya	1
	Dieffenbachia camilla	Camille	1
	Digitaria	Crabgrass	1, 2
	Dracaena reflexa	Song of India	1
	Dracaena trifasciata	Snake plant	1
	Drynaria quercifolia	Oakleaf fern	1
	Duranta repens	Golden dewdrop	1
	Ehretia microphylla	Fukien tea tree	1
	Euphorbia heterophylla	Japanese poinsettia	1
	Ficus pseudopalma	Philippine fig	1, 2, 4
	Ficus ulmifolia	Ficus	1, 2, 4
	Helianthus sp.	Sunflower	1
	Heliconia rostrata	Lobster claw plant	1
	Heliconia psittacorum	Lady Di	1
	Heliconia indica "Spectabilis"	Spectabilis	1
	Hibiscus	Gumamela	1

(continued)

Table 10.1 (continued)

Inventory of plant species by use category			
Use category	Scientific name	Common name	Zone 0—House 1—Garden 2—Grazing 3—Cash Crops 4—Food Forest 5—Wilderness (Flores and Buot 2021)
	Hippobroma longiflora	Star of Bethlehem	1
	Impatiens balsamina	Kamantigi	1
	Ipomoea triloba	Littlebell	1
	Ixora coccinea	Jungle geranium	1
	Jasminum sambac	Sampaguita	1
	Laportea interrupta (L.)	Lipang-aso	1, 2
	Lindernia procumbens	Common false pimpernel	1
	Maranta arundinacea	Arrowroot plant	1
	Monochoria hastata	Gabi-gabihan	1
	Musa ornata	Flowering banana	1
	Myosotis	Forget-me-not	1
	Nelumbo nucifera	Indian lotus	1
	Nephrolepis obliterata	Kimberly queen fern	1
	Opuntia	Prickly pear cactus	1
	Orchidaceae	Orchid	1
	Paspalum conjugatum	Carabao grass	1, 2
	Passiflora foetida	Stinking passion flower	1
	Philodendron Ceylon	Philodendron golden plant	1
	Pilea nummulariifolia	Creeping Charlie	1
	Rhapis excelsa	Broadleaf lady palm	1
	Rosa	Rose	1
	Solanum diphyllum	Two-leaf nightshade	1
	Spathiphyllum wallisii	Peace lily	1

(continued)

Table 10.1 (continued)

Inventory of plant species by use category			
Use category	Scientific name	Common name	Zone 0—House 1—Garden 2—Grazing 3—Cash Crops 4—Food Forest 5—Wilderness (Flores and Buot 2021)
	Strelitzia reginae	Bird of paradise	1
	Syngonium podophyllum	Arrowhead plant	1
	Tabernaemontana pandacaqui	Banana bush	1
	Tagetes	Marigold	1
	Taraxacum	Dandelion	1
	Tradescantia zebrina	Wandering Jew	1
	Trichanthera gigantea	Madre de agua	1, 2
	Tridax procumbens	Tridax daisy	1
	Turnera diffusa	Damiana	1
	Vinca	Periwinkle	1
	Zephyranthes sp.	Fairy lily	1
	Zingiber zerumbet	Bitter ginger	1
Herbs	*Allium schoenoprasum*	Chives	1
	Anethum graveolens	Dill	1
	Artemisia dracunculus	Tarragon	1
	Carmona retusa	Fukien tea	1
	Centella asiatica	Gotu kola	1
	Chamaecostus cuspidatus	Insulin plant	1
	Chromolaena odorata	Hagonoy	1, 2, 3
	Clitoria ternatea	Butterfly pea	1
	Corchorus olitorius	Saluyot	1
	Coriandrum sativum	Coriander	1
	Cymbopogon citratus	Lemongrass	1, 2, 3
	Eryngium foetidum	Sawtooth coriander	1
	Euphorbia hirta	Asthma plant	1
	Gynura procumbens	Gynura	1

(continued)

Table 10.1 (continued)

Inventory of plant species by use category			
Use category	Scientific name	Common name	Zone 0—House 1—Garden 2—Grazing 3—Cash Crops 4—Food Forest 5—Wilderness (Flores and Buot 2021)
	Hibiscus sabdariffa	Roselle	1, 3
	Ipomoea indica	Blue morning glory	1
	Mentha	Mint	1
	Mentha × piperita	Peppermint	1
	Mentha spicata	Spearmint	1
	Mimosa diplotricha	Giant sensitive plant	1
	Ocimum basilicum	Sweet basil	1
	Ocimum tenuiflorum	Holy basil	1
	Origanum vulgare	Oregano	1
	Pandanus amaryllifolius	Pandan	1
	Paspalidium flavidum	Watercrown grass	1
	Peperomia pellucida	Pansit-pansitan	1
	Petroselinum crispum	Parsley	1
	Portulaca	Purslane	1
	Ocimum basilicum "Purpurascens"	Red Rubin basil	1
	Salvia rosmarinus	Rosemary	1
	Talinum fruticosum	Talinum	1
Feed crops	*Azolla pinnata*	Azolla	1
	Broussonetia papyrifera	Paper mulberry	1, 2, 4
Weeds	*Calopogonium mucunoides*	Calopo	1, 2, 3
	Cleome rutidosperma	Fringed spider flower	1, 2
	Commelina diffusa	Climbing dayflower	1, 2
	Cyperus kyllingia	White kyllinga	1, 2
	Galinsoga parviflora	Potato weed	1, 2, 3
	Imperata cylindrica	Cogon grass	1, 2

(continued)

Table 10.1 (continued)

Inventory of plant species by use category			
Use category	Scientific name	Common name	Zone 0—House 1—Garden 2—Grazing 3—Cash Crops 4—Food Forest 5—Wilderness (Flores and Buot 2021)
	Melochia concatenata	Bankalanan	1, 2
	Mimosa pudica	Shameplant	1, 2
	Sida acuta	Broom weed	1, 2, 3

as landscape components which provide food and habitat for nonhuman species as well.

Inventoried plants were categorized into six "use" categories: vegetable/cereal crops, tree/fruit-bearing crops, ornamentals, herbs, feed crops, and weeds. Each category is discussed as follows.

Vegetable/Cereal Crops This category refers to either annual or perennial plants that are mainly used for household food security and livelihood. A total of 45 species were recorded. Plants belonging to this category were the most abundant on the landscape in 50% of the sites.

Tree/Fruit-Bearing Crops This category refers to multifunctional plant species that produce either fruit, fiber, or fuelwood for personal consumption and commercial production. A total of 61 species were inventoried. Plants in this category dominated the landscape in 20% of the sites.

Ornamentals This category broadly refers to plants that are used for aesthetic and sociocultural purposes. A total of 75 species were observed. Ornamental plants were commonly found planted near houses as hanging plants, hedges, green walls, or small garden patches.

Herbs This category refers to generally aromatic plant species particularly used for culinary, medicinal, and pest management purposes. A total of 32 species were documented. It is found near house structures and intercropped in vegetable gardens.

Feed Crops This category refers to plant species for nonhuman consumption. A total of two species were documented. *A. pinnata* was observed in man-made ponds used for aquaponic systems. *Broussonetia papyrifera* (paper mulberry) was observed growing abundantly in farm edges. Its leaves were said to be used as forage for livestock.

Weeds This category refers to plant species that were not propagated intentionally but were left to grow on the landscape. A total of only eight species were found in the sampling plots mostly due to the density and diversity of other crops in the shared space (Pakeman et al. 2019). Some weeds were used as either cover crops, living mulch, sacrificial crops, animal forage, or biomass material for making compost.

Table 10.2 shows the most common plant species found in at least 50% of permaculture sites. *Colocasia esculenta* (taro) and *Capsicum frutescens* (chili pepper) were the most common crops found in 83% of sites. Both species were said to be easily propagated either intentionally or unintentionally for household consumption according to interviews. Taro was observed in low-elevation waterlogged areas while chili pepper was reported to be propagated by droppings from chickens and birds according to respondents.

Annona muricata (soursop) was the most common fruit-bearing tree found in 75% of sites. Soursop was a multifunctional tree used by practitioners that they claim to provide many health benefits. The fruit eaten is fresh or made into juice concentrate while the leaves and bark were steeped for tea. *Bambusoideae sp.* (bamboo) was found in 67% of sites with its young shoots (called *labong* in Tagalog) commonly consumed as a household meal while its stems were used as construction materials for building houses and furniture. The plant species also functioned as living fences and windbreaks.

Abelmoschus esculentus (okra) is an easy-to-grow vegetable that was found in 58% of sites. Okra's resilient characteristics as a perennial crop make it a favored high-value crop for household consumption and livelihood. It is the most preferred vegetable food source among all plant species according to interviews. *Carica papaya* (papaya), *Curcuma longa* (turmeric), and *Moringa oleifera* (malunggay) were also documented in the same number of sites. According to respondents, papaya and moringa were preferred food sources ranking fourth and fifth, respectively, when asked which crops within the farm they consumed regularly.

Ipomoea aquatica (water spinach) and *Solanum lycopersicum* (tomato) were the only two plant species found in 50% of sites. Both the crops were reported to be a regular part of the daily household diet along with previous crops mentioned. This demonstrates that the link between crop diversity and dietary diversity (Singh et al. 2020; Madzorera et al. 2021) is well established in permaculture (Fig. 10.2).

10.4 Conclusion

The design emphasis on biodiversity is a significant departure from the common practice of monoculture in conventional farming, thus creating heterogeneous landscapes that mimic the appearance of forests more than farms. However, the scale of permaculture design is still limited to smallholder farm and home garden landscapes.

In theory, permaculture is scalable but examples of whole communities benefiting from the collective design efforts of multiple practitioners originating from a common locale have yet to be documented in the country. In addition to scale, annual crop yield from mixed cropping and agroforestry systems and the income derived

Table 10.2 Ten plant species mostly belonging to the vegetable/cereal crops category were found in at least 50% of permaculture sites

Plant species	Common name	Use category	Site												
			1	2	3	4	5	6	7	8	9	10	11	12	Total (%)
Colocasia esculenta	Taro	Vegetable/cereal crops	✓	✓		✓	✓	✓	✓		✓	✓	✓	✓	83
Capsicum frutescens	Chili pepper	Vegetable/cereal crops	✓	✓	✓	✓		✓	✓	✓	✓		✓	✓	83
Annona muricata	Soursop	Tree/fruit-bearing crops	✓			✓	✓	✓	✓	✓	✓	✓		✓	75
Bambusoideae	Bamboo	Tree/fruit-bearing crops	✓	✓	✓	✓	✓	✓		✓	✓				67
Abelmoschus esculentus	Okra	Vegetable/cereal crops	✓	✓	✓		✓	✓				✓		✓	58
Carica papaya	Papaya	Tree/fruit-bearing crops	✓		✓	✓		✓		✓		✓	✓		58
Curcuma longa	Turmeric	Herbs	✓	✓				✓	✓			✓	✓	✓	58
Moringa oleifera	Moringa	Tree/fruit-bearing crops	✓	✓		✓			✓	✓		✓	✓		58
Ipomoea aquatica	Water spinach	Vegetable/cereal crops	✓				✓	✓			✓		✓	✓	50
Solanum lycopersicum	Tomato	Vegetable/cereal crops	✓				✓	✓			✓		✓	✓	50

Fig. 10.2 Taro (**a**), chili pepper (**b**), soursop (**c**), bamboo (**d**), okra (**e**), papaya (**f**), turmeric (**g**), and moringa (**h**) are six of the ten common crops in at least 50% of permaculture sites visited

from diversified revenue streams generated by farm multifunctionality were not included in the study.

Permaculture has yet to receive enough attention in order to attract institutional support to promote its practice. But with more research, especially on local examples of successful practitioners and their landscape designs, permaculture can provide the necessary design framework for restructuring our agricultural landscapes into diverse and productive ecosystems and redefining our relationship with nature.

References

Cohen D, Crabtree B (2006) Qualitative Research Guidelines Project. http://www.qualres.org/HomeMaxi-3803.html

Crossley MS, Burke KD et al (2020) Recent collapse of crop belts and declining diversity of US agriculture since 1840. Glob Chang Biol 27(1):151–164. https://doi.org/10.1111/gcb.15396

Flores JJM, Buot IE (2021) The structure of permaculture landscapes in the Philippines. Biodiversitas 22(4):2032–2044. https://doi.org/10.13057/biodiv/d220452

Garibaldi LA, Perez-Mendez N (2019) Positive outcomes between crop diversity and agricultural employment worldwide. Ecol Econ 164:5. https://doi.org/10.1016/j.ecolecon.2019.106358

Grant T, Madden EM et al (2004) Monitoring native prairie vegetation: the belt transect method. Ecol Restor 22(2):106–111. https://doi.org/10.3368/er.22.2.106

Hirschfeld S, Van Acker R (2019) Permaculture farmers consistently cultivate perennials, crop diversity, landscape heterogeneity and nature conservation. Renew Agric Food Syst 35(3): 342–351. https://doi.org/10.1017/S1742170519000012

Holmgren D (2002) Permaculture: principles & pathways beyond sustainability. Permanent Publications

Lovell ST, Desantis S, Nathan CA, Olson MB, Mendez VE, Kominami HC, Erickson DL, Morris KS, Morris WB (2010) Integrating agroecology and landscape multifunctionality in Vermont: an evolving framework to evaluate the design of agroecosystems. Agric Syst

Madzorera I, Blakstad MM et al (2021) Food crop diversity, women's income-earning activities, and distance to markets in relation to maternal dietary quality in Tanzania. J Nutr 151(1): 186–196. https://doi.org/10.1093/jn/nxaa329

Meldrum G, Mijatovic D et al (2018) Climate change and crop diversity: farmers' perceptions and adaptation on the Bolivian Altiplano. Environ Dev Sustain, 20, 703–730 doi:https://doi.org/10.1007/s10668-016-9906-4

Pakeman RJ, Brooker AJ et al (2019) Increased crop diversity reduces the functional space available for weeds. Weed Res 60(2):121–131. https://doi.org/10.1111/wre.12393

Putro RH, Miyaura R (2020) Indonesian permaculture: factors shaping permaculture farm systems in humid tropical Indonesia. Trop Agric Develop 64(33):113–124

Radeschall CA, Bommarco R et al (2021) Landscape crop diversity and semi-natural habitat affect crop pollinators, pollination benefit and yield. Agric Ecosyst Environ 306:1. https://doi.org/10.1016/j.agee.2020.107189

Renard D, Tilman D (2019) National food production stabilized by crop diversity. Nature 571:257–260. https://doi.org/10.1038/s41586-019-1316-y

Roberts T (2017) Perennial polycultures and the richness of diversity. https://permaculturenews.org/

Singh S, Jones AD et al (2020) The association between crop and income diversity and farmer intra-household dietary diversity in India. Food Sec 12:369–390. https://doi.org/10.1007/s12571-020-01012-3

Zhao H, Li J et al (2021) Crop diversity at the landscape level affects the composition and structure of the vegetation-dwelling arthropod communities in naked oat (Avena chinensis) fields. Int J Environ Res Public Health 18(1):30. https://doi.org/10.3390/ijerph18010030

Conservation of Musa Diversity and Germplasm Management at the National Plant Genetic Resources Laboratory in the Philippines

11

Lavernee S. Gueco, Maila S. Tejano, Lyka A. Yanos, Michelle Lyka V. Descalsota, Jonathan C. Descalsota, and Rachel C. Sotto

Keywords

Banana · Conservation · Gene bank

11.1 Introduction

The loss of biodiversity caused by vegetative clearing, habitat destruction, industrialization, chemical pollution, introduction of exotic species and high-yielding varieties, and natural calamities has resulted in genetic erosion and extinction of many species. Genetic uniformity had also caused catastrophic events in the past; including the famine in Ireland due to potato late blight (Berkeley 1846) and elimination of corn hybrids carrying t-cytoplasm caused by Southern corn blight (Ullstrup 1972) to mention a few. In banana, the occurrence of fusarium wilt caused by a fungal pathogen has led to the devastation of susceptible Gros Michel variety in Panama and were replaced by Cavendish varieties in the 1950s (Promusa 2021). A new race of Fusarium tropical race 4 is currently a threat for the Cavendish banana which occupies almost 50% of banana production worldwide. These prompted many institutions worldwide to conserve plant genetic resources over the last 50 years (Yong-Bi Fu 2017).

In 2016, the global strategy for the conservation and use of Musa (banana) genetic resources was documented and prepared by the members of the Global Musa

L. S. Gueco (✉) · M. S. Tejano · L. A. Yanos · M. L. V. Descalsota · J. C. Descalsota
National Plant Genetic Resources Laboratory, Institute of Plant Breeding, College of Agriculture and Food Science, University of the Philippines, Los Baños, Laguna, Philippines

R. C. Sotto
Institute of Biological Sciences, College of Arts and Sciences, University of the Philippines, Los Baños, Laguna, Philippines

S. Ramamoorthy et al. (eds.), *Plant Genetic Resources, Inventory, Collection and Conservation*, https://doi.org/10.1007/978-981-16-7699-4_11

Genetic Resources Network. Its vision is the security of the world's Musa genetic diversity, valued and utilized to support the livelihood and improve food and nutrition (MusaNet 2016). Based on a global survey carried out from 2012 to 2015, there are 9051 accessions of *Musa* germplasm conserved in the field gene bank, 4507 accessions in vitro, 898 accessions in the greenhouse, and 926 accessions in cryopreservation collections (MusaNet 2016). In the Philippines, the National Plant Genetic Resources at the University of the Philippines-Los Baños, Laguna, and the Bureau of Plant Industry of the Department of Agriculture in Davao City have been identified as key institutions in conserving the national as well as regional banana collections.

Banana is one of the most important fruit crops in the Philippines both for domestic and export markets. In 2019, the total area planted with banana was estimated at 449,030 ha with the total volume of production estimated at 9.157 million metric tons (Philippine Statistics Authority 2021). The Cavendish, which is the export variety, accounts for more than 52% of the total volume of banana produced in the Philippines with 4.85 MT produced in 2019. In terms of the area planted, the Saba variety occupies more than 180,000 ha representing 41% of the total banana hectarage. Saba is the most important local banana cultivar, which is planted in all the provinces in the Philippines. Saba is mostly eaten boiled/steamed, processed into chips, or cooked into banana cue, "turon," and "maruya" (Gueco 2020). Lakatan variety is the most popular desert cultivar, which commands a premium price in the market.

11.1.1 Diversity of Musa Genetic Resources in the Philippines

According to Valmayor et al. (2002), the Philippine archipelago lies within the area of greatest *Musa* diversity. Indigenous *Musa species* include *Musa balbisiana Colla, Musa acuminata Colla*, and *Musa textilis* with several interspecific hybrids that occur naturally, enriching the diversity of *Musa* germplasm in the country. The wild *M. balbisiana* and wild *M. acuminata* are recognized as the parents of edible, commercial banana cultivars. It is reported that there are 91 cultivars in the Philippines that are primarily of local origin. It is composed of 47 dessert types, 34 cooking bananas, and 7 dual-purpose types, i.e., can serve as dessert and cooking bananas.

Aside from Saba, Cavendish, and Lakatan, some of the known indigenous and introduced banana cultivars include Latundan, Morado, Sulay Baguio, Pitogo, Pastilan, Tindok, and Pisang Seribu among others which still remain in the gardens and backyards of some indigenous people in the country. Latundan is also considered as one of the most common dessert bananas, which are grown extensively in backyards and commercial farms. Morado is a popular backyard mutant dessert cultivar with red, sweet, and aromatic fruits. The red color is very attractive and is believed to bring good luck in some cultures. There are also other orange/yellow-colored flesh bananas which have a high beta-carotene and vitamin C content than other banana cultivars. Sulay Baguio is a typhoon-resistant cultivar that stands about

1.5 m in height. Pitogo and Pastilan are two of the most distinct and curious cultivars found in the country. Pitogo is a cultivar with roundish fruits and very compact bunch while Pastilan usually produces two or more bunches in a single plant. Another cultivar known as the most popular plantain in the country is Tindok. It has no male bud with very large fruits of highest quality. On the other hand, Pisang Seribu is known for its extended fruit bunch that nearly touches the ground. It is also called as the thousand-fruited banana. Further studies should be done in order to unlock the potential use of these cultivars and to increase its utilization.

Another *Musa* species, the abaca (*Musa textilis*) known internationally as Manila hemp, is a native plant species in the Philippines. The close relative of banana is the most important fiber and export product of the country. It is endemic to the Philippines being the top producer of abaca fiber, the strongest among the existing natural fibers. It is reported that there are about 400 cultivars of abaca found in the country but only 20 are commercially important. Different abaca varieties were produced according to geographic regions. Corresponding produce of abaca varieties by regions are the following: in Bicol Region (Tinawagan Pula, Lausigun, Sogmad), in Eastern Visayas (Inosa, Linawaan, Laylay), and in Mindanao (Tangongon, Bungolanon, Maguindano). In 2017, PHILFIDA registered the first three varieties of abaca with the National Seed Industry Council (NSIC), namely Abuab, Inosa, and Tangongon. These newly approved varieties are the ones recommended for commercial planting for abaca production (PhilFIDA 2017). Recent breeding techniques are being employed to develop varieties with desirable traits like high yield, good quality, and resistance to pests and diseases that are important for the strengthening of the security of genetic resources available in the country.

11.1.2 Conservation of Musa Genetic Resources in the Philippines

The Bureau of Plant Industry - Davao National Crop Research, Development and Production Support Center (BPI-DNCRDPSC) of the Department of Agriculture (DA) in Bago-Oshiro, Davao City maintains the SouthEast Asian Banana and Plantain Germplasm Resource Center while NPGRL holds the duplicate collection of the SouthEast Asian Banana and Plantain Germplasm. The BPI-DNCRDPSC established the South East Asian (SEA) banana field gene bank in 1977 which contained accessions assembled from the Philippines, Malaysia, Thailand, Indonesia, as well as Papua New Guinea and Belgium.

National efforts in Musa conservation in the University of the Philippines Los Baños started when the National Plant Genetic Resources Laboratory (NPGRL) was established on November 12, 1976, through Presidential Decree No. 1046-A. The NPGRL was mandated to serve as the national repository of important and potentially useful agricultural crops. Presently, NPGRL is one of the divisions of the Institute of Plant Breeding, College of Agriculture and Food Science, University of the Philippines Los Baños. The NPGRL's gene bank ensures availability of diverse, economically, and potentially important crops through collecting missions and conservation in situ and ex situ. Collected plant species are being conserved and characterized to identify accessions with superior quality for selection and breeding purposes for specific trait. Selected accessions may be used in improving and developing new varieties that can withstand the changing climate patterns with resistance to insect pests and diseases. Breeders are now focusing not only on increasing yield but also on superior quality, resistance to diseases and pests, and better adaptation to abiotic stresses.

The conservation of Musa germplasm by NPGRL focuses on three strategies: field gene bank, in vitro conservation, and seeds.

Edible bananas do not produce seeds and are usually propagated vegetatively such as suckers and corms. However, wild Musa species and other related families produce viable seeds. Most of the plant germplasm conserved in field gene banks are those that are either difficult to conserve as seeds (orthodox) or asexually propagated crops (Reed et al. 2004). Field gene banks are considered the more practical and cost-effective way of clonal crop germplasm conservation specifically when resources and skills are limiting. Bananas are one of the most expensive plant germplasms to conserve, which annually costs around $650 per accession (Yong-Bi Fu 2017). One of the advantages of banana field gene bank conservation is the preservation of genetic integrity. Plants maintained in the field over a long period of time have lower risks of losing genetic integrity.

Asexually propagated crops including root crops and *Musa* species are conventionally conserved ex situ in field gene banks and pot collections; however, conditions in the field are not always secure, hence threatening the safety of these germplasms. Tissue culture-based conservation which is considered as an ex situ form of plant germplasm conservation has been established and implemented worldwide to complement field and seed-bank conservation (Oliveira et al. 2000).

Some Musa species such as the wild *M. balbisiana* and wild *M. acuminata* contain viable seeds. These seeds are classified as orthodox which can be kept viable by drying at lower moisture content and stored under a low-temperature storage. This is a very simple method of conserving germplasm collection and is only applicable for Musa species with viable seeds. However, regeneration of the genetic materials once seed viability decreases at a critical level is difficult.

11.1.3 Status of Musa Genetic Resources Conserved at NPGRL

The *Musa* field germplasm collections of NPGRL consist of various *Musa* accessions collected from the different regions of the country as well as donations from other research institutions in the Philippines and from other countries. The list of accessions, conservation site, and genomic composition of edible bananas of local and foreign origin, wild *Musa balbisiana*, and other *Musa* species conserved at NPGRL is presented in Tables 11.1, 11.2, 11.3, and 11.4, respectively.

A total of 320 *Musa* accessions are currently conserved at NPGRL. These are composed of 216 edible accessions (171 local and 45 foreign) and 104 accessions

Table 11.1 List of accessions, conservation, and genomic composition of local edible banana conserved at NPGRL

No.	PHL/GB number	Accessions	Conservation site[a]	Genome group
1	PHL32728	Amas	TCL	AA
2	PHL32731	Bata-Bata	FGB	AA
3	GB71364	Binaktong	TCL	AA
4	PHL32741	Cuarenta Dias	FGB	AA
5	PHL32744	Eda-an	FGB, TCL	AA
6	GB71378	Ga-o	FGB	AA
7	PHL32753	Inarnibal	FGB, TCL	AA
8	PHL32757	Kinamay Dalaga	TCL	AA
9	GB71398	Lansing	FGB	AA
10	GB71402	Manang	FGB	AA
11	PHL32766	Mintal	FGB, TCL	AA
12	GB71359	Morong Prinsesa	FGB, TCL	AA
13	PHL32779	Rawari	TCL	AA
14	PHL32785	Señorita	FGB, TCL	AA
15	GB71416	Señorita Tagaytay	FGB, TCL	AA
16	GB70027	Suyak	FGB	AA
17	PHL32793	Tudlo Datu	FGB, TCL	AA
18	PHL32727	Alaswe	TCL	AA
19	PHL20621	Gloria	FGB, TCL	AA
20	GB71397	Laku	TCL	AA
21	GB71413	Pusit	TCL	AA
22	PHL32788	Suyak	TCL	AA
23	GB71419	Talip	TCL	AA
24	GB71360	Binawe	FGB, TCL	AAA
25	PHL32736	Bungulan	FGB, TCL	AAA
26	GB71365	Bungulan (BL.) Quezon	FGB, TCL	AAA
27	GB71372	Cavendish	FGB, TCL	AAA
28	PHL32740	Cavendish	FGB	AAA
29	GB60154	Dugong Baka	FGB, TCL	AAA
30	PHL32748	Gros Michel	FGB, TCL	AAA
31	GB66553	Morado	FGB	AAA
32	GB66559	Morado	FGB	AAA
33	GB66611	Morado Pula	FGB	AAA
34	PHL32768	Morado Puti	FGB	AAA
35	GB71403	Moradong Puti	FGB	AAA
36	GB71404	Moradong Puti	FGB, TCL	AAA
37	GB70035	Murado	FGB	AAA
38	PHL32771	Oramos	FGB, TCL	AAA
39	PHL32775	Pastilan	FGB	AAA
40	GB60153	Saba Pula	FGB	AAA
41	PHL32691	Sulay Baguio	FGB, TCL	AAA
42	PHL32692	Sulay Baguio	FGB	AAA

(continued)

Table 11.1 (continued)

No.	PHL/GB number	Accessions	Conservation site[a]	Genome group
43	PHL32795	Tudok	FGB, TCL	AAA
44	GB66555	Tumok	FGB	AAA
45	GB69720	Tumok	FGB	AAA
46	PHL32796	Umalag	FGB, TCL	AAA
47	PHL32734	Bodles Altafort	FGB, TCL	AAA
48	PHL32747	Grand (G.) Naine	FGB, TCL	AAA
49	PHL32780	Robusta	FGB, TCL	AAA
50	PHL32797	Valery	FGB, TCL	AAA
51	GB71393	Lakatan Cavite	FGB, TCL	AAA/AA
52	PHL32758	Lakatan Davao	FGB	AAA/AA
53	GB71394	Lakatan Palawan	FGB, TCL	AAA/AA
54	GB71395	Lakatan Tagaytay	TCL	AAA/AA
55	PHL32738	Calypso	FGB, TCL	AAAA
56	GB70024	Bata-Bata 1226	FGB	AAB
57	GB70025	Bungaoisan	FGB	AAB
58	PHL32743	Duhoy	FGB, TCL	AAB
59	PHL32746	Galamay Señora	FGB, TCL	AAB
60	GB71396	Laknaw/Laknau	FGB	AAB
61	GB69717	Latundan	FGB	AAB
62	GB71399	Latundan	FGB	AAB
63	PHL32759	Latundan Davao	FGB	AAB
64	GB70029	Maramag	FGB	AAB
65	PHL32764	Maramag	FGB	AAB
66	GB70026	Muracho	FGB, TCL	AAB
67	GB71405	Muracho	FGB	AAB
68	GB61042	P. Seribu	FGB, TCL	AAB
69	GB70031	Radja	FGB	AAB
70	GB71415	Radja	FGB, TCL	AAB
71	GB67552	Saksik	FGB	AAB
72	GB71421	Tindok	FGB, TCL	AAB
73	PHL32798	Waggi Muwab	FGB, TCL	AAB
74	GB63082	Binato	FGB	AAB
75	PHL32733	Binato	FGB, TCL	AAB
76	PHL20600	K. Nang-nuan	FGB, TCL	AAB
77	GB67557	Isabelo	FGB	AAB
78	GB71389	Kabai	FGB	AAB
79	PHL20577	Kapas	FGB, TCL	AAB
80	GB71375	Daydayawen	FGB	ABB
81	PHL32655	Dippig/Tiparot	FGB	ABB
82	PHL32755	Kalapagnon	FGB	ABB
83	PHL20685	Katali	FGB, TCL	ABB
84	GB71391	Katsila	FGB, TCL	ABB
85	GB69725	Lagkitan	FGB	ABB

(continued)

Table 11.1 (continued)

No.	PHL/GB number	Accessions	Conservation site[a]	Genome group
86	GB71400	Madurangga	TCL	ABB
87	PHL32765	Matavia	FGB, TCL	ABB
88	GB61396	Pakawan	FGB	ABB
89	GB63067	Pelipita	FGB	ABB
90	PHL32775	Pilipa	FGB	ABB
91	GB71412	Pitogo	FGB	ABB
92	PHL32786	Siusok	FGB, TCL	ABB
93	GB71373	Dadanugen	FGB	ABB
94	PHL32791	Tiparot	FGB, TCL	ABB
95	PHL20683	Gubao	FGB	ABB
96	PHL32749	Gubao	FGB	ABB
97	GB69703	Inabaniko	FGB	ABB
98	PHL20627	Inabaniko	FGB, TCL	ABB
99	PHL32772	Pa-a Dalaga	TCL	ABB
100	GB71383	Saba Hapon	FGB	ABB
101	PHL32782	Saba Hapon	FGB, TCL	ABB
102	PHL32783	Saba Kastila	FGB, TCL	ABB
103	PHL32665	Sabang Barako	FGB	ABB
104	PHL32666	Sabang Barako	FGB	ABB
105	PHL32664	Sabang Binong	FGB	ABB
106	PHL32660	Susong Dalaga	FGB	ABB
107	GB71390	Kastila	FGB	ABB
108	PHL32638	Abuhon	FGB	BB/ABB
109	PHL32684	Cadisnon/Cardaba	FGB	BBB/ABB
110	PHL32685	Cadisnon/Cardaba	FGB	BBB/ABB
111	GB61946	Calibo	FGB	BBB/ABB
112	GB71366	Calibo 1	FGB	BBB/ABB
113	GB71367	Calibo 2	FGB	BBB/ABB
114	GB71368	Calibo 3	FGB	BBB/ABB
115	GB71370	Cardaba	FGB	BBB/ABB
116	GB71371	Cardaba	FGB, TCL	BBB/ABB
117	PHL32630	Cardaba	FGB	BBB/ABB
118	PHL32639	Cardaba	FGB	BBB/ABB
119	PHL32644	Cardaba	FGB	BBB/ABB
120	GB71369	Cardaba 6	FGB	BBB/ABB
121	GB71361	Dali-an	FGB, TCL	BBB/ABB
122	GB71374	Dali-an	FGB, TCL	BBB/ABB
123	PHL32622	Dali-an	FGB, TCL	BBB/ABB
124	PHL32637	Dali-an	FGB	BBB/ABB
125	PHL32742	Dali-an	FGB	BBB/ABB
126	PHL32646	Damilig	FGB	BBB/ABB
127	GB71376	Dippig	FGB, TCL	BBB/ABB
128	GB71377	Dippig	FGB	BBB/ABB

(continued)

Table 11.1 (continued)

No.	PHL/GB number	Accessions	Conservation site[a]	Genome group
129	PHL32629	Dippig	FGB	BBB/ABB
130	PHL32649	Dippig	FGB	BBB/ABB
131	PHL32651	Dippig	FGB	BBB/ABB
132	PHL32652	Dippig	FGB	BBB/ABB
133	PHL32653	Dippig	FGB	BBB/ABB
134	PHL32657	Dippig/Luyluy	FGB, TCL	BBB/ABB
135	PHL32656	Dippig/Luyluyan	FGB, TCL	BBB/ABB
136	PHL32625	Dwarf Cardaba	FGB	BBB/ABB
137	PHL32640	G. Cardaba	FGB	BBB/ABB
138	PHL32643	Giant Cardaba	FGB, TCL	BBB/ABB
139	PHL32624	Giant Cardaba	FGB	BBB/ABB
140	PHL32642	Giant Cardaba	FGB	BBB/ABB
141	PHL32756	Kalimpos	FGB, TCL	BBB/ABB
142	GB69735	Mundo	FGB	BBB/ABB
143	PHL32769	Mundo	FGB, TCL	BBB/ABB
144	GB63094	Sab-a	FGB	BBB/ABB
145	PHL32712	Sab-a	FGB	BBB/ABB
146	PHL32714	Sab-a	FGB	BBB/ABB
147	PHL32694	Sab-a-Manila	FGB	BBB/ABB
148	PHL32658	Saba	FGB	BBB/ABB
149	PHL32661	Saba	FGB	BBB/ABB
150	PHL32662	Saba	FGB	BBB/ABB
151	PHL32662	Saba	FGB	BBB/ABB
152	PHL32662	Saba	FGB	BBB/ABB
153	PHL32663	Saba	FGB	BBB/ABB
154	PHL32671	Saba	FGB	BBB/ABB
155	PHL32672	Saba	FGB	BBB/ABB
156	PHL32674	Saba	FGB	BBB/ABB
157	PHL32675	Saba	FGB	BBB/ABB
158	PHL32676	Saba	FGB	BBB/ABB
159	PHL32677	Saba	FGB	BBB/ABB
160	PHL32678	Saba	FGB	BBB/ABB
161	GB71385	Saba Puti	FGB	BBB/ABB
162	PHL32680	Saba Puti	FGB	BBB/ABB
163	PHL32784	Saba Puti	TCL	BBB/ABB
164	PHL32638	Saba/Dippig	FGB	BBB/ABB
165	PHL32648	Saba/Dippig	FGB	BBB/ABB
166	PHL32669	Sabang Pilos	FGB	BBB/ABB
167	PHL32668	Sabang Pulpol	FGB	BBB/ABB
168	GB60150	Sabang Puti	FGB	BBB/ABB
169	GB60151	Sabang Puti	FGB	BBB/ABB
170	GB60152	Sabang Puti	FGB	BBB/ABB
171	GB71432	Dwarf Cardaba	FGB	BBB/ABB

[a] *FGB* field gene bank, *TCL* tissue culture, *SS* seed storage

which belong to the wild *Musa* and related families, majority of which belong to the wild *Musa balbisiana* species.

11.1.3.1 Conservation Strategies of Musa Accessions at NPGRL

Figure 11.1 shows the number of accessions conserved in the field gene bank (FGB), in vitro, and in the seed storage (SS) area. More than 92% (296 accessions) of the total *Musa* germplasm collections at NPGRL are conserved in the field gene bank. One hundred ninety-four accessions (60%) are conserved only in the FGB, 77 accessions (24%) are conserved both in the FGB and in vitro, while 25 accessions (7.81%) are conserved in the FGB and SS. These germplasm collections are currently being documented, characterized, and evaluated in the field for identification of germplasm materials with potential for direct utilization or as parents for breeding materials with sources of desirable traits in the future. The current *Musa* field germplasm collection is situated in the experimental area located along Pili Drive, University of the Philippines at Los Baños, College, Los Baños, Laguna. The soil in the *Musa* field germplasm collection was identified to be Lipa clay loam and is characterized to be well-drained fertile loam with high water-holding capacity and organic matter content (Fernandez and De Jesus 1980).

The in vitro *conservation* unit is currently maintaining 92 accessions of local and foreign bananas (Table 11.1 and Fig. 11.1). Seventy-seven accessions are conserved both in vitro and in the field while 15 accessions are conserved only in vitro. The entire banana collections include the duplicate collection of Southeast Asia Banana Genetic Resources Center-Bureau of Plant Industry, some accessions from *Musa* Transit Center (MTC), and some *Musa balbisiana* collections from the field gene bank. The cultures were maintained in 25 × 150 mm test tubes containing MS basal medium or MS basal medium +3 mg/L BAP and covered with cotton plug. For germplasm conservation, 5–10 culture vessels per accession were maintained. Using this system of maintenance, the subculture interval ranged from 4 to 6 months. Cultures were subcultured onto fresh medium when signs of culture degeneration due to excessive browning of media, leaf senescence, and browning of explants were evident, and there was already insufficient culture media due to usage and desiccation. Regular subculture of the whole collection onto fresh medium generally required about 3–4 weeks.

Ideally, all the *Musa* collections in the field should be duplicated in vitro to serve as a backup in case something wrong happens to the field collections. However, some accessions could not be easily maintained in the laboratory. Some of the problems in the laboratory are culture degeneration especially for the *Musa balbisiana,* occurrence of severe browning, limited or slow shoot proliferation, and fungal and bacterial contamination.

The current studies on in vitro germplasm conservation are the identification of medium-term conservation storage for *Musa* germplasm applicable in NPGRL in vitro unit's condition. This method had been devised to limit growth by modifying the basal growth media either by increasing of the osmotic potential of the culture media by using sugars or sugar alcohols coupled with reduced temperature or addition of osmoticum and modification of culture vessel cover (Ko et al. 1991;

Table 11.2 List of accessions, conservation, and genomic composition of edible banana from foreign origin, conserved at NPGRL

No.	PHL/GB number	Cultivar	Conservation site[a]	Genome group
1	GB71363	AACV rose13–16	FGB	AA
2	GB71362	AACV rose s3 728	FGB	AA
3	GB70019	Malaysian blood	FGB	AA
4	GB70028	Pisang Rojo Uter	FGB	AA
5	GB71433	Domie Leite	FGB, TCL	AAA
6	PHL32745	F26 Honduras	FGB, TCL	AAA
7	GB71379	GCTCV 105	FGB, TCL	AAA
8	GB71357	GCTCV 106	FGB, TCL	AAA
9	GB71351	GCTCV 119	FGB, TCL	AAA
10	GB71380	GCTCV 215	FGB, TCL	AAA
11	GB71381	GCTCV 218	FGB, TCL	AAA
12	GB71382	GCTCV 219	FGB	AAA
13	GB71358	GCTCV 247	FGB, TCL	AAA
14	GB71386	HPSDC105	FGB, TCL	AAA
15	GB71392	Komargh	FGB, TCL	AAA
16	GB71406	OT 09–1	FGB, TCL	AAA
17	GB71423	William	TCL	AAA
18	GB71349	Fhia 17	FGB, TCL	AAAA
19	GB71356	Fhia 25	FGB	AAAA
20	GB71353	SH 3640	FGB, TCL	AAAA
21	GB70020	CRBP	FGB	AAAB
22	GB71346	Fhia 1	FGB, TCL	AAAB
23	GB71354	Fhia 18	FGB, TCL	AAAB
24	GB71347	Fhia 2	FGB, TCL	AAAB
25	GB71355	Fhia 21	FGB, TCL	AAAB
26	GB71350	Fhia 23	FGB	AAAB
27	GB71418	Simili Radjah	FGB, TCL	AAAB
28	GB71387	Humwey	FGB	AAB
29	GB69732	Katimor	FGB	AAB
30	GB70023	MAA plantain	FGB	AAB
31	GB69733	Obubit N'tang GM	FGB	AAB
32	GB71408	P. Keling	FGB, TCL	AAB
33	GB71409	P. Lingi	FGB, TCL	AAB
34	GB69734	Hender Neyargh	FGB	AAS
35	GB71414	Calcutta 4	FGB, TCL	*M. accuminata* ssp. Burmanica
36	GB71410	Pahang	FGB, TCL	M. accuminata ssp. Malaccensis
37	GB71348	Cachaco	FGB	ABB
38	PHL32754	Java	FGB	ABB
39	GB71388	K. Namwa Khom	FGB	ABB

(continued)

Table 11.2 (continued)

No.	PHL/GB number	Cultivar	Conservation site[a]	Genome group
40	PHL20597	K. Namwa Luang	FGB	ABB
41	GB71401	P.K. Malacacina	FGB	ABB
42	GB71407	P.K. Malacacina	TCL	ABB
43	GB71352	TMBX 1378	FGB, TCL	ABBB
44	GB70160	Lyon (selfed)	FGB, SS	
45	GB70161	Lyon (x rose)	FGB, SS	

[a] *FGB* field gene bank, *TCL* tissue culture, *SS* seed storage.

Bhat and Chandel 1993). Medium-term storage of banana plantlets in vitro is best achieved at 16 °C under low-intensity illumination (Banerjee and De Langhe 1995). The use of sodium alginate as an encapsulating agent for shoot tip has also been used for short- to medium-term storage and transportation purposes (Rao et al. 1993).

Maintenance of banana collections in vitro remains labor intensive and demands regular replanting of subculturing. Even with reduced growth, there remains the risk of losing collections because of latent contamination by microorganisms and human error. Genetic variations may also occur naturally during both conventional and in vitro propagation of bananas (Vroh-Bi et al. 2010). Incidence of somatic mutations is low in bananas propagated conventionally. It is observed to be frequent in tissue-cultured materials (Stover 1988; Robinson 1996) and often induced by somatic embryogenesis (Strosse et al. 2006). Somatic mutations are limited to nonreproductive cells, with somaclonal variation often being exhibited by plants regenerated from cultured cells, particularly those regenerated via a callus phase. Mutations have been associated with somaclonal variation, including point mutations, gene duplication, chromosomal rearrangements, and changes in chromosome complements. Chromosome instability is one of the most common causes of tissue culture-induced variations in bananas (Larkin 2004; Msogoya et al. 2008). Thus, in vitro collections require monitoring of genetic integrity every 10 years to ensure that accessions are not undergoing somaclonal variation.

Thirty-four wild *Musa* accessions (10.6%) have seeds conserved, 9 accessions in seeds only, while 25 accessions are conserved in both the FGB and SS. Majority of accessions with seeds stored are wild *Musa balbisiana* and with some wild *Musa accuminata*, abaca, Canton, lyon selfed, and lyon hybrid.

11.1.3.2 Genomic Composition of Musa Accessions at NPGRL

The genomic composition of Musa germplasm accessions at NPGRL is either pure *M. acuminata* (AA, AAA, AAAA) or *M. balbisiana* (BB, BBB) or a combination of both *M. acuminata* and *M. balbisiana* (AB, AAB, ABB, AAAB, ABBB). Figure 11.2 shows the genomic classifications of the *Musa* germplasm conserved at NPGRL.

In terms of edible bananas, 63 accessions have genomic classifications of ABB/BBB. These cultivars mostly belong to the Saba subgroup which was classified

Table 11.3 List of accessions and conservation wild *Musa balbisiana* collection conserved at NPGRL

No.	PHL/GB number	Local name	Conservation site[a]	Genome group
1	PHL31770	Butuhan	FGB	BBw
2	PHL31773	Butuhan	FGB	BBw
3	PHL31778	Butuhan	FGB	BBw
4	PHL31779	Butuhan	FGB	BBw
5	PHL31780	Butuhan	FGB	BBw
6	PHL31781	Butuhan	FGB	BBw
7	PHL31782	Butuhan	FGB	BBw
8	PHL31784	Butuhan	FGB, SS	BBw
9	PHL31786	Butuhan	FGB	BBw
10	PHL31787	Butuhan	FGB	BBw
11	PHL31789	Butuhan	FGB	BBw
12	PHL31790	Butuhan	FGB	BBw
13	PHL31792	Butuhan	FGB	BBw
14	PHL31793	Butuhan	FGB	BBw
15	PHL31796	Pik-iw	FGB	BBw
16	PHL31799	Butuhan	FGB	BBw
17	PHL31801	Butuhan	FGB	BBw
18	PHL31803	Butuhan	FGB	BBw
19	PHL31804	Butuhan	FGB	BBw
20	PHL31807	Butuhan	FGB	BBw
21	PHL31809	Butuhan	FGB	BBw
22	PHL31810	Butuhan	FGB	BBw
23	PHL31816	Butuhan	FGB	BBw
24	PHL31820	Butuhan	FGB	BBw
25	PHL31821	Butuhan	FGB	BBw
26	PHL31825	Butuhan	FGB	BBw
27	PHL31826	Balayang	FGB	BBw
28	PHL31828	Balayang	FGB	BBw
29	PHL31832	Pacol	FGB	BBw
30	PHL31836	Butuhan	FGB	BBw
31	PHL31838	Pacol	FGB	BBw
32	PHL31841	Butuhan	FGB	BBw
33	PHL31846	Butuhan	FGB	BBw
34	PHL31851	Pacol	FGB	BBw
35	PHL31852	Pacol	FGB	BBw
36	PHL31855	Pacol	FGB	BBw
37	PHL31856	Pisgan	FGB	BBw
38	PHL31857	Pacol	FGB	BBw
39	PHL31859	Pacol	FGB	BBw
40	PHL31861	Pik-iw	FGB	BBw
41	PHL31862	Pik-iw	FGB, SS	BBw
42	PHL31866	Butuhan	FGB	BBw

(continued)

Table 11.3 (continued)

No.	PHL/GB number	Local name	Conservation site[a]	Genome group
43	PHL31873	Butuhan	FGB	BBw
44	PHL31874	Butuhan	FGB	BBw
45	PHL31875	Butuhan	FGB	BBw
46	PHL31877	Paguha	FGB	BBw
47	PHL32673	Banana	FGB	BBw
48	GB61360	Badiating	SS	BBw
49	GB61941	Pacol	FGB, SS	BBw
50	GB61942	Butuhan	SS	BBw
51	GB61978	Calibo	FGB	BBw
52	GB61996	Moko	FGB, SS	BBw
53	GB61977	Butuhan	SS	BBw
54	GB62000	Moko	FGB	BBw
55	GB66554	Balbisiana	FGB	BBw
56	GB66558	Balayang	FGB	BBw
57	GB66560	Balbisiana	FGB	BBw
58	GB66562	Black Banana	FGB	BBw
59	GB66565	Balbisiana	FGB	BBw
60	GB66566	Balayang	FGB, SS	BBw
61	GB66568	Balayang	FGB, SS	BBw
62	GB66583	Pacol	FGB	BBw
63	GB66584	Butuhan	FGB	BBw
64	GB66585	Pacol	FGB	BBw
65	GB66586	Butuhan	FGB, SS	BBw
66	GB66587	Butuhan	FGB, SS	BBw
67	GB66588	Butuhan	FGB	BBw
68	GB66591	Balayang	FGB	BBw
69	GB66592	Moko	FGB	BBw
70	GB67541	Pik-iw	FGB	BBw
71	GB67548	Pik-iw	FGB	BBw
72	GB67550	Pik-iw	FGB, SS	BBw

[a] *FGB* field gene bank, *SS* seed storage

by Valmayor et al. (2002) as BBB. This was supported by a study conducted by Sales et al. (2011) using flow cytometry and molecular markers. Using the same techniques, Christelová et al. (2017) reclassified Saba (S14 ITC1745) genome as ABB. This is the reason why ABB/BBB is used to classify accessions belonging to Saba subgroup which is similar to that used by Dela Cruz et al. (2020). One of the reasons for the numerous accessions for this subgroup is due to a project funded by PCAARRD-DOST on the development/selection of Saba with short stature. Thirty-four accessions belong to ABB "Bluggoe sub-group" which includes Gubao, Matavia, Lagkitan, Katali, Inabaniko, and Pelipita, among others. Tiparot, Dadanugen, and Pakawan are interesting accessions because of their unstable male

Table 11.4 List of accessions, conservation, and genomic composition of other Musa species and related families conserved at NPGRL

No.	PHL/GB number	Cultivar	Conservation site[a]	Genome group
1	GB66589	Abaca	FGB	*Musa textilis*
2	GB70151	Negro	FGB	*Musa textilis*
3	GB70152	Bagwisan	FGB	*Musa textilis*
4	GB70153	Minc	FGB	*Musa textilis*
5	GB70154	PSU no.3	FGB	*Musa textilis*
6	GB70155	TG 12	FGB	*Musa textilis*
7	GB70156	TG 13	FGB	*Musa textilis*
8	GB70157	TG 15	FGB	*Musa textilis*
9	GB70158	BRA 33	FGB	*Musa textilis*
10	GB67554	Agutay	Seed storage (SS)	*Musa acuminata spp. Errans*
11	GB61407	Alinsanay	FGB	*M. alinsanaya*
12	GB66593	Alinsanay	FGB	*M. alinsanaya*
13	GB68602	False birds of paradise	FGB	*Heliconia rostrata*
14	GB66581	Cantong	FGB	*M, balbisiana x M. textilis*
15	GB66582	Cantong	FGB, SS	*M, balbisiana x M. textilis*
16	GB71424	Cantong	FGB	*M, balbisiana x M. textilis*
17	GB68595	Giant Ornata	FGB	*Musa ornata*
18	GB68598	Jaquinie Orange	FGB	*Heliconia bihai*
19	GB71425	Lateriflora	FGB	*Musa laterita*
20	GB71426	Laterita	FGB	*Musa laterita*
21	GB66564	Ornata	FGB	*Musa ornata*
22	PHL32770	Ornata	FGB	*Musa ornata*
23	GB68601	Red Caribbean	FGB	*Heliconia caribaea*
24	GB68593	Yellow Caribbean	FGB	*Heliconia caribaea*
25	GB68599	Yellow Heliconia	FGB	*Heliconia psittacorum*
26	GB66561	Zebrina	FGB	*Musa acuminata spp. Zebrina*
27	GB71427	Zebrina	FGB	*Musa acuminata spp. Zebrina*

[a] *FGB* field gene bank, *SS* seed storage

bud characteristics which sometimes have one or two male bud per pseudostem, and in some cases none.

Forty accessions of triploid *M. acuminata* (AAA) where Cavendish, Bungulan, Morado, and Tumok belong are conserved at NPGRL. Interesting cultivars that also belong to this group include Sulay Baguio, a dwarf local Cavendish clone, and Pastilan, which sometimes produces two or more bunches per pseudostem. Lakatan has been previously identified by Valmayor et al. (2002) as diploid AA but recent reports have classified Lakatan as AAA (Dela Cruz et al. 2020). For this chapter, AA/AAA was used as a classification of Lakatan.

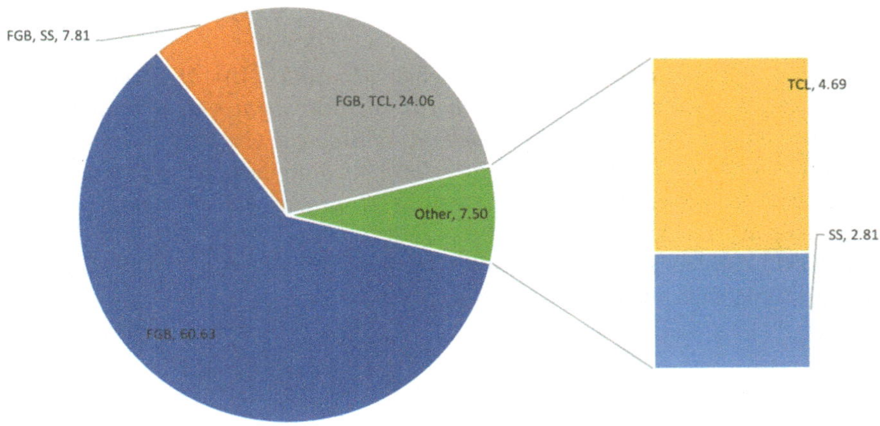

Fig. 11.1 Number of accessions conserved in the field gene bank, in vitro, and in seed storage area. *FGB* field gene bank, *TCL* tissue culture, *SS* seed storage

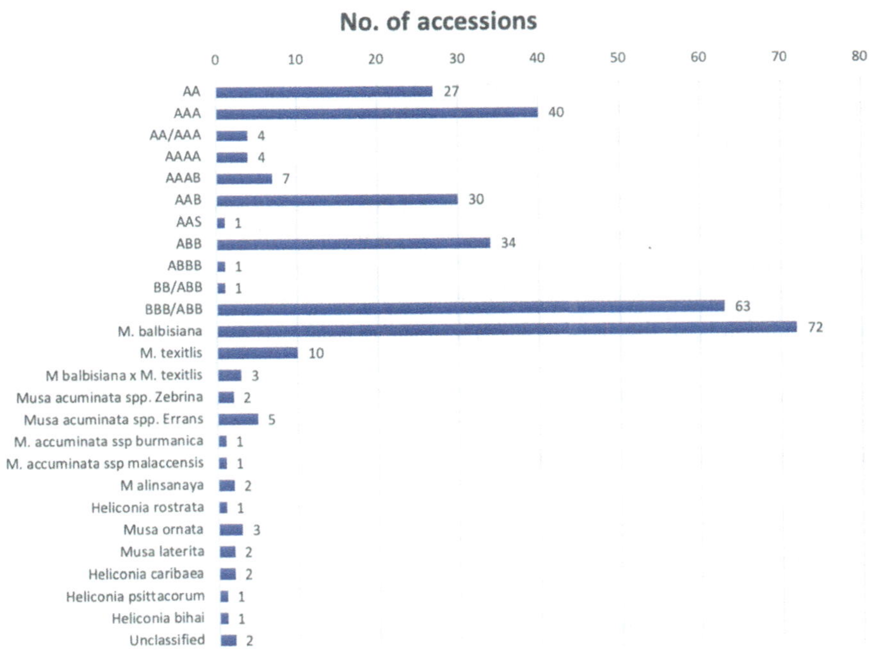

Fig. 11.2 Genomic classifications of the Musa germplasm conserved at NPGRL

In terms of the wild *Musa* species, the wild *Musa balbisiana* had the most number of accessions (72). A DOST-PCAARRD-funded project had initiated a field exploration throughout the Philippines which was able to collect 97 wild *Musa balbisiana* accessions (Sotto and Rabara 2002). At present, only 45 accessions remain from the original collection. The additional 27 accessions of wild *Musa balbisiana* have been recently collected.

Other wild *Musa* species conserved include *M. textilis* (10), *M. acuminata* ssp. *Errans* (5), *M. acuminata* ssp. *Zebrina* (2), *M. acuminata* ssp. *Burmanica* (1), *M. acuminata* ssp. malaccensis (1), and *M. alinsanaya* (2). Canton is a natural hybrid between *Musa textilis* and *Musa balbisiana*, with three accessions currently conserved. Ornamental *Musa* species such as *M. ornata* and *M. laterita* are also maintained at the field gene bank.

NPGRL also has some species under the order Zingiberales, where the *Musa* family also belongs, including *Heliconia* species.

11.2 Conclusion

Banana germplasm conservation is a real challenge in countries like the Philippines where diseases are prevalent. *Musa* accessions maintained in field gene bank are lost due to pest and diseases as well as natural calamities such as typhoons. In vitro conservation may result in somaclonal variation due to continuous subculturing in an artificial medium and prolonged storage at low temperature. The possibility of losing materials arises due to unreliable electrical supply and high cost of maintenance. A specialized skill for laboratory staff is necessary for the work to be efficiently performed. Ideally, the in vitro conservation will serve as a backup in case some accessions in the field are lost or infected with viruses. An alternative strategy is to maintain the germplasm collections in large pots under an insect-proof screen cage. Insect-proof cage prevents insects/vectors from infecting the plants inside the cage, hence maintaining virus-free plants. At the same time, the risk of somaclonal variations is minimized since they are planted on pots with normal soil. The Musa diversity existing in the field may contain the desirable genes that we need in the future, hence it is important to sustainably conserve them both *in-situ* and *ex-situ*. These irreplaceable germplasm materials may also provide a broad base of genetic diversity that can be used by breeder's in developing improved banana varieties for the present and the generations to come.

References

Banerjee N, De Langhe E (1995) A tissue culture technique for rapid clonal propagation and storage under minimal growth conditions of *Musa* (banana and plantain). Plant Cell Rep 4:351–354

Bhat SR, Chandel KPS (1993) *In vitro* conservation of Musa germplasm: effects of mannitol and temperature on growth and storage. J Hortic Sci 68:841–846

Berkeley MJ (1846) Observations, botanical and physiological on the potato murain. J Hort Soc London 1:9–34

Christelová P, De Langhe E, Hřibová E, Čížková J, Sardos J, Hušáková M, Sutanto A, Kepler A, Swennen R, Roux N, Doležel J (2017) Molecular and cytological characterization of the global Musa germplasm collection provides insights into the treasure of banana diversity. Biodivers Conserv 26(4):801–824

Dela Cruz FR, Laude RP, Sandoval CMC, Gueco LS, Huelgas VC, Garcia RN, Mendoza EMT (2020) Genetic Characterization of Philippine Saba Germplasm Collection using Microsatellite markers. Philipp J Crop Sci 149(1):189–197

Fernandez NC, de Jesus JC (1980) Philippine soils: their distribution, general land-use and parent materials. NFAC/UPLB countryside action. Dent Prog 229:5

Gueco LS, Posada IB, Corpuz FV, Yanos LA, dela Cruz FS Jr, Damasco OP, Descalsota MLV, Molina AB Jr (2020) Characterization and! agronomic evaluation naturally occuring short statured Saba banana in the Philippines. Philipp J Crop Sci 149(3Wa):981–992

Ko WH, Hwang SC, Ku FM (1991) A new technique for storage of meristem tip cultures of 'Cavendish' banana. Plant Cell Tiss Org 25:179–183

Larkin P (2004) Somaclonal variation: origin and causes. In: Goodman RM (ed) Encyclopedia of plant and crop science. Marcel Dekker, New York

Msogoya T, Grout B, Roberts A (2008) Karyotypic and 2C nuclear DNA size instability *in vitro* induced off-types of East African Highland banana (*Musa* AAA East Africa). Biotechnol (Pak) 7:578–581

MusaNet (2016) In: Chase R, Laliberté B (eds) Summary of the global strat egy for the conservation and use of musa genetic resources. Bioversity International, Montpellier

Oliveira RP, de Oliveira S, de Silva KM, Silveira DDG (2000) *In vitro* conservation of diploid bananas accessions. Sci Agric 57:245–249

Philippine Fiber Industry Development Authority (PhilFIDA) (2017) Philippine abaca helps in global environment conservation. http://www.philfida.da.gov.ph/index.php/archived-articles/19-philippine-abaca-helps-in-global-environment-conservation

Philippine Statistics Authority (2021) OpenStat database: agriculture, forestry, fisheries. Other Crops: Volume of Production and Area Planted, Year and Item. https://openstat.psa.gov.ph/PXWeb/pxweb/en/DB/DB__2E__CS/?tablelist=true&rxid=bdf9d8da-96f1-4100-ae09-18cb3eaeb313. Accessed 16 December 2021

Promusa (2021) Fusarium wilt of Banana. https://www.promusa.org/Fusarium+wilt. Accessed 15 December 2021

Rao PS, Ganapathi TR, Suprasanna P, Bapat VA (1993) Encapsulated shoot tips of banana - a new propagation and delivery system. InfoMusa 2:4–5

Reed BM, Engelmann F, Dullo ME, Engels JMM (2004) Technical guidelines for the management of field and in vitro germplasm collections. IPGRI handbook for Genebanks no.7. IPGRI, Rome, p 116

Robinson JC (1996) Bananas and plantains. CABI, Oxfordshire

Sales EK, Butardo NG, Paniagua HG, Jansen H, Dolozel J (2011) Assessment of ploidy and genome constitution of some Musa balbisiana cultivars using DArT markers. Philipp J Crop Sci 36:11–18

Sotto R, Rabara R (2002) Morphological diversity of Musa balbisiana colla in the Philippines. Info Musa 9(2):28–30

Stover RH (1988) Variation and cultivar nomenclature in *Musa*, AAA group, Cavendish subgroup. Fruits 43:353–356

Strosse H, Schoofs H, Panis B, Andre E, Reyniers K, Swennen R (2006) Development of embryogenic cell suspensions from shoot meristematic tissue in bananas and plantains (*Musa* spp.). Plant Sci 170:104–112

Ulstrup AJ (1972) The impacts of Southern corn leaf blight epidemics of 1970-1971. Annu Rev Phytopathol 10:37–50

Valmayor RV, Espino RRC, Pascua OC (2002) The wild and cultivated bananas of the Philippines. PARRFI and BAR, Los Baños, p 242

Vroh-Bi I, Anagbogu C, Nnandi S, Tenkouano A (2010) Genomic characterization of natural and somaclonal variations in bananas (*Musa* spp.). Plant Mol Biol Rep 29:440–448

Yong-Bi F (2017) The vulnerability of plant genetic resources conserved ex-situ. Crop Sci 57:2314–2328

Diversity and Management of Indoor Plants at Urban Dwellings in Bangladesh: A Case Study from Halishahar of Chattogram Metropolitan Area

12

Md. Akhter Hossain, Abul Hayat Poyal, Mohammed Kamal Hossain, Mohammad Mosharraf Hossain, Mohammed Iqram Uddin Al Amran, and Md. Oliur Rahman

Abstract

Integrating greeneries into the indoor dwelling environment boosts work performance and relieves stress to add to the overall psychological well-being especially in deserted urban settings. In addition to mental soothing, thermal regulation, air purification, aesthetics, public health, and comfort, the addition of plants to indoor settings may also contribute to the conservation of dwindling floral biodiversity. Despite authorities' pledge for sustainable urban management, the status of indoor gardening has hitherto remained unexplored in the emerging megapolis of Bangladesh—Chattogram—the second largest urban center of the country. In addressing that gap, this study aims to explore the composition, diversity, and management of indoor plants in urban dwellings at Halishahar of Chattogram based on interviews on 48 households selected through multistage random sampling. Data from all selected households were collected by using a semi-structured questionnaire through physically visiting the households. Almost half of the households (48%) living at Halishahar had indoor plants in their dwellings. The study recorded a handsome 120 indoor plant species belonging to 108 genera from 60 families. While the diversity was in no way comparable to the tropical ecosystem of the country, in consideration of the strict set of

M. A. Hossain (✉) · A. H. Poyal · M. K. Hossain · M. M. Hossain
Institute of Forestry and Environmental Sciences, University of Chittagong, Chittagong, Bangladesh
e-mail: akhter.hossain@cu.ac.bd; mosharraf@cu.ac.bd

M. I. U. A. Amran
International Union for Conservation of Nature, Dhaka, Bangladesh
e-mail: iqram.amran@iucn.org

M. O. Rahman
Department of Botany, University of Dhaka, Dhaka, Bangladesh
e-mail: oliur.bot@du.ac.bd

S. Ramamoorthy et al. (eds.), *Plant Genetic Resources, Inventory, Collection and Conservation*, https://doi.org/10.1007/978-981-16-7699-4_12

requirements for plants to be suitable for an indoor setting, the diversity seemed excellent as evident from four diversity indices. Soil mixed with compost, sand, and *surki* at different ratios is used as potting media. Pests were identified as the major challenge in managing the indoor plants. Application of domestic manure with the potting media was common as a means to maintain the nutrient flow. Bruised tea leaf is the most frequently added nutrient supplement. Apart from the aesthetic values, urban dwellers from Halishahar reported the immense potential of indoor gardening in supplementing daily nutrition and in mitigating the impacts of climate change. The lessons from this study can be used in informed policymaking for the promotion of biodiversity conservation and other benefits from indoor greening among urban dwellers in Bangladesh.

Keywords

Indoor plants · Ornamental plants · Urban greening · Diversity indices · Chattogram

12.1 Introduction

Living plants, if used properly, make houses cozy, environmentally friendly, and aesthetically pleasing. Besides, plants in indoor environment freshen the air, make passing leisure time a retreat to nature, help to make vacant interior spaces elements of aesthetics, improve cognitive states, contribute towards mitigating the urban heat island, ameliorate microclimate, add to human comfort and happiness, boost work performance, elevate recovery from illness, and reduce stress (Bringslimark et al. 2009; Smith and Pitt 2010; Dover 2015; Deng and Deng 2018; Ambrose et al. 2020; Young et al. 2020). Additionally, indoor plants can add life to a sterile office, provide privacy, and abate noise levels (Burton 2001). Compared to conventional air purification techniques, indoor plants offer an environment-friendly cost-effective purification of indoor air by eliminating harmful toxins and particulate matter (Gawronska and Bakera 2015) as proven by systematic evaluation of 28 commonly used ornamental indoor plant species for their ability to remove volatile indoor pollutants: aromatic hydrocarbons (benzene and toluene), aliphatic hydrocarbon (octane), halogenated hydrocarbon, and terpene (α-pinene) (Yang et al. 2009; Treesubsuntorn and Thiravetyan 2012; Cruz et al. 2014; Parseha et al. 2018). On the other hand, declining green spaces in the megapolises increasingly confining the urban dwellers inside concrete walls (Weng 2001) was a stressful phenomenon felt acutely during the COVID-19 pandemic. Evidently, during the pandemic-forced confinement at home, people exposed to greenery at home experienced better health (Dzhambov et al. 2021). In addition, indoor plants can contribute to biodiversity conservation and provide ecosystem service by fulfilling multiple policy goals of cities (Young et al. 2020). It is no surprise, therefore, that greening of balconies, terraces, and stairs, as well as the indoor spaces, is becoming a common practice in urban-dwelling design and development besides traditional roof gardening (Bruins 2005;

Xu et al. 2007; Lin et al. 2018). Indoor plants consist of a wide variety of native and introduced herbs, shrubs, dwarf trees, climbers, and succulents. The potential role of the indoor plants largely varies with their genetic variation, species, and habit (Park et al. 2010; Aydogan and Montoya 2011; Treesubsuntorn and Thiravetyan 2012; Gawronska and Bakera 2015). It is essential to know the composition and preference of species mix of the indoor greeneries as practiced by the city people to assess the means to augment benefits from indoor plants especially to use it as a means to attain SDG 11—Sustainable Cities and Communities (Moran 2019).

Chattogram, the port city of Bangladesh, is one of the most densely populated cities globally (Mia et al. 2015) with 14,200 people per square kilometer (Jafrin and Beza 2018). The city became dull and unattractive despite its excellent environmental settings with hills, rivers, and sea due to faulty zoning, unplanned constructions, and shortage of green spaces (Uddin and Firoj 2013). Currently, the city has only 0.18 square meters of open space per capita compared to the standard 9 square meters per capita that cities all around the world are trying to maintain (Paul et al. 2020). Accommodating the rapidly increasing urban populace in Chattogram leads to faster urban sprawl engulfing the green fringes of the city resulting in a fragile ecology for the urban dwellers (Hossain 2008). The increase in the built-up environment is exacerbating the urban heat island effect, and indoor gardening has a great potential to mitigate it (Safayet et al. 2017). Though the use of indoor plants is gaining popularity in Chattogram, information on the diversity and composition of species used in the indoor spaces in Chattogram is sparse. In this backdrop, this study was conducted by taking the Halishahar residential area of Chattogram city as a case to assess the composition and diversity of indoor plants and their management practices by the city dwellers. The outcome of the study may assist the policymakers in informed policymaking to promote a sustainable model for indoor greening in pursuing SDG 11.

12.2 Materials and Methods

12.2.1 Study Area

Chattogram, the second largest and the principal port city of Bangladesh, is located on the North bank of the river Karnaphuli at the south-eastern coast of Bangladesh between 22°14′ and 22°24′30" N Latitude and 91°-46′ and 91°-53′ E Longitude (Fig. 12.1). Chattogram, with a land area of 157 km^2 (Mia et al. 2015), is considered the national economic hub of Bangladesh (BBS 2020). The small town rapidly grew into a megacity due to rapid industrialization and development leading to its ascension as Chattogram City Corporation (CCC) in 1990 outstripping the old municipality. The topography is characterized by its hills, river, canals, and lake with a spatial and temporal trajectory of urban expansion extending over 40 years. Over the period, 56% of the land cover has changed—mainly because of the expansion of built-up areas and other human activities (Hassan and Nazem 2016). The city environment is characterized by an average annual temperature of 25.57 °C

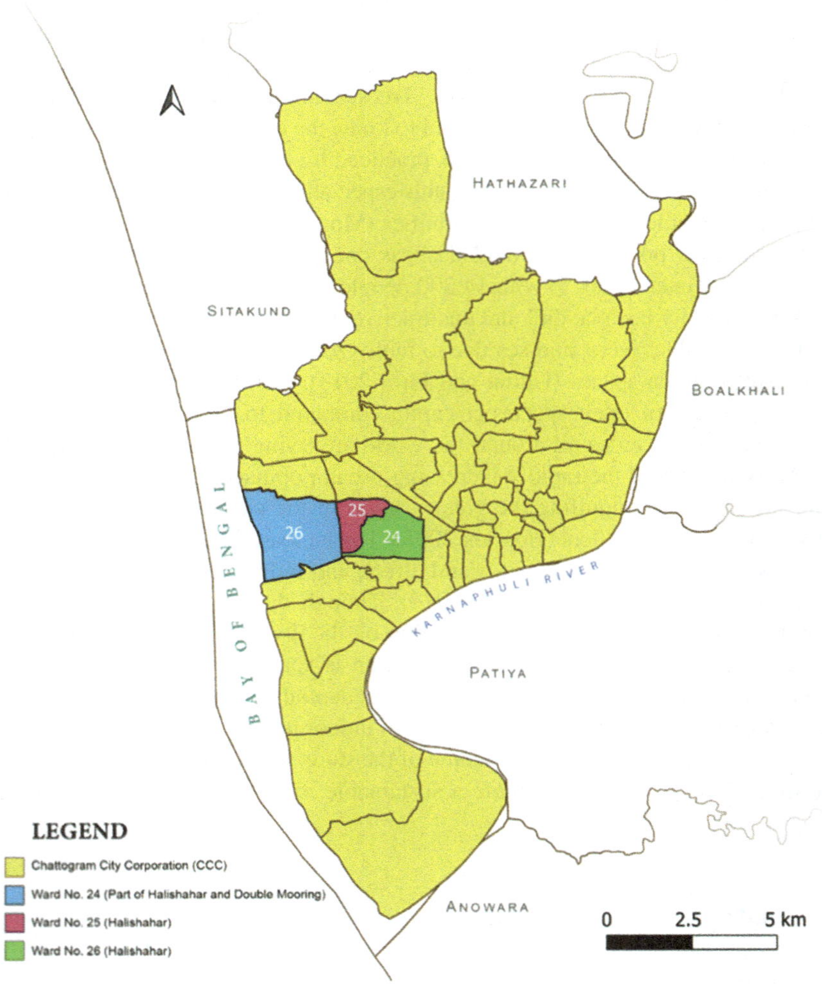

Fig. 12.1 Location of the studied sites in the map of Chattogram City Corporation

(Yousuf et al. 2016), higher nighttime temperature, comparatively lower mean diurnal temperature, and an average rainfall of 2841.3 mm with large seasonal variations in monthly rainfall (Shahid et al. 2016). The rainy period lasts for 9.8 months between February and December. Shahid et al. (2016) reported an increasing trend in the average annual temperature of Chattogram city due to the urban heat island effect.

12.2.2 Study Design

The study was conducted through observation of the indoor plants from the selected households and interviews of the household heads of the Halishahar area of CCC. Three wards, out of 41 wards of CCC, namely North Agrabad, Rampur, and North Halishahar that compose the Halishahar area, were demarcated on the map. In that area, the households were selected following a multistage sampling method. First, all the major roads were listed from the three wards from which two prominent roads were selected randomly for each of the three wards. From each road, eight residential buildings were selected systematically taking one building after each three buildings located along the road. Most of the buildings were multistoried housing complexes with families living in separate flats. Each flat was considered as an individual household. All the households of the selected buildings were asked if they have indoor plants. After that, two households having indoor plants from each building were finally selected randomly and interviewed. Indoor plants were observed in each household and information thereof was recorded. In total, 48 households were interviewed and surveyed by using a semi-structured questionnaire. A data collection form for indoor plants was used to take details about the plants.

12.2.3 Field Data Collection

Field data were collected between August 2019 and March 2020. The local name, number of individuals, and place of occurrence for all the indoor plant species were recorded in the datasheet. During the interview, data related to respondent's background, source of seedlings, potting media, maintenance operation, perceived benefits received from the indoor plants, etc. were collected. For unknown species, herbarium specimen was prepared and used for identification. In cases where the respondents did not permit plant sample collection, photographs of the plants were taken and used for identification.

12.2.4 Identification of the Indoor Plants

Some of the indoor plants were identified during the data collection. For others, voucher specimens for future reference were prepared using the standard herbarium technique (Singh and Subramaniam 2008). Relevant taxonomic materials, i.e., Encyclopedia of Flora and Fauna of Bangladesh (Ahmed et al. 2009a, b, c; Ahmed et al. 2008a, b, c; Siddiqui et al. 2007a, b; Uddin and Hassan 2018), were consulted for identification of the collected and photographed specimens. Since many of the indoor plants were not native to Bangladesh and they are largely introduced for their aesthetic values, Web-based platforms of the ornamental plants, i.e., Flora of Bangladesh and Bonayon (http://www.floraofbangladesh.com/ and http://www.bonayon.com/), were used for identification. All the identified plants were taxonomically arranged under different families according to the Cronquist's system of classification (Cronquist 1981).

12.2.5 Calculating Diversity Indices

The diversity of indoor plants recorded from the city dwellings was expressed by using four diversity indices. Shannon-Wiener's diversity index (SWDI) and maximum value of Shannon's diversity index were computed following Shannon and Wiener (1963) following Eqs. (12.1) and (12.2). Typically, the value of the SWDI indicates low diversity when it is less or close to 1.5 and high species diversity when it is high or close to 3.5. Shannon's equitability index, also known as species evenness index (E), Simpson's dominance index (D), and Margalef's species richness index (R), is defined following Pielou (1966), Simpson (1949), and Margalef (1958), respectively, as indicated below by Eqs. (12.3) through (12.5):

$$H = -\sum_{i=1}^{s} P_i \ln(P_i) \tag{12.1}$$

$$H_{max} = ln(s) \tag{12.2}$$

$$E = \frac{H}{H_{max}} \tag{12.3}$$

$$D = \sum_{i=1}^{s} P_i^2 \tag{12.4}$$

$$R = \frac{(S-1)}{ln(N)} \tag{12.5}$$

Here, $P_i = \frac{k_i}{\sum_{i=1}^{s} k_i}$; k_i is the number of individuals of ith species and s is the total number of species recorded from the study area, and $N = \sum_{i=1}^{s} k_i$.

The diversity indices were compared with the similar tree diversity indices reported from different forests of Bangladesh. Though the bioclimatic condition, management techniques, and environmental characteristics are not identical with the present study, the comparison nevertheless was made to evaluate the potentiality of harboring diverse plants in indoor conditions. Diversity indices from 10 ecosystems of Bangladesh were gathered to compare the mean with the respective diversity index value from the present study (Table 12.1). The null hypothesis as tested using the t-statistic was that the mean of a certain diversity index reported from the 10 ecosystems did not significantly differ from the respective diversity index of the present study (Eq. (12.6)):

Table 12.1 Values of the four important diversity indices from ten different forest ecosystems of Bangladesh

SN	Diversity indices Shannon-Wienner	Simpson's dominance	Margalef's species richness	Species evenness	References
1	4.45	0.02	23.46	0.85	Hossain et al. (2013)
2	3.22	0.07	7.91	0.47	Chowdhury et al. (2019)
3	1.40	0.30	2.19	0.42	GoB (2019), Henry et al. (2021)
4	3.73	0.04	12.67	0.83	Hossen et al. (2021)
5	4.01	0.28	13.21	0.90	Jannat et al. (2019)
6	0.87	0.63	1.48	0.44	Rahman et al. (2019a)
7	3.35	0.09	14.20	0.73	Mamun et al. (2015)
8	3.18	0.07	7.57	0.79	Rahman et al. (2019b)
9	1.16	0.46	1.63	0.21	Uddin et al. (2020a)
10	0.90	0.37	2.03	0.62	Rahman et al. (2016), Dey and Akhter (2020)

$$t = \frac{x_i - \mu_i}{\sigma_i} \tag{12.6}$$

where x_i is the mean of diversity index i obtained from the studies conducted in 10 different ecosystems of Bangladesh; μ_i is the value diversity index i from the present study; and σ_i is the standard error of the mean of the diversity index i obtained in the studies from different ecosystems.

12.3 Results and Discussion

12.3.1 Species Composition of Indoor Plants

The study identified the occurrence of 120 indoor plant species belonging to 108 genera of 60 families (Table 12.2). Among the genera, *Mammillaria* was represented with the maximum four species each followed by *Euphorbia* and *Jasminum* genera with three species each. Apocynaceae and Cactaceae were the most dominant families with seven species. Presence of seven species under the Cactaceae family indicates the popularity of succulents in urban households. These two families were followed by Araceae with six species and Acanthaceae, Asteraceae, and Euphorbiaceae with five species each. Most of the plants were herbs (63%) and shrubs (33%) with some (9%) dwarf trees, e.g., *Cinnamomum tamala* and *Phoenix dactylifera*. Indoor plants were largely maintained for ornamentation purposes (82% of the species); besides, some of them were used for fruit (7%), medicine (7%), and vegetable (4%). Jaman et al. (2017) reported a comparatively higher number of plant species (221 species) in Dhaka south city corporation. Akber et al. (2011) mentioned 67 medicinal plant species from Khulna city. Moreover, the

Table 12.2 Composition of indoor plants recorded from the study area

SN.	Scientific name	Local name	Family	Habit
1	*Acalypha wilkesiana* Müll.Arg.	Seedham	Euphorbiaceae	Shrub
2	*Acer macrophyllum* Pursh	Bigleaf maple	Sapindaceae	Tree
3	*Adenium obesum* (Forssk.) Roem. & Schult.	Adenium	Apocynaceae	Shrub
4	*Aerides odorata* Lour.	Sukhphul	Orchidaceae	Herb
5	*Allamanda cathartica* L.	Golden trumpet	Apocynaceae	Shrub
6	*Aloe vera* (L.) Burm.f.	Aloe vera	Asphodelaceae	Herb
7	*Anigozanthos flavidus* DC.	Tall kangaroo paw	Haemodoraceae	Herb
8	*Antigonon leptopus* Hook. & Arn.	Ononto Lota	Polygonaceae	Herb
9	*Aphelandra squarrosa* Nees	Zebra plant	Acanthaceae	Herb
10	*Asystasia gangetica* (L.) T. Anderson	Chinese violet	Acanthaceae	Herb
11	*Azadirachta indica* A. Juss.	Neem	Meliaceae	Tree
12	*Basella alba* L.	Pui shak	Basellaceae	Herb
13	*Bougainvillea glabra* Choisy	Bagan bilash	Nyctaginaceae	Shrub
14	*Bryophyllum pinnatum* (lam.) Oken	Pathor kuchi	Crassulaceae	Herb
15	*Caesalpinia pulcherrima* (L.) Sw.	Radachura	Fabaceae	Tree
16	*Calendula arvensis* M. Bieb.	Calendula flower	Asteraceae	Herb
17	*Capsicum frutescens* L.	Morich	Solanaceae	Herb
18	*Carissa carandas* L.	Korom cha	Apocynaceae	Shrub
19	*Catharanthus roseus* (L.) G. Don	Noyon tara	Apocynaceae	Herb
20	*Centella asiatica* (L.) urban	Thankuni	Apiaceae	Herb
21	*Chlorophytum comosum* (Thunb.) Jacq.	Spider plant	Asparagaceae	Herb
22	*Chrysanthemum morifolium* (Ramat.) Hemsl.	Chandramallika	Asteraceae	Herb
23	*Cinnamomum tamala* (Buch.-ham.) T. Nees & C.H. Eberm	Tezpata, pulao leaf	Lauraceae	Tree
24	*Citrus Limon* (L.) Osbeck	Lebu	Rutaceae	Shrub
25	*Clitoria ternatea* L.	Aparajita	Fabaceae	Herb
26	*Coleus scutellarioides* (L.) Benth.	Pata bahar	Lamiaceae	Herb
27	*Colocasia gigantea* (Blume) Hook. f.	Elephant ear	Araceae	Herb
28	*Combretum indicum* (L.) DeFilipps	Madhuri Lota	Combretaceae	Herb
29	*Convallaria majalis* L.	Lily of the valley	Asparagaceae	Herb
30	*Costus woodsonii* (LSID)	Red button ginger	Costaceae	Herb
31	*Crassula ovata* (Mill.) Druce	Jade plant	Crassulaceae	Shrub
32	*Cryptanthus bromelioides* Otto & Dietrich	Cryptanthus	Bromeliaceae	Herb
33	*Cryptostegia grandiflora* Roxb. ex R. Br.	Rubber vine	Apocynaceae	Herb
34	*Cyathea medullaris* (G. Forst) Sw.	Black tree fern	Cyatheaceae	Tree
35	*Cymbopogon citratus* (DC.) Stapf	Thai pata	Poaceae	Herb

(continued)

Table 12.2 (continued)

SN.	Scientific name	Local name	Family	Habit
36	*Delosperma cooperi* (Hook. f.) L. Bolus	Trailing Iceplant	Aizoaceae	Herb
37	*Dendrobium nobile* Lindl.	Green leaf orchid	Orchidaceae	Herb
38	*Dendrobium teretifolium* R. Br.	Pencil orchid	Orchidaceae	Herb
39	*Desmodium intortum* (Mill.) Urb.	Greenleaf Desmodium	Fabaceae	Herb
40	*Dischidia vidalii* Becc.	Kangaroo pocket plant	Apocynaceae	Herb
41	*Dracaena trifasciata* (Prain) Mabb.	Snake plant	Asparagaceae	Herb
42	*Dypsis lutescens* (H.Wendl.) Beentje & J. Dransf.	Areca palm	Arecaceae	Herb
43	*Epipremnum aureum* (Linden & André) G.S. Bunting	Marvel queen pothos, Neon pothos	Araceae	Herb
44	*Eremurus robustus* (Regel) Regel	Giant desert candle	Asphodelaceae	Herb
45	*Eryngium foetidum* L.	Bilati dhonia	Apiaceae	Herb
46	*Euphorbia milii* Des Moul.	Kata mukut	Euphorbiaceae	Herb
47	*Euphorbia tirucalli* L.	Pencil cactus	Euphorbiaceae	Shrub
48	*Euphorbia tithymaloides* L.	Devil's backbone	Euphorbiaceae	Shrub
49	*Evolvulus nuttallianus* L.	Blue daze	Convolvulaceae	Herb
50	*Ficus benghalensis* L.	Bot	Moraceae	Tree
51	*Fittonia albivenis* (Lindl. ex Veitch) Brum.	Mosaic plant	Acanthaceae	Herb
52	*Fragaria* × *ananassa* (Duchesne ex Weston) Duchesne ex Rozier	Garden strawberry	Rosaceae	Herb
53	*Gardenia jasminoides* J. Ellis	Gandhraj	Rubiaceae	Shrub
54	*Gladiolus italicus* Mill.	Gladiolus flower	Iridaceae	Herb
55	*Gynura procumbens* (Lour.) Merr.	Gynura	Asteraceae	Herb
56	*Hedychium coronarium* J. Koenig	Dolan chapa	Zingiberaceae	Herb
57	*Hibiscus rosa-sinensis* L.	Joba	Malvaceae	Shrub
58	*Hylocereus undatus* (Haw.) Britton & Rose	Dragon fol	Cactaceae	Shrub
59	*Impatiens balsamina* L.	Dopati	Balsaminaceae	Herb
60	*Ipomoea purpurea* (L.) Roth	Morning glory	Convolvulaceae	Herb
61	*Ipomoea quamoclit* L.	Kunjo Lota	Convolvulaceae	Herb
62	*Iris versicolor* L.	Blue flag	Iridaceae	Herb
63	*Jasminum multiflorum* (Burm. f.) Andrews	Star jasmine	Oleaceae	Shrub
64	*Jasminum officinale* L.	Aromatic jui	Oleaceae	Shrub
65	*Jasminum sambac* L.	Beli, Arabian jasmine	Oleaceae	Shrub
66	*Juniperus chinensis* L.	Chinese juniper	Cupressaceae	Tree
67	*Justicia adhatoda* L.	Basak	Acanthaceae	Shrub
68	*Kalanchoe blossfeldiana* Poelln.	Christmas kalanchoe, Lal Pathorkuchi	Crassulaceae	Herb

(continued)

Table 12.2 (continued)

SN.	Scientific name	Local name	Family	Habit
69	*Lactuca sativa* L.	Lettuce	Asteraceae	Herb
70	*Lawsonia inermis* L.	Mehendi	Lythraceae	Shrub
71	*Lilium longiflorum* Thunb.	Easter lily	Liliaceae	Herb
72	*Mammillaria bombycina* Quehl	Cactus	Cactaceae	Shrub
73	*Mammillaria microhelia* Werderm.	Cactus	Cactaceae	Shrub
74	*Mammillaria spinosissima* Lem.	Cactus	Cactaceae	Shrub
75	*Mammillaria vetula* Mart.	Cactus	Cactaceae	Shrub
76	*Mansoa alliacea* (Lam.) A.H. Gentry	Lota parul	Bignoniaceae	Shrub
77	*Mentha spicata* L.	Spearmint	Lamiaceae	Herb
78	*Mirabilis jalapa* L.	Shondha maloti	Nyctaginaceae	Herb
79	*Morus nigra* L.	Black mulberry	Moraceae	Tree
80	*Murraya koenigii* (L.) Sprengel	Kaari pata	Rutaceae	Shrub
81	*Murraya paniculata* (L.) Jack	Kamini plant	Rutaceae	Shrub
82	*Nymphaea nouchali* Burm. f.	Sapla	Nymphaeaceae	Herb
83	*Ocimum tenuiflorum* L.	Kalo tulsi	Lamiaceae	Herb
84	*Oncidium altissimum* (Jacq.) Sw.	Dancing lady orchid	Orchidaceae	Herb
85	*Origanum vulgare* L.	Wild marjoram	Lamiaceae	Herb
86	*Othonna capensis* L.H. Bailey	Rubi necklace	Asteraceae	Herb
87	*Oxalis triangularis* A.St.-Hil.	False shamrock	Oxalidaceae	Herb
88	*Peperomia pellucida* Kunth	Luchi pata	Piperaceae	Herb
89	*Pereskia grandifolia* Haw.	Rose cactus	Cactaceae	Shrub
90	*Petroselinum crispum* (Mill.) Fuss	Dhone pata	Apiaceae	Herb
91	*Philodendron quinquenervium* Miq.	Philodendron	Araceae	Herb
92	*Phoenix dactylifera* L.	Date palm	Arecaceae	Tree
93	*Pilea cadierei* Gagnep. & Guillaumin	Aluminum plant	Urticaceae	Herb
94	*Piper betle* L.	Betel leaf	Piperaceae	Herb
95	*Plukenetia volubilis* L.	Mountain peanut	Euphorbiaceae	Shrub
96	*Plumeria alba* L.	Kat golap	Apocynaceae	Tree
97	*Portulaca grandiflora* hook.	Moss rose	Portulacaceae	Herb
98	*Portulaca oleracea* L.	Portulaca	Portulacaceae	Herb
99	*Rosa gallica* L.	Gallic rose	Rosaceae	Shrub
100	*Rosa rubiginosa* L.	China rose	Rosaceae	Shrub
101	*Ruellia simplex* C. Wright	Mexican petunia	Acanthaceae	Herb
102	*Russelia equisetiformis* Schltdl. & Cham.	Russelia	Plantaginaceae	Shrub
103	*Schlumbergera truncata* (Haw.) Moran	Christmas cactus	Cactaceae	Shrub
104	*Scindapsus aureus* (Linden & André) G.S. bunting	Golden pothos	Araceae	Herb
105	*Sedum acre* L.	Sedum	Crassulaceae	Herb
106	*Senna alexandrina* Mill.	Senokot plant	Caesalpiniaceae	Shrub
107	*Serissa japonica* (Thunb.) Thunb.	Snowbush	Rubiaceae	Shrub

(continued)

Table 12.2 (continued)

SN.	Scientific name	Local name	Family	Habit
108	*Solanum melongena* L.	Begun	Solanaceae	Shrub
109	*Soleirolia soleirolii* (Req.) Dandy	Baby tears	Urticaceae	Herb
110	*Spathiphyllum wallisii* Regel	Peace lily	Araceae	Herb
111	*Spondias mombin* L.	Amra	Anacardiaceae	Tree
112	*Syngonium podophyllum* Schott	Arrowhead plant	Araceae	Herb
113	*Tagetes erecta* L.	Gada phul	Asteraceae	Herb
114	*Talinum fruticosum* (L.) Juss.	Water leaf	Talinaceae	Herb
115	*Tradescantia spathacea* Sw.	Boat lily	Commelinaceae	Herb
116	*Trientalis borealis* (Raf.) U. Manns & Anderb.	Star flower	Primulaceae	Herb
117	*Tropaeolum majus* L.	Garden nasturtium	Tropaeolaceae	Herb
118	*Verbena bonariensis* L.	Purpletop vervain	Verbenaceae	Herb
119	*Zephyranthes rosea* Lindl.	Rain lily	Amaryllidaceae	Herb
120	*Zingiber spectabilis* Griff.	Maraca	Zingiberales	Herb

composition of indoor plant species in the households of the study area was comparable to 125 species from Daulatpur Upazila of Kushtia (Moniruzzaman et al. 2012) but less than those reported from many other studies from different Upazilas (Islam et al. 2009; Mahmudah et al. 2017; Rahman and Alam 2013; Rahman et al. 2012; Rahman et al. 2013; Rahman et al. 2019a; Sajib et al. 2014). In comparison to the floristic composition of protected areas (PA), the indoor plants of Halishahar, Chattogram City Corporation, were richer than only Khadimnagar National Park (Sobuj and Rahman 2011) but the composition was much less than that of several PAs reported by Dutta et al. (2014), Harun-ur-Rashid et al. (2018), Hossain et al. (2014), Rahman et al. (2017), and Sobuj and Rahman (2011). However, the number of indoor plant species was higher than several Village Common Forests of Bandarban and Khagrachhari (Chowdhury et al. 2019; Jannat et al. 2019; Uddin et al. 2020b); some distinguished forest ecosystems like Durgapur Hill Forest, Rajeshpur Sal Forest, and Ratargul Swamp Forest (Rahman et al. 2019b; Choudhury et al. 2004a; Choudhury et al. 2004b); and tree species composition of some reserve forests and forest beats like Idgaon Forest Reserve, Kamalachari natural forest, and Shitalpur Forest Beat (Biswas and Misbahuzzaman 2008; Hossain et al. 2015; Nur et al. 2016). Therefore, the study indicated a strong role of indoor green spaces in conserving diverse plant species.

12.3.1.1 Diversity of the Indoor Plant Species

The diversity of indoor plants in urban dwellings was high as indicated by the SWDI (4.207) which was remarkably close to Shannon's maximum possible value of diversity index, 4.79, for the recorded species in the study area. Moreover, the evenness index (0.879), species richness index (19.301), and dominance index (0.026) indicated higher evenness in the distribution of individuals of different species, species richness of the indoor plants, and less dominance of few species, respectively. Indirectly, all these indices indicated higher diversity of plants. Plant diversity indices from similar studies and contexts are very scarce. We compared the diversity indices with the same reported from different forest ecosystems. It has been observed that the SWDI was less than only compared to Dudhpukuria-Dhopachari Wildlife Sanctuary (DDWS) and selected VCFs of Rangamati (Hossain et al. 2013; Rudra et al. 2021) but the value of SWDI was much higher than Fasiakhali Wildlife Sanctuary, Himchari National Park, Kaptai National Park, Madhupur National Park, Medha Kachhapia National Park, Betchori VCF, Dhonpata VCF, Komolchari VCF, Maijchari VCF, and Shantivila VCF (Das et al. 2018; Hossen et al. 2021; Rahman et al. 2016; Rahman et al. 2019a; Uddin et al. 2020a, b). The comparison of the Simpson dominant index also indicated a similar trend to SWDI. Only a few ecosystems such as DDWS indicated a higher diversity than the present study with a comparatively lower value (0.019) of Simpson's dominant index. However, many other ecosystems, including several PAs, i.e., Durgapur Hill Forest, Fasiakhali Wildlife Sanctuary, Kaptai National Park, and several VCFs with a comparatively higher value of Simpson's dominance index, showed lower species diversity (Das et al. 2018; Rahman et al. 2016; Rahman et al. 2019b; Uddin et al. 2020b). The species evenness index and species richness index were also found to be higher than several protected areas, i.e., Himchari National Park, Madhupur National Park, and Medha Kachhapia National Park (Hossen et al. 2021; Rahman et al. 2019a; Uddin et al. 2020a), and several VCFs, i.e., Komolchari and Dhonpata (Chowdhury et al. 2019; Uddin et al. 2020b). However, both the species richness and evenness were comparatively lower than DDWS and Renikhayong Bagan Para VCF (Hossain et al. 2013; Jannat et al. 2019).

The t-test indicated a significantly higher SWDI ($t = -3.597, p = 0.005$), species richness index ($t = -4.642, p = 0.001$), and evenness index ($t = -3.445, p = 0.007$) of indoor plants in urban dwellings of Halishahar than the means of respective diversity indices from the studies of 10 forest ecosystems. The test also showed that the Simpson's dominance index from the present study was significantly lower ($t = 3.129, p = 0.012$) than its means from other studies indicating a little dominance of few species over others. Overall, the t-test confirmed the significantly higher diversity indices for urban indoor plants than other forest ecosystems.

12.3.1.2 Distribution of Indoor Plant Species per Household

The study attempted to reveal the extent of the presence of indoor plants in urban dwellings. About half (48%) of the households in the study area reported indoor plant species at their homes. People maintain the plants generally at the balcony or other suitable spaces of their houses. There was at least one family member in surveyed families who were carefully maintaining and taking care of the plants. Among the households, 2% had >9, 9% had 7–8, 38% had 3–4, 27% had 5–6, and 23% had ≤ 2 indoor plant species (Fig. 12.2). The variation was linked mainly to the availability of space and affluence.

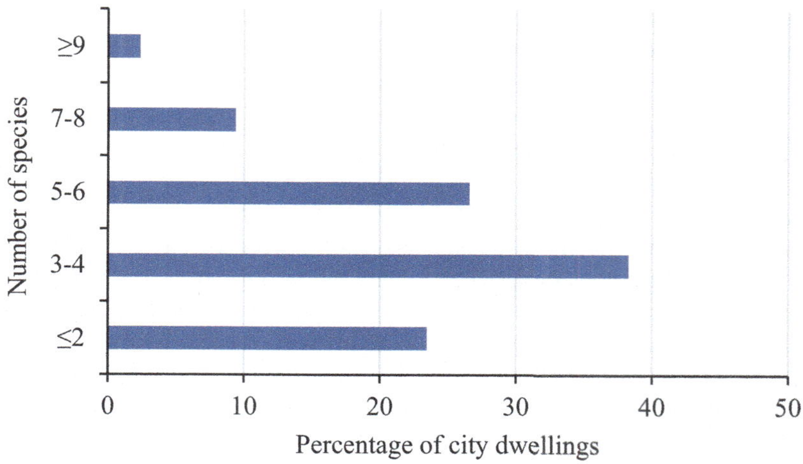

Fig. 12.2 Distribution of indoor plant species into the households

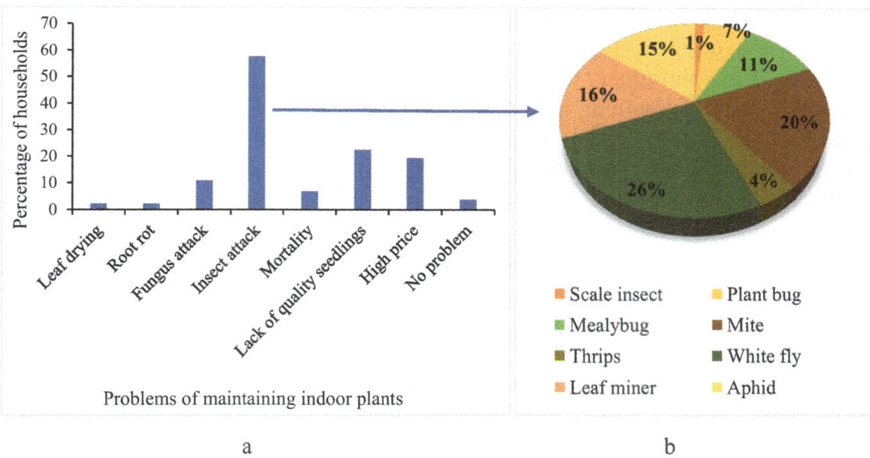

Fig. 12.3 (**a**) Problems in maintaining the indoor plants; (**b**) percent of households reported attack by different insect

12.3.2 Problems in Managing Indoor Plants

Insect attack (58% HH) was the main challenge in managing the indoor plants followed by the lack of quality seedlings and high price. Seven percent of respondents mentioned mortality where they are not clear about the reasons (Fig. 12.3a). However, 4% people do not face any managerial problem. Whitefly, mites, aphids, plant bug, mealybugs, leaf miners, thrips, and scale insects were the major pests with whitefly being more frequent followed by mites (Fig. 12.3b). The respondents did not adopt any intensive control measures except removing the affected leaves and branches or clearing off the affected plant.

12.3.3 Potting Media and Nutrition Supplement

In Bangladesh use of soil with cow dung, compost, sand, brick powder, wood chips, etc. is a common potting media to gardeners. People generally buy the potting media from the nearby nurseries of the city. Among the respondents, 28% bought plants with pots and media from nursery without much concern about the composition of potting media. However, a handsome 82% of them used a mixture of soil and compost at 3:1 ratio as the potting media for their indoor plants. On the other hand, 16% of them used a gross mixture of soil, sand, and compost while only 2% used brick powder with soil and compost (Fig. 12.4a). The study also explored the nutrient supplement for the indoor plants' potting media. Most (46%) of the households used a range of domestically produced organic waste materials (i.e., tea leaves, biodegradable kitchen wastage, banana peel) to supplement nutrients in the potting medium. The majority applied bruised tea leaves (46%), followed by kitchen wastage (30%) (Fig. 12.4b), while 11% used chemical fertilizers including urea, triple superphosphate, muriate of potash, etc.

Growing and maintaining indoor plants is tough compared to not growing plants in gardens. The success of maintaining indoor plants depends largely on the selection of suitable species, suitable containers, and potting media, and regular care. Application of appropriate cultural methods, regular maintenance, and protection from pest diseases are some of the primary needs for indoor plants. Most of the people had been maintaining indoor greens as a hobby despite the risk of increased insects (i.e., ants, mosquitoes) and humidity in the interior environment. People considered indoor plants as an element of beautification (52%), a means of air

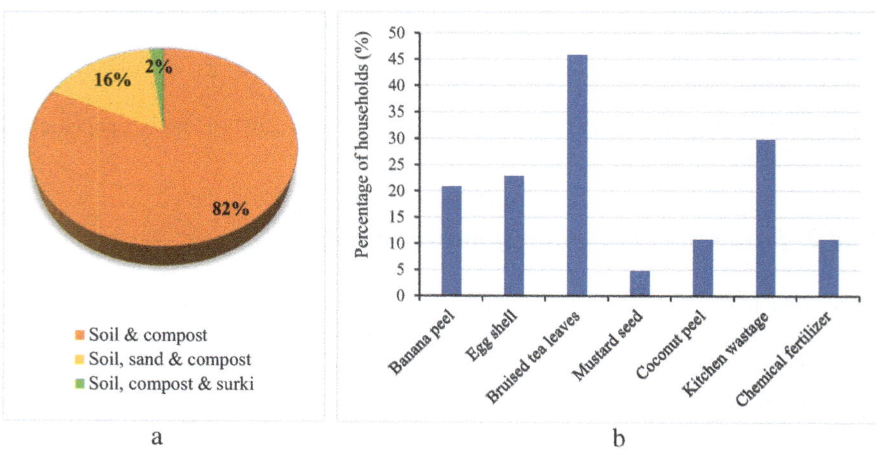

Fig. 12.4 (a) Percentage of different types of potting media; (b) maintenance of the potting media

purification (45%), and a good way to pass leisure time (17%). In addition to enriching the plant diversity, indoor plants can supplement the need for continuously decreasing urban green space. The green indoor concept could be most effective in urban areas where the lack of green open spaces is a common phenomenon.

12.4 Conclusion

The study assessed the composition and diversity of the indoor plants along with their management as practiced at the urban dwellings in the Halishahar—the north-western part of the Chattogram City. About half of the households had indoor plants and overall the study recorded 120 indoor plant species belonging to 60 families which indicated a handsome diversity of the species. The diversity indices showed significantly higher values compared to different forest ecosystems indicating the role that the urban green space can play in the conservation of herbs and shrubs. The study indicated the lack of technical know-how among the households in managing indoor plants. Observations indicated the scope to improve pot designs to make them suitable for inhouse placement (i.e., living rooms, balcony); to improve potting media to make it more nutrient rich, species specific, and less bulky; to improve strategies of insect and pest controls, etc.

Maintaining green inside the house has a soothing impact with a variety of benefits, i.e., aesthetics, air purification, and positive psychological impacts, as mentioned by the respondent families. However, increased insect occurrences inside the home and humidification of the indoor environment were the negative collaterals as identified by the respondents. The lack of quality seedlings, high cost, pest attack, and rots are some of the problems associated with the indoor plants. Despite some maintenance by the households (i.e., enriching the potting media, fertilizing, watering) the lack of technical knowledge was evident which needs to be addressed to enhance the impact of urban indoor greening. Indoor gardening has immense potential as a component of urban greening and should be given more emphasis amid the shrinkage of available space of urban green spaces (Homer-Dixon 2010), especially to attain the SDG 11.

Acknowledgement The authors are thankful to the Institute of Forestry and Environmental Sciences, University of Chittagong, for supporting the research through allowing unlimited use of the laboratories and library.

Declaration The authors declare that there is no conflict of interest.

References

Ahmed ZU, ZNT B, Hassan MA, Khondker M, SMH K, Ahmad M, ATA A, AKA R, Haque EU (2008a) Encyclopedia of Flora and Fauna of Bangladesh. In: Angiosperms: Dicotyledons (Acanthaceae – Asteraceae), vol 6. Asiatic Society of Bangladesh, Dhaka, Bangladesh

Ahmed ZU, ZNT B, Hassan MA, Khondker M, SMH K, Ahmad M, ATA A, AKA R, Haque EU (2008b) Encyclopedia of Flora and Fauna of Bangladesh. In: Angiosperms: Dicotyledons (Balsaminaceae–Euphorbiaceae), vol 7. Asiatic Society of Bangladesh, Dhaka, Bangladesh

Ahmed ZU, ZNT B, Hassan MA, Khondker M, SMH K, Ahmad M, ATA A, AKA R, Haque EU (2008c) Encyclopedia of Flora and Fauna of Bangladesh. In: Angiosperms: Monocotyledons (Orchidaceae–Zingiberaceae), vol 12. Asiatic Society of Bangladesh, Dhaka, Bangladesh

Ahmed ZU, Hassan MA, ZNT B, Khondker M, SMH K, Ahmad M, ATA A (2009b) Encyclopedia of Flora and Fauna of Bangladesh. In: Angiosperms: Dicotyledons (Magnoliaceae–Punicaceae), vol 9. Asiatic Society of Bangladesh, Dhaka, Bangladesh

Ahmed ZU, Hassan MA, ZNT B, Khondker M, SMH K, Ahmad M, ATA A (2009c) Encyclopedia of Flora and Fauna of Bangladesh. In: Angiosperms: Dicotyledons (Ranunculaceae–Zygophyllaceae), vol 10. Asiatic Society of Bangladesh, Dhaka, Bangladesh

Ahmed ZU, Hassan MA, Begum ZNT, Khondker M, Kabir SMH, Ahmad M, Ahmed ATA, Rahman AKA (2009a) Encyclopedia of Flora and Fauna of Bangladesh. In: Angiosperms: Dicotyledons (Fabaceae–Lythraceae), vol 8. Asiatic Society of Bangladesh, Dhaka, Bangladesh

Akber M, Seraj S, Islam F, Ferdausi D, Ahmed R, Nasrin D, Nahar N, Ahsan S, Jamal F, Rahmatullah M (2011) A survey of medicinal plants used by the traditional medicinal practitioners of Khulna city. Bangladesh American-Eurasian J Sustain Agric 5(2):177–195

Ambrose G, Das K, Fan Y, Ramaswami A (2020) Is gardening associated with greater happiness of urban residents? A multi-activity, dynamic assessment in the twin-cities region, USA. Landsc Urban Plan 198:103776. https://doi.org/10.1016/j.landurbplan.2020.103776

Aydogan A, Montoya LD (2011) Formaldehyde removal by common indoor plant species and various growing media. Atmos Environ 45:2675–2682. https://doi.org/10.1016/j.atmosenv.2011.02.062

BBS (2020) Statistical Yearbook Bangladesh 2019. Bangladesh Bureau of Statistics, Ministry of Planning, Government of the People's Republic of Bangladesh, Bangladesh, pp. 570

Biswas SR, Misbahuzzaman K (2008) Tree species diversity and regeneration traits of the dominant species in a dipterocarp forest in Bangladesh: implications for conservation. Int J Biodiversity Sci Management 4(2):81–91. https://doi.org/10.3843/Biodiv.4.2:2

Bringslimark T, Hartig T, Patil GG (2009) The psychological benefits of indoor plants: a critical review of the experimental literature. J Environ Psychol 29(4):422–433. https://doi.org/10.1016/j.jenvp.2009.05.001

Bruins CK (2005) Borrowing from heaven and earth. Contemporary balcony gardens of Wuhan, China, in their historical context. University of Cincinnati. https://etd.ohiolink.edu/pg_10?0::NO:10:P10_ACCESSION_NUM:ucin1115955282. Accessed 1 Oct 2020

Burton S (2001) Energy efficient office refurbishment. Earthscan. James & James (Science Publishers) Ltd., London

Choudhury JK, Biswas SR, Islam MS, Rahman O, Uddin SN (2004a) Biodiversity of Rajeshpur Sal forest. IUCN, Comilla, Bangladesh, p 21

Choudhury JK, Biswas SR, Islam MS, Rahman O, Uddin SN (2004b) Biodiversity of Ratargul swamp forest. IUCN, Sylhet, Bangladesh, p 24

Chowdhury MA, Islam KN, Hafiz N, Islam K (2019) Diversity of trees in a community managed forest: the case of Komolchari VCF, Khagrachari, Bangladesh. Geol Ecol Landscapes 3(2): 95–103. https://doi.org/10.1080/24749508.2018.1508980

Cronquist A (1981) An integrated system of classification of flowering plants. Columbia University Press, New York, p 1262

Cruz MD, Muller R, Svensmark B, Pedersen JS, Christensen JH (2014) Assessment of volatile organic compound removal by indoor plants—a novel experimental setup. Environmental science. Pollut Res 21:7838–7846. https://doi.org/10.1007/s11356-014-2695-0

Das SC, Alam MS, Hossain MA (2018) Diversity and structural composition of tree species in remnant dipterocarp forests: a study from Fasiakhali wildlife sanctuary. Bangladesh J Forestry Res 29(5):1241–1249. https://doi.org/10.1007/s11676-017-0548-7

Deng L, Deng Q (2018) The basic roles of indoor plants in human health and comfort. Environ Sci Pollut Res 25(36):36087–36101. https://doi.org/10.1007/s11356-018-3554-1

Dey A, Akther A (2020) Tree species composition and natural regeneration status of southeast region of Bangladesh. J Trop Biodiver Biotechnol 5(1):27–34. https://doi.org/10.22146/jtbb.49988

Dover JW (2015) Green infrastructure: incorporating plants and enhancing biodiversity in buildings and urban environments. Routledge, Abingdon, UK, p 352. https://doi.org/10.4324/9780203121993

Dutta S, Hossain MK, Hossain MA, Chowdhury P (2014) Floral diversity of Sitakunda botanical garden and Ecopark in Chittagong. Bangladesh Indian J Trop Biodiversity 22(2):106–118

Dzhambov AM, Lercher P, Browning MHEM, Stoyanov D, Petrova N, Novakov S, Dimitrova DD (2021) Does greenery experienced indoors and outdoors provide an escape and support mental health during the COVID-19 quarantine? Environ Res 196:110420. https://doi.org/10.1016/j.envres.2020.110420

Gawronska H, Bakera B (2015) Phytoremediation of particulate matter from indoor air by *Chlorophytum comosum* L. plants. Air Qual Atmos Health 8:265–272. https://doi.org/10.1007/s11869-014-0285-4

GoB (2019) Tree and forest resources of Bangladesh: report on the Bangladesh Forest inventory. Forest Department, Ministry of Environment, Forest and Climate Change, Government of the People's Republic of Bangladesh, Dhaka, Bangladesh, pp. 213

Harun-ur-Rashid M, Islam S, Kashem SB (2018) Floristic diversity (Magnoliids and Eudicots) of Baraiyadhala National Park, Chittagong, Bangladesh. Bangladesh J Plant Taxon 25(2):273–288. https://doi.org/10.3329/bjpt.v25i2.39532

Hassan MM, Nazem MNI (2016) Examination of land use/land cover changes, urban growth dynamics, and environmental sustainability in Chittagong city, Bangladesh. Environ Dev Sustain 18(3):697–716. https://doi.org/10.1007/s10668-015-9672-8

Henry M, Iqbal Z, Johnson K, Akhter M, Costello L, Scott C, Jalal R, Hossain MA, Chakma N, Kuegler O, Mahmood H, Mahamud R, Siddique MRH, Misbahuzzaman K, Uddin MM, Al Amin M, Ahmed FU, Sola G, Siddiqui MB, Birigazzi L, Rahman M, Animon I, Ritu S, Rahman LM, Islam A, Hayden H, Sidik F, Kumar MF, Mukul RH, Nishad H, Belal AH, Anik AR, Khaleque A, Shaheduzzaman M, Hossain SS, Aziz T, Rahaman MT, Mohaiman R, Meyer P, Chakma P, Rashid AZMM, Das S, Hira S, Jashimuddin M, Rahman MM, Wurster K, Uddin SN, Azad AK, Islam SMZ, Saint-André L (2021) A multi-purpose National Forest Inventory in Bangladesh: design, operationalisation and key results. Forest Ecosystems 8:12. https://doi.org/10.1186/s40663-021-00284-1

Homer-Dixon TF (2010) Environment, scarcity, and violence. Princeton University Press, Princeton, NJ, p 272

Hossain MA, Hossain MK, Alam MS, Bhuiyan MK (2014) Floristic composition of a protected area of Dudhpukuria-Dhopachori Wildlife Sanctuary, Chittagong, Bangladesh. Bangladesh Agriculture 6(1):38–70

Hossain MA, Hossain MK, Alam MS, Uddin MM (2015) Composition and diversity of tree species in Kamalachari Natural Forest of Chittagong South Forest Division, Bangladesh. J Forest Environ Sci 31(3):192–201. https://doi.org/10.7747/JFES.2015.31.3.192

Hossain MA, Hossain MK, Salam MA, Rahman S (2013) Composition and diversity of tree species in Dudhpukuria-Dhopachori Wildlife Sanctuary of Chittagong (South) Forest Division, Bangladesh. Res J Pharm Biol Chem Sci 4(2):1447–1457

Hossain S (2008) Rapid urban growth and poverty in Dhaka City. Bangladesh e-J Sociol 5(1):1–24

Hossen S, Hossain MK, Hossain MA, Uddin MF (2021) Quantitative assessment of tree species diversity of Himchari National Park (HNP) in Cox's Bazar. Bangladesh Asian J Forestry 5(1):1–7. https://doi.org/10.13057/asianjfor/r050101

Islam MR, Uddin MZ, Hassan MA (2009) An assessment of the angiospermic flora of Ramgarh Upazila of Khagrachhari district, Bangladesh. Bangladesh J Plant Taxon 16(2):115–140

Jafrin M, Beza BB (2018) Developing an open space standard in a densely populated city: a case study of Chittagong city. Inf Dent 3(3):1–25. https://doi.org/10.3390/infrastructures3030040

Jaman MS, Jahan I, Jamil M, Hossain MF (2017) Structure and composition of plant across different urban land vegetation in Dhaka South City, Bangladesh. J Agric Ecol Res Int 13(2): 1–14. https://doi.org/10.9734/JAERI/2017/38562

Jannat M, Kamruzzaman M, Hossain MA, Hossain MK (2019) Tree species diversity in the forest of Renikhayong Para Village in Bandarban, Bangladesh: a case study. J Biodivers Conserv Bioresour Manag 5(2):115–126. https://doi.org/10.3329/jbcbm.v5i2.44922

Lin BB, Egerer MH, Ossola A (2018) Urban gardens as a space to engender biophilia: evidence and ways forward. Front Built Environ 4:79. https://doi.org/10.3389/fbuil.2018.00079

Mahmudah Z, Islam M, Haque T, Uddin MZ (2017) Taxonomic enumeration of angiosperm flora of Sreenagar Upazila, Munshiganj, Dhaka, Dhaka, Bangladesh. J Asiatic Soc Bangladesh, Science 43(2):161–172

Mamun A, Hossain MA, Hossain MK, Alam S (2015) Quantifying diversity and composition of tree species in secondary hill forests of Chunati forest, Chittagong, Bangladesh. Indian Forester 141(5):566–572

Margalef R (1958) Information theory in ecology. General Systematics 3:36–71

Mia MA, Nasrin S, Zhang M, Rasiah R (2015) Chittagong, Bangladesh. Cities 48:31–41. https://doi.org/10.1016/j.cities.2015.05.011

Moniruzzaman M, Hassan MA, Rahman MM, Layla S, Islam MR (2012) A preliminary checklist of the angiospermic flora of Daulatpur Upazila in Kushtia district, Bangladesh. J Asiatic Soc Bangladesh Sci 38(1):53–65

Moran C (2019) Using plants in conjunction with permaculture design principles to provide an effective and affordable way to address air pollution in urban areas. In the proceeding of the 1st International Conference on Sustainable Energy and Climate in African Municipalities–SECAM 2019 22nd - 25th of October 2019, ENSTP, Yaoundé, Cameroon, p 1–5

Nur A, Nandi R, Jashimuddin M, Hossain MA (2016) Tree species composition and regeneration status of Shitalpur Forest beat under Chittagong north Forest division, Bangladesh. Adv Ecol 2016:5947874. https://doi.org/10.1155/2016/5947874

Park S, Kim M, Yoo M, Oh M, Son K (2010) Comparison of indoor CO_2 removal capability of five foliage plants by photosynthesis. Korean J Horticultural Sci Technol 28(5):864–870

Parseha I, Hakimeh T, Hajizadehc Y, Ebrahimpourc K (2018) Phytoremediation of benzene vapors from indoor air by *Schefflera arboricola* and *Spathiphyllum wallisii* plants. Atmos Pollut Res 9(6):1083–1087. https://doi.org/10.1016/j.apr.2018.04.005

Paul A, Nath TK, Noon SJ, Islam MM, Lechner AM (2020) Public open space, green exercise and well-being in Chittagong, Bangladesh. Urban Forestry Urban Greening 55:126825. https://doi.org/10.1016/j.ufug.2020.126825

Pielou EC (1966) Species diversity and pattern diversity in the study of ecological succession. J Theor Biol 10:370–383

Rahman MM, Mahmud MAA, Shahidullah M, Nath TK, Jashimuddin M (2016) The competitiveness of the phytosociological attributes of the protected areas in Bangladesh with that in the other tropical countries. J Sustain For 35(6):431–450

Rahman MO, Alam MT (2013) A taxonomic study on the angiosperm Flora of Trishal Upazila, Mymensingh Dhaka University. J Biol Sci 22(1):63–74

Rahman MO, Antara RT, Begum M, Hassan MA (2012) Floristic diversity of Dhamrai Upazila of Dhaka with emphasis on medicinal plants. Bangladesh J Bot 41(1):71–85

Rahman MO, Begum M, Ullah MW (2013) Angiosperm flora of Sadar Upazila of Munshiganj district, Bangladesh. Bangladesh J Plant Taxon 20(1):213–231

Rahman MR, Hossain MK, Hossain MA (2019a) Diversity and composition of tree species in Madhupur National Park, Tangail, Bangladesh. J Forest Environ Sci 35(3):159–172. https://doi.org/10.7747/JFES.2019.35.3.159

Rahman MR, Hossain MK, Hossain MA, Haque MS (2017) Floristic composition of Madhupur National Park (MNP), Tangail, Bangladesh. Bangladesh Agriculture 7(1):27–45

Rahman MR, Rahman MM, Chowdhury MA, Akhter J (2019b) Tree species diversity and structural composition: the case of Durgapur Hill Forest, Netrokona, Bangladesh. Asian J Forest 3(1):10–19

Rudra S, Islam KN, Rahman MM, Uddin SB (2021) Medicinal plant diversity and their therapeutic uses in selected village common forests in Chittagong Hill tracts, Bangladesh. J Herbs Spices Med Plants 27(1):1–25. https://doi.org/10.1080/10496475.2020.1786874

Safayet M, Arefin MF, Hasan MMU (2017) Present practice and future prospect of rooftop farming in Dhaka city: a step towards urban sustainability. J Urban Management 6(2):56–65. https://doi.org/10.1016/j.jum.2017.12.001

Sajib NH, Uddin SB, Islam MM (2014) Angiospermic plant diversity of Subarnachar Upazila in Noakhali, Bangladesh. J Asia Soc Bangladesh Sci 40(1):39–60

Shahid S, Wang HSB, Shamsudin SB, Ismail T, Minhans A (2016) Climate variability and changes in the major cities of Bangladesh: observations, possible impacts and adaptation. Reg Environ Chang 16(2):459–471. https://doi.org/10.1007/s10113-015-0757-6

Shannon CE, Wiener W (1963) The mathematical theory of communities. University of Illinois Press, Champaign, IL, pp 111–117

Siddiqui KU, Islam MA, Ahmed ZU, ZNT B, Hassan MA, Khondker M, Rahman MM, SMH K, Ahmad M, ATA A, AKA R, Haque EU (2007a) Encyclopedia of Flora and Fauna of Bangladesh. In: Bryophyta, pteridophyta and gymnosperms, vol 5. Asiatic Society of Bangladesh, Dhaka, Bangladesh

Siddiqui KU, Islam MA, Ahmed ZU, ZNT B, Hassan MA, Khondker M, Rahman MM, SMH K, Ahmad M, ATA A, AKA R, Haque EU (2007b) Encyclopedia of Flora and Fauna of Bangladesh. In: Angiosperm: Agavaceae–Najadaceae, vol 11. Asiatic Society of Bangladesh, Dhaka, Bangladesh

Simpson EM (1949) Measurement of diversity. Nature 163:688

Singh HB, Subramaniam B (2008) Field manual on herbarium techniques. National Institute of Science Communication and Information Resources, pp 1-297

Smith A, Pitt M (2010) Healthy workplaces: plantscaping for indoor environmental quality. Facilities 29(3):169–187. https://doi.org/10.1108/02632771111109289

Sobuj NA, Rahman M (2011) Assessment of plant diversity in Khadimnagar National Park of Bangladesh. Int J Environ Sci 2(1):79–91

Treesubsuntorn C, Thiravetyan P (2012) Removal of benzene from indoor air by Dracaena sanderiana: effect of wax and stomata. Atmos Environ 57:317–321. https://doi.org/10.1016/j.atmosenv.2012.04.016

Uddin M, Chowdhury FI, Hossain MK (2020a) Assessment of tree species diversity, composition and structure of Medha Kachhapia National Park, Cox's Bazar, Bangladesh. Asian J Forest 4(1): 15–21. https://doi.org/10.13057/asianjfor/r040104

Uddin MN, Firoj M (2013) Causes and consequences of rural–urban migration in Bangladesh: an empirical study in Chittagong City. Int J Ethics Social Sci 1(1):89–104

Uddin MN, Hossain MM, Karim MS, Siriwong W, Boonyanuphap J (2020b) The phytosociological attributes of village common forests in Chittagong Hill tracts, Bangladesh. Songklanakarin J Sci Technol 42(4):819–829

Uddin SN, Hassan MA (eds) (2018) Vascular Flora of Chittagong and the Chittagong Hill tracts, vol 1-3. Bangladesh National Herbarium, Dhaka, Bangladesh

Weng Q (2001) A remote sensing-GIS evaluation of urban expansion and its impact on surface temperature in the Zhujiang Delta, China. Int J Remote Sens 22:1999–2014

Xu JG, Liao BG, Shen Q, Zhang F, Mei AX (2007) Urban spatial restructuring in transitional economy-changing land use pattern in Shanghai. Chin Geogr Sci 17:19–27

Yang DS, Pennisi SV, Son K, Kays SJ (2009) Screening indoor plants for volatile organic pollutant removal efficiency. Hort Science 44(5):1377–1381. https://doi.org/10.21273/HORTSCI.44.5.1377

Young C, Hofmann M, Frey D, Moretti M, Bauer N (2020) Psychological restoration in urban gardens related to garden type, biodiversity and garden-related stress. Landsc Urban Plan 198: 103777. https://doi.org/10.1016/j.landurbplan.2020.103777

Yousuf AHM, Ahmed MK, Kundu G, Hoque S (2016) Characteristics of temperature and rainfall in Bangladesh: a historical perspective 1948-2011. Dhaka University. J Earth Environ Sci 4:1–20

Exploration, Identification, Multiplication, and Conservation of Rare Forest Genetic Resources in Chittagong University Campus, Bangladesh

13

Mohammed Kamal Hossain, Mohammed Danesh Miah, Md. Akhter Hossain, and Mohammed Saifullah

Abstract

Trees play a vital role as key structural components of forest ecosystems, sources of timber and non-timber products, and providers of many nonmarket ecosystem services. Rapid loss and degradation of natural forests have brought an alarming rate of forest genetic resource (FGR) depletion in Bangladesh. Considering the importance of conserving FGR, we need to explore, identify, and collect propagules (i.e., fruits, seeds) of the selected tree species from the remnant natural forests; develop propagation techniques for multiplication; and establish ex situ conservation in the University of Chittagong campus. This study has been conducted under the project titled "Exploration, Identification, Characterization, Multiplication and ex situ Conservation of Endangered Forest Genetic Resources Including Medicinal Plants of Bangladesh." The remnant natural forests of Chattogram, Chittagong Hill Tracts, Cox's Bazar, and Tangail were surveyed for identifying rare native tree species. The study identified 74 rare tree species for multiplication and conservation under this program. The selected species were observed in the field for phenological events. Fruits and seeds of the species were collected for seed germination experiments. Seeds with pre-sowing treatments were sown in the propagator house, polypots, seedbed, and seed germination trays to raise seedlings. Seedlings of the species were planted in the conservation stand of the University of Chittagong campus. In addition, seedlings of rare tree

M. K. Hossain (✉) · M. D. Miah · M. A. Hossain
Institute of Forestry and Environmental Sciences, University of Chittagong, Chittagong, Bangladesh
e-mail: akhter.hossain@cu.ac.bd

M. Saifullah
Bangladesh Agriculture Research Council, Farm gate, Dhaka, Bangladesh
e-mail: m.saif@barc.gov.bd

S. Ramamoorthy et al. (eds.), *Plant Genetic Resources, Inventory, Collection and Conservation*, https://doi.org/10.1007/978-981-16-7699-4_13

species were distributed to different organizations for ex situ conservation programs. The conservation stands will conserve the threatened tree species and become a potential source of fruits and seeds to propagate the species in future.

Keywords

Rare tree species · Deforestation · Multiplication · Nursery · Conservation

13.1 Introduction

The natural forests of Bangladesh were rich with diverse tree species (Hossain et al. 2013; Mukul et al. 2018). Trees are very important as key structural components of forest ecosystems, sources of timber and non-timber products, and sources of vital ecosystem services (Hossain et al. 2019). Like the global trend, the natural forest with its rich biodiversity is undergoing rapid loss and degradation in Bangladesh (Reza and Hasan 2020). The number of angiospermic species in Bangladesh is estimated to be 5700 (Firoz et al. 2004), whereas Ahmed et al. (2009a, b, c), Ahmed et al. (2008a, b, c), and Siddiqui et al. (2007b) recorded 3611 angiosperms in the Encyclopedia of Flora and Fauna of Bangladesh. However, there was no complete enumeration of the flora across the country due to which the species composition is just an approximation. As a result, it is difficult to make a precise estimate about the number of species being lost from the forest due to deforestation and degradation of forests across the country (UNREDD Bangladesh National Program 2016). It is evident from the field observation that the disappearance of native Forest Genetic Resources (FGR) in Bangladesh was more severe than the expectation. Conservation of the FGR is a pressing need in this situation but very little special conservation efforts are found. Hossain et al. (2017) reported a conservation effort which studied the germination techniques and established conservation plots of 60 tree species. Considering the importance of conserving rare FGR from natural forests of Bangladesh, we need to explore, identify, and collect propagules (i.e., fruits, seeds) of the rare tree species from the remnant natural forests and develop propagation techniques for multiplication and ex situ conservation. This situation leads to an initiative titled "Exploration, Identification, Characterization, Multiplication and ex-situ Conservation of Endangered Forest Genetic Resources Including Medicinal Plants of Bangladesh," which was implemented by the Institute of Forestry and Environmental Sciences, University of Chittagong (IFESCU), with funding support from Bangladesh Agricultural Research Council (BARC). The program identified rare tree species from the forests for studying their best multiplication techniques, establishing conservation plots in the University of Chittagong campus, and evaluating their growth performance in the conservation stand. This chapter explains the process of selecting the rare tree species, seed germination, and initial growth performance of the seedlings in the plantations. It is expected that the contents will be helpful to the conservationists and relevant policy makers for implementing forest restoration and floral conservation programs.

13.2 Materials and Methods

13.2.1 Study Area

The study is conducted in the University of Chittagong; more specifically, nursery and seed research laboratory of IFESCU was used for germination experimentation and seedling raising. Administratively, this area falls under Hathazari Upazila of Chattogram district, Bangladesh. The seedlings were then planted in the conservation plots located at the southern hills of Science Faculty. The conservation plots are medium hills characterized by sandy loam soils and bushy vegetation.

13.2.2 Study Design

The study has started with the exploration and mapping of the status of the FGR followed by studying their phenology, collection of propagules at seeding season, seed germination treatments, seedling raising, and finally establishment of conservation plots by planting and maintaining the plantations.

13.2.3 Exploring the Status of FGR and Selection of Species

Field visits in different forests of Bangladesh including Bangabandhu Safari Park at Dulhazra of Cox's Bazar district, Hazarikhil Wildlife Sanctuary (HWS) of Chattogram district, Srimai forest of Chattogram district, and Himchari National Park of Cox's Bazar district were conducted for assessing the status of FGR. In addition to this, areas such as Barshijura Seed Orchard Centre of Moulvibazar, Ukhiya Seed Orchard Centre, and Madhupur National Park of Tangail were explored. Transect walks were conducted in the forests using the foot trails. The native or naturalized tree species apparently seemed rare in the natural forests were selected for recolonization through multiplication. Locality of the selected species is recorded by its administrative location and a handheld global positioning system (GPS) machine.

13.2.4 Studying the Selected Species

Preliminarily selected species were observed periodically for recording their phenology. Propagules (i.e., matured seeds, ripen fruits) were collected during the fruiting or seeding period. Separated orthodox seeds were dried in open sun to an appropriate level in case the mature fruit is collected. Morphological characters (i.e., length, breadth, and weight) of a representative sample of seeds were recorded and another sample is preserved in the Seed Research Laboratory of IFESCU.

13.2.5 Seedling Raising

Seeds of the selected species went through pre-sowing treatments (if needed) before sowing in the nursery bed, propagator house, and indoor condition of laboratory. The pre-sowing treatments varied with species to suit it with the type of seeds and nature of seed coats of different species. Seeds are generally sown in the polybags and propagator house beds but very small seeds were sown in the germination trays. Sandy loam soil was used in the polybags and nursery trays, whereas propagator house beds were composed of coarse sands. Decomposed cow dung was added with the sandy loam soil at 1:3 proportion.

13.2.6 Care and Maintenance of the Seedlings

Seedlings raised in propagator and seedbed were transferred to polybag within 1–2 months of germination. The polybags, nursery trays, and propagator house beds were regularly watered and cleaned off weeds. Partial shade was provided to protect the seedlings from strong sunlight.

13.2.7 Conservation Stand Establishment

The 1-year-old seedlings were planted in the conservation site prepared on the hills located at the south of Faculty of Science Building and adjacent to the hill bottom colony of the University of Chittagong campus. The site was cleared off bushes and weeds. A total of 25–36 seedlings for each species were planted in the sufficiently sized pits filled with soil and cow dung. A uniform spacing, 2 m × 2 m, was maintained in the plantations between the seedlings. The conservation plots were surrounded by biodegradable fence material to protect them from grazing. Weeding was conducted regularly with a frequency of three times a year.

13.2.8 Seedling Data Collection

Average height of the seedlings (cm) from ground when 6 and 18 months old was collected for each of the species to assess the growth performance. From each of the species and age range, a total of 25 seedlings were measured for total height with a centimeter-scaled ranging rod.

13.2.9 Data Analysis

Height of the seedlings from the conservation plots was analyzed per species to indicate the survival percentage and growth performance in the nursery. Species with significantly higher height growth were identified through ANOVA and Duncan's multiple range test (DMRT) post hoc test at $P < 0.05$.

Conservation status of the species is available from the Encyclopedia of Flora and Fauna of Bangladesh (Ahmed et al. 2009a, b, c; Ahmed et al. 2008a, b, c; Siddiqui et al. 2007a, b). Since the IUCN red list status for the selected plant species is not available in Bangladesh perspective, the status from the Encyclopedia of Flora and Fauna of Bangladesh is compared with the selected species that seemed to be rare in the natural forests of Bangladesh.

13.3 Results and Discussion

13.3.1 Status of the Selected Rare Species

The study selected 74 rare indigenous tree species from different parts of Bangladesh for experimentation, multiplication, and conservation (Table 13.1). Taxonomically, the species belong to 65 genera and 39 families. It seemed that the highest four species were from *Terminalia* genus followed by *Dipterocarpus*, *Mallotus*, *Miliusa*, *Sterculia*, and *Vitex* genera with two species each. Among the family's representation, Euphorbiaceae was the highest with nine species followed by Anacardiaceae, Combretaceae, Meliaceae, and Sterculiaceae, each with four species.

13.3.2 Status of the Selected Species

IUCN red list of threatened species is the world's most comprehensive source of information about the conservation status of plants, animals, and fungi. The first red listing of species was done in 2000 for selected animals of Bangladesh which was repeated with more species in 2016. IUCN red listing of the plant species is yet to be done in Bangladesh. However, conservation status of the plants of Bangladesh is provided in the Encyclopedia of Flora and Fauna of Bangladesh by Ahmed et al. (2009a, b, c), Ahmed et al. (2008a, b, c), and Siddiqui et al. (2007a, b). The 0.5-1.5 years old status indicated that as much as 41 species were Least Concern, followed by 10 species Not Evaluated and 6 species under Conservation Dependent and Vulnerable each (Fig. 13.1). Moreover, there are two Data Deficient, one Lower Risk, one Near Threatened, and one Endangered species. In this discussion, we want to focus on the Least Concern plants that have become rare nowadays in the natural forests of Bangladesh. *Adenanthera pavonina, Antidesma ghaesembilla, Aphanamixis polystachya, Aquilaria malaccensis, Artocarpus lacucha, Baccaurea ramiflora, Bauhinia acuminata, Bischofia javanica, Carallia brachiata,* and *Caryota urens* are some of the then Least Concern species that have seemed to become rare at present and are conserved in the University of Chittagong campus.

Table 13.1 Composition of the rare tree species selected for experimentation, multiplication, and conservation at Chittagong University campus

No.	Botanical name	Vernacular name	English name	Family	Conservation status
1	***Adenanthera pavonina*** L.	**Raktakambal,** Raktakanchan, Ranjan	Red bead tree	Mimosaceae	LC
2	***Anisoptera scaphula*** (Roxb.) Pierre	**Boilam**, Boilshora, **Boilsur**, Sada boilam	Mascal wood	Dipterocarpaceae	CD
3	***Antidesma ghaesembilla*** Gaertn.	**Chutkigota**, **Elena,** Khudijam, Tendera	-----	Euphorbiaceae	LC
4	***Aphanamixis polystachya*** (Wall.) Parker	Boddiraj, Berirata, **Pitraj**, rata, Rohina, **Roina**	Amoora	Meliaceae	LC
5	***Aquilaria malaccensis*** Lamk.	Agar, Agaru	Agar tree	Thymelaeaceae	LC
6	***Artocarpus lacucha*** Buch.-Ham.	Ban Kanthal, **Barta**, Deophal, **Dewa**, Dewa Cham	Monkey jack	Moraceae	LC
7	***Baccaurea ramiflora*** Lour.	**Bhubi**, Harfata, Kangra gula, Latka, **Latkon**	Burmese grape	Euphorbiaceae	LC
8	***Bauhinia acuminata*** L.	Sada kanchan, **Swet-kanchan**	**Mountain ebony**, white bauhinia	Caesalpiniaceae	LC
9	***Bischofia javanica*** Bl.	**Kanjal badhi**, Kehura bhadi, laubhadi	Java cedar, West Indian cedar	Euphorbiaceae	LC
10	***Bouea oppositifolia*** (Roxb.) Meissn.	Ban aam, Maila, **Miriam**, Moyam, Uriam	Burmese plum, Marian plum	Anacardiaceae	NE but seems rare
11	***Carallia brachiata*** (Lour.) Merr.	Raskao, Lotkao, **Kiabong**, Kemal gach	Not known	Rhizophoraceae	LC
12	***Careya arborea*** Roxb.	Bidipata, **Gadila, Kumbi**, Tendu pata	Slow-match tree	Lecythidaceae	VU
13	***Caryota urens*** L.	**Chau supari**, Gol sago, Chau-gota	Hill palm, Indian sago-palm	Arecaceae	LC
14	***Cassia fistula*** L.	**Bandar Lathi**, Honalu, **Sonalu**	Golden shower	Caesalpiniaceae	LC
15	***Chukrasia tabularis*** A. Juss.	**Chikrassi**, Chabarassy, **Haithra-poma**, Paba	Chittagong wood, Indian mahogany	Meliaceae	LC
16	***Crateva magna*** (Lour.) DC.	**Barun**, Banya, Gota burna, **Pitagola**	Three-leaved caper	Capparaceae	LC
17	***Croton tiglium*** L.	**Jaypal**, jamalgota, Bish khagor	Purging croton	Euphorbiaceae	CD

18	***Cryptocarya amygdalina*** Nees	Bhuia gach, **Sutrong**	----	Lauraceae	NE but seems rare
19	***Dehaasia kurzii*** King	Bograj, **Modon mosta**	---	Lauraceae	NE but seems rare
20	***Diospyros malabarica*** **(Desr.) Kostel.**	Deshi gab	Indian persimmon	Ebenaceae	LC
21	***Dipterocarpus alatus*** Roxb. *ex* G. Don	Dahlia garjan, **Dhullya garjan, Sada Garjan**, Mashkhaliya garjan, Sil garjan	Garjan tree	Dipterocarpaceae	LC
22	***Dipterocarpus costatus*** Gaertn.	**Baittya garjan**, Guti garjan, Kessho garjan	Garjan tree	Dipterocarpaceae	CD
23	***Dysoxylum binectariferum*** (Roxb.) Hook. f. et. Bedd.	Bandar ratal, **bara rata**, Hota rata	---	Meliaceae	CD
24	***Ehretia serrata*** Roxb.	Kalahuja, **Kalauja,** Kat goa, Paban	Heliotrope tree	Boraginaceae	LC
25	***Elaeocarpus floribundus*** Bl.	**Belphoi**, ban belphoi, Jalpai	Indian olive, rugged oil fruit	Elaeocarpaceae	LC
26	***Elaeocarpus tectorius*** (Lour.) Poir.	Belfoi, **Jalpai**	Indian olive	Elaeocarpaceae	LC
27	***Engelhardtia spicata*** **Leschen** *ex* Bl.	Chorkata lej, Dhala rata, **Jhumka bhadi**, Kichra-bhadi, Loha bhadi	Great Malay beam	Juglandaceae	NE
28	***Fernandoa adenophylla*** (Wall. ex G. Don) vn Steenis	**Bar patta**, bon Sal, bon segun, Chilana, Koira aswal, Pahari hijal, Sil parul	Karen wood	Bignoniaceae	LC
29	***Ficus hispida*** L. f.	Dumur, **Kak dumur**, Khosaka, **Khuksa**, Kudura	Opposite-leaved fig, rough-leaved fig	Moraceae	LC
30	***Firmiana colorata*** (Roxb.) R. Br.	**Faisa udal**, Mula, Pata gota, Udal, Ujal	The colored sterculia	Sterculiaceae	LC
31	***Flacourtia jangomas*** (Lour.) Raeusch	**Lok luki**, Paniala, **Paina gola**, Peala gota, **pela gota**	Indian cherry	Flacourtiaceae	LC
32	***Gardenia coronaria*** Buch.-Ham.	Connari, **Kannyari**	Golden gardenia	Rubiaceae	VU
33	***Grewia nervosa*** (Lour.) Panigr.	**Achar gulla**, Assar, **Datoi**, Patka, Sitki	Phalsa	Tiliaceae	LC
34	***Haldina cordifolia*** (Roxb.) Ridsdale	Dakrum, **Haldu, Kaika**, Kali kadam	Yellow teak	Rubiaceae	CD

(continued)

Table 13.1 (continued)

No.	Botanical name	Vernacular name	English name	Family	Conservation status
35	***Hymenodictyon orixense*** (Roxb.) Mabberley	**Bhutum, Buikadam**, Puti kadam, Kali kadam, Srikadam	Bridal couch tree	Rubiaceae	VU
36	***Lagerstroemia parviflora*** Roxb.	**Sidha jarul**, tilla-jarul		Lythraceae	LC but seems rare
37	***Lepisanthes rubiginosa*** (Roxb.) Leenhouts	Bara harina, **Chagal guti, Chagaler Bori**, Chagaler leda, Harina gula	Rusty sapindus	Sapindaceae	LC
38	***Lithocarpus acuminata*** (Roxb.) Rehder	**Dholi Batna**, Gola Batna, kali batna, Kanta batna, Lota Batna	Indian chestnut	Fagaceae	EN
39	***Litsea monopetala*** (Roxb.) Pers.	**Lalkhori, Manda,** Meda, Pipul gach, **Sukurja**	Common Grey mango laurel	Lauraceae	NE
40	***Lophopetalum wightianum*** Arn.	**Raktan**, rattan, Sutrong	Wight's lophopetalum	Celastraceae	NT
41	***Macaranga denticulata*** (Bl.) Muell.-Arg.	**Bura, Bura Kochi**, Burna, rata bura	Blistery Macaranga	Euphorbiaceae	LC
42	***Madhuca longifolia*** (Koenig) MacBride	Mahua, Mohwa	Butter tree, the mohua tree	Sapotaceae	NE
43	***Mallotus philippensis*** (Lamk.) Muell.-Arg.	**Kamela**, Sindur, **Sinduri**, Singara	Red berry, monkey face tree	Euphorbiaceae	LC
44	***Miliusa longiflora*** (Hook. f. & Thom.) Finet & Gagnep.	Phesikau, Kuchu kao	----	Annonaceae	CD
45	***Miliusa velutina*** (Dunal) Hook. f. et Thoms.	Bal gozari, Gandhi-gazari, Gobra Sal, Pora Sal, Senia Sal	----	Annonaceae	LC but seems rare
46	***Ormosia robusta*** (Roxb.) Baker	Ormosi, **Hokkanali**	Horse-eye bean	Fabaceae	DD but seems rare
47	***Oroxylum indicum*** (L.) Benth. *Ex* Kurz	Hona, **Kanaidinga, Khona**, Thona	Indian trumpet flower	Bignoniaceae	LC
48	***Protium serratum*** (Wall. *ex* Colebr.)	**Gutgutya**, Hiliabhadi, Lohabhadi, Neur, Niyar	Indian red pear	Burseraceae	LC

49	***Pterocarpus indicus*** Willd.	Andaman padauk, **Padak**, Paduak, Rakta Chandan	Andaman red wood	Fabaceae	NE
50	***Pterygota alata*** (Roxb.) R. Br.	**Buddha narikel**, Jangli badam	The Buddha's coconut tree	Sterculiaceae	LC
51	***Putranjiva roxburghii*** Wall.	**Jiaputa**, Putranjib		Euphorbiaceae	LC
52	***Quercus gomeziana*** A. Camus	**Dhoilla Batna**, Goorja Batna, Khooisa Batna, **Tal Batna**	Oak tree	Fagaceae	DD
53	***Saraca asoca*** (Roxb.) de Wilde	**Asok**, Monea, Owsad tree	Asoka tree	Caesalpiniaceae	LC
54	***Schima wallichii*** (DC.) Korth.	**Banak,** bon Champa, **Kanak,** Makra	Schima	Theaceae	LC
55	***Schleichera oleosa*** (Lour.) Oken.	**Kusum, Joyna**, Lakkha	Ceylon oak, gum tree, honey tree	Sapindaceae	NE
56	***Semecarpus anacardium***	**Behula**		Anacardiaceae	NE
57	***Spondias pinnata*** (L.f.) Kurz.	Amra, **Deshi-amra, bon-amra**, Pial	Hog plum	Anacardiaceae	LC
58	***Sterculia foetida*** L.	Jangli badam, **box-badam**, Udal badam	Wild almond tree	Sterculiaceae	NE
59	***Sterculia villosa*** Roxb. ex. Smith.	Bara ujal, **Fashya udal**, Udal, Ujal	Elephant rope tree	Sterculiaceae	LC
60	***Stereospermum colais*** (Buch.-Ham. ex Dillw.) Mabberley	**Dharmara**, Pahari awal	Yellow snake tree	Bignoniaceae	LC
61	***Suregada multiflora*** (A. Juss.) Bail.	**Ban naranga**, Chagal ladha, Maricha	False lime	Euphorbiaceae	NE
62	***Swintonia floribunda*** Griff.	**Am-chandul**, Bailsur, **Civit**	Civit	Anacardiaceae	VU
63	***Syzygium firmum***	**Dhakai jam**	Sea apple	Myrtaceae	LC
64	***Terminalia arjuna*** (Roxb. *ex* DC.) Wt. et Arn.	**Arjun**, Arjuna	The arjuna myrobalan, white murdah	Combretaceae	VU
65	***Terminalia bellirica*** (Gaertn.) Roxb.	Bahera, Bayra, Boira, bora gach	Bastard myrobalan, Belleric myrobalan	Combretaceae	LC
66	***Terminalia chebula*** Retz.	**Haritaki**, Horttal, Hartoki	Black myrobalan, Chebulic myrobalan	Combretaceae	VU
67	***Terminalia citrina*** (Gaertn) Roxb. ex. Fleming	Hartiki, **Hatiyal**, Hartaki	Citrina myrobalan	Combretaceae	LC

(continued)

Table 13.1 (continued)

No.	Botanical name	Vernacular name	English name	Family	Conservation status
68	***Tetrameles nudiflora*** R. Br.	**Chandul**, Mainakat, Tirol, **Toirol**, Taru	Tetrameles	Datiscaceae	NE
69	***Trewia nudiflora*** L.	Batul, **Pitali**, lattoo, gotagamar, Mera gota, pitagola	False white teak	Euphorbiaceae	NE
70	***Uvaria cordata*** (Dunal) Alston	**Bagh-runga**, Gagh-ranga	Carabao	Annonaceae	LC
71	***Vitex glabrata*** R.Br.	**Arsal**, Asphal**, Harina**, Tokra gach	Smooth chaste tree	Verbenaceae	LC
72	***Vitex peduncularis*** Wall. ex Schauer	Awal, **Arsol**, **Goda, Horina**, Aswal	Long-spike chaste tree	Verbenaceae	NE but seems rare
73	***Walsura robusta*** Roxb.	**Ataligula, ban lichi**, Lal batta	------	Meliaceae	LC
74	***Zanthoxylum rhetsa*** (Roxb.) DC.	Badrang, **Bajna, Basbarana**	------	Rutaceae	LC

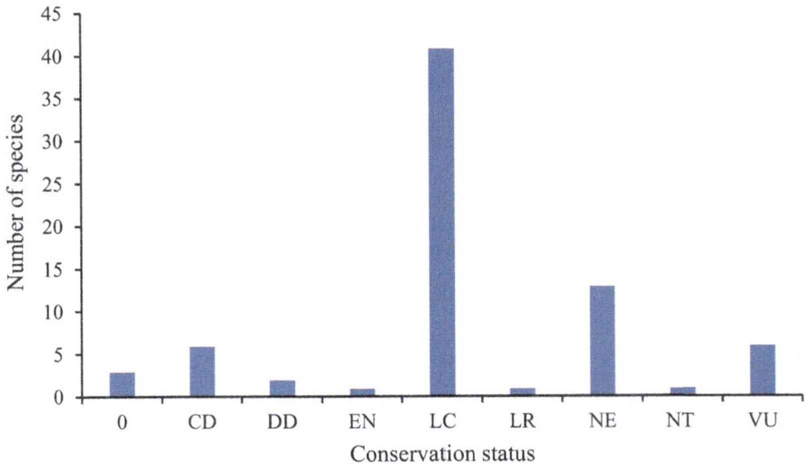

Fig. 13.1 Number of species under different conservation status [according to Ahmed et al. (2009a, b, c); Ahmed et al. (2008a, b, c); Siddiqui et al. (2007a, b); here, *CD* conservation dependent, *DD* data deficient, *EN* endangered, *LC* least concern, *LR* lower risk, *NE* not evaluated, *NT* near threatened, *VU* vulnerable]

13.3.3 Growth Performance

Survival percentage and height (cm) measurement were taken when the plantations were 6 months (34 species) and 18 months old (40 species). A total of 36 seedlings were planted for more than 60 species followed by 50 seedlings planted for 10 species. Number of seedlings from the remaining 18 species varied from 25 to 60. The average survival percentage of the 6-month-old seedlings was 98% which varied from 83% to 100% (Table 13.2). On the other hand, the average survival percentage is 97%, ranging from 65% to 100%, for 18-month-old seedlings. It is to be noted that 83% of species showed 100% survival percentage indicating a very successful seedling stand which may be attributed to an intensive care and maintenance. Overall, the mean survival percentage is 97% varying from 65% to 100% where 54 species (72% species) showed the highest 100% survival percentage and only 5% species showed less than 90% survival percentage portraying a very good success rate of the plantations up to this stage.

At 6 months old, the average height of the seedlings was 74.7 cm which ranged from 13 cm to 276 cm. The average height crossed 100 cm for only seven species (21%). *S. firmum* attained the maximum height (276.8 cm) followed by *T. nudiflora* (257.5 cm) and *S. foetida* (167.0) (Table 13.2). Minimum height was found for *Caryota urens* (12.3 cm) followed by *Diospyros malabarica* (18.3 cm) and *Firmiana colorata* (18.6 cm).

Average height of the seedlings at the age of 18 months was 163.1 cm which ranged from 19 cm to 292 cm. About 73% of species showed a mean height of more than 100 cm. Pitali (*Trewia nudiflora*) attained the maximum height (392.3 cm)

Table 13.2 Scientific name, source of the fruits/seeds, survival percentage, and height growth of the tree species in the conservation plot

No.	Botanical name	Fruit/seed source	Seedlings planted (no.)	Survival %	Age (months)	Average ht. (cm)
1	***Adenanthera pavonina*** L.	BTRI	36	100	18	185.4
2	***Anisoptera scaphula*** (Roxb.) Pierre	HWS	36	100	18	117.3
3	***Antidesma ghaesembilla*** Gaertn.	HNP	36	100	18	183.9
4	***Aphanamixis polystachya*** (Wall.) Parker	HWS	36	100	18	119.7
5	***Aquilaria malaccensis*** Lamk.	LNP	36	100	6	60.1
6	***Artocarpus lacucha*** Buch.-Ham.	HWS	30	93	6	21.3
7	***Baccaurea ramiflora*** Lour.	KNP	36	94	6	51.7
8	***Bauhinia acuminata*** L.	BDNP	36	100	18	186.6
9	***Bischofia javanica*** Bl.	KNP	60	100	6	115.8
10	***Bouea oppositifolia*** (Roxb.) Meissn.	Boalkhali	36	83	6	21.1
11	***Carallia brachiata*** (Lour.) Merr.	CWS	36	100	6	87.4
12	***Careya arborea*** Roxb.	MNP	36	100	6	24.1
13	***Caryota urens*** L.	CU campus	35	100	6	12.3
14	***Cassia fistula*** L.	CWS	36	100	18	266.1
15	***Chukrasia tabularis*** A. Juss.	HWS	36	100	18	252.4
16	***Crateva magna*** (Lour.) DC.	Guman Mardan, Hathazari	48	100	18	188.3
17	***Croton tiglium*** L.	MNP	36	100	18	295.0
18	***Cryptocarya amygdalina*** Nees	HWS	36	100	18	60.3
19	***Dehaasia kurzii*** King	HWS	36	100	18	24.4
20	***Diospyros malabarica*** **(Desr.) Kostel.**	BDNP	30	93	6	18.3
21	***Dipterocarpus alatus*** Roxb. *ex* G. Don	Silkhali Forest Range	36	100	18	203.0
22	***Dipterocarpus costatus*** *Gaertn.*	Silkhali Forest Range	40	85	18	23.3
23	***Dysoxylum binectariferum*** (Roxb.) Hook. f. et. Bedd.	HWS	36	100	18	64.8
24	***Ehretia serrata*** Roxb.	Bahaddarhat, Chattogram	36	100	18	245.5

25	***Elaeocarpus floribundus*** Bl.	HWS	36	100	18	210.3
26	***Elaeocarpus tectorius*** (Lour.) Poir.	HWS	36	100	18	232.4
27	***Engelhardtia spicata*** Leschen *ex* Bl.	BFRI campus	36	86	18	111.1
28	***Fernandoa adenophylla*** (Wall. ex G. Don) vn Steenis	HNP	50	100	6	41.9
29	***Ficus hispida*** L. f.	HWS	36	94	6	144.3
30	***Firmiana colorata*** (Roxb.) R. Br.	HWS	36	100	6	18.6
31	***Flacourtia jangomas*** (Lour.) Raeusch	KNP	25	100	6	67.3
32	***Gardenia coronaria*** Buch.-Ham.	HWS	36	100	6	73.4
33	***Grewia nervosa*** (Lour.) Panigr.	KNP	36	100	6	73.4
34	***Haldina cordifolia*** (Roxb.) Ridsdale	KNP	30	100	6	76.4
35	***Hymenodictyon orixense*** (Roxb.) Mabberley	MNP	36	83	18	18.9
36	***Lagerstroemia parviflora*** Roxb.	MNP	36	100	18	116.7
37	***Lepisanthes rubiginosa*** (Roxb.) Leenhouts	HWS	50	98	6	66.3
38	***Lithocarpus acuminata*** (Roxb.) Rehder	CWS	36	100	18	204.1
39	***Litsea monopetala*** (Roxb.) Pers.	KNP	30	100	6	33.5
40	***Lophopetalum wightianum*** Arn.	HWS	30	100	6	47.3
41	***Macaranga denticulata*** (Bl.) Muell.-Arg.	HWS	30	100	6	41.8
42	***Madhuca longifolia*** (Koenig) MacBride	HWS	50	96	6	64.4
43	***Mallotus philippensis*** (Lamk.) Muell.-Arg.	BDNP	50	93	6	27.5
44	***Miliusa longiflora*** (Hook. f. & Thom.) Finet & Gagnep.	BDNP	50	98	6	48.1
45	***Miliusa velutina*** (Dunal) Hook. f. et Thoms.	MNP	36	93	6	23.5
46	***Ormosia robusta*** (Roxb.) Baker	HWS	36	100	18	191.6
47	***Oroxylum indicum*** (L.) Benth. *ex* Kurz	HWS	36	100	18	278.0
48	***Protium serratum*** (Wall. *ex* Colebr.)	Mirpur Botanical Garden	50	100	18	53.8
49	***Pterocarpus indicus*** Willd.	CU campus	36	100	18	84.4

(continued)

Table 13.2 (continued)

No.	Botanical name	Fruit/seed source	Seedlings planted (no.)	Survival %	Age (months)	Average ht. (cm)
50	***Pterygota alata*** (Roxb.) R. Br.	Mirpur Botanical Garden	36	100	18	204.1
51	***Putranjiva roxburghii*** Wall.	CU campus	36	100	6	80.9
52	***Quercus gomeziana*** A. Camus	KNP	25	100	6	36.9
53	***Saraca asoca*** (Roxb.) de Wilde	MNP	36	94	6	83.2
54	***Schima wallichii*** (DC.) Korth.	MNP	25	94	6	124.1
55	***Schleichera oleosa*** (Lour.) Oken.	HWS	36	100	18	342.8
56	***Semecarpus anacardium***	KNP	60	100	18	86.4
57	***Spondias pinnata*** (L.f.) Kurz.	HWS	30	100	18	230.8
58	***Sterculia foetida*** L.	HWS	36	100	6	167.0
59	***Sterculia villosa*** Roxb. ex. Smith.	HWS	36	100	18	198.5
60	***Stereospermum colais*** (Buch.-Ham. ex Dillw.) Mabberley	TWS	36	97	18	87.1
61	***Suregada multiflora*** (A. Juss.) Bail.	BDNP	25	92	18	111.9
62	***Swintonia floribunda*** Griff.	HWS	50	100	18	96.0
63	***Syzygium firmum***	FWS	36	100	6	276.8
64	***Terminalia arjuna*** (Roxb. *ex* DC.) Wt. et Arn.	Banskhali Eco-Park	36	65	18	142.5
65	***Terminalia bellirica*** (Gaertn.) Roxb.	KNP	50	100	6	155.2
66	***Terminalia chebula*** Retz.	Guman Mardan, Hathazari	50	100	18	192.3
67	***Terminalia citrina*** (Gaertn) Roxb. ex. Fleming	Banskhali	36	100	18	128.0
68	***Tetrameles nudiflora*** R. Br.	KNP	36	100	6	257.7
69	***Trewia nudiflora*** L.	Guman Mardan, Hathazari	36	100	18	392.3
70	***Uvaria cordata*** (Dunal) Alston	Bangabhandu Safari Park, Chakaria	36	94	18	83.8
71	***Vitex glabrata*** R.Br.	FWS	30	100	18	105.7
72	***Vitex peduncularis*** Wall. ex Schauer	HWS	50	98	6	66.3

73	***Walsura robusta*** Roxb.	KNP	36	100	18	204.1
74	***Zanthoxylum rhetsa*** (Roxb.) DC.	KNP	30	100	6	33.5

**BDNP* Baraiyadhala National Park, *BTRI* Bangladesh Tea Research Institute, *CWS* Chunati Wildlife Sanctuary, *FWS* Fasiakhali Wildlife Sanctuary, *HWS* Hazarikhil Wildlife Sanctuary, *KNP* Kaptai National Park, *MNP* Madhupur National Park followed by Kusum (342.8 cm) and jamalgota (295.0) and minimum height was for Sada garjan (23.3 cm) followed by Modon mosto (24.4 cm) (Table 13.2). The growth performance may be influenced by the slow-growing nature of species along with the edaphic and climatic variations attributed to the sites.

13.4 Conclusion

Forest landscape restoration is one of the mostly talked programs nowadays for wildlife protection, biodiversity conservation, maintenance of environmental balance, and sustainable management of the ecosystems. Afforestation is one of the key tools for forest landscape restoration. But due to rapid loss of the mature individuals and mother trees of many native tree species, the seed sources of rare trees are becoming scarcer day by day. Moreover, though there are scattered plantation programs across Bangladesh, very little is known about the conservation programs dedicated for rare native tree species except the study of Hossain et al. (2017). The present program explored, multiplicated, and established conservation plots for 74 tree species which can also be considered as the potential stand for future seed source of these rare native tree species.

Acknowledgements The authors are highly grateful to the NATP Phase 2 Project ID # 074 IFESCU Component supported by the Natural Resources Division of Bangladesh Agricultural Research Council (BARC) for providing financial support and necessary suggestions under Project, "*Exploration, Identification, Characterization, Multiplication and Ex-Situ Conservation of Endangered Forest Genetic Resources Including Medicinal Plants of Bangladesh.*" The authors would like to thank all staff of "Seed Research Laboratory" and nursery of the Institute of Forestry and Environmental Sciences, University of Chittagong.

References

Ahmed ZU, Begum ZNT, Hassan MA, Khondker M, Kabir SMH, Ahmad M, Ahmed ATA, Rahman AKA, Haque EU (2008a) Encyclopedia of Flora and Fauna of Bangladesh. In: Angiosperms: Dicotyledons (Balsaminaceae–Euphorbiaceae), vol 7. Asiatic Society of Bangladesh, Dhaka, Bangladesh

Ahmed ZU, Begum ZNT, Hassan MA, Khondker M, Kabir SMH, Ahmad M, Ahmed ATA, Rahman AKA, Haque EU (2008b) Encyclopedia of Flora and Fauna of Bangladesh. In: Angiosperms: Dicotyledons (Acanthaceae–Asteraceae), vol 6. Asiatic Society of Bangladesh, Dhaka, Bangladesh

Ahmed ZU, Begum ZNT, Hassan MA, Khondker M, Kabir SMH, Ahmad M, Ahmed ATA, Rahman AKA, Haque EU (2008c) Encyclopedia of Flora and Fauna of Bangladesh. In: Angiosperms: monocotyledons (Orchidaceae–Zingiberaceae), vol 12. Asiatic Society of Bangladesh, Dhaka, Bangladesh

Ahmed ZU, Hassan MA, Begum ZNT, Khondker M, Kabir SMH, Ahmad M, Ahmed ATA (2009a) Encyclopedia of Flora and Fauna of Bangladesh. In: Angiosperms: Dicotyledons (Ranunculaceae–Zygophyllaceae), vol 10. Asiatic Society of Bangladesh, Dhaka, Bangladesh

Ahmed ZU, Hassan MA, Begum ZNT, Khondker M, Kabir SMH, Ahmad M, Ahmed ATA (2009b) Encyclopedia of Flora and Fauna of Bangladesh. In: Angiosperms: Dicotyledons (Magnoliaceae–Punicaceae), vol 9. Asiatic Society of Bangladesh, Dhaka, Bangladesh

Ahmed ZU, Hassan MA, Begum ZNT, Khondker M, Kabir SMH, Ahmad M, Ahmed ATA, Rahman AKA (2009c) Encyclopedia of Flora and Fauna of Bangladesh. In: Angiosperms: Dicotyledons (Fabaceae–Lythraceae), vol 8. Asiatic Society of Bangladesh, Dhaka, Bangladesh

Firoz R, Mobasher SM, Waliuzzaman M, Alam MK (2004) Proceedings of the regional workshops on National Biodiversity Strategy and Action Plan. IUCN Bangladesh Country Office, Dhaka, Bangladesh

Hossain MA, Hossain MK, Salam MA, Rahman S (2013) Composition and diversity of tree species in Dudhpukuria-Dhopachori wildlife sanctuary of Chittagong (south) Forest division, Bangladesh. Res J Pharm, Biol Chem Sci 4(2):1447–1457

Hossain MA, Pavel MA, Al Harada K, Beierkuhnlein C, Jentsch A, Uddin MB (2019) Tree species diversity in relation to environmental variables and disturbance gradients in a northeastern forest in Bangladesh. J For Res 30(6):2143–2150. https://doi.org/10.1007/s11676-018-0786-3

Hossain MK, Alam MS, Hossain MA (2017) Conservation of threatened tree species in Chittagong University campus. Arannayk Foundation, Dhaka, Bangladesh, p 183

Mukul SA, Biswas SR, Rashid AZMM (2018) Biodiversity in Bangladesh. In: Global biodiversity: selected countries in Asia, vol 1. Apple Academic Press, Williston, VT, pp 93–107. https://doi.org/10.1201/9780429487743-3

Reza AA, Hasan MK (2020) Forest biodiversity and deforestation in Bangladesh: the latest update. In: Forest degradation around the world. Intechopen, New York. https://doi.org/10.5772/intechopen.86242

Siddiqui KU, Islam MA, Ahmed ZU, Begum ZNT, Hassan MA, Khondker M, Rahman MM, Kabir SMH, Ahmad M, Ahmed ATA, Rahman AKA, Haque EU (2007a) Encyclopedia of flora and fauna of Bangladesh. In: Bryophyta, pteridophyta and gymnosperms, vol 5. Asiatic Society of Bangladesh, Dhaka, Bangladesh

Siddiqui KU, Islam MA, Ahmed ZU, Begum ZNT, Hassan MA, Khondker M, Rahman MM, Kabir SMH, Ahmad M, Ahmed ATA, Rahman AKA, Haque EU (2007b) Encyclopedia of flora and fauna of Bangladesh. In: Angiosperm: Agavaceae–Najadaceae, vol 11. Asiatic Society of Bangladesh, Dhaka, Bangladesh

UN-REDD Bangladesh National Program (2016) Drivers of deforestation and Forest degradation in Bangladesh: final report. UN-REDD Bangladesh National Program p 250

The Chemistry of Antidiabetic Plants 14

Emmanuel Adikwu Orgah, Mbaoji Camillus, Ifeoluwa Margaret Idowu, Olaniyan Kayode Olaoluwa, Chimezie Esther Uchechukwu, Luke Michael Ifreke, and Ajaegbu Henry Chukwudi

Abstract

A great many plant species that demonstrate activity against hyperglycemia are listed in literature. They are known to contain classes of compounds with biological activities including terpenoids, carotenoids, flavonoids, and others such as alkaloids and glycosides. These plants are variously reported to show antidiabetic effects. Some of the plant sources commonly found around various continents include aloe vera, *Amaranthus tricolor, Anacardium occidentale, Momordica charantia (bitter gourd), Moringa oleifera, Gongronema latifolia, Vernonia amygdalina, Tridax procumbens, Phyllanthus amarus, Phyllanthus niruri, Piper longum, Salvia miltiorrhiza, Ocimum gratissimum, Mangifera indica,* etc.

E. A. Orgah (✉) · M. Camillus · C. E. Uchechukwu
Nigeria Natural Medicine Development Agency, 9 Kofo Abayomi Street, Victoria Island, Lagos, Nigeria

School of Clinical Research, Texila American University (TAU), Georgetown, Guyana

I. M. Idowu
Nigeria Natural Medicine Development Agency, 9 Kofo Abayomi Street, Victoria Island, Lagos, Nigeria

Tianjin University of Traditional Chinese Medicine (TUTCM), Tianjin, China

O. K. Olaoluwa
Nigeria Natural Medicine Development Agency, 9 Kofo Abayomi Street, Victoria Island, Lagos, Nigeria

Federal University of Agriculture Abeoakuta Department of Veterinary Parasitology and Entomology, Abeokuta, Nigeria

L. M. Ifreke · A. H. Chukwudi
Nigeria Natural Medicine Development Agency, 9 Kofo Abayomi Street, Victoria Island, Lagos, Nigeria

S. Ramamoorthy et al. (eds.), *Plant Genetic Resources, Inventory, Collection and Conservation*, https://doi.org/10.1007/978-981-16-7699-4_14

However, these plants do not exert their effects by their mere beauty and form but by their constituents of chemical entities in structural orders that behooves on them these medicinal capabilities. This review attempts to highlight some classes of plant natural products or secondary metabolites and some specificities that render them the hidden medicines for the management and treatment of diabetes in the floral factories, found everywhere in our environment.

Keywords

Diabetes mellitus · Antidiabetic medicinal plants · Phytochemicals · Bioactive natural products · Herbal medicines clinical research · Metabolic pathways

14.1 Introduction

Over 800 plants of wonderful aesthetic beauty have been categorized as having antidiabetic properties. However, knowledge on the systematic empirical evaluation remains minimal; hence, it is important to acquire the knowledge about their chemistry on the one hand and the bioactivity in a stepwise manner of clinical efficacy, safety, and mechanism of action on the other hand. The creator, in effecting these wonderful designs, has clearly passed a message to man to study and discover the secret of the interrelatedness of nature to our lives.

How is this vast wealth of medicinal factory useful for the management and control of diabetes? The chemistry of antidiabetic plants like that of all medicinal plants is a knowledge that not only unwraps the beauty of creation but also reveals the key to solving health puzzles cast by the metabolic syndrome known as diabetes. What are the physical qualities of plant compounds that render them effective in ameliorating diabetes and what are the predictable pathways for such structure–activity relation of the natural products specifically targeting the antidiabetic clinical characteristics of these compounds?

This review aims to identify and catalogue specific natural products known to possess antidiabetic properties and the current understanding of the clinical pathways by which they exert these properties.

14.2 The Past

Old literatures of the late twentieth century (Ivorra et al. 1989; Alarcon et al. 1993) noted clearly that quite a few of these plants had been chemically and clinically assessed to ascertain efficacy.

At this period, principal chemical entities isolated, characterized, and identified were chiefly glycans, proteins, and mucilage. Also, steroids, flavonoids, alkaloids, and triterpenoids were demonstrated to possess hypoglycemic activity (Maries and Farnsworth 1995; Raman and Lau 1996). Mucilage of the Malvaceae family, of the seed of *Plantago asiatica*, demonstrated hypoglycemic activity as was noted by

Tomoda et al. (1987). Glycans of *Aloe arborescens* and *D. japonica* were reported by Hikino et al. (1986a) and Hikino et al. (1986b) to show similar properties. Hypoglycemic effects of the peptidoglycans "ganoderan A, B, and C" were reported by Hikino et al. (1988), in normoglycemic and hyperglycemic mice.

An aqueous extract of rice bran (*Oryza sativa* L.), source of vitamin B, was shown (Hikino et al. 1988) to reduce the levels of plasma glucose in experimental animal (mice). This activity-guided fraction of rice bran yielded the glycans oryzabrans, A, B, C, and D and demonstrated distinctive antidiabetic effects on healthy and alloxan-induced diabetic mice.

The juice of the inflorescence stalk of *Musa sapientum* yielded a pectin with antidiabetic properties. This extract was reported (Gomathy et al. 1990) to increase the concentration of hepatic glycogen and glycogenesis leading to a spontaneous rise in the activity of the enzyme glycogen synthetase with subsequent addition of glucose into hepatic glycogen in nondiabetic rats.

An important discovery was that some of the plant compounds have structural and physiological resemblance to insulin. Some of these insulin-like principles include peptidoglycan-isolated Panax ginseng and other glycans (panaxans A to P and ginsenoside Rb1).

14.3 Current Observations

From the previous works, empirical data have given scientists the knowledge that plants possess natural compounds of specific activities for the clinical management of diabetes. Some specific phytochemicals/natural products with such antidiabetic potential include flavonoids, saponins, alkaloids, glycosides, phenolic acids, tannins, stilbenes, polysaccharides, etc.

These classes of compounds have been found to exhibit antidiabetic effects manifested through various mechanisms and clinical pathways like:

- Glucose regulation
- Potentiating insulin secretion
- Mimicking insulin-like actions
- Lipid metabolism
- Stimulating cellular regeneration
- Signaling NF-kB pathway
- Gluconeogenic enzyme inhibition
- Protection against reactive oxygen species' (ROS) destructive actions

14.4 Specific Compounds and Their Characteristics

The compound **adenosine**, a peptide, was shown to have anti-lipidemic activity (Ng and Yeung 1985). **p-Insulin**, a polypeptide from *Momordica charantia*, only lacked methionine to equate with bovine insulin and p-insulin exhibited hypoglycemic activity when administered in juvenile diabetic patients, in a comparable manner to bovine insulin (Pushpa et al. 1981). The bitter gourd *Momordica charantia* appears to be a special plant created for diabetes going by its rich natural product constituents targeting different pathways and mechanisms of diabetes. Raman and Lau (1996) listed some of these compounds to include "saponins, sitosterol and stigmastadienol glucosides, proteins, p-insulin, steroidal glycosides, momordicines, momordicosides, and pyrimidine nucleoside vicine," which leads to fall in blood glucose and reduction in pancreatectomy according to Lotlikar and Rajarama (1966). Hii and Howell (1984), Chakravarthy et al. (1981), and Kolb et al. (1982) reported the regeneration of β-cells in the pancreatic islets of Langerhans, by these compounds. However, these laudable properties of *M. charantia* are not without the downside as it is also reported by Raman and Lau (1996) to show hepatic and reproductive system toxicity.

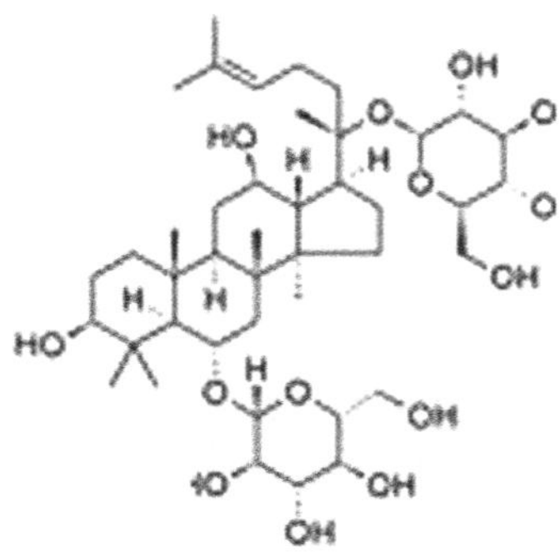

Ginsenoside

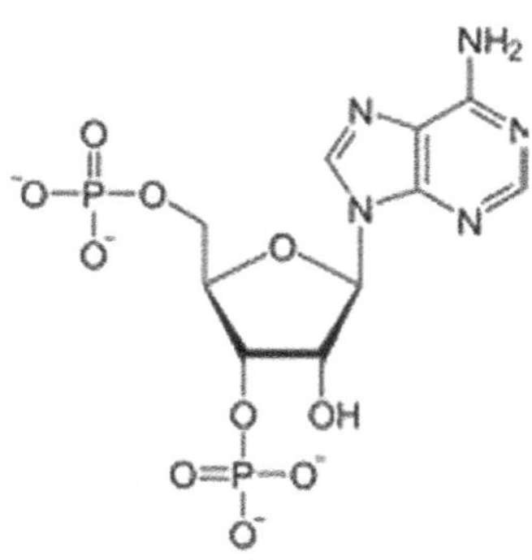

Adenosine,

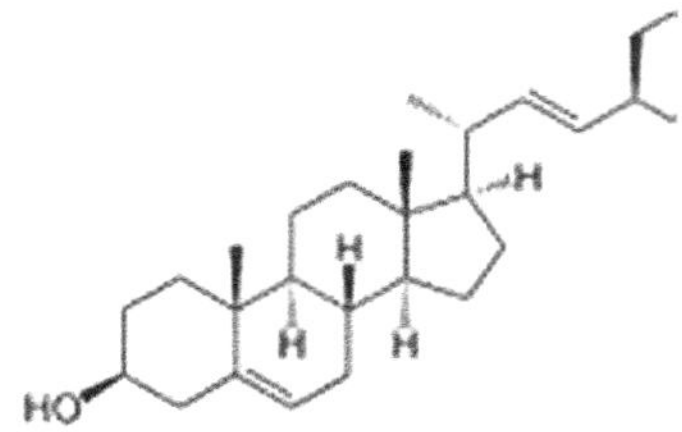

Sitosterol

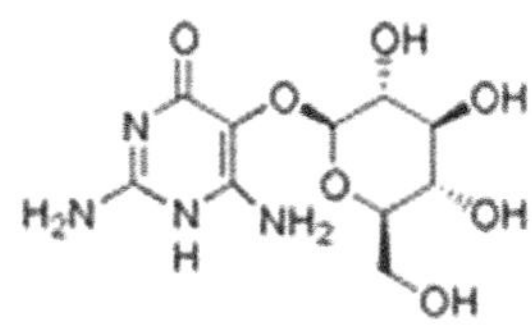

Vicine

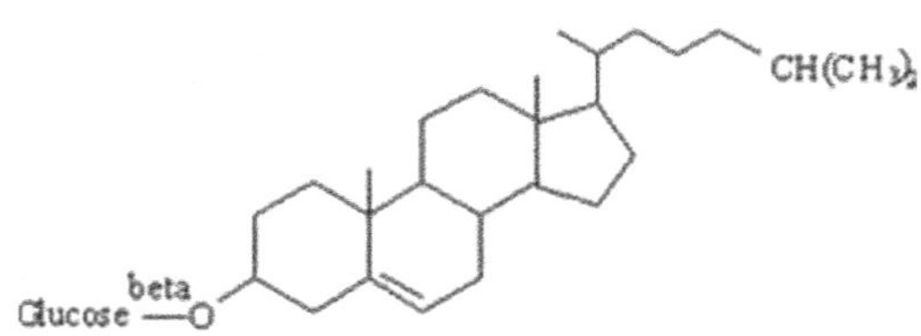

Charantin

Coterill et al. (1976) reported that the compound **kolaflavanone** demonstrated significant antidiabetic property in nondiabetic and alloxan-induced diabetic mice. Kolaflavanone is an anti-glycation agent, a biflavonoid obtained from *Garcinia kola* seed and found to exhibit protective effects against liver toxicity and blood glucose reduction properties.

Jalil et al. (1986) reported that the pentacyclic triterpenoids, **oleanolic acids and ursolic acid**, from *C. officinalis* decrease the volume of consumed water and urination in diabetic rats.

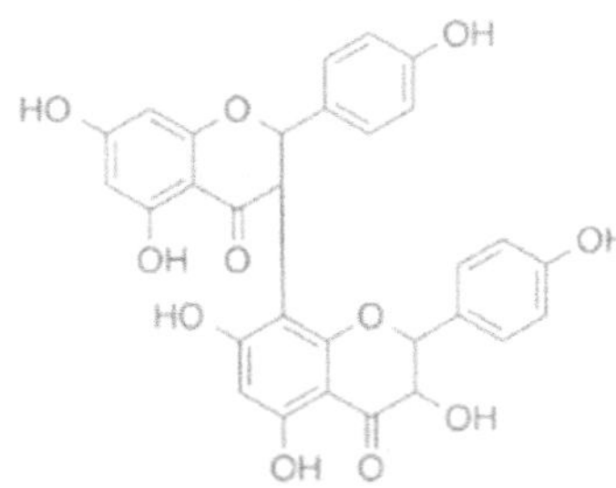

Kolaflavone

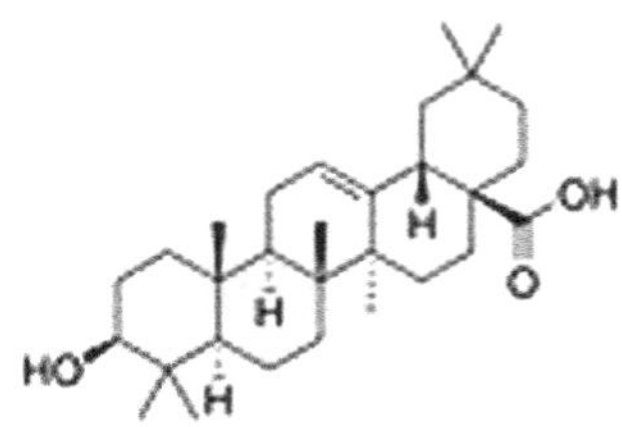

Oleanolic acid

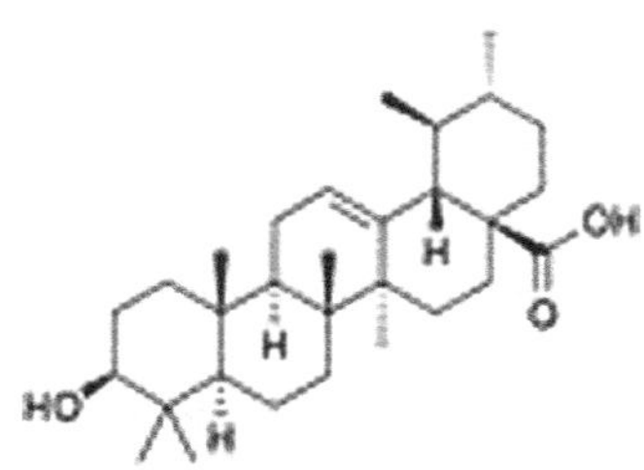

ursolic acid

Dioscoretine, an extract of *Dioscorea dumetorum*, tropical yam, *reported by* Iwu et al. (1990), was found to produce hypoglycemic effects. **Capsaicin** from *Capsicum annuum* L. was reported by Monsereenunurson (1980) to inhibit intestinal glucose transport. Like piperine, this brings up the value of the pepper family as antidiabetic plants. Its action is said to be associated with the breakdown of glucose to lactic acid/ or inhibitory effect on the ATPase-dependent sodium pump.

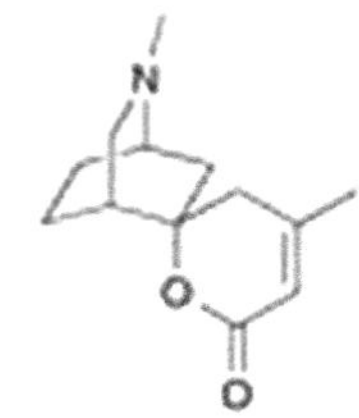

Dioscoretine

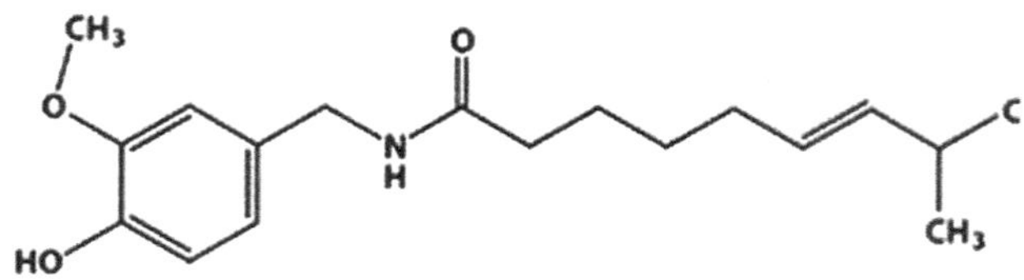

Capsaicin

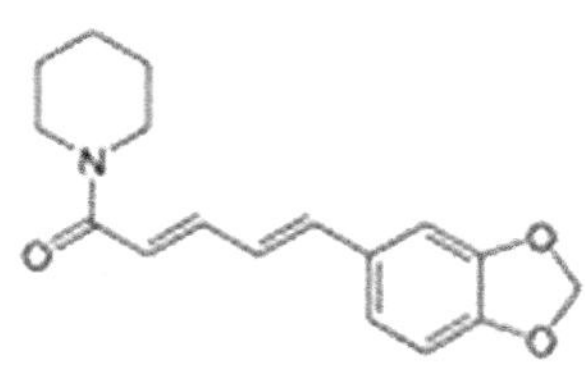

piperine

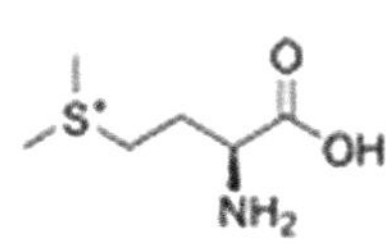

Smethylmethionine

Extracts of onion **diphenylamine** showed hypoglycemic effect in rabbits and performed better than tolbutamide as a hypoglycemic agent (Karawya et al. 1984), while **S-methylmethionine** demonstrated "an insulin-like effect on the β-cells of the pancreas" in some diabetic rats. Maries and Farnsworth (1995) suggest that allyl propyl disulfide and diallyl oxide (**allicin**) exert their antidiabetic effects by competing with insulin.

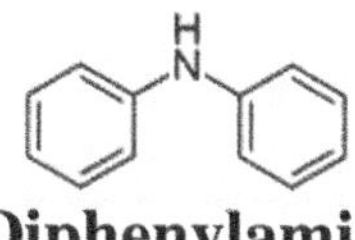

Diphenylamine

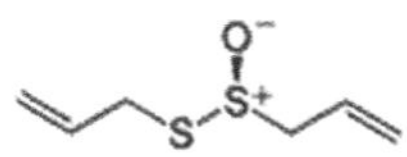

Allicin

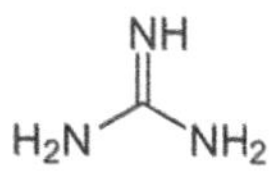

Guanidine

The derivative of the compound **guanidine** from *Ganoderma lucidum* Fruit Karst. (Polyporaceae) was reported by Sterne (1982) to have a marked hypoglycemic effect.

14.5 Alkaloids

Alkaloids have valuable roles in intervening in the insulin pathway. Alkaloids such as piperine from the common pepper and other piper species were utilized in the dietary formula of many ancient cultures but have been lately abandoned due to changing lifestyles and scientific claims that they lacked any functional food benefits, and similar ones like **oxymatrine, sanguinarine, boldine, neferin, berberine, and lupanine** have all been found to have antidiabetic effects through various molecular targets and signaling of specific pathways of insulin. Atal et al. (2016) showed that piperine has "bio-enhancing effects with metformin" in reducing plasma glucose of diabetic rats.

Berberine, an isoquinoline alkaloid, from Berberis was shown to ameliorate diabetic syndrome by slowing down the absorption of glucose from the stomach into the bloodstream by its inhibition of the enzyme glycosidase (Pan et al. 2003). In the same way, Cicero and Baggioni (2016) pointed to its glucose-lowering, lipid-lowering, antioxidant, and anti-inflammatory effects.

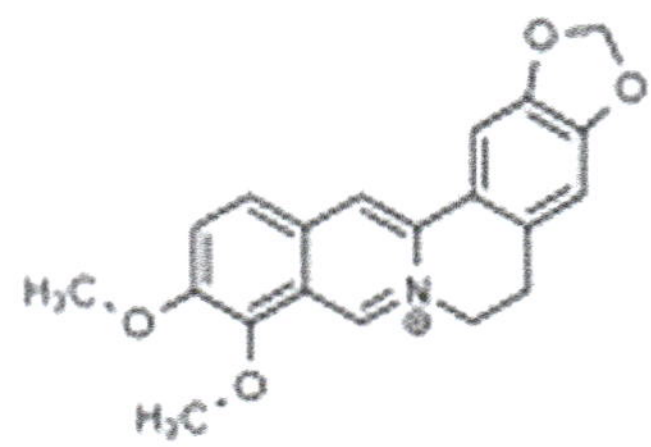

Berberine

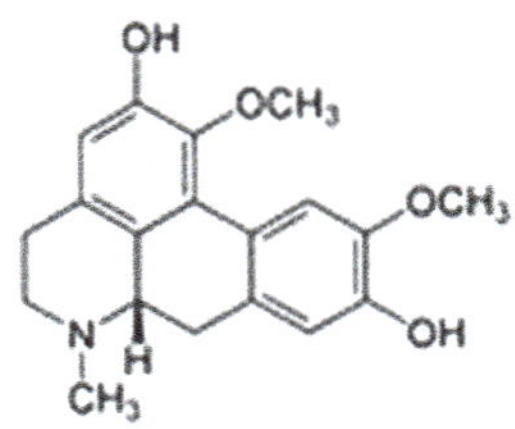

Boldine

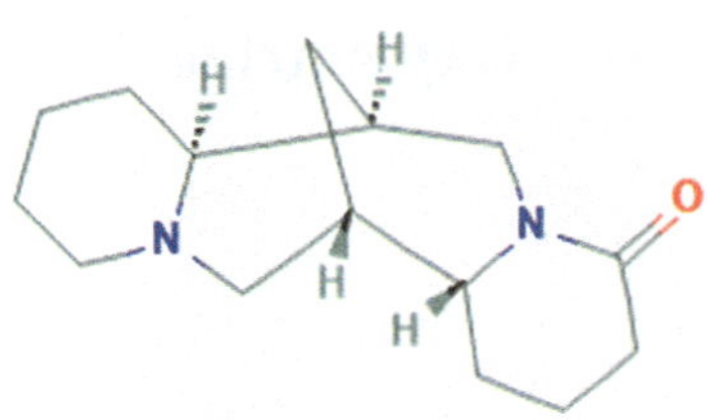

Lupanine

The benzylisoquinoline alkaloid, **boldine**, from the plant *Peumus boldus* Molina, was reported by Oza and Kulkarni (2016) to ameliorate endothelial dysfunction in diabetic mice by inhibiting oxidative stress cascade through the "angiotensin II-mediated BMP4 expression" pathway. Lau et al. (2013) state that this arises from downregulating overproduction of reactive oxygen species.

The plant *Lupinus perennis* produces the quinolizidine alkaloid **lupanine**. This compound enhances the secretion of insulin (Lopez et al. 2004). Wiedemann et al. (2015) suggest that this influences "ATP-sensitive potassium (KATP) channels and insulin genes" to improve homeostasis of glucose.

The plant Nelumbo yields the molecule **neferin,** a bisbenzylisoquinoline alkaloid. This compound was reported by Li et al. (2013a) to reduce total cholesterol

(TChol.), body weight, fasting blood glucose (FBG), blood pressure (BP), and triglyceride, while high-density lipoprotein cholesterol (HDLc) increased. Also CCL5 and CCR5 mRNA expression was reduced in hyperglycemic rats treated with the compound (Li et al. 2013b).

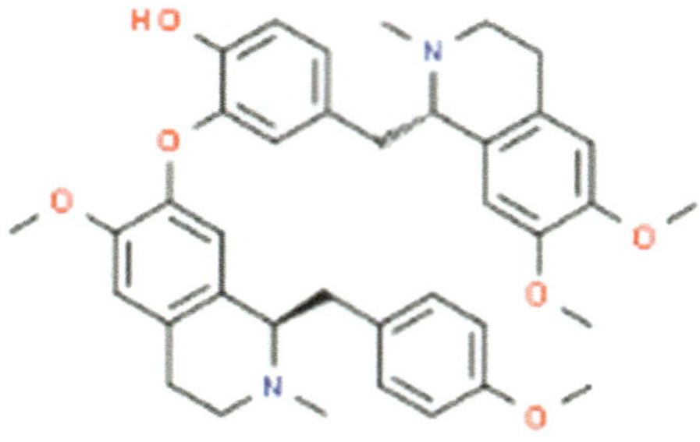

Neferine

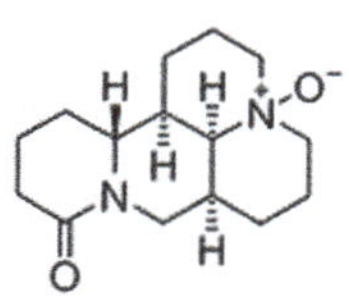

Oxymatrine

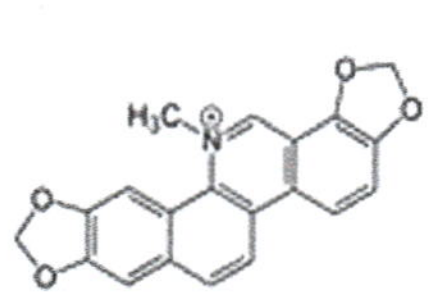

Sanguinarine

The alkaloid **oxymatrine**, another quinolizidine alkaloid extracted from *Sophora flavescens* root, reduced albumin excretion, blood glucose, serum creatinine, and urinary protein including urea nitrogen in an animal model for type 2 diabetes mellitus' evaluation of nephropathy and protective effect of the compound (Guo et al. 2014; Oza and Kulkarni 2016).

Current medicinal chemistry approaches like computational bioinformatics have expounded the knowledge of the natural product scientists via modeling approaches to drug discovery. Wang et al. (2014), through this approach, reported that the alkaloidal candidate agent **sanguinarine,** a benzophenanthridine, from the plant *Sanguinaria canadensis*, was an intercalator of DNA and RNA in type 2 diabetes mellitus treatment.

14.6 Flavonoids

Flavonoids are favored compounds of plants not just for their aromatic essence but also for their wide-ranging medicinal properties for internal and external applications. As secondary metabolites they are found in various plant parts especially in flowers, leaves, fruits, stem barks, and sometimes roots, rhizomes, tubers, etc. The key value of flavonoids lies in their structure of hydroxyl ends. This specific property allows flavonoids to mimic and blend with physiological compounds of the human system. Their end hydroxyl groups and skeleton of aromatic rings make them favored candidates for resolving the reactive oxygen species (ROS) oxidative stress cascade at play in cellular damage and inflammation and as such they serve as natural antioxidants. Oxidative stress, being a major villain of diabetes and its comorbid disorders and other predisposing risk factors, makes plant flavonoids an attractive candidate for evaluation as prospective drug materials. Many herbs in our culinary formula as spices are chosen for their attractive contents of natural flavonoids.

Flavonoids that have been evaluated for antidiabetic effects include **quercetin, kaempferol, genistein, naringenin, chrysin rutin, catechins, baicalein, luteolin, diosmin**, etc. Den Hartogh and Tsiani (2019) detailed the effect of **naringenin** as an antidiabetic agent which included its capacity to decrease intestinal absorption of glucose and renal reabsorption, while at the same time improving muscular uptake and utilization of glucose. Other actions of this compound as noted by the same author include reduction in triglyceride production and gluconeogenesis, thereby ameliorating hyperglycemia and hyperlipidemia.

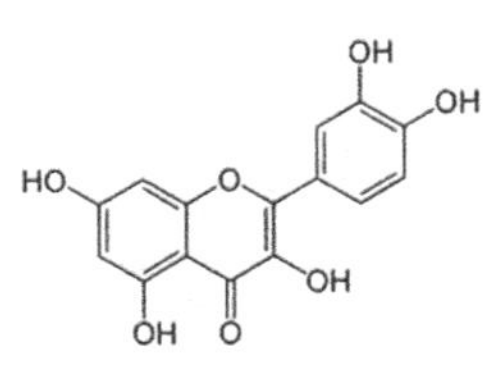

Quercetin,

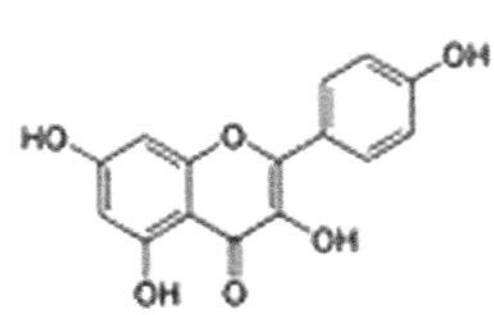

Kaempferol

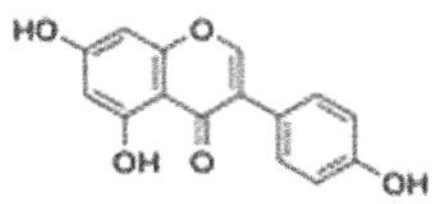

Genistein,

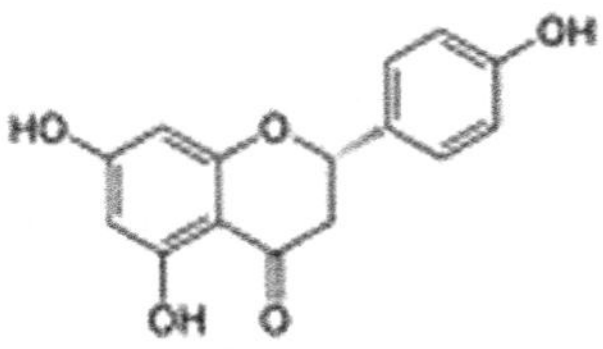

Naringenin,

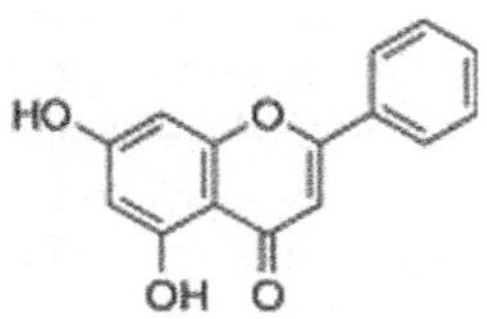

Chrysin

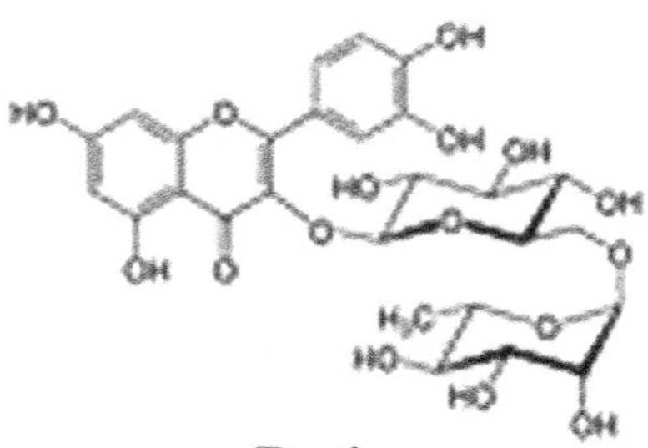

Rutin,

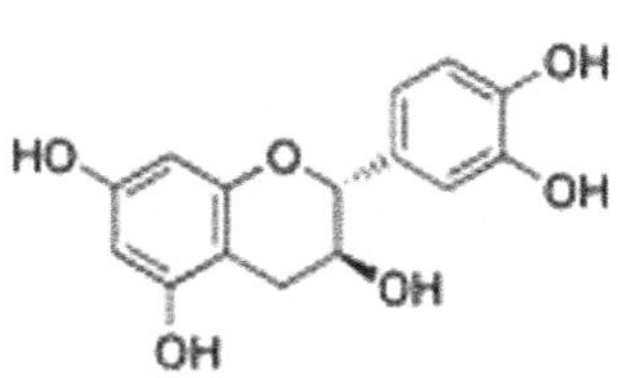

Catechin

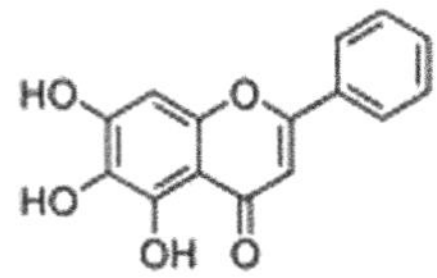

Baicalein,

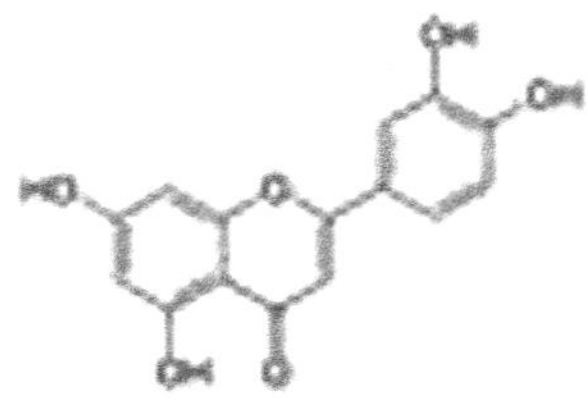

Luteolin,

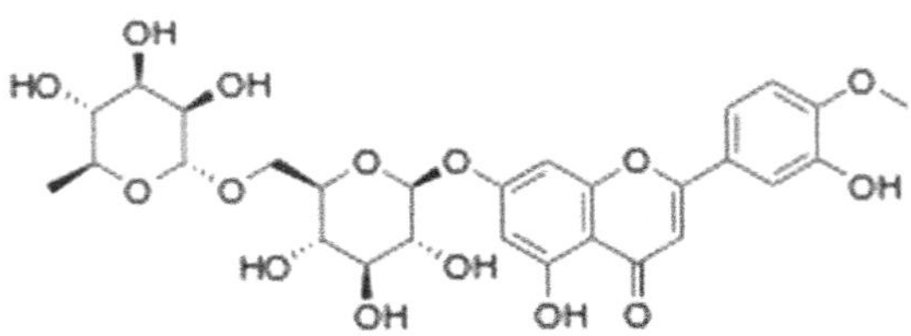

Diosmin

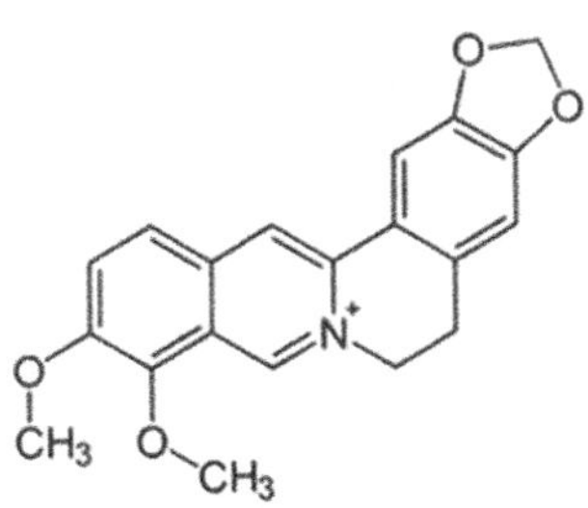

Berberine

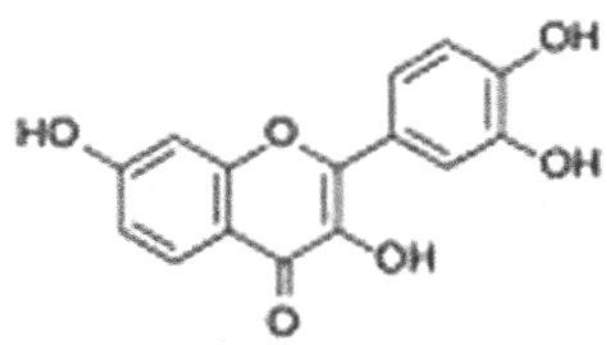

Fisetin

The catechins noted as flavonoids in cacao, tea, and other related beverage herbs include various derivatives like epicatechin and epigallocatechin gallate, which have been demonstrated to exhibit protective effects against oxidative damage which, as noted previously (Orgah et al. 2020), may be a critical process at play in metabolic syndrome like diabetes, atherosclerosis, hypertension, renal disorder disease, and cancer.

Fisetin, a bioflavonoid of strawberries, was shown by several authors like Prasath et al. (2014), Prasath and Subramanian (2014), and Maher et al. (2011) to cut back blood glucose significantly; inhibit gluconeogenic enzymes, thus leading to improvement of glucose homeostasis; and improve the glyoxalase 1 level and activity. By this way, Maher et al. (2011) suggest that it limits the complications of diabetes.

Alkhalidy et al. (2015) reported that **kaempferol** promotes insulin sensitivity and preserved pancreatic β-cell mass in middle-aged obese diabetic mice. This compound and its derivative, a flavonol, are well distributed in many plants and naturally designed to target ROS and reverse oxidative stress and the author further notes that it preserves pancreatic β-cell mass.

The flavone **luteolin** was reported by Wang et al. (2011) to have ameliorative effects on diabetic nephropathy of diabetic rats. It was thus suggested as a compound for treating diabetic nephropathy. Further, Wang et al. (2012a) noted that **luteolin** ameliorates cardiac arrest in type 1 diabetes mellitus cardiomyopathy.

In a study to evaluate the compound naringenin, a flavanone common in grapefruit, Tsai et al. (2012) found that the compound possessed anti-inflammatory and anti-fibrotic properties in diabetic mice. By this means it attenuates diabetic nephropathy and was also found by the same author to "decrease the expression of interleukin (IL)-1, IL-6, type IV collagen, fibronectin, and transforming growth factor 1."

Pentamethylquercetin, a derivative of **quercetin, was** investigated by Li et al. (2013a, b) for its protective effects in "diabetes-related cognitive deficits in diabetic Goto-Kakizaki rats," and Tang et al. (2011) on the other hand investigated the "in vitro suppression of quercetin on hypertrophy and extracellular matrix accumulation in rat glomerular mesangial cells cultured by high glucose." These workers found that quercetin and its derivative:

- Have protective effects in diabetes.
- Decreased the cellular amount of "G(0)/G(1) phase, Smad 2/3 expression, laminin, and type IV collagen and TGF-(1) mRNA levels."
- "Activate the Akt/cAMP response element-binding protein pathway."

Rutin, the flavonoid glycoside, found in some plants, has been indicated to alter glycolytic and gluconeogenic enzymes (Prince and Kamalakkannan 2006), through which it improves glucose homeostasis. Hsu et al. (2014) later reported that rutin enhanced insulin-dependent glucose transporter 4 translocation by potentiating insulin receptor kinase while Kappel et al. (2013) discussed "the stimulatory effect of rutin on glucose uptake in rat muscle." These all point to the various mechanisms by which this flavonoid improves diabetic conditions.

Morin was reported by several authors as an insulin mimetic (Paoli et al. 2013); a rescuer of "endothelial dysfunction by activating the Akt/eNOS pathway" (Taguchi et al. 2016); "downregulator of mir-29a to exert antidiabetic effects in human hepg2 cells" (Razavi et al. 2018); and attenuator of ER stress (Pandey et al. 2019).

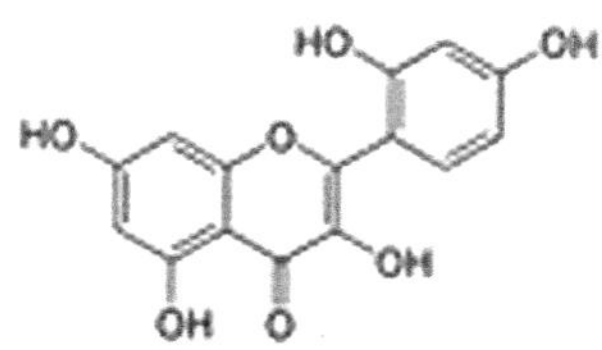

Morin

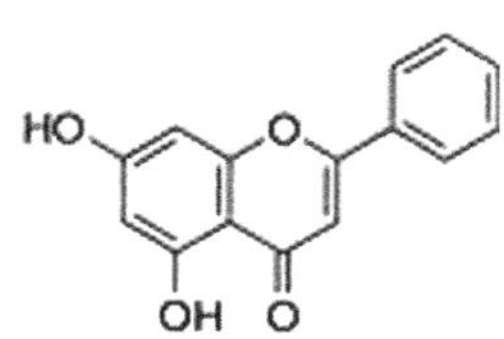

Chrysin

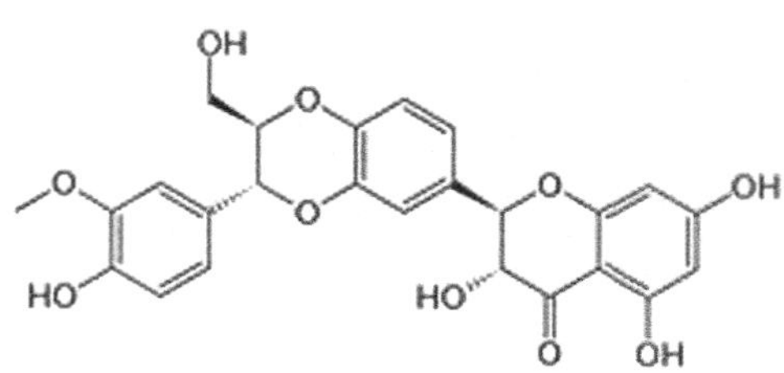

silymarin

Milk thistle (*Silybum marianum*) has been put forward as a herbal treatment for several metabolic systemic disorders and as an anti-infective. The main compound of activity in this plant is a lipophilic complex extracted from the seeds and is made of three flavonoid isomers known collectively as **silymarin,** a flavonolignan comprising silybin, silydianin, and silychristin (Bijak 2017). This complex flavonoid has been reported by various authors to have "nephroprotective effects in type 2 diabetes mellitus" (Bijak 2017; Jaggi and Singh 2016; Abenavoli et al. 2018). Sheela et al. (2013) had previously indicated that the compound could decrease hyperglycemia. Silymarin ameliorates diabetic cardiomyopathy through the inhibition of TGF-1/ Smad signaling; thus, it is useful in the treatment of diabetic cardiomyopathy.

The flavone, **chrysin**, is a type of flavonoid found in various plants and is noted to suppress TGF-β, a transforming growth factor, fibronectin, and expression of protein collagen IV in renal tissues. Amjid et al. (2014) noted its anti-inflammatory property and demonstrated its capacity to abrogate renal dysfunction challenges in diabetic rats induced with type 2 diabetes. Similarly, Taslimi (2019) evaluated the activities of some metabolic enzymes and noted that chrysin demonstrated "antidiabetic and anticholinergic effects on cyclophosphamide-induced multiple-organ toxicities" and ameliorates these toxicities in the liver, kidney, heart, testis, and the brain in rats.

Baicalein has been reported (Ahad et al. 2014) to mitigate oxidative stress, and could be used as a prophylactic against diabetic renal dysfunction as it "suppresses the activation of NF-B, decreases the expression of iNOS and TGF-1, and normalizes the levels of serum pro-inflammatory cytokines and liver function enzymes."

14.7 Terpenoids

Four **cucurbitane** glycosides of *Momordica charantia* were reported to exhibit beneficial effects in diabetes and obesity in "both L6 myotubes and 3T3-L1 adipocytes" (Tan et al. 2015). These authors further stated that the compounds "stimulated GLUT4 translocation to the cell membrane—an essential step for inducible glucose entry into cells." They explained that this was "a key pathway mediating glucose uptake and fatty acid oxidation, being associated with AMP-activated protein kinase (AMPK) increased inactivity." The role of these compounds (momordicosides) in enhancing fatty acid oxidation and glucose disposal was noted in both insulin-responsive and insulin-resistant mice.

A detailed review on the chemistry of tetracyclic triterpenoids from various plants was carried out by Hamid et al. (2015), in which the following compounds were listed as the main important members: "dammarane, cucurbitane, cycloartane, lanostane, and protostane groups."

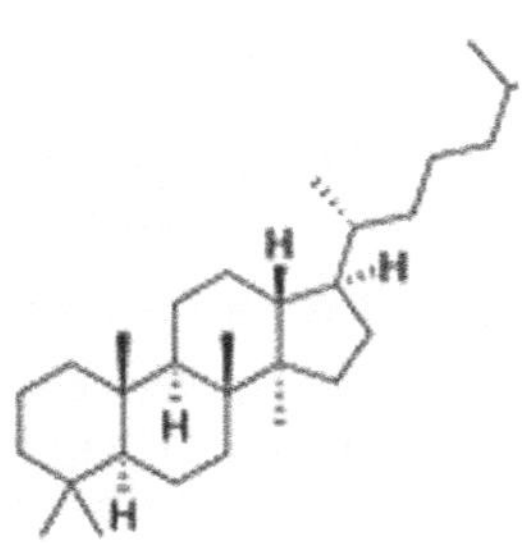

dammarane,

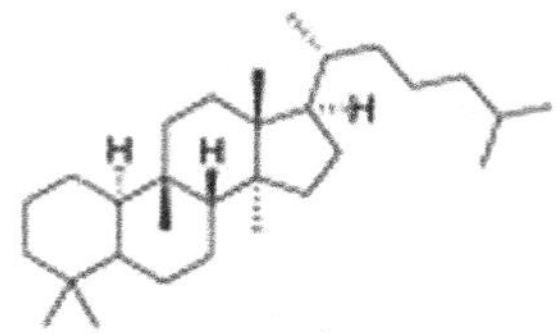

cucurbitane,

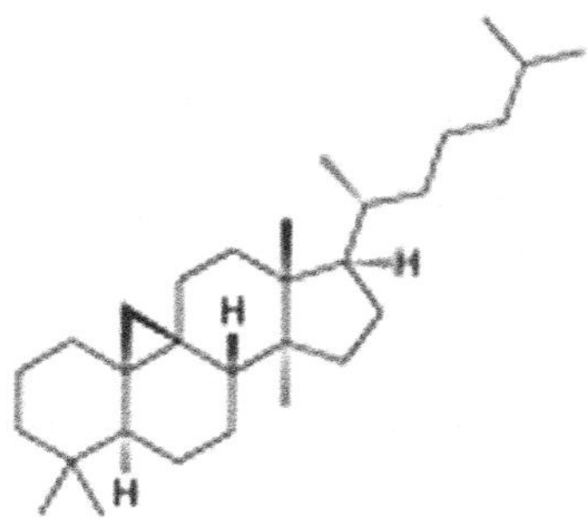

cycloartane,

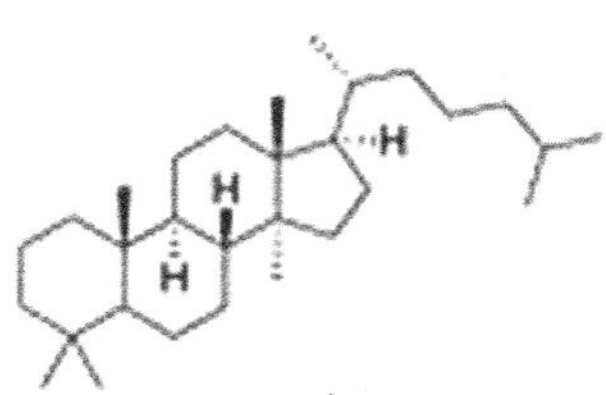

anostane,

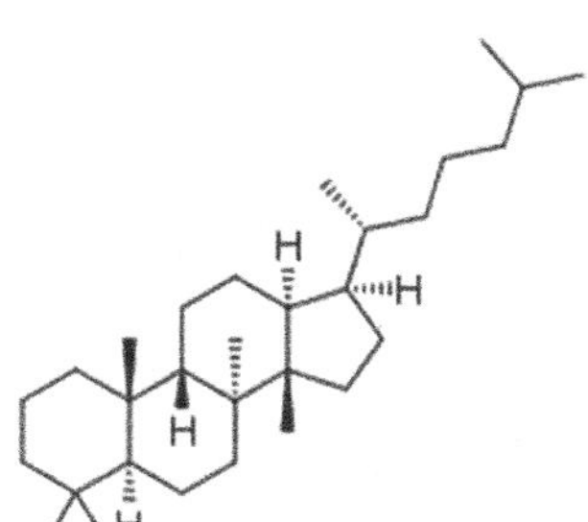

protostane skeleton

These workers noted that plants such as "*Panax ginseng, P. quinquefolium, P. notoginseng, Gynostemma pentaphyllum, Astragalus membranaceus, Momordica charantia*, and *Ganoderma lucidum*" widely used as herbal anti-diabetes treatments contain these tetracyclic triterpenoids though with unproven efficacies through classical clinical research evaluation approach. In comparison to pentacyclic triterpenoids, Hamid et al. (2015) further noted that tetracyclic triterpenoid plants appear to be more exploited clinically for anti-diabetes, but largely as plant extracts rather than pure compounds in which form they serve as blood glucose-reducing potions.

Their overall mechanism of action appears to be "glucose absorption, glucose uptake, insulin secretion, diabetic vascular dysfunction, retinopathy, and nephropathy" (Hamid et al. 2015), including anti-inflammatory. Previously reported similar activities of pentacyclic triterpenoid compounds such as "oleanolic acid, glycyrrhizin, glycyrrhetinic acid, ursolic acid, betulin, betulinic acid, and lupeo." In the view of these authors, structure–activity relationship (SAR) in pharmacological effects appears to be at play since they have structural similarity to steroid hormones, thus facilitating their intercalation into plasma membranes.

Performing well as safe and efficacious extracts is a positive value for the holistic philosophy of herbal and natural medicine and since traditional use of herbs does not suggest adverse effects. However, systematic clinical evaluation remains an important step.

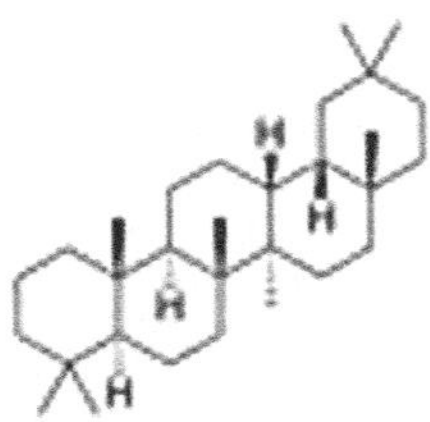

Oleananes,

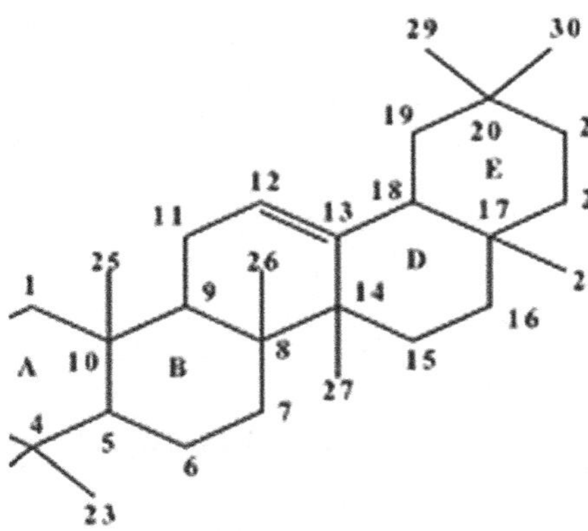

Ursanes Skeleton

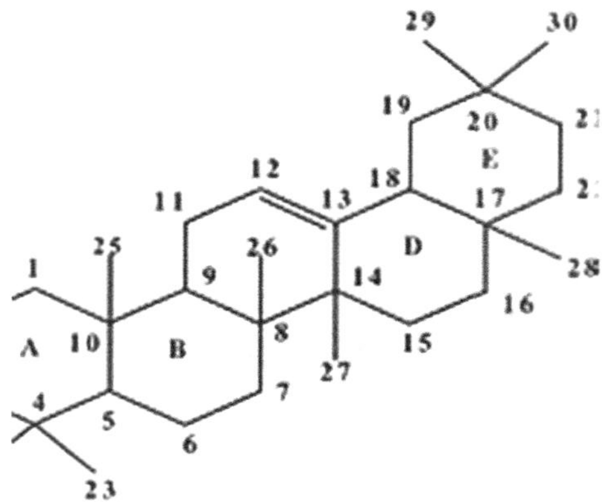

Lupanes Skeleton

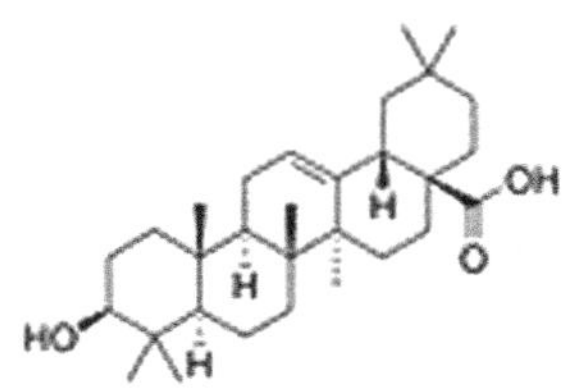

Oleanolic acid,

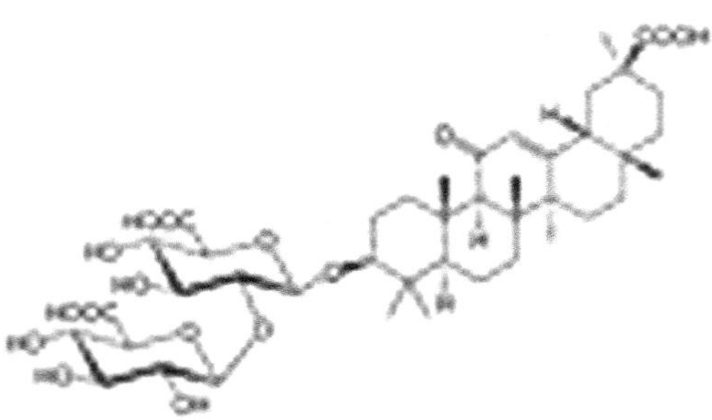

Glycyrrhizin,

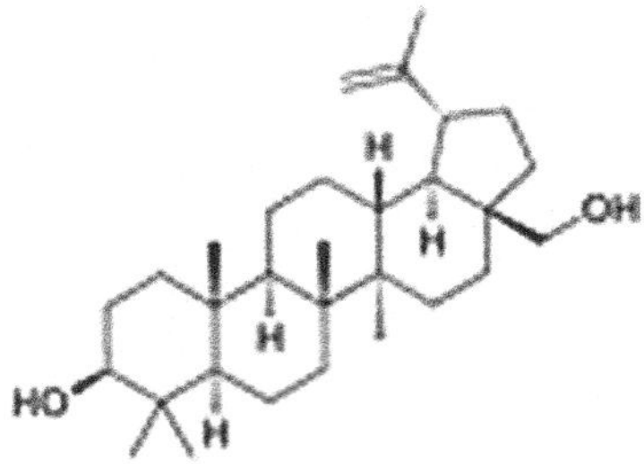

Betulin,

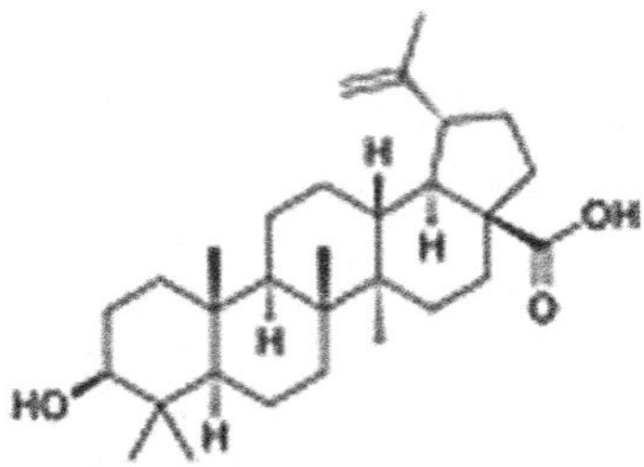

betulinic acid

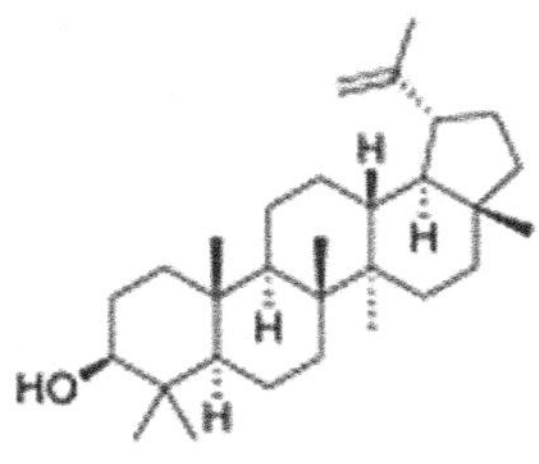

Lupeo

The activity of **boswellic acid**, a pentacyclic triterpene of Boswellia species, has been attributed to the stimulation of cellular release of insulin, and thus their usage mechanism is the target of damaged and inflamed islet of Langerhans and antidiabetic prophylaxis (Ammon 2012); Jadhav and Puchchakala 2011).

Celastrol, a triterpene of different plant sources, has been shown (Han et al. 2016; Wang et al., 2015) to attenuate renal injury, aiding lipid metabolism and rendering antioxidant activities, thus protecting against diabetes-related liver injury.

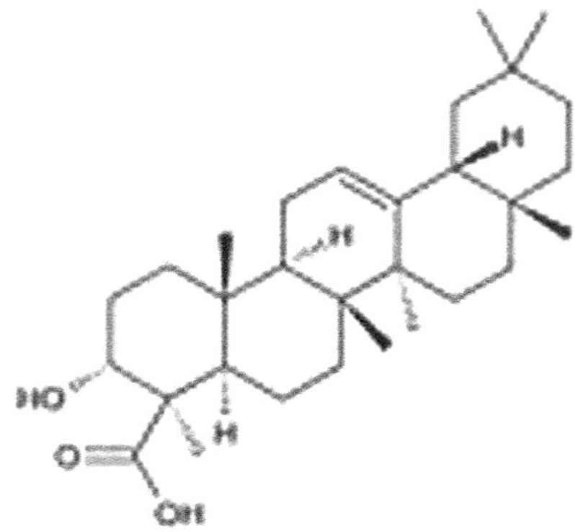

Boswellic acid

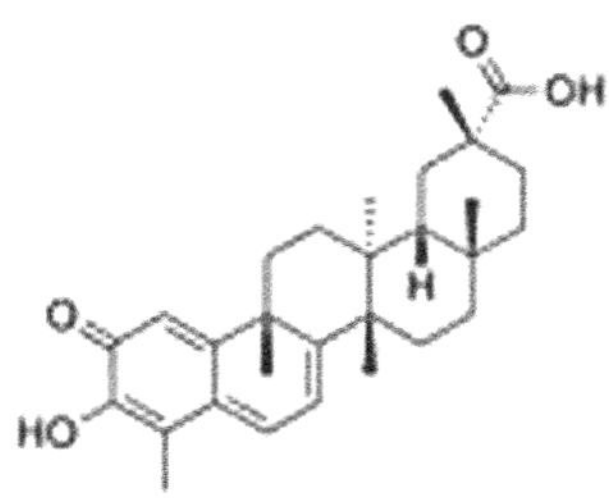

Celastrol

The pentacyclic triterpenoids, **oleanolic acid** and **ursolic acid**, are also reported as prophylactic and management compounds against type 2 diabetes mellitus (Camer et al. 2014, Mancha-Ramirez and Slaga 2016). Further, ursolic acid has been reported (Ling et al. 2013) to ameliorate insulin sensitivity and plasma glucose levels including the protection of the kidney of diabetic rats.

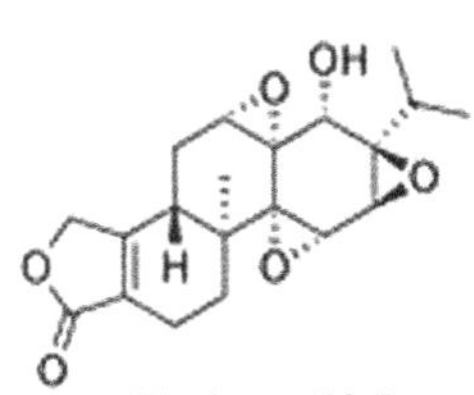

Triptolide

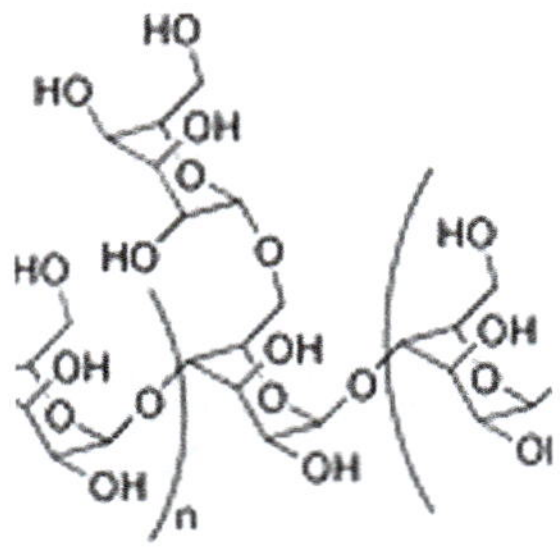

Galactomannan

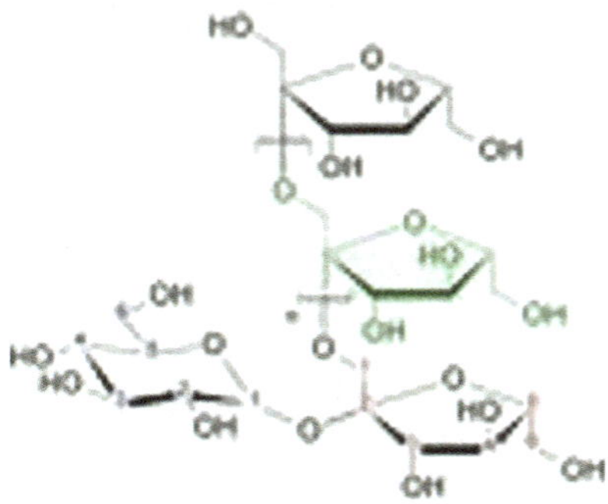

Inulin

The diterpenoid, **triptolide**, from *Tripterygium wilfordii* ameliorates glomerular hypertrophy and reduces "the levels of phosphorylated protein kinase B and phosphorylated inhibitor of kappa B and increases caspases 3, 8, and 9" (Huang et al. 2013). However, its toxicity to reproductive, urogenital, and circulatory systems and liver has also been noted (Li et al. 2014).

Galactomannan, a polysaccharide from *Amorphophallus konjac* and *Cyamopsis tetragonolobus*, delays glucose absorption rate and reduces postprandial hyperglycemia. Pinto Vierira et al. (2018) reported that the compound "decreased glycemia, total cholesterol, and triacylglycerol." However, the report indicated "toxic galactomannan effects" due to the histopathological changes observed in the pancreas in streptozotocin-induced diabetic animal.

The polysaccharide **inulin** (carbohydrate) from *Helianthus tuberosus was* reported (Dehghan et al. 2013) to cause significant decrease in "fasting blood sugar (FBS), glycated hemoglobin (HbA1c), and fasting insulin," and the author concluded that "inulin supplementation appears to modulate inflammation and metabolic endotoxemia in type 2 diabetes."

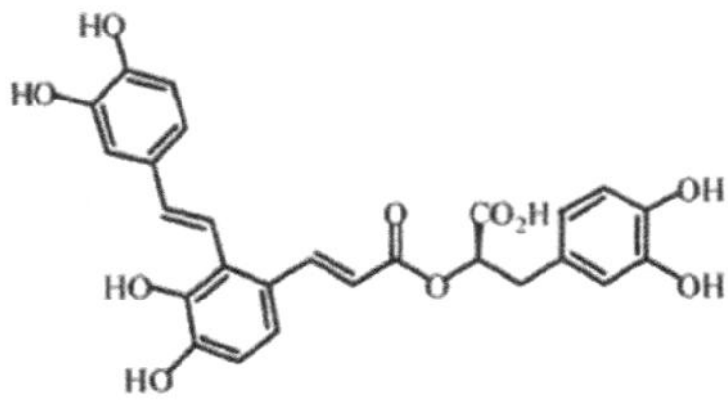

Salvianolic acid A

Molecular Formula: $C_{26}H_{22}O_{10}$
Molecular Weight: 494.45

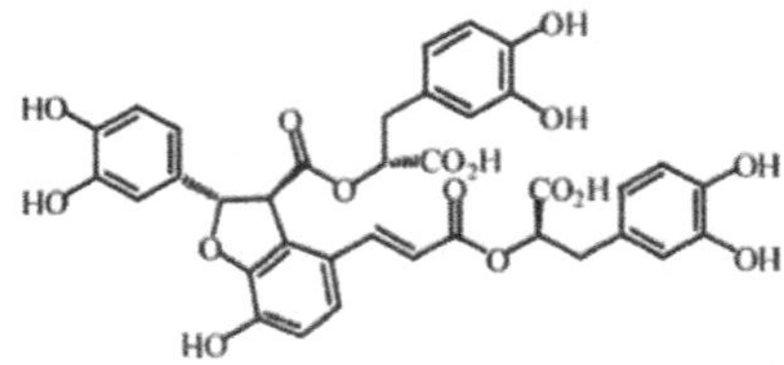

Salvianolic acid B

Molecular Formula: $C_{36}H_{30}O_{16}$
Molecular Weight: 718.6138

Salvianolic acid A and B

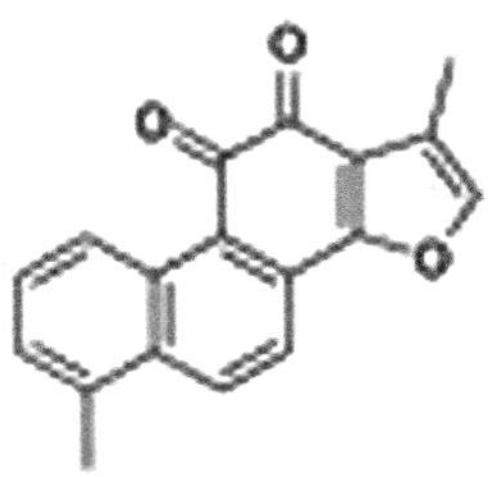

Tanshinone

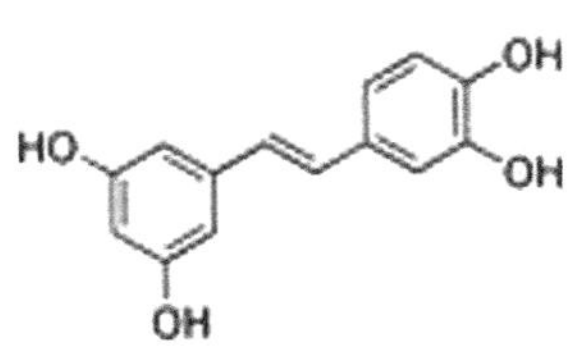

Piceatannol

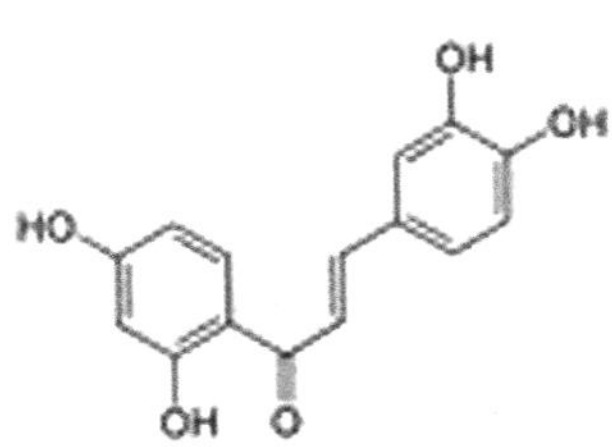

Butein

The derivative of resveratrol, **piceatannol**, was reported by Minakawa et al. (2012) as a suppressor of blood glucose levels in type 2 diabetes mellitus mice model; Jeong et al. (2015) stated that it restores the "palmitic acid-induced impairment of insulin signaling." In the overview of the activities of **resveratrol** class of compounds in diabetes written by Szkudelski and Szkudelska (2015), based on diabetic animal models, they stated that **resveratrol** "improved glucose homeostasis, reduced resistance to insulin, protected pancreatic cells, and stimulated and increased the secretion of insulin including general intervention against metabolic disorders."

Butein, a phenolic chalcone, isolated from various plant species, such as *Cyclopia subternata, Semecarpus anacardium, and Coreopsis tinctoria,* and **tocotrienol** were reported by Kuhad and Chopra (2009) and Benzler et al. (2015), respectively, to inhibit central NF-kB signaling, leading to attenuation of diabetic nephropathy, oxidative stress, and glucose tolerance.

Curcumin has been reported by various authors (Kunwar and Priyadarsini 2016) to decrease end-point complications of diabetes such as weight loss, neuropathy fatty liver, vascular diseases, islet regeneration, and nephropathy.

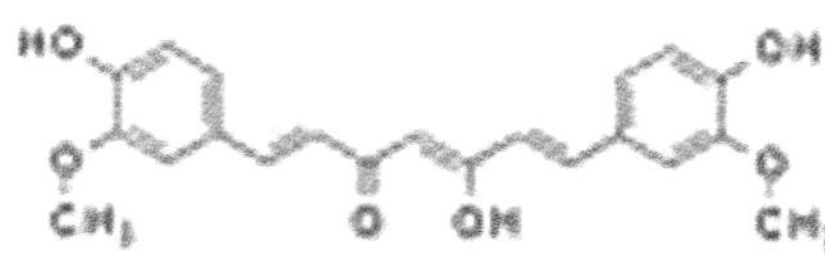

Curcumin

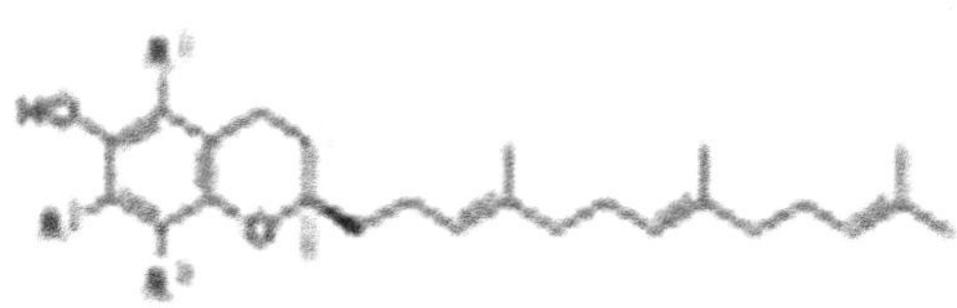

tocotrienol

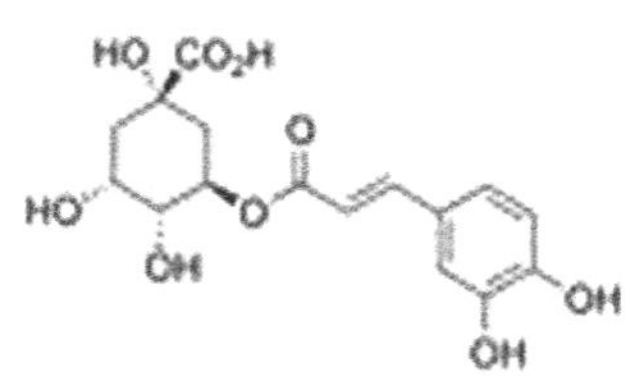

Chlorogenic acid

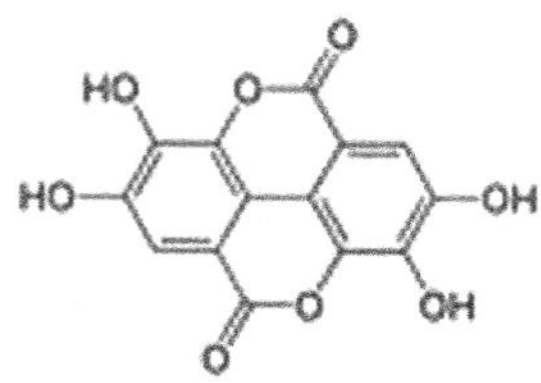

Elagic acid

Chlorogenic acid, a polyphenol, stimulates glucose transport in skeletal muscle and is also reported to affect hepatic glucose release and glycemia (Ong et al. 2012) **Ellagic acid**, a dilactone acid on the other hand, exerts its antidiabetic action by stimulating insulin secretion through the β-cells of the pancreas and decreasing glucose intolerance including being an antioxidant (Fatima et al. 2017).

Embelin, a hydroxyl benzoquinone, was reported by Durg et al. (2017) to "reduce glycosylated hemoglobin and pro-inflammatory mediators (interleukin 6 and tumor necrosis factor)." Yu et al. (2016) reported that **erianin**, a phenolic compound obtained from *Dendrobium chrysotoxum*, "inhibits high glucose-induced retinal angiogenesis via blocking the ERK1/2-regulated HIF-1-VEGF/VEGFR2 signaling pathway." Cui et al. (2018) stated that **the compound gambogic acid** found in Garcinia species ameliorates "diabetes-induced proliferative retinopathy," while **garcinol** in the same species was reported by Madhuri and Naik (2017) to modulate various functionalities in diabetic conditions such as lipids and high-density lipoprotein cholesterol, activities of antioxidant enzymes, glycosylated hemoglobin, and levels of blood glucose generally.

The polyphenol lignan **honokiol**, isolated from *Magnolia officinalis*, was reported by Li et al. (2018) to "protect pancreatic β-cell against high glucose and intermittent hypoxia-induced injury" via increasing "phosphorylation and insulin signaling factors."

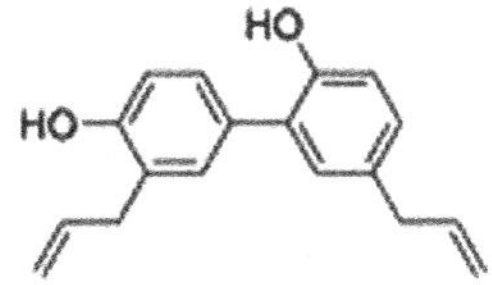

Honokiol,

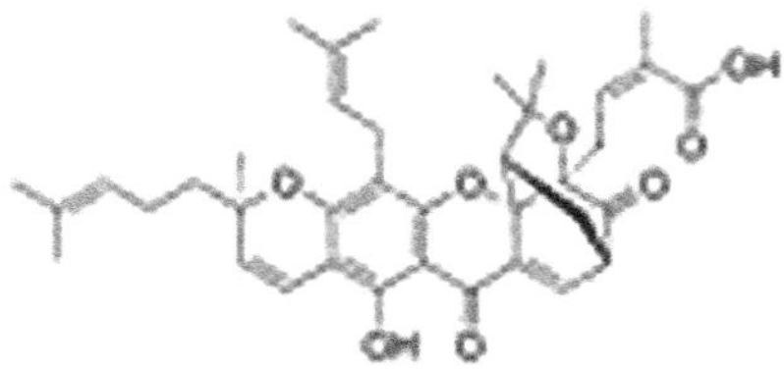

Gambogic acid

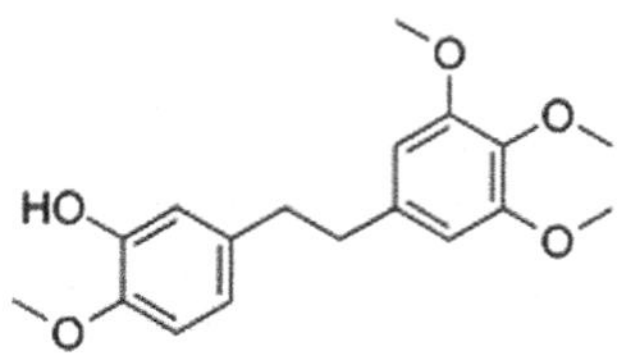

Erianin

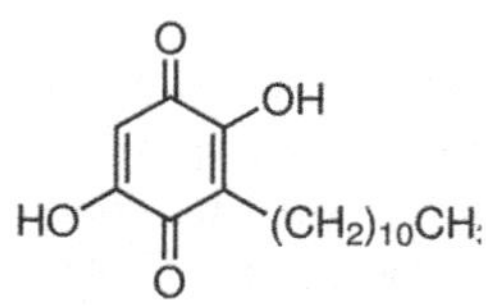

Embelin

Fang et al. (2011) reported that the levels of fasting blood glucose (FBG), serum triglyceride, "total cholesterol, low-density lipoprotein cholesterol, and nonesterified fatty acid (NEFA)" were significantly decreased in **salvianolic acid A**-treated diabetic rats. Also, superoxide dismutase activity was significantly increased and malondialdehyde content was significantly decreased. Salvianolic acid A thus decreases hyperlipidemia and improves glycometabolism.

Baojian et al. (2018) reported that **tanshinone IIA** treatment "significantly attenuated mechanical allodynia and thermal hyperalgesia" in diabetic rats. Also, Baojian et al. (2018) further noted that "the expression of the pro-inflammatory cytokines IL-1β, IL-6, and TNF-α was inhibited, and the level of the anti-inflammatory cytokine IL-10 was increased" with **tanshinone IIA**.

Orgah et al. (2020) noted that liver PPARγ increases insulin sensitivity while decreasing inflammation. Hence, "promoting PPARγ-adiponectin pathway improves insulin sensitivity and prevents type 2 diabetes." The importance of this observation in this review is the fact that upregulating the PPARγ signal pathway was the mechanism by which the phytochemicals of the Chinese patent medicine "Danhong injection ameliorated kidney injury as well as proteinuria and serum lipid levels" in a diabetic kidney disease rat model reported by Orgah et al. (2020).

Further, bioactive compounds of interest variously reported by other authors in the antidiabetic cascade were isolated from this product by Liu et al. (2013). These include polyphenolic acids such as rosmarinic acid, caffeic acid, protocatechuic aldehyde, 5-hydroxymethyl-2-furfural, ferulic acid, protocatechuic acid, Safflower yellow A, lithospermic acid, salvianolic acid A, salvianolic acid B, salvianolic acid D, and danshensu.

The multiplicity of the activity of combined natural products was further demonstrated in the various other pathways noted in this single product (Danhong injection) which according to He et al. (2016) also demonstrated activity "through VEGF/VEGFR-2 and PPARδ signaling pathway to regulate angiogenesis and glucose homeostasis, lower the high blood glucose levels, improve glucose tolerance, and accelerate perfusion recovery of the ischemic hind limb."

Orgah et al. (2020) further reported that **salvianolic acid B**, a component of this medicine, (Danhong injection), reduced oxidative stress in in vitro human aortic endothelial cells and "in vivo cholesterol-fed rabbits, inhibited low-density lipoprotein (LDL) oxidation, and reduced oxidized LDL-induced cytotoxicity in the human aortic endothelial cells." They further noted that the injection reduced the deposition of lipid in the thoracic aorta, intimal thickness of the aorta arch/thoracic aorta, and neointimal formation in the abdominal aorta of the rabbits.

Tanshinone IIA, in a work by Li et al. (2015), was found to exert protection against impairment of endothelium-dependent vaso-relaxation through the eNOS and posttranscriptional mechanism caused by diabetes. Previously, this compound was reported by Hwang et al. (2013) to improve glucose metabolism by aiding insulin sensitivity via the AMPK pathway.

Norris et al. (2009) had reported the improvement of glycemia in obese postmenopausal women and suggested that this observation was due to an increased level of serum "alpha tocopherols and plasma adiponectin, consequently causing a decreased adipose mass" in these type 2 diabetics.

14.8 Conclusion

Early on, scientists focused on the identification of plants that show antidiabetic properties as well as plants that ameliorate the state of various diseases. At that time, research was focused on identifying plant families and their relatives that exert such desirable properties to ameliorate the conditions of disease such as diabetes. It was believed that plants of same genus would be capable of exerting similar bioactivities.

However, plants do not exert their medicinal effects by their mere beauty and form but by their constituents of chemical entities in structural orders that behooves on them these medicinal capabilities. These plant natural compounds are believed to arise from their responses to stress and metabolic challenges.

Identifying plants with antidiabetic properties and the natural products at play in the disease cascade brings the question of how these compounds actually achieve the target objective of ameliorating the health conditions observed in the pathophysiology of the disease. This study became the later concern of scientists as they braced up to study the physical qualities of plants' natural products making them effective in ameliorating diabetes. The knowledge is incomplete in bringing plants to the bedside except that the natural plant products are purified, isolated, characterized, identified, and evaluated medically. Scientists are thus interested in the predictable pathways for which known structures exert activity on human biomolecules which for the purpose of this review is specifically targeting the antidiabetic clinical characteristics of plant compounds. Predicting biological pathways aids the drug development process by bringing to fore compounds with potentials for management of a pathological condition.

This review has brought to fore the plants/families, the compounds/structures, and the mechanisms/pathways of some antidiabetic plants' natural products. As previously noted by Wang et al. (2012b), the factor of synergy in the holistic philosophy of plant medicines is quite obvious from the different pathways by which the multicomponents of some polyherbal medicines exert their activities, as they appear to interact in order to attain a combined efficacy greater than that of their individual capacities which improves the medicines' efficacy to target complexities found in metabolic syndromes like diabetes mellitus and related cardiovascular diseases. Some chemical families are notable in their ability to align favorably with human biomolecules to mimic or modulate their activities positively such as to ameliorate diabetes.

While the level of current knowledge is quite advanced, as long as this disease remains with us, future research must continue to seek ways to modify these natural compounds to effectively reverse the diabetic scourge.

Declaration There is no conflict of interest.

References

Abenavoli L, Izzo AA, Milić N, Cicala C, Santini A, Capasso R (2018) Milk thistle (Silybum marianum): A concise overview on its chemistry, pharmacological, and nutraceutical uses in liver diseases. Phytother Res 32:2202–2213

Ahad A, Mujeeb M, Ahsan H, Siddiqui WA (2014) Prophylactic effect of baicalein against renal dysfunction in type 2 diabetic rats. Biochimie 106:101–110

Alarcon E, Roman RR, Floress SJ (1993) Plantas medicinales usadas en el control de la diabetes mellitus. Ciencia 44:363–381

Alkhalidy H, Moore W, Zhang Y, McMillan R, Wang A, Ali M, Suh K-S, Zhen W, Cheng Z, Jia Z (2015) Small molecule kaempferol promotes insulin sensitivity and preserved pancreatic b-cell mass in middle-aged obese diabetic mice. J Diabetes Res 2015:532984

Amjid A, Ajaz AG, Mohd M, etc. (2014) Chrysin, an anti-inflammatory molecule, abrogates renal dysfunction in type 2 diabetic rats. Toxicol Appl Pharm 279:1–7

Ammon HPT (2012) Use of boswellic acids for the prophylaxis and/or treatment of damage to and/or inflammation of the islets of Langerhans

Atal S, Atal S, Vyas S, Phadnis P (2016) Bio-enhancing effect of piperine with metformin on lowering blood glucose level in alloxan induced diabetic mice. Pharm Res 8:56–60

Benzler J, Ganjam GK, Pretz D, Oelkrug R, Koch CE, Legler K, Stöhr S, Culmsee C, Williams LM, Tups A (2015) Central inhibition of ikkβ/nf-κb signaling attenuates high-fat diet-induced obesity and glucose intolerance. Diabetes 64:2015–2027

Bijak MS (2017) A major bioactive component of milk thistle (Silybum marianum l. Gaernt.)—chemistry, bioavailability, and metabolism. Molecules 22:1942

Camer D, Yu Y, Szabo A, Huang X (2014) The molecular mechanisms underpinning the therapeutic properties of oleanolic acid, its isomer and derivatives for type 2 diabetes and associated complications. Mol Nutr Food Res 58:1750–1759

Chakravarthy BK, Cupta S, Gambhir SS, Gode KD (1981) Pancreatic beta-cell regeneration in rats by (−) epicatechin. Lancet 2:759

Cicero AFG, Baggioni A (2016) Berberine and its role in chronic disease. Springer, Cham, Switzerland, p 928

Coterill P, Scheinmannan E, Stenhouse I (1976) 1976 Kolaflavanone a new biflavanone from *Garcinia kola*. J Chem Soc Perkin Trans 6:532–539

Cui J, Gong R, Hu S, Cai L, Chen L (2018) Gambogic acid ameliorates diabetes-induced proliferative retinopathy through inhibition of the hif-1α/VEGF expression via targeting pi3k/Akt pathway. Life Sci 192:293–303

Dehghan P, Gargari BP, Jafar-Abadi MA, Aliasgharzadeh A (2013) Inulin controls inflammation and metabolic endotoxemia in women with type 2 diabetes mellitus: A randomized-controlled clinical trial. Int J Food Sci Nutr 65:117–123. https://doi.org/10.3109/09637486.2013.836738

Den Hartogh DJ, Tsiani E (2019) Antidiabetic properties of naringenin: A citrus fruit polyphenol. Biomol Ther 9:99

Durg S, Veerapur VP, Neelima S, Dhadde SB (2017) Antidiabetic activity of *Embelia ribes*, embelin and its derivatives: A systematic review and meta-analysis. Biomed Pharmacother 86:195–204

Fang L-H, He G-R, Du G-H (2011) Anti-diabetic effect of salvianolic acid A and the possible mechanisms in rats with diabetes mellitus. Chinese J New Drugs 20(21):2063–2068

Fatima N, Hafizur RM, Hameed A, Ahmed S, Nisar M, Kabir N (2017) Ellagic acid in *Emblica officinalis* exerts anti-diabetic activity through the action on β-cells of pancreas. Eur J Nutr 56: 591–601

Guo C, Han F, Zhang C, Xiao W, Yang Z (2014) Protective effects of oxymatrine on experimental diabetic nephropathy. Planta Med 80:269–276

Hamid K, Alqahtani A, Kim M-S, Cho J-L, Cui PH, Guang Li C (2015) Tetracyclic triterpenoids in herbal medicines and their activities in diabetes and its complications. Curr Top Med Chem 15: 2406–2430

Han L, Li C, Sun B, Xie Y, Guan Y, Ma Z, Chen L (2016) Protective effects of celastrol on diabetic liver injury via tlr4/myd88/NF-κB signaling pathway in type 2 diabetic rats. J Dia Res 2016: 2641248. https://doi.org/10.1155/2016/2641248

He S, Zhao T, Guo H, Meng Y, Qin G, Goukassian DA, Han J, Gao X, Zhu Y (2016) Coordinated activation of VEGF/VEGFR-2 and PPAR-delta pathways by a multi-component Chinese medicine DHI accelerated recovery from peripheral arterial disease in type 2 diabetic mice. PLoS One 11(12):e0167305. https://doi.org/10.1371/journal.pone.0167305

Hii CST, Howell SL (1984) Effects of epicatechin on rat islets of Langerhans. Diabetes 33:291–296

Hikino H, Konno C, Takahashi M, Murakami M, Kato Y, Karikura M, Hayashi T (1986b) Isolation and hypoglycemic activity of dioscorans A, B, C, D, E and F; glycans of *Dioscorea [japonica rhizophors]*. Planta Med 52:168–171

Hikino H, Takahashi M, Murakam IM, Konno C, Mirin Y, Karikura M, Hayashi T (1986a) Isolation and hypoglycemic activity of arborans A and B, glycans of *Aloe arborescens* var natalensis leaves. Int Crude Drug Res 24:183–186

Hikino H, Takahashi M, Oshima Y (1988) Konno C: Isolation and hypoglycemic activity of oryzabrans A, B, C and D, glycans of *Oryza sativa* bran. Planta Med 54:1–3

Hsu CY, Shih HY, Chia YC, Lee CH, Ashida H, Lai YK, Weng CF (2014) Rutin potentiates insulin receptor kinase to enhance insulin-dependent glucose transporter 4 translocation. Mol Nutr Food Res 58:1168–1176

Huang SH, Lin GJ, Chu CH, Yu JC, Chen TW, Chen YW, Chien MW, Chu CC, Sytwu HK (2013) Triptolide ameliorates autoimmune diabetes and prolongs islet graft survival in nonobese diabetic mice. Pancreas 42:442–451

Hwang SL, Yang JH, Jeong YT, Kim YD, Li X, Lu Y, Chang YC, Son KH, Chang HW (2013) Tanshinone IIA improves endoplasmic reticulum stress-induced insulin resistance through AMP-activated protein kinase. Biochem Biophys Res Commun 430(4):1246–1252. https://doi.org/10.1016/j.bbrc.2012.12.066
Ivorra MD, Paya M, Villar A (1989) A review of natural products and plants as potential antidiabetic drugs. Ethnopharmacal 27:243–275
Iwu M, Okunji CO, Akah PT, Corley D (1990) Dioscoretine the hypoglycemic principle of *Dioscorea dumetorum*. Planta Med 56:119–120
Jadhav R, Puchchakala G (2011) Hypoglycemic and antidiabetic activity of flavonoids: Boswellic acid, ellagic acid, quercetin, rutin on streptozotocin-nicotinamide induced type 2 diabetic rats. Int J Pharm Pharm Sci 4:251–256
Jaggi AS, Singh N (2016) Silymarin and its role in chronic diseases. In: Drug discovery from mother nature. Springer, Cham, pp 25–44
Jalil U, Jabbar A, Hasan C (1986) Hypoglycemic activities of the glycosides of Momordica cochinchinensis. J Bangladesh Acad Sci 10:25–30
Jeong SO, Son Y, Lee JH, Cheong YK, Park SH, Chung HT, Pae HO (2015) Resveratrol analog piceatannol restores the palmitic acid-induced impairment of insulin signaling and production of endothelial nitric oxide via activation of anti-inflammatory and antioxidative heme oxygenase-1 in human endothelial cells. Mol Med Rep 12:937–944
Kappel VD, Cazarolli LH, Pereira DF, Postal BG, Zamoner A, Reginatto FH, Silva FRMB (2013) Involvement of glut-4 in the stimulatory effect of rutin on glucose uptake in rat soleus muscle. J Pharm Pharmacol 65:1179–1186
Karawya MS, Wahab SM, Farrag NM (1984) Diphenylamine, an antihyperglycemic agent from onion and tea. Nat Prod 47:775–780
Kolb H, Kiesel H, Greulich B, Bosh V (1982) Lack of antidiabetic effect of (−) epicatechin. Lancet 1:13031304
Kuhad A, Chopra K (2009) Tocotrienol attenuates oxidative-nitrosative stress and inflammatory cascade in experimental model of diabetic neuropathy. Neuropharmacology 57:456–462
Kunwar A, Priyadarsini KI (2016) Curcumin and its role in chronic diseases. In: Gupta SC, Prasad S, Aggarwal BB (eds) Advances in experimental medicine and biology. Anti-inflammatory nutraceuticals and chronic diseases, vol 928. Springer, Cham, pp 1–26
Lau YS, Tian XY, Mustafa MR, Murugan D (2013) Boldine improves endothelial function in diabetic db/db mice through inhibition of angiotensin II-mediated BMP4oxidative stress cascade. Br J Pharmacol 170:1190–1198
Li CG, Ni CL, Yang M, Tang YZ, Li Z, Zhu YJ, Jiang ZH, Sun B, Li CJ (2018) Honokiol protects pancreatic β-cell against high glucose and intermittent hypoxia-induced injury by activating Nrf2/are pathway in vitro and in vivo. Biomed Pharmacother 97:1229–1237
Li G, Xu H, Zhu S et al (2013a) Effects of neferin on ccl5 and ccr5 expression in SCG of type 2 diabetic rats. Brain Res Bull 90:79–87
Li J, Shen F, Guan C, Wang W, Sun X, Fu X et al (2014) Activation of Nrf2 protects against triptolide-induced hepatotoxicity. PLoS One 9(7):e100685. https://doi.org/10.1371/journal.pone.0100685
Li XH, Xin X, Wang Y et al (2013b) Pentamethylquercetin protects against diabetes-related cognitive deficits in diabetic Goto-Kakizaki rats. J Alzheimers Dis 34:755–767
Li YH, Xu Q, Xu WH, Guo XH, Zhang S, Chen YD (2015) Mechanisms of protection against diabetes-induced impairment of endothelium-dependent vasorelaxation by Tanshinone IIA. Biochim Biophys Acta 1850(4):813–823. https://doi.org/10.1016/j.bbagen.2015.01.007
Ling C, Jinping L, Xia L, Renyong Y (2013) Ursolic acid provides kidney protection in diabetic rats. Curr Res 75:59–63
Liu H-T, Wang Y-F, Olaleye O, Zhu Y, Gao X-M, Kang L-Y, Zhao T (2013) Characterization of in vivo antioxidant constituents and dual-standard quality assessment of Danhong injection. Biomed Chromatogr 27(5):655–663. https://doi.org/10.1002/bmc.2842

Lopez PM, Mora PG, Wysocka W, Maiztegui B, Alzugaray ME, Zoto HD, Borelli MI (2004) Quinolizidine alkaloids isolated from Lupinus species enhance insulin secretion. Eur J Pharmacol 504:139–142

Lotlikar MM, Rajarama R (1966) Pharmacology of a hypoglycaemic principle isolated from the fruits of *Momordica charantia* Linn. Indian Pharm 28:129–133

Madhuri K, Naik PR (2017) Modulatory effect of garcinol in streptozotocin-induced diabetic Wistar rats. Arch Physiol Biochem 123:322–329

Maher P, Dargusch R, Ehren JL, Okada S, Sharma K, Schubert D (2011) Fisetin lowers methylglyoxal dependent protein glycation and limits the complications of diabetes. PLoS One 6:e21226

Mancha-Ramirez AM, Slaga TJ (2016) Ursolic acid and chronic disease: an overview of UA's effects on prevention and treatment of obesity and cancer. In: Gupta SC, Prasad S, Aggarwal BB (eds) Advances in experimental medicine and biology. Anti-inflammatory nutraceuticals and chronic diseases, vol 928. Springer, Cham

Maries RJ, Farnsworth NR (1995) Antidiabetic plants and their active constituents. Phytomedicine 2:137–189

Minakawa M, Miura Y, Yagasaki K (2012) Piceatannol, a resveratrol derivative, promotes glucose uptake through glucose transporter 4 translocation to plasma membrane in l6 myocytes and suppresses blood glucose levels in type 2 diabetic model db/db mice. Biochem Biophys Res Commun 422:469–475

Monsereenunurson R (1980) Effect of *Capsicum annuum* on blood glucose level. Quar J Crude Drug Res 18:1–7

Ng TB, Yeung HW (1985) Hypoglycemic constituents of Panax ginseng. Gen Pharmacol 16(6): 549–552. https://doi.org/10.1016/0306-3623(85)90140-5

Norris LE, Collene AL, Asp ML, Hsu JC, Liu L-F, Richardson JR, Li D, Bell D, Osei K, Jackson RD, Belury MA (2009) Comparison of dietary conjugated linoleic acid with safflower oil on body composition in obese postmenopausal women with type 2 diabetes mellitus. Am J Clin Nutr 90(3):468–476. https://doi.org/10.3945/ajcn.2008.27371

Ong KW, Hsu A, Tan BKH (2012) Chlorogenic acid stimulates glucose transport in skeletal muscle via ampk activation: A contributor to the beneficial effects of coffee on diabetes. PLoS One 7: e32718

Orgah JO, Wanga SHY, Jianga M, Wanga Y, Orgah EA, Duand Y, Zhaoa B, Zhanga B, Hand J, Zhua Y (eds) (2020) Pharmacological potential of the combination of Salvia miltiorrhiza (Danshen) and Carthamus tinctorius (Honghua) for diabetes mellitus and its cardiovascular complications. Pharmacol Res 153:104654

Oza MJ, Kulkarni YA (2016) Phytochemical and complication in type 2 diabetes—an update. Int J Pharm Sci Res 7:14–24

Pan GY, Huang ZJ, Wang GJ, Fawcett JP, Liu XD, Zhao XC, Sun JG, Xie YY (2003) The antihyperglycaemic activity of berberine arises from a decrease of glucose absorption. Planta Med 69:632–636

Pandey VK, Mathur A, Khan MF, Kakkar P (2019) Activation of perk-eif2α-atf4 pathway contributes to diabetic hepatotoxicity: attenuation of er stress by morin. Cell Signal 59:41–52

Paoli P, Cirri P, Caselli A, Ranaldi F, Bruschi G, Santi A, Camici G (2013) The insulin-mimetic effect of morin: A promising molecule in diabetes treatment. Biochim Biophys Acta 1830: 3102–3111

Pinto Vierira IG, Mendes FN, da Silva SC, Paim RT, da Silva BB, Benjamin SR, Florean EP, Florindo Guedes MI (2018) Antidiabetic effects of galactomannans from Adenanthera pavonina L. in streptozotocin-induced diabetic mice. Asian Pac. J Trop Med 11:116–122

Prasath GS, Pillai SI, Subramanian SP (2014) Fisetin improves glucose homeostasis through the inhibition of gluconeogenic enzymes in hepatic tissues of streptozotocin induced diabetic rats. Eur J Pharmacol 740:248–254

Prasath GS, Subramanian SP (2014) Antihyperlipidemic effect of fisetin, a bioflavonoid of strawberries, studied in streptozotocin-induced diabetic rats. J Biochem Mol Toxicol 28:442–449. https://doi.org/10.1002/jbt.21583

Prince P, Kamalakkannan N (2006) Rutin improves glucose homeostasis in streptozotocin diabetic tissues by altering glycolytic and gluconeogenic enzymes. J Biochem Mol Toxicol 20:96–102

Pushpa K, Jain S, c., Panagariva, A., Dixit, V. P. (1981) Hypoglycemic activity of polypeptide from a plant source. J Nat Prod 44:648–655

Raman A, Lau C (1996) Anti-diabetic properties and phytochemistry of *Momordica charantia* L. (Cucurbitaceae). Phytomedicine 2:349–362

Razavi T, Kouhsari SM, Abnous K (2018) Morin exerts anti-diabetic effects in human hepg2 cells via down-regulation of mir-29a. Exp Clin Endocrinol Diabetes 127:615–622

Sheela N, Jose MA, Sathyamurthy D, Kumar BN (2013) Effect of silymarin on streptozotocin-nicotinamide-induced type 2 diabetic nephropathy in rats. Iran J Kidney Dis 7:117–123

Sterne J (1982) Pharmacology and mode of action of the hypoglycemic guanidine derivatives. In: Campbell GD (ed) Oral hypoglycemic agents. Academic Press, New York, pp 193–245

Szkudelski T, Szkudelska K (2015) Resveratrol and diabetes: from animal to human studies. Biochim Et Biophys Acta (BBA)-Mol Basis Dis 1852:1145–1154

Taguchi K, Hida M, Hasegawa M, Matsumoto T, Kobayashi T (2016) Dietary polyphenol morin rescues endothelial dysfunction in a diabetic mouse model by activating the Akt/eNOS pathway. Mol Nutr Food Res 60:580–588

Tan M-J, Ye J-M, Turner N, Hohnen-Behrens C, Ke C-Q, Tang C-P, Chen T, Weiss H-C, Gesing E-R, Rowland A, James DE, Ye Y (2015) Antidiabetic activities of triterpenoids isolated from bitter melon associated with activation of the AMPK pathway. Chem Biol 15(5):520

Tang DQ, Wei YQ, Yin XX, al., e. (2011) In vitro suppression of quercetin on hypertrophy and extracellular matrix accumulation in rat glomerular mesangial cells cultured by high glucose. Fitoterapia 82:920–926

Taslimi P, Kandemir FM, Demir Y, İleritürk M, Temel Y, Caglayan C, Gulçin İ (2019) The antidiabetic and anticholinergic effects of chrysin on cyclophosphamide-induced multiple organ toxicity in rats: pharmacological evaluation of some metabolic enzyme activities. J Biochem Mol Toxicol:e22313

Tomoda M, Shimizu N, Oshima Y, Takahashi M, Murakami M, Hikino H (1987) Hypoglycemic activity of twenty plant mucilages and three modified products. Planta Med 53:8–12

Tsai SJ, Huang CS, Mong MC, al., e. (2012) Anti-inflammatory and antifibrotic effects of naringenin in diabetic mice. J Agric Food Chem 60:514–521

Wang C, Shi C, Yang X, Yang M, Sun H, Wang C (2015) Celastrol suppresses obesity process via increasing antioxidant capacity and improving lipid metabolism. Eur J Pharmacol 744:52–58

Wang G, Li W, Lu X, Bao P, Zhao X (2012a) Luteolin ameliorates cardiac failure in type i diabetic cardiomyopathy. J Diabetes Complicat 26:259–265

Wang GG, Lu XH, Li W, Zhao X, Zhang C (2011) Protective effects of luteolin on diabetic nephropathy in STZ-induced diabetic rats. Evid-Based Complement Altern Med 2011:323171

Wang Q, Zhao Z, Shang J, Xia W (2014) Targets and candidate agents for type 2 diabetes treatment with computational bioinformatics approach. J Diabetes Res 2014:763936

Wang SYH, Tan W, Wu X, Chen R, Cao J, Chen M, Y. (2012b) Wang, compatibility art of traditional Chinese medicine: from the perspective of herb pairs. J Ethnopharmacol 143(2): 412–423. https://doi.org/10.1016/j.jep.2012.07.033

Wiedemann M, Gurrola-Díaz C, Vargas-Guerrero B, Wink M, García-López P, Düfer M (2015) Lupanine improves glucose homeostasis by influencing KATP channels and insulin gene expression. Molecules 20:19085–19100

Yu Z, Zhang T, Gong C, Sheng Y, Lu B, Zhou L, Ji L, Wang Z (2016) Erianin inhibits high glucose-induced retinal angiogenesis via blocking erk1/2-regulated hif-1α-VEGF/VEGFR2 signaling pathway. Sci Rep 6:34306

Conservation of Medicinal Plant Genetic Resources in Southern Africa

15

Michael Pillay, Samkeliso Takaidza, Lerato N. Madike, and Unisa Terblanche

Abstract

Various diseases and ailments have been treated with medicinal plants for over 60,000 years. The easy accessibility and affordability of these plants compared to the high cost of pharmaceutical drugs are, perhaps, the most compelling reasons for the extensive use of ethnomedicine. New research findings on the phytochemicals present in medicinal plants are needed to test the veracity of some of the indigenous knowledge. One of the serious problems with ethnomedicine is the lack of studies on the toxicity of the secondary compounds in the plants. This chapter presents information on three medicinal plants (*Tulbaghia violacea*, *Carpobrotus edulis*, and *Cotyledon orbiculata*) that are used widely in Southern Africa. The three plants contain a number of common phytochemicals including alkaloids, flavonoids, flavonols, phenolic compounds, proanthocyanidins, saponins, tannins, and triterpenes. In general, studies on different cell lines showed that these plants are useful in traditional medicine. However, most of the extracts were toxic at high concentrations. There is no information on the extent that these plants are being used or on the availability of these plants in the wild. It is envisaged that once more scientific information on these plants becomes available the danger of overexploitation of these species is possible. Therefore, this chapter also presents a review of some of the various ways in which medicinal plants are conserved for future use.

Keywords

Carpobrotus edulis · *Cotyledon orbiculata* · *Tulbaghia violacea* · Conservation · Ethnomedicine · Genetic resources

M. Pillay (✉) · S. Takaidza · L. N. Madike · U. Terblanche
Department of Biotechnology, Vaal University of Technology, Private Bag X021, Vanderbijlpark, South Africa

S. Ramamoorthy et al. (eds.), *Plant Genetic Resources, Inventory, Collection and Conservation*, https://doi.org/10.1007/978-981-16-7699-4_15

15.1 Introduction

Various diseases and ailments have been treated with medicinal plants from about 60,000 years ago (Solecki 1975). One of the main reasons for the extensive use of medicinal plants centers around the accessibility, affordability, and perceived safety of medicinal plants for millions of people that either do not have access to modern-day healthcare facilities (Singh and Singh 1981; Dar et al. 2017) or prefer treatment from traditional healers. Many of the current pharmaceutical, agricultural, and cosmetic products have been derived from plants and particularly with leads from indigenous knowledge. Many medicinal plants have provided precursors for the synthesis of novel compounds with medicinal properties (chemo-pharmaceutical semisynthesis) (Doughari 2012). For example, one of the most important medicines for treating childhood leukemia, which was derived from the rosy periwinkle, is an old remedy that was used by witch doctors in Madagascar (Foster 2010). However, there has been no recognition of indigenous knowledge in the development of modern medicines and many people are still fighting for their right to receive compensation.

The exact chemical constituents of many of these plants are unknown primarily because very little scientific research has been conducted on them. There has been a steady rise in research on medicinal plants especially in South Africa when the government recognized the need for traditional medicine as being part of primary healthcare since 1994. This was also promoted, perhaps, to relieve the pressure on the limited healthcare facilities that were available at the inception of a new government. It is becoming increasingly accepted that medicinal plants should be integrated into modern medicine since current research is validating the traditional use of plants in several scientific disciplines and with the likelihood of developing new medicines (Abdullahi 2011). Our laboratory at the Vaal University of Technology started work on medicinal plants in 2015. We have selected a few plants that included *Tulbaghia violacea, Carpobrotus edulis*, and *Cotyledon orbiculata* mainly because the plants were used for traditional medicines by a large percentage of the population and meagre technical evidence was available to verify or support the use of these plants.

In this chapter we review some of the literature on the medicinal proprieties of these plants and some of the new findings from our studies. We envisage that as new knowledge of these plants becomes available there is the danger that these plants will be exploited for use indiscriminately. It is believed that over 3000 flowering plants in South Africa are used for medicinal reasons (Louw et al. 2002; Van Wyk and Gericke 2007). There is evidence to support the view that about 82 species (0.4% of the total national flora of South Africa) are close to extinction at a national level in the short and medium terms and a further 100 species are of conservation concern (including two species already extinct in the wild) (Williams et al. 2013). Therefore, this chapter also considers how these genetic resources could be conserved for future use and how to prevent their extinction.

15.2 Botanical Description and Distribution of Plants

15.2.1 *Carpobrotus edulis*

Carpobrotus edulis L. of the family Aizoaceae is an indigenous South African plant commonly known as "sour fig" or "Cape fig" (Fig. 15.1) (Van Wyk et al. 2009; Rocha et al. 2017). It is a prostrate creeping perennial plant with evergreen succulent leaves and edible fruits (Campoy et al. 2018).

The plant grows predominantly on sandy coastal dunes and inland on sandy or well-drained slopes. Several characteristics of the plant such as its ability for controlling soil erosion on dunes and embankments, its ability to grow in a wide range of soils and climates, and its edible fruits have led to the plant's dispersal to several parts of the world including California, USA (Albeit 1995); Portugal (Martins et al. 2010); Mediterranean countries (Roiloa and Retuerto 2016); and Europe and Australia (Rocha et al. 2017).

15.2.2 *Cotyledon orbiculata*

Cotyledon orbiculata L. (family Crassulaceae) is endemic to South Africa and is also known as "pig's ear" (Van Wyk et al. 2009). It is a small evergreen, low-growing to erect shrub with woody branches and large decussate succulent leaves (Fig. 15.2).

The fleshy leaves have a waxy layer and vary in color from green to grey with a red line around the leaf margins. The shape, size, and color of the leaves are influenced by environmental factors. *Cotyledon orbiculata* is widespread in Southern Africa and other parts of the world (Maroyi 2019).

Fig. 15.1 *Carpobrotus edulis* (photograph by Dr. W. Terblanche)

Fig. 15.2 *Cotyledon orbiculata* (photograph by Dr. W. Terblanche)

Fig. 15.3 *Tulbaghia violacea*

15.2.3 *Tulbaghia violacea*

Tulbaghia violacea (family Amaryllidaceae) is a perennial bulbous herb with rhizomes, and narrow, hairless, and strap-shaped dark green and leathery leaves (Van Wyk et al. 2009). The plant ranges in height from 30 to 120 cm depending on growth conditions (Fig. 15.3).

The plant produces an umbel of about ten mauve or lilac flowers at the tip of a pedicel and blooms from summer to autumn (Watson and Dallwitz 1992; Kindersley 2008). It grows in dry to moist soils either in partial shade or in full sunny areas (Benham 1993). *Tulbaghia violacea* is also known as "wild garlic or society garlic" due to its garlic-like odor (Van Wyk et al. 2009; George et al. 2010). The sulfur compounds present in various *Allium* spp. are similar to those identified in *T. violacea* (Jacobsen et al. 1968) and consequently the plants have similar medicinal properties. *Tulbaghia violacea* is mainly distributed in the KwaZulu-Natal, the Eastern Cape, and Northern Gauteng in South Africa. The plant has also been located in the rocky grasslands of Zimbabwe (Dyson 1998).

15.3 Ethnomedicinal Uses of the Plants

15.3.1 *Carpobrotus edulis*

Carpobrotus edulis has been used widely as a traditional medicine for the treatment of mouth and throat infections, as a diuretic, or for the treatment of diphtheria, dysentery, diarrhea, and tuberculosis. The plant is renowned for its external use to treat wounds, burns, skin infections, and eczema (Van Wyk et al. 2009; De Beer and Van Wyk 2011). Aqueous and alcoholic extracts of the plant are used to treat infections arising from acquired immunodeficiency syndrome (AIDS) as well as diabetes mellitus, high blood pressure, intestinal worms, and constipation (Omoruyi et al. 2012). In Tunisia, *Carpobrotus edulis* is traditionally used for wound healing, to alleviate inflammation of the mucous membranes of the respiratory and urinary systems, and to treat water retention and painful urination (Meddeb et al. 2017).

15.3.2 *Cotyledon orbiculata*

Cotyledon orbiculata has been used to treat numerous ailments (Amabeoku et al. 2007; Van Wyk et al. 2009; Semenya et al. 2013). For instance, the peeled leaves have been used to facilitate easy removal of corns, warts, or plantar warts. Warmed peeled leaves have been used as a poultice for boils, skin eruptions, and abscesses. The consumption of fresh leaves has been used as a vermifuge. Fresh leaf juice of *Cotyledon orbiculata* has been used to treat epilepsy when taken orally and as a lotion for acne, earache, or toothache. Lastly, a leaf decoction, used as an enema, has also been used to treat syphilis (Terblanche 2021). *Cotyledon orbiculata* has not been the subject of many in vitro or in vivo studies despite its popularity as a traditional medicinal plant (SANBI 2004).

15.3.3 *Tulbaghia violacea*

The plant has been used as a traditional medicine for the treatment of various ailments including gastrointestinal problems, asthma, fever, colds, pulmonary tuberculosis, seizures/fit, rheumatoid arthritis (Burton 1990; Dyson 1998; Ebrahim and Pool 2010; Ncube et al. 2011; Kubec et al. 2013), and constipation/purgative and as an anti-helminthic (Watt and Breyer-Brandwijk 1962). Traditional uses in the Eastern Cape province of South Africa include treatment of colic, wind, restlessness, headache, and fever, primarily in young children (Dyson 1998). Crushed leaves of the plant are used to treat sinus headaches, repel fleas, ticks, as well as mosquitoes (Lim 2013). A concoction made from soaking the leaves and bulbs of *T. violacea* in water is used for rheumatism and arthritis and to reduce fever. The plant is also used as an ornamental and the flowers and leaves are edible (Dyson 1998).

15.4 Review of the Phytochemistry and Pharmacological Properties of *Carpobrotus edulis, Cotyledon orbiculata,* and *Tulbaghia violacea*

15.4.1 *Carpobrotus edulis*

Carpobrotus edulis contains a variety of phytochemicals including alkaloids, flavonoids, flavonols, phenolic compounds, proanthocyanidins, saponins, tannins, and triterpenes (Omoruyi et al. 2012). Many studies have reported on the antimicrobial activity of different *Carpobrotus edulis* extracts (Van Der Watt and Pretorius 2001; Ordway et al. 2003; Chokoe et al. 2008; Martins et al. 2011; Meddeb et al. 2017). Crude extracts of the plant were effective against both gram-negative and -positive bacteria (Van der Watt and Pretorius 2001; Martins et al. 2011). The latter study reported that several isolated compounds from *Carpobrotus edulis* showed moderate activity against *Mycobacterium tuberculosis*. The antibacterial properties of *Carpobrotus edulis* were shown to vary seasonally in South Africa (Chokoe et al. 2008). Van der Watt and Pretorius (2001) demonstrated the efficacy of a crude methanolic extract as well as five isolated flavonoids, namely, rutin, neohesperidin, hyperoxide, catechin, and ferulic acid, against the gram-negative bacterium, *Moraxella catarrhalis*, and gram-positive bacteria, *Staphylococcus epidermidis* and *S. aureus*. Essential oil extract inhibited the growth of some pathogenic fungal isolates including *Candida albicans*, *C. krusei*, *C. rugose*, *C. glabrata*, and *Cryptococcus neoformans*, with MICs ranging from 0.02 to 0.31 mg/mL (Omoruyi et al. 2014). The antioxidant activity of *Carpobrotus edulis* extracts has been extensively researched (Chokoe et al. 2008; Falleh et al. 2011; Omoruyi et al. 2012; Meddeb et al. 2017; Hafsa et al. 2018).

Higher antioxidant activities were reported for the stem and leaf extracts, which correlated with the observed higher flavonoid, phenolic, and tannin contents of these extracts (Falleh et al. 2011). A number of phenolic compounds including sinapic acid, chlorogenic acid, gallic acid, quercetin, luteolin-7-o-glucoside, hyperoside, isoquercitrin, and ellagic acid with strong antioxidant activity were identified in the leaf extracts of *Carpobrotus edulis* (Omoruyi et al. 2012; Hafsa et al. 2018; Meddeb et al. 2017; Rocha et al. 2017; Castañeda-Loaiza et al. 2020).

Several compounds isolated from the leaves of *Carpobrotus edulis*, including uvaol, ß-amyrin, oleanolic acid, catechin, epicatechin, and monogalactosyldiacylglycerol, exhibited an antiproliferative effect in parental mouse lymphoma cells and human MDR1-transfected mouse lymphoma cells (Martins et al. 2010). Several studies have reported on the cytotoxicity of different *Carpobrotus edulis* extracts (Cock and van Vuuren 2014; Meddeb et al. 2017; Rocha et al. 2017; Hafsa et al. 2018; Mulaudzi et al. 2019).

15.4.2 *Cotyledon orbiculata*

The death of fowls that were accidently fed *Cotyledon orbiculata* showed that the plant is toxic (Kamerman 1926). In addition, further studies showed that *Cotyledon orbiculata* caused an acute or chronic condition known as "krimpsiekte" in livestock. A variety of other phytochemicals including flavonoids, phenolic compounds, reducing sugars, saponins, tannins, and triterpene steroids have been identified in *Cotyledon orbiculata* with different solvent extractions (Amabeoku et al. 2007; Molefe et al. 2013; Xaba 2016). *Cotyledon orbiculata* extracts have been reported to show antimicrobial activity (Van der Watt and Pretorius 2001; Ordway et al. 2003; Chokoe et al. 2008; Martins et al. 2011; Meddeb et al. 2017). Methanolic extracts of *Carpobrotus edulis* inhibit the MDR efflux pumps, promote the killing of phagocytosed *S. aureus*, and support immune modulation. Methanolic extract as well as five isolated flavonoids, namely, rutin, neohesperidin, hyperoxide, catechin, and ferulic acid, acted against the gram-negative bacterium, *Moraxella catarrhalis*, and gram-positive bacteria, *Staphylococcus epidermidis* and *S. aureus*. Aqueous and acetone extracts of the shoots of *Cotyledon orbiculata* displayed no antitumor activity against the breast (MCF-7) cancer cell line, whereas some antitumor activity was observed for both extracts against the cervical cancer cell line (HeLa) after 24 h of treatment (Molefe et al. 2013). However, the antitumor activity was less pronounced after 48 h of treatment. Water and acetone extracts of *Cotyledon orbiculata* suppressed hatching and larval development of gastrointestinal parasitic nematodes by 82.63% and 85.32%, respectively (Molefe et al. 2013). The water extract was more effective than the acetone extract. Amabeoku et al. (2007) showed that both the aqueous and methanolic extracts of *Cotyledon orbiculata* significantly reduced induced seizures in mice. Low doses (25 mg/kg) of a methanolic extract of *Cotyledon orbiculata* produced analgesic effects on the nervous system (Kabatende 2005). Studies by Roux (2012) and Aremu et al. (2010) demonstrated that extracts showed significant antioxidant activity and anti-inflammatory activity. The cytotoxicity of an acetone and aqueous extract of *Cotyledon orbiculata* was evaluated against a mammalian bovine kidney cell line (MDBK), using the MTT and lactate dehydrogenase (LDH) assays. The brine shrimp lethality test was used to test the cytotoxicity of the extracts in the same study. The study reported that both the acetone and aqueous extract exhibited significant cytotoxicity at the highest concentration tested (2.5 mg/mL) after prolonged treatment (72 h). The acetone and aqueous extract at 0.2, 0.4, and 0.6 mg/mL killed all larvae within 72 h; however, no larval mortality was observed for the first 24 h of treatment. Based on the observations, it was concluded that the acetone extract had greater toxicity than the aqueous extract and that, although cytotoxicity was observed, particularly after prolonged treatment (72 h), *Cotyledon orbiculata* appeared to be nontoxic (Molefe et al. 2013). Roux (2012) reported that a methanolic extract showed significant toxicity at 2 and 10 mg/ml against a neuroblastoma cell line whereas treatment with 0.08 and 0.2 mg/ml displayed no toxicity.

15.4.3 *Tulbaghia violacea*

Phytochemicals from different organs of *T. violacea* possess antibacterial, antifungal, anticancer, antioxidant, and anthelmintic activities (McGaw et al. 2000; Ncube et al. 2011; Aremu and van Staden 2013). Three studies described the anticancer properties of *T. violacea* (Lyantagaye and Rees 2003; Bungu et al. 2006; Nöthlings et al. 2008). Bungu et al. (2008) also reported on the antithrombotic and anticoagulant properties. Kaempferol has been isolated from *T. violacea* (Bate-Smith 1968) and helps in the reduction of heart disease (Park et al. 2006). Crude methanol extracts of *T. violacea* were effective against a broad spectrum of bacterial species such as *Clavibacter michiganensis*, *Ralstonia solanacearum*, and *Xanthomonas campestris* (Nteso and Pretorius 2006). Methanolic extracts from the rhizomes of *T. violacea* showed high antioxidant activities as well as a high degree of cytotoxic activity (Olorunnisola et al. 2011). Dichloromethane extracts of *T. violacea* were effective against the parasites *Trypanosoma brucei* and *Leishmania tarentolae* (Krstin et al. 2018).

Tulbaghia violacea also appears to have cardiovascular properties (Duncan and van Staden 1999; Mackraj et al. 2008). High phenolic and flavonoid contents were identified in the aqueous extracts of dried *T. violacea* leaves (Madike et al. 2017) and acetone extracts of fresh leaves (Takaidza et al. 2018a). Takaidza et al. (2018b) also reported that the acetone extract of fresh *T. violacea* leaves had the highest antioxidant activity among eight *Tulbaghia* species.

15.5 Recent Findings on the Biological Activities of *Carpobrotus edulis*, *Cotyledon orbiculata*, and *Tulbaghia violacea*

Several medicinal plants have been shown to contain bioactive compounds that include phenolic compounds, flavonoids, tannins, triterpenes, alkaloids, and saponins that have several properties including (1) antimicrobial activity, (2) antioxidant and anti-inflammatory activity, and (3) cell-stimulating properties and other properties unique to some plants. In this section of the chapter we provide some recent research results on three plants identified for this study. In all the studies whole-plant extracts were used to verify the biological properties of the plants. Many studies have shown that pure drugs isolated from plants rarely have the same degree of activity as the whole-plant extract at comparable concentrations or dosage of the active component (Cravotto et al. 2010; Rasoanaivo et al. 2011; Gallily et al. 2015). It was shown that CBD together with other plant constituents improved the dose-response and is supported by some recent reports showing that CBD in a standardized *Cannabis sativa* extract is more potent or efficacious than pure CBD (Gallily et al. 2015). One of the advantages of herbal medicines is their multiplex composition. Their components have multiple activities that produce a greater total action (Carmona and Pereira 2013).

15.5.1 *Carpobrotus edulis* and *Cotyledon orbiculata*

Carpobrotus edulis and *Cotyledon orbiculata* have been used in traditional medicine for the treatment of various skin conditions and wounds. This is achieved through wound re-epithelialization, where keratinocytes migrate, proliferate, and differentiate until a new epithelium is formed. Using the HaCaT cells that are used for studies of skin biology and differentiation it was shown in various tests that *Carpobrotus edulis* showed potential to improve the rate of keratinocyte migration and proliferation and could positively influence the wound-healing process (Terblanche 2021). On the contrary, crude extracts of *Cotyledon orbiculata* inhibited the migration, proliferation, and differentiation when introduced to keratinocytes (HaCaT) in vitro. Phytochemical screening showed that reducing sugars, phenols, tannins, flavonoids, cardiac glycosides, diterpenes, phytosteriods, and triterpenoids were present in most of the aqueous and all methanolic extracts of *Carpobrotus edulis*. Reducing sugars, phenols, tannins, flavonoids, cardiac glycosides, diterpenes, phytosteriods, and triterpenoids were present in all the crude extracts of *Cotyledon orbiculata* (Terblanche 2021).

One of the highlights of the study on *Carpobrotus edulis* and *Cotyledon orbiculata* was the method used for the extraction of phytochemicals from the plants. Most studies on medicinal plants have used a single-factor-at-a-time approach to ascertain the optimal extraction from plants. The disadvantages of this approach are that the method is expensive, time consuming, and laborious. It also limits the interpretation of the results since the interactions among different factors (parameters or variables) in extraction are not considered (Nkuzinna et al. 2014). A factorial design where many input variables, usually at two levels for each variable, can be investigated simultaneously is a valuable screening methodology. Our studies have found that for *Carpobrotus edulis* aqueous extract, extraction time, temperature, and interactions between them were most influential in optimizing the extraction yield. For the *Carpobrotus edulis* methanolic extract, the solvent pH, extraction time, and temperature as well as their three-way interaction strongly influenced the extraction yield (Terblanche 2021). It is highly recommended that factorial designs are used for efficient extraction of compounds from plants.

15.5.2 *Tulbaghia violacea*

Screening of *Tulbaghia violacea* for phytochemicals showed that flavonoids, glycosides, tannins, terpenoids, saponins, protein, phenols, carbohydrates, amino acids, and cardiac glycosides were present (Madike et al. 2017; Takaidza et al. 2018a). Acetone and water extracts of *T. violacea* showed antioxidant and antibacterial activity against *Bacillus subtilis, Serratia marcescens, Staphylococcus aureus,* and *S. epidermidis* and antifungal activity. Acetone extracts also showed both bacteriostatic/fungistatic and bactericidal/fungicidal activity depending on the incubation time and concentration of the extract. These activities may be due to rich source of phytochemicals and broad range of pharmaceutically important compounds in *T. violacea.*

Acetone and water crude extracts of *T. violacea* showed antiproliferative activity on two human cancer cell lines, Hkesc-1 cells with IC50 values of 0.4 mg/mL and 1.625 mg/mL, respectively, and the tumor KB cells with IC50 values of 0.2 mg/mL and 1 mg/mL, respectively. These extracts also induced apoptotic activity in these cells as was evidenced by the morphological changes such as blebbing, cell shrinkage, and rounding of the cells. Acridine orange-ethidium bromide staining showed that the level of apoptosis was influenced by the extract concentrations. The multicaspase assay that determines cellular plasma membrane permeabilization and cell death showed that apoptosis occurred in both the Hkesc-1 and KB-treated cells and cell cycle arrest was induced at the G2/M phase (Takaidza et al. 2018b). The immunomodulatory activity of acetone and water crude extracts from *T. violacea* on LPS-stimulated RAW264.7 cells indicated that the extracts had strong immuno-stimulative properties (Takaidza 2019).

The presence of cardiac glycosides in *T. violacea* allowed us to initiate some studies on the potential of the plant to play a role in cardiovascular diseases. Cardiac glycosides occur as secondary metabolites in several plants but also in some animals such as the milkweed butterflies. They are drugs used for the treatment and management of congestive heart failure and cardiac arrhythmia (Denwick 2002). Cardiac glycosides are responsible for increased levels of sodium ions in the myocytes, promoting an increase of calcium ions which enables contraction of the heart muscle, thus improving cardiac output while reducing distention of the heart (Chukwuebuka and Chinenye 2015). Water extracts of *T. violacea* showed antithrombogenic properties by delaying blood clotting on polystyrene discs. However, clotting was dependent on the concentrations of the extract with a lower concentration being more effective (Madike et al. 2020). When the anticlotting properties of *T. violacea* were further tested on polycaprolactone-synthesized scaffolds it was also shown that lower extract concentrations were more effective in inhibiting platelet adhesion and activation. Scaffolds containing water extracts of *T. violacea* were evaluated for their toxicity, viability, adhesion, proliferation, morphology, and differentiation of adipose-derived stem cells into osteoblasts (Madike et al. 2019). It was shown that scaffolds fabricated with a low concentration of *T. violacea* extract promoted cell adhesion, proliferation, and differentiation. *Tulbaghia violacea* may be useful for tissue engineering applications and bone repair with prospects of preventing cardiovascular diseases associated with bone defects (Madike 2020).

15.6 Conservation of Medicinal Plants

The medicinal plants described in this chapter are well known for their traditional healing properties. However, none of them have been listed as being in danger of being overexploited. The scientific information that was recently gathered on these plants has not been widely dispersed as yet. There is a danger that as this information becomes published there may be increased interest in these plants by both local users and pharmaceutical companies. Therefore, it becomes pertinent to review some of the ways in which these plants could be conserved as genetic resources for the future. There are several reviews on the conservation and sustainable use of medicinal plants (Chen et al. 2016; van Wyk and Prinsloo 2018; Kadam and Pawar 2020; Cahyaningsih et al. 2021). We provide a summary of the main ideas from these reviews as well as other literature. Despite the advances in chemical sciences and development of synthetic drugs the need for natural materials and modern medicine still depends on the availability of biological materials (Hawkins 2008). The harvesting of plants for medicinal uses depends on the part of a plant that is used. Several plant parts such as the leaves, bark, roots, fruits, seeds, or even whole plant are harvested for traditional medicine. Therefore, the pressure for harvesting medicinal plants is different (Andel and Havinga 2008). Overexploitation, indiscriminate collection, uncontrolled deforestation, and habitat destruction invasive species are some of the main reasons for the conservation of medicinal plants (Chen et al. 2016).

There are several strategies that are used in the conservation of medicinal plants. The type of approach is determined by the biological characteristics and geographic distribution the species (Chen et al. 2016). The common recommendation is that both in situ and ex situ methods should be used in conserving medicinal plants (Mestanza-Ramon et al. 2020, Asgher et al. 2020). Ex situ conservation is conservation of medicinal plants outside their natural habitat (Kadam and Pawar 2020) and includes procedures such as seed storage, DNA storage, field gene banks, and botanical gardens. In situ conservation deals with the "on-site conservation" of the wild genetic diversity in the natural habitat. In situ conservation such as natural reserves and wild nurseries is an efficient way to retain the phytochemical composition of plants in their natural habitats (Chen et al. 2016).

15.7 In Situ Conservation

The habitat of most medicinal plants determines the secondary metabolites that they harbor. It is considered that medicinal plants harvested from its natural settings are more effectual than cultivated plants (Joshi and Joshi 2014, Raimondo 2015). These phytochemicals may not be produced when the plants are cultured (Chen et al. 2016). Conservation of medicinal plants in their ecosystems via in situ conservation of whole communities is advantageous for protecting indigenous plants and maintaining natural communities (Chen et al. 2016). In situ conservation has focused on the establishment of protected areas that mimic the ecosystem-oriented, rather than species-oriented, approach (Zhang et al. 2012).

15.8 Nature Reserves

Natural reserves protect the biodiversity within a given topographical area (Cantu et al. 2004). Nature reserves are managed with the aim of preserving and restoring biodiversity. Nature reserves greatly increase the proportion of protected lands in a country. Unfortunately, in many countries, nature reserves face high rates of deforestation and landscape fragmentation—factors which add urgency to their protection (Menon et al. 2001). Nature reserves provide opportunities for research, especially to show the advantages of reserves in preserving biodiversity and investigating the extent of protecting specific resources within the existing reserve system (Cantu et al. 2004).

15.9 Wild Nurseries

Nurseries are required to maintain vegetatively propagated species and species with poor seed germination, and other materials for difficult plants that cannot be maintained in gene banks (Engels 2003). Wild nurseries are generally established for the cultivation and domestication of specific species that are endangered. These nurseries are usually close to the sites where the plants grow naturally. Cost considerations and competing land uses make it impossible to establish wild nurseries for every natural wild plant habitat (Chen et al. 2016). Wild nurseries are considered advantageous for the in situ conservation of medicinal plants that are endemic, threatened, and in need (Chang et al. 2011).

15.10 Ex Situ Conservation

Both ex situ and in situ conservations are complementary ways for overexploited and endangered medicinal plants that grow slowly, are scarce, and are highly prone to replanting diseases (Lakshman 2016; Chen et al. 2016). Ex situ conservation aims to grow and establish vulnerable species to enable their survival. It is also a way to produce much planting material for drug production (Chen et al. 2016). Many wild medicinal plants cannot maintain their effectiveness when they are produced in spaces that are different from their natural habitats, but their reproductive structures can be stored in seed banks for future use (Chen et al. 2016).

15.11 Botanic Gardens

Globally, botanic gardens have been playing a role in studying and cultivating medicinal plants for over 500 years (Hawkins 2008). Botanic gardens were initiated with the main purpose to grow and study medicinal plants based on the theories, beliefs, and experiences of indigenous people (Hawkins 2008). These plants were used in the general maintenance of health as well as in the prevention, diagnosis, improvement, or treatment of physical and mental illness. Botanical gardens are important for the exhibition of ornamental plants and the preservation of plants of a region, especially those that are rare and endangered. These repositories may provide growing material in cases of disasters such as fires that may destroy plants in their natural ecosystems. Botanic gardens can enhance medicinal plant conservation by developing techniques for propagation and cultivation, as well as starting programs for domestication and breeding (Maunder et al. 2001). Botanic gardens are ideal for conservation of species which cannot be kept in seed banks (Kadam and Pawar 2020).

15.12 Seed Banks

Seed banks can also be used in integrated plant conservation (Sharrock et al. 2014). Plants stored as seeds is a common way of preserving germplasm and is regarded as one way of ex situ conservation (Rivière and Müller 2018). Seed banks are a crucial component of conserving medicinal plants especially under unpredictable conditions such as climate change and habitat loss (Rajjou and Debeaujon 2008). It is believed that current extinction rates are much higher than during the recent geological past mainly due to wide-scale destruction of natural habitats (Pimm et al. 1995). Seed banks are facilities where seeds are stored under cold and dry conditions. This prolongs seed viability and thereby preserves plants for future use. The main advantages of seed banks are that they provide immediate access to plant samples, they facilitate scientific studies for conserving a species, and plants conserved in seed banks are protected from habitat destruction, diseases, and predators (Schoen and Brown 2001). Seeds are useful to reinstate species into existing, suitable habitats where they were once present, or to increase the diversity of small, genetically homogeneous populations (Riviere and Muller 2018). The storage of seeds is easier due to reduced space required, low labor demands, and ability to conserve large samples at a low cost (Riviere and Muller 2018). In addition, seeds are a convenient means of long-term storage of genetic diversity, as the samples are small in size, are easily handled, require low maintenance, and frequently remain viable for long periods (Riviere and Muller 2018). Seed banks offer easier ways of storing the genetic diversity of many medicinal plants ex situ compared to botanic gardens, and are recommended to help preserve the biological and genetic diversity of wild plant species (Schoen and Brown 2001). The challenging tasks of seed banking are how to reintroduce the plant species back into the wild and how to actively assist in the restoration of wild populations (Li and Pritchard 2009).

15.13 Cultivation of Medicinal Plants

There are few or no regulations which control the collection of material from the wild in many countries. Therefore, there is growing anxiety about diminishing populations, loss of genetic diversity, extinction of species, and habitat degradation with regard to medicinal plants (Canter et al. 2005). Although adequate protection may be provided to protect some species, many countries are adopting methods for long-term cultivation of medicinal plants. At present, increasing attention is given to the cultivation of medicinal plant cultivation as a means to combine biodiversity conservation and poverty alleviation. In many countries, cultivation of medicinal plants is being promoted to provide a source of income and employment (Wiersum et al. 2006). Medicinal plants also form an important aspect for international trade (Fabricant and Farnsworth 2001).

The value of the global trade in plants used for medicinal purposes is estimated to be in access of US$ 2.5 billion, and is increasingly driven by industry demand (UN Comtrade 2013).

The global market for traditional medicine was estimated at US$ 83 billion annually (WHO 2011). Cultivation of medicinal plants provides the opportunity to use new techniques to solve problems encountered in their production including the presence of toxic components, pesticide contamination, low levels of active ingredients, and misclassification of botanical origin (Raina et al. 2011). Controlled growth conditions of medicinal plants can improve the yields of active compounds and ensure production stability (Chen et al. 2016). Cultivation practices are designed to provide optimal levels of water, nutrients, optional additives, and environmental factors including temperature, light, and humidity to obtain improved yields of target products (Chen et al. 2016). Several advantages of cultivating medicinal plants are discussed by Canter et al. (2005). They include cultivation to control the content of the active, traditional breeding of medicinal plants, genetic transformation, and genetic engineering. Recently, synthetic products have been developed and they are popular due to their direct effect, easy production, and stringent controls (Veeresham 2012). Despite this, crude extracts of medicinal plants continue to be the main form used by the public.

Due to space constraints, we have not addressed the role of biotechnology including tissue culture for the conservation of medicinal plants and the large-scale production of desirable bioactive compounds.

15.14 Conclusion

The stigma associated with using medicinal plants for healthcare has dwindled all over the world. There is an increase in the use of traditional medicine, also known as ethnomedicine, folk medicine, native healing, or complementary or alternative medicine (Abdullahi 2011). The more prosperous countries such as Australia, France, Canada, the USA, the UK, and Germany are also seeing an increased use of traditional medicine. There is evidence to show that traditional medicine has contributed to the promotion of health behavior and referral to the modern healthcare system (Abdullahi 2011). As more research is done on medicinal plants and evidence emerges for their therapeutic properties the appeal for ethnic medicine will broaden. There is an average of 5000 papers published on medicinal plants per annum. But research on medicinal plants in Africa is underdeveloped compared to that in China and India (Salmerón-Manzano et al. 2020).

Despite the existence of various sets of recommendations for the conservation and sustainable use of medicinal plants, only a small portion of these have achieved adequate protection of medicinal plant resources through conventional conservation in natural reserves or botanic gardens. The daily demand of medicinal plants is increasing exponentially and there is danger that many useful plants will become extinct. Therefore, it is critical that the focus on conserving medicinal plants should receive high priority to prevent the loss of the genetic diversity of these plants. The conservation methods discussed in this chapter will provide sustainable ways to conserve medicinal plants for human healthcare needs. Conservation strategy defines conservation as "the management of human use of the biodiversity so that it may yield the greatest sustainable benefit to present generation while maintaining its potential to meet the needs and aspirations of future generations," cited by Lakshman (2016).

Declaration The authors declare no conflict of interest.

References

Abdullahi AA (2011) Trends and challenges of traditional medicine in Africa. African Journal of Traditional, Complementary and Alternative Medicine 8(5 Suppl):115–123

Albeit ME (1995) The ecology and management of *Carpobrotus edulis*. CalEPPC News, p 11

Amabeoku GJ, Green I, Kabatende J (2007) Anticonvulsant activity of *Cotyledon orbiculata* L. (Crassulaceae) leaf extract in mice. J Ethnopharmacol 112(1):101–107

Andel TV, Havinga R (2008) Sustainability aspects of commercial medicinal plant harvesting in Suriname. For Ecol Manag 256(8):1540–1545

Aremu AO, Ndhlala AR, Fawole OA, Light ME, Finnie JF, Van Staden J (2010) *In vitro* pharmacological evaluation and phenolic content of ten South African medicinal plants used as anthelmintics. S Afr J Bot 76(3):558–566

Aremu AO, van Staden J (2013) The genus *Tulbaghia* (Alliaceae)--a review of its ethnobotany, pharmacology, phytochemistry and conservation needs. J Ethnopharmacol 149:387–400

Asgher M, Verma S, Khan NA, Vyas D, Kumari P, Rashid S, Khan S, Qadir S, Ajmal Ali M, Ahmad P (2020) Physiological, biochemical and reproductive studies on *Valeriana wallichii*, a critically endangered medicinal Plant of the Himalayan Region Grown under *in-situ* and *ex-situ* conditions. Plan Theory 9(2):131

Bate-Smith E (1968) The phenolic constituents of plants and their taxonomic significance: II. Monocotyledons. Bot J Linn Soc 60(383):325–356

Benham S (1993) *Tulbaghia*. A survey of the species cultivation. Plantsman 15(2):89–110

Bungu L, Frost CL, Brauns CS, van de Venter M (2006) *Tulbaghia violacea* inhibits growth and induces apoptosis in cancer cells in vitro. Afr J Biotechnol 5(20):1936–1943

Bungu L, Van De Venter M, Frost C (2008) Evidence for an *in vitro* anticoagulant and antithrombotic activity in *Tulbaghia violacea*. Afr J Biotechnol 7(6):58496

Burton SG (1990) A chemical investigation of Tulbaghia violacea. MSc Thesis. Rhodes University, South Africa

Cahyaningsih R, Brehm JM, Maxted N (2021) Gap analysis of Indonesian priority medicinal plant species as part of their conservation planning. Global Ecol Conserv 26:e01459

Campoy JG, Acosta ATR, Affre L, Barreiro R, Brundu G, Buisson E, González L, Lema M, Novoa A, Retuerto R, Roiloa SR, Fagúndez J (2018) Monographs of invasive plants in Europe: *Carpobrotus*. Botany Letters 165(3–4):440–475

Canter PH, Thomas H, Ernst E (2005) Bringing medicinal plants into cultivation: opportunities and challenges for biotechnology. Trends Biotechnol 23(4):180–185

Cantu CR, Wright G, Scott JM, Strand E (2004) Assessment of current and proposed nature reserves of Mexico based on their capacity to protect geophysical features and biodiversity. Biol Conserv 115:411–417

Carmona IF, Pereira AMS (2013) Herbal medicines: old and new concepts, truths and misunderstandings. Revista Brasileira de Farmaco 23(2):379–385

Castañeda-Loaiza V, Placines C, Rodrigues MJ, Pereira C, Zengin G, Uysal A, Jeko J, Cziáky Z, Reis CP, Gaspar MM, Custódio L (2020) If you cannot beat them, join them: exploring the fruits of the invasive species *Carpobrotus edulis* (L.) N.E. Br as a source of bioactive products. Ind Crop Prod 144:112005

Chang L, Hua YU, Chen S-L (2011) Framework for sustainable use of medicinal plants in China. Plant Diversity 33:65–68

Chen SL, Yu H, Luo HM, Wu Q, Li C-H, Steinmetz A (2016) Conservation and sustainable use of medicinal plants: problems, progress, and prospects. Chin Med 11:37

Chokoe PK, Masoko P, Mokgotho MP, Howard RL, Mampuru LJ (2008) Does seasonal variation influence the phytochemical and antibacterial properties of *Carpobrotus edulis*? Afr J Biotechnol 7(22):4164–4171

Chukwuebuka E, Chinenye IJ (2015) Biological functions and anti-nutritional effects of phytochemicals in living system. J Pharm Biol Sci 10(2):10–19

Cock IE, Van Vuuren SF (2014) Anti-Proteus activity of some South African medicinal plants: their potential for the prevention of rheumatoid arthritis. Inflammopharmacology 22(1):23–36

Cravotto G, Boffa L, Genzini L, Garella D (2010) Phytotherapeutics: an evaluation of the potential of 1000 plants. J Clin Pharm Ther 35:11–48

Dar RA, Shahnawaz M, Qazi PH (2017) General overview of medicinal plants: a review. J Phytopharmacol 6(6):349–351

De Beer JJJ, Van Wyk B-E (2011) An ethnobotanical survey of the Agter–Hantam, Northern Cape Province, South Africa. S Afr J Bot 77(3):741–754

Denwick PM (2002) Natural products: a biosynthetic approach. Wiley, London

Doughari JH (2012) Phytochemicals: extraction methods, basic structures and mode of action as potential chemotherapeutic agents, phytochemicals. In: Rao V (ed) A global perspective of their role in nutrition and health. IntechOpen, New York. https://doi.org/10.5772/26052

Duncan AC, van Staden JAK (1999) Screening of Zulu medicinal plants for angiotensin converting enzyme (ACE) inhibitors. J Ethnopharmacol 68(1):63–70

Dyson A (1998) Discovering Indigenous healing plants of herb and fragrance gardens at Kirstenbosch National Botanical Garden. National Botanical Institute, Kirstenbosch, South Africa

Ebrahim M, Pool EJ (2010) The effect of *Tulbaghia violacea* extracts on testosterone secretion by testicular cell cultures. J Ethnopharmacol 132(1):359–361

Engels JMM (2003) Plant genetic resources management and conservation strategies: problems and progress. Acta Hortic 623:179–192

Fabricant DS, Farnsworth NR (2001) The value of plants used in traditional medicine for drug discovery. Environ Health Perspect 109(Suppl 1):69–75

Falleh H, Ksouri R, Medini F, Guyot S, Abdelly C, Magné C (2011) Antioxidant activity and phenolic composition of the medicinal and edible halophyte *Mesembryanthemum edule* L. Ind Crop Prod 34(1):1066–1071

Foster S (2010) From herbs to medicines: the Madagascar Periwinkle's impact on childhood leukemia: a serendipitous discovery for treatment. Alternat Complement Therapies 16(6):16609

Gallily R, Yekhtin Z, Hanuš LO (2015) Overcoming the bell-shaped dose-response of Cannabidiol by using cannabis extract enriched in Cannabidiol. Pharmacol Pharm 6:75–85

George S, Bhalero SV, Lidstone EA, Ahmad IS, Abbasi A, Cunningham BT, Watkin KL (2010) Cytotoxicity screening of Bangladesh medicinal plant extracts on pancreatic cancer cells. BMC Complement Altern Med 10(52):1–12

Hafsa J, Hamm KM, Cerf DL, Limem K, Majdoub H, Charfeddine B (2018) Characterization, antioxidant and antiglycation properties of polysaccharides extracted from the medicinal halophyte *Carpobrotus edulis* L. Int J Biol Macromol 107:833–842

Hawkins B (2008) Plants for life: medicinal plant conservation and botanic gardens. Botanic Gardens Conservation International, Richmond, UK

Jacobsen JV, Yamaguchi Y, Mann LK, Howard F, Bernhard RA (1968) An alkyl-cysteine sulfoxide lyase in *Tulbaghia violacea* and its relation to other alliinase-like enzymes. Phytochemistry 7(7):1099–1108

Joshi BC, Joshi RK (2014) The role of medicinal plants in livelihood improvement in Uttarakhand. Int J Herbal Med 1:55–58

Kabatende J (2005) Pharmacological evaluation of some central nervous system effects of Cotyledon orbiculata. MSc Thesis. University of the Western Cape, Bellville, South Africa

Kadam ST, Pawar AD (2020) Conservation of medicinal plants: a review. Int Ayurvedic Med J 8(7):3890–3895

Kamerman P (1926) The toxicity of the genus *Cotyledon* cause of the disease krimpsiekte. S Afr J Sci 23:185

Kindersley D (2008) RHS A-Z encyclopedia of garden plants, vol 1136. Dorling Kindersley Ltd, London, UK

Krstin S, Sobeh M, Braun M, Wink M (2018) *Tulbaghia violacea* and *Allium ursinum* extracts exhibit anti-parasitic and antimicrobial activities. Molecules 23(2):313

Kubec R, Krejčová P, Mansur L, García N (2013) Flavor precursors and sensory-active sulfur compounds in *Alliaceae* species native to South Africa and South America. J Agric Food Chem 60(1):1335–1342

Lakshman CD (2016) Bio-diversity and conservation of medicinal and aromatic plants. Adv Plants Agric Res 5(4):561–566

Li DZ, Pritchard HW (2009) The science and economics of *ex situ* plant conservation. Trends Plant Sci 14(11):614–621

Lim T (2013) Edible medicinal and non-medicinal plants. Springer, Amsterdam

Louw CAM, Regnier TJC, Korsten L (2002) Medicinal bulbous plants of South Africa and their traditional relevance in the control of infectious diseases. J Ethnopharmacol 82:147–154

Lyantagaye SSL, Rees DJG (2003) Screening *Tulbaghia violacea* extracts for the presence of apoptotic compounds. S Afr J Bot 69:256–257

Mackraj I, Ramesar S, Singh M, Govender T, Baijnath H, Singh R, Gathiram P (2008) The *in vivo* effects of *Tulbaghia violacea* on blood pressure in a salt-sensitive rat model. J Ethnopharmacol 117(2):263–269

Madike LN, Pillay M, Popat K (2019) *In vitro* cell adhesion, proliferation and differentiation of adipose derived stem cell on *Tulbaghia violacea* loaded polycaprolactone (PCL) nanofibers. J Biomater Tissue Engineering 9:1485–1498

Madike LN, Pillay M, Popat K (2020) Antithrombogenic properties of *Tulbaghia violacea* loaded polycaprolactone (PCL) nanofibers. J Bioact Compat Polym 35:1–15

Madike LN, Takaidza S, Pillay M (2017) Preliminary phytochemical screening of crude extracts from the leaves, stems and roots of *Tulbaghia violacea*. Int J Pharmacogn Phytochem Res 9: 1300–1308

Madike NL (2020) The effect of crude water extracts of *Tulbaghia violacea* on scaffolds for cardiovascular applications. PhD Thesis, Vaal University of Technology, Vanderbijlpark, South Africa

Maroyi A (2019) A review of botany, medicinal uses, phytochemistry and biological activities of *Cotyledon orbiculata*. J Pharm Sci Res 11(10):3491–3496

Martins A, Vasas A, Schelz ZS, Viveiros M, Molnar J, Hohmann J, Amaral L (2010) Constituents of *Carpobrotus edulis* inhibit P-glycoprotein of MDR1-transfected mouse lymphoma cells. Anticancer Res 30(3):829–835

Martins A, Vasas A, Viveiros M, Molnár J, Hohmann J, Amaral L (2011) Antibacterial properties of compounds isolated from *Carpobrotus edulis*. Int J Antimicrob Agents 37(5):438–444

Maunder M, Higgens S, Culham A (2001) The effectiveness of botanic garden collections in supporting plant conservation: a European case study. Biodivers Conserv 10:383–401

McGaw LJ, Jäger AK, van Staden J (2000) Antibacterial, anthelmintic and anti-amoebic activity in South African medicinal plants. J Ethnopharmacol 72(1–2):247–263

Meddeb E, Charni M, Ghazouani T, Cozzolino A, Fratianni F, Raboudi F, Nazzaro F, Fattouch S (2017) Biochemical and molecular study of *Carpobrotus edulis* bioactive properties and their effects on *Dugesia sicula* (Turbellaria, Tricladida) regeneration. Appl Biochem Biotechnol 182(3):1131–1143

Menon S, Pontius G Jr, Rose J, Khan ML, Bawa KS (2001) Identifying conservation-priority areas in the tropics: a land-use change modelling approach. Conserv Biol 15:501–512

Mestanza-Ramón C, Henkanaththegedara SM, Vásconez Duchicela P, Vargas Tierras Y, Sánchez Capa M, Constante Mejía D, Jimenez Gutierrez M, Charco Guamán M, Mestanza Ramón P (2020) *In-situ* and *ex-situ* biodiversity conservation in Ecuador: a review of policies. Actions Challenges Diversity 12(8):315

Molefe NI, Tsotetsi AM, Ashafa AOT, Thekisoe OMM (2013) *In vitro* anthelmintic activity of *Cotyledon orbiculata*, *Hermannia depressa* and *Nicotiana glauca* extracts against parasitic gastrointestinal nematodes of livestock. J Med Plants Res 7(9):536–542

Mulaudzi RB, Aremu AO, Rengasamy KRR, Adebayo SA, McGaw LJ, Amoo SO, Van Staden J, Du Plooy CP (2019) Antidiabetic, anti-inflammatory, anticholinesterase and cytotoxicity determination of two *Carpobrotus* species. S Afr J Bot 125:142–148

Ncube B, Ngunge VNP, Finnie J, van Staden J (2011) A comparative study of the antimicrobial and phytochemical properties between outdoor grown and micropropagated *Tulbaghia violacea* Harv. plants. J Ethnopharmacol 134:775–780

Nkuzinna OC, Menkiti MC, Onukwuli OD, Mbah GO, Okolo BI, Egbujor MC (2014) Application of factorial design of experiment for optimization of inhibition effect of acid extract of *Gnetum africanum* on copper corrosion. Nat Resour 5:299–307

Nöthlings U, Murphy SP, Wilkens LR, Boeing H, Schulze MB, Bueno-de-Mesquita HB, Michaud DS, Roddam A, Rohrmann S, Tjønneland A (2008) A food pattern that is predictive of flavonol intake and risk of pancreatic cancer. Am J Clin Nutr 88(6):1653–1662

Nteso L, Pretorius JC (2006) *Tulbaghia violacea* L. in vitro antimicrobial properties towards plant pathogens. Aust J Agric Res 57(5):511–516

Olorunnisola OS, Bradley G, Folayan AJ (2011) Antioxidant properties and cytotoxic evaluation of methanolic extract of dried and fresh rhizomes of *Tulbaghia violacea*. Afr J Pharm Pharmacol 5(22):2490–2497

Omoruyi BE, Afolayan AJ, Bradley G (2014) The inhibitory effect of *Mesembryanthemum edule* (L.) bolus essential oil on some pathogenic fungal isolates. BMC Complement Altern Med 14:168

Omoruyi BE, Bradley G, Afolayan AJ (2012) Antioxidant and phytochemical properties of *Carpobrotus edulis* (L.) bolus leaf used for the management of common infections in HIV/AIDS patients in Eastern Cape Province. BMC Complement Altern Med 12(1):1186

Ordway D, Hohmann J, Viveiros M, Viveiros A, Molnar J, Leandro C, Arroz MJ, Gracio MA, Amaral L (2003) *Carpobrotus edulis* methanol extract inhibits the MDR efflux pumps, enhances killing of phagocytosed *S. aureus* and promotes immune modulation. Phytother Res 17(5): 512–519

Park JS, Rho HS, Kim DH, Chang IS (2006) Enzymatic preparation of kaempferol from green tea seed and its antioxidant activity. J Agric Food Chem 54(8):2951–2956

Pimm SL, Russel GJ, Gittleman JL, Brooks TM (1995) The future of biodiversity. Science 269: 347–350

Raimondo D (2015) South Africa's strategy for plant conservation. South African National Biodiversity Institute and the botanical Society of South Africa, Pretoria, South Africa

Raina R, Chand R, Sharma YP (2011) Conservation strategies of some important medicinal plants. Int J Med Arom Plants 1:342–347

Rajjou L, Debeaujon I (2008) Seed longevity: survival and maintenance of high germination ability of dry seeds. C R Biol 331:796–805

Rasoanaivo P, Wright CW, Willcox ML, Gilbert B (2011) Whole plant extracts versus single compounds for the treatment of malaria: synergy and positive interactions. Malaria J 10:S4

Rivière S, Müller J (2018) Contribution of seed banks across Europe towards the 2020 global strategy for plant conservation targets, assessed through the ENSCONET database. Oryx 52(3): 464–470

Rocha MI, Rodrigues MJ, Pereira C, Pereira H, Da Silva MM, da Rosa NN, Nogueira JMF, Varela J, Barreira L, Custódio L (2017) Biochemical profile and *in vitro* neuroprotective properties of *Carpobrotus edulis* L., a medicinal and edible halophyte native to the coast of South Africa. S Afr J Bot 111:222–231

Roiloa SR, Retuerto R (2016) Effects of fragmentation and seawater submergence on photochemical efficiency and growth in the clonal invader *Carpobrotus edulis*. Flora 225:45–51

Roux W (2012) The antioxidant properties of the methanol extract of Cotyledon orbiculata L. var orbiculata (Haw.) DC. Leaves. MSc Thesis. North-West University, Potchefstroom, South Africa

Salmerón-Manzano E, Garrido-Cardenas JA, Manzano-Agugliaro F (2020) Worldwide research trends on medicinal plants. Int J Environ Res Public Health 17(10):3376

SANBI (2004) *Cotyledon orbiculata*|PlantZAfrica-SANBI pza.sanbi.org › cotyledon-orbiculata. Accessed 31 Mar 2021

Schoen DJ, Brown AHD (2001) The conservation of wild plant species in seed banks: attention to both taxonomic coverage and population biology will improve the role of seed banks as conservation tools. Bioscience 51(11):960–966

Semenya SS, Potgieter MJ, Erasmus LJC (2013) Indigenous plant species used by Bapedi healers to treat sexually transmitted infections: their distribution, harvesting, conservation and threats. S Afr J Bot 87:66–75

Sharrock S, Oldfield S, Wilson O (2014) Plant Conservation Report 2014: A Review of Progress in Implementation of the Global Strategy for Plant Conservation 2011–2020. Secretariat of the convention on biological diversity, Montréal, Canada, and Botanic Gardens Conservation International, Richmond, UK

Singh P, Singh C (1981) Chemical investigations of *Clerodendrum fragrans*. J Indian Chem Soc 58(6):626–627

Solecki RS (1975) Shanidar IV, a Neanderthal flower burial in northern Iraq. Science 190(4217): 880–881

Takaidza S (2019) Phytochemical analysis and biological activities of crude extracts from selected Tulbaghia species. PhD Thesis. Vaal University of Technology, Vanderbijlpark, South Africa

Takaidza S, Madan Kumar A, Ssemakalu CC, Natesh NS, Karanam G, Pillay M (2018b) Anticancer activity of crude acetone and water extracts of *Tulbaghia violacea* on human oral cancer cells. Asian Pac J Trop Biomed 8(2)

Takaidza S, Mtunzi FM, Pillay M (2018a) Analysis of the phytochemical contents and antioxidant activity of crude from Tulbaghia species. J Tradit Chin Med 39:272–279

Terblanche U (2021) The effect of *Carpobrotus edulis* and *Cotyledon orbiculata* on the migration, proliferation and differentiation of keratinocytes *in vitro*. PhD Vaal University of Technology, Vanderbijlpark, South Africa

UN Comtrade (2013) International trade statistics yearbook II. United Nations New York. https:// comtrade.un.org

Van der Watt E, Pretorius JC (2001) Purification and identification of active antibacterial components in *Carpobrotus edulis* L. J Ethnopharmacol 76(1):87–91
Van Wyk AS, Prinsloo G (2018) Medicinal plant harvesting, sustainability and cultivation in South Africa. Biol Conserv 227:335–342
Van Wyk BE, Gericke N (2007) People's plants. A guide to useful plants of Southern Africa. Briza Publications, Pretoria, South Africa, p 351
Van Wyk BE, Van Oudsthoorn B, Gericke N (2009) Medicinal plants of South Africa, 2nd edn. Briza Publications, Pretoria, South Africa
Veeresham C (2012) Natural products derived from plants as a source of drugs. J Adv Pharm Technol Res 3(4):200–201
Watson L, Dallwitz M (1992) Onwards. The families of flowering plants: descriptions, illustrations, identification, and information retrieval Version: 14th December 2000
Watt M, Breyer-Brandwijk MG (1962) The medicinal and poisonous plants of southern and Eastern Africa. E & S Livingstone Ltd., Edinburgh, Great Britain
WHO (2011) The world medicines situation 2011 traditional medicines: global situation, issues and challenges. WHO, Geneva, Switzerland
Wiersum KA, Dold AP, Husselman M, Cocks M (2006) Cultivation of medicinal plants as a tool for biodiversity conservation and poverty alleviation in the Amatola region, South Africa. In: Bogers RJ, Craker LE, Lange D (eds) Medicinal and aromatic plants. Springer, Amsterdam, The Netherlands, pp 43–57
Williams VL, Victor JE, Crouch NR (2013) Red listed medicinal plants of South Africa: status, trends, and assessment challenges. S Afr J Bot 86:23–35
Xaba VM (2016) Pharmacological screening of traditional medicinal plants used against skin ailments in the Free State, South Africa. MSc Thesis. University of the Free State, Bloemfontein, South Africa
Zhang MJ, Rong K, Cheng K (2012) Research and practice on biodiversity in situ conservation in China: progress and prospect. Biodivers Sci 20(5):551–558

Chloroplast DNA Haplotype Specific to Genetic Resources of *Dalbergia oliveri* in Thailand

16

Suchitra Changtragoon and Kittiya Singthong

Abstract

The chloroplast DNA in three regions, with a total length of 1812 bp, *trn*S-*trn*G (504 bp), *trn*V-*trn*M (702 bp), and *trn*LF-*trn*LR (606 bp), was sequenced in 15 *Dalbergia oliveri* populations of protected areas in Thailand. The results showed that 32 haplotypes were found. Out of 182 loci of variation, 37 were SNP loci and 145 indel loci. The total haplotype diversity (h) was 0.857 and nucleotide diversity (π) was 0.00637, indicating that there was a genetic diversity within the *D. oliveri* of 15 populations with high haplotype diversity (h) and low nucleotide diversity (π). Each haplotype appears in different proportions in their population. The most common haplotype is H25, which is found in 4 out of 15 populations, accounting for 27.03%, and the 29th haplotype was found only in one population. The haplotype of *D. oliveri* populations in the four regions was completely different. In addition, haplotypes that were specific to each population group were found, making it possible to distinguish between provinces and population groups, with the exception of the four northern populations, Lampang and Chiang Mai, which were yet unable to differentiate. The results of this study are expected to be used to identify future populations of illegally logged wood as well as in situ and ex situ gene conservation design of *D. oliveri* in Thailand.

Keywords

Dalbergia oliveri · Chloroplast DNA · Haplotype · Genetic resources · Thailand

S. Changtragoon (✉) · K. Singthong
Department of National Parks, Wildlife and Plant Conservation, Forest and Plant Conservation Research Office, 61 Phaholyothin, Bangkok, Thailand

S. Ramamoorthy et al. (eds.), *Plant Genetic Resources, Inventory, Collection and Conservation*, https://doi.org/10.1007/978-981-16-7699-4_16

16.1 Introduction

Dalbergia oliveri is distributed in deciduous, semi-deciduous, and non-deciduous forests at a moderate depth of 1000 m and is often found in humid areas near water bodies and on hills (Aerts et al. 2009). *D. oliveri* has been found in many provinces of Cambodia and in conservation areas in southern Vietnam (CTSP 2004). It has also been reported to be found in some provinces of Laos and Myanmar (Van Sam et al. 2004, USNH 2003). In Thailand, it is often found in the Northern, North Eastern, and Eastern regions. Since there are not many studies on *D. oliveri* genetic variation, Hartvig et al. (2017) studied genetic variation in *D. cochinchinensis* and use of *D. oliveri* as microsatellite in nucleus genome. Changtragoon et al. (2018) and Changtragoon et al. (2019) evaluated genetic variation in nucleus and chloroplast genome in *D. cochinchinensis*. In other timber species, Petit et al. (2002) studied chloroplast DNA variation in European white oak phylogeography and patterns of diversity. Deguilloux et al. (2003) checked geographical origin of oak wood by using molecular and statistic tools. In addition, Deguilloux et al. (2004) used DNA-based method to control oak wood geographic origin in the context of the cooperage industry. In other tropical timber species, Tnah et al. (2009) analyzed haplotypes from chloroplast DNA to trace the source of *Neobalanocarpus heimii* in Peninsular Malaysia. Ng et al. (2020) analyzed haplotypes from chloroplast DNA for geographic origin identification of *Intsia palembanica* in Peninsular Malaysia. Ng et al. (2017) generated a population identification database of *Shorea platyclados* for geographic origin identification and controlling illegal logging as well as to manage the sustainability of this valuable timber species in Malaysia. Ng et al. (2016) developed a timber tracking tool for *Gonystylus bancanus* in the context of illegal logging and chain-of-custody certification in Indonesia and Malaysia. Therefore, to prepare for forensic law enforcement, the genetic variation chloroplast genome is studied. Differentiation of the DNA haplotypes in different areas can be used as a preliminary genetic information for identifying the origin of confiscated wood, which will provide accurate and effective forensic evidence for genetic investigation as well as in situ and ex situ gene conservation design.

16.2 Materials and Methods

DNA was extracted from *D. oliveri* leaf and wood samples as shown in Table 16.1 using the Doyle and Doyle (1990) method, and then the DNA content (PCR) of the noncoding regions is increased from a polymer-containing primer. Three pairs of genome area of the intron and the interactive spacers of the chloroplast were *trn*S-*trn*G (Hamilton 1990), *trn*V-*trn*M (Demesure et al. 1995), and *trn*LF-*trn*LR (Kress et al. 2005, Viglietti et al. 2019). The universal primers were selected to increase the DNA content. The DNA concentration was 20–50 ng/μl volume of 1 μl, a primer solution, both forwarded primer and reversed primer, at a concentration of 10 pmol/μ l each volume of 1 μl, buffer with a concentration of ten times (10x reaction buffer: 100 mM Tris–HCL pH 8.3, 15 mM $MgCL_2$, 500 mM KCL), volume 2.5 μl, dNTP

Table 16.1 Investigated populations and location of *D. oliveri* in Thailand

Populations	Regions	Name of populations	Samples/ populations
1. Ramkhamhaeng National Park, Sukhothai	Central	RKH	10
2. Namtok Sam Lan National Park, Saraburi		SSL	10
3. Huai Sala Wildlife Sanctuary, Sisaket	North eastern	HSL	10
4. Pha Taem National Park, Ubon Ratchathani		PT	10
5. Phu Chong Na Yoi National Park, Ubon Ratchathani		PJ	10
6. Phu Lan Kha National Park, Chaiyaphum		PLC	10
7. Na Yung-Nam Som National Park, Udon Thani		NYN	10
8. Phu Hin Chom Tat-Phu Phrabat National Park, Udon Thani		PUD	10
9. Phu Sa Dok Bua National Park, Mukdahan		PSD	10
10. Tham Pha Thai National Park, Lampang	Northern	TPT	10
11. Mae Wang National Park, Chiang Mai		MAV	10
12. Si Lanna National Park, Mae Taeng Chiang Mai		LNM	10
13. Si Lanna National Park, Chiang Dao Chiang Mai		LND	10
14. Namtok Khlong Kaeo National Park, Trat	Eastern	KTR	10
15. Namtok Phlio National Park, Chanthaburi		NTP	8
Total			148

(dNTP mix: dATP, dCTP, dGTP, and dTTP, each at a concentration of 2 mM), volume 1.5 μl and 1 unit of Taq DNA polymerase/μl was adjusted with 25 μl of distilled water and mixed together and increased the DNA content by PCR. The DNA fragment size was examined using agarose gel electrophoresis and the gel was dyed with ethidium bromide. The DNA size from PCR product was compared with standard DNA (1 kb DNA ladder). The nucleotides from those PCR products were sequenced by ABI 3500 Genetic Analyzer (Applied) Biosystem, USA. The combined data of the cpDNA sequences of this study was analyzed. The obtained sequences were aligned and edited by using Bioedit 7.0.5.3 software (Hall 1999). The cpDNA haplotype diversity (h) and nucleotide diversity (π) were analyzed and defined from the obtained sequences by using DnaSP 5.10.01 (Librado and Rozas 2009) software.

16.3 Results and Discussion

The result of nucleotide sequence analysis in three chloroplast regions from 148 samples of 15 *D. oliveri* populations showed that there were 1812 bp in total length: *trn*S-*trn*G (504 bp), *trn*V-*trn*M (702 bp), and *trn*LF-*trn*LR (606 bp). It was found out that there were 182 positions of variation, 37 of which were SNP positions and 145 indel positions. The highest variation was found in *trn*S-*trn*G as shown in Figs. 16.1 and 16.2. For all SNP loci in those three chloroplast regions, there were 4 transitions and 33 transversions. The proportion of AT nucleotide sequences was higher than CG which is in line with the evolution of noncoding chloroplast DNA as mentioned by Kelchner (2000).

The results showed that haplotype 1 (H1) and haplotype 2 (H2) were found in Ramkhamhaeng National Park, Sukhothai Province; haplotype 3 (H3), haplotype 4 (H4), haplotype 5 (H5), and haplotype 6 (H6) were found in Namtok Sam Lan National Park, Saraburi Province; haplotype 7 (H7), haplotype 8 (H8), haplotype 9 (H9), and haplotype 10 (H10) were found in Huai Sala Wildlife Sanctuary, Sisaket Province; haplotype 11 (H11) was found in Pha Taem National Park, Ubon Ratchathani Province; haplotype 10 (H10) and haplotype 12 (H12) were found in Phu Chong Na Yoi National Park, Ubon Ratchathani Province; haplotype 13 (H13) was found in Phu Lan Kha National Park, Chaiyaphum Province; haplotype 14 (H14), haplotype 15 (H15), haplotype 16 (H16), haplotype 17 (H17), haplotype 18 (H18), haplotype 19 (H19), and haplotype 20 (H20) were found in Na Yung-Nam Som National Park, Udon Thani Province; haplotype 21 (H21), haplotype 22 (H22), and haplotype 23 (H23) were found in Phu Hin Chom Tat-Phu Phrabat National Park, Udon Thani Province; haplotype 23 (H23) and haplotype 24 (H24) were found in Phu Sa Dok Bua National Park, Mukdahan Province; haplotype 26 (H26), haplotype 27 (H27), and haplotype 23 (H28) were found in Namtok Khlong Kaeo National Park, Trat Province; haplotype 29 (H29), haplotype 30 (H30), haplotype 31 (H31), and haplotype 32 (H32) were found in Namtok Phlio National Park, Chanthaburi Province; and haplotype 25 (H25) was found in the northern regions: Tham Pha Thai National Park, Lampang Province; Mae Wang National Park, Chiang Mai Province; and Si Lanna National Park (Mae Taeng and Chiang Dao), Chiang Mai Province as shown in Table 16.2 and Fig. 16.3.

This study has shown that there was a total of 32 chloroplast DNA haplotypes in 15 *D. oliveri* populations of 13 national parks and one wildlife sanctuary in four regions (Central, North Eastern, Northern, and Eastern regions) of Thailand as shown in Table 16.2 and Fig. 16.3.

The results of the genetic diversity analysis by combining the nucleotide sequences of the *trn*S-*trn*G, *trn*V-*trn*M, and *trn*LF-*trn*LR segments showed that the h-values and the π-values in each population were between 0–0.511 and 0–0.00046, respectively, and all samples were analyzed with values of 0.857 and 0.00637, respectively. The population with the highest h- and π-values was found in Huai Sala Wildlife Sanctuary, Sisaket Province; Na Yung-Nam Som National Park, Udon Thani Province; and Phu Sa Bua National Park, Mukdahan Province, followed by Ramkhamhaeng National Park, Sukhothai Province, and Namtok Khlong Kaeo National Park, Trat Province, which were in line with the variation of haplotypes found in those populations. The populations with h- and π-values as 0.000, respectively, were found in Pha Taem National Park, Ubon Ratchathani Province; Phu Chong Na Yoi National Park, Ubon Ratchathani Province; Phu Laen Kha National Park, Chaiyaphum Province; Tham Pha Thai National Park, Lampang Province; Mae Wang National Park, Chiang Mai Province; and Sri Lanna National Park,

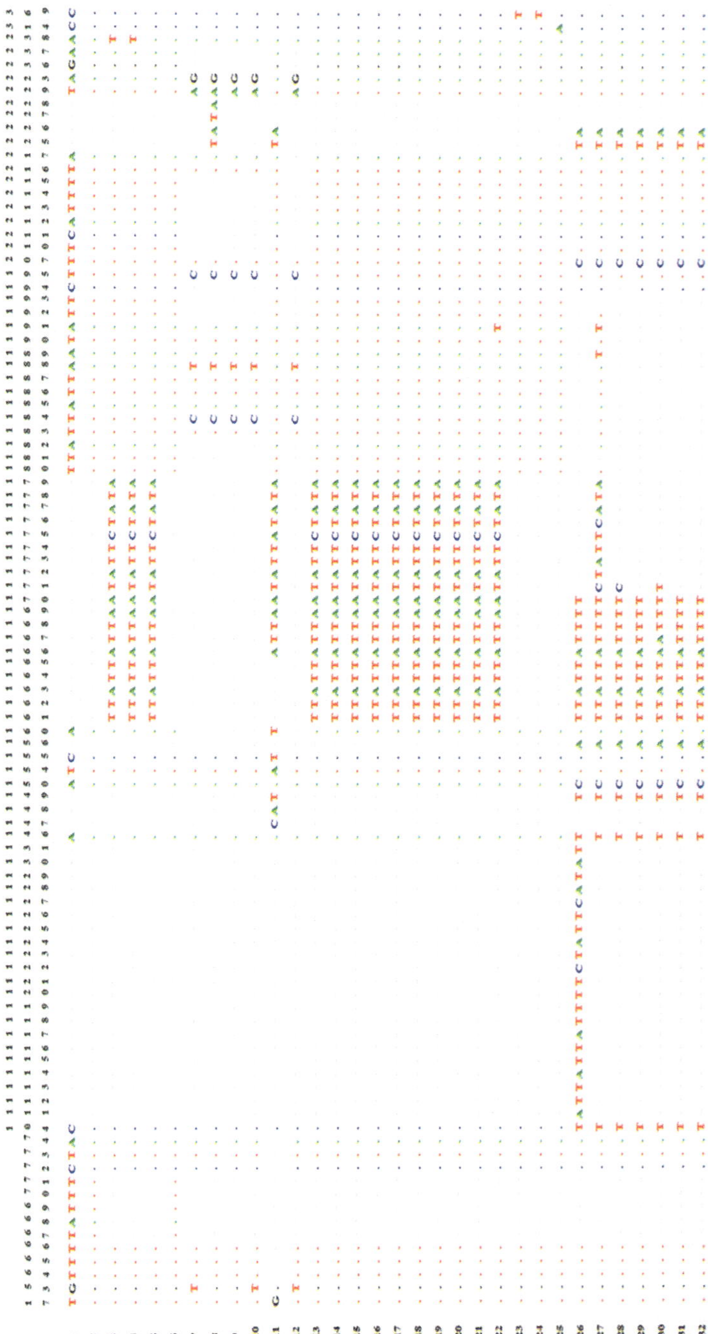

Fig. 16.1 SNP polymorphic sites based on the sequences of *trnS-trnG* in *Dalbergia oliveri*

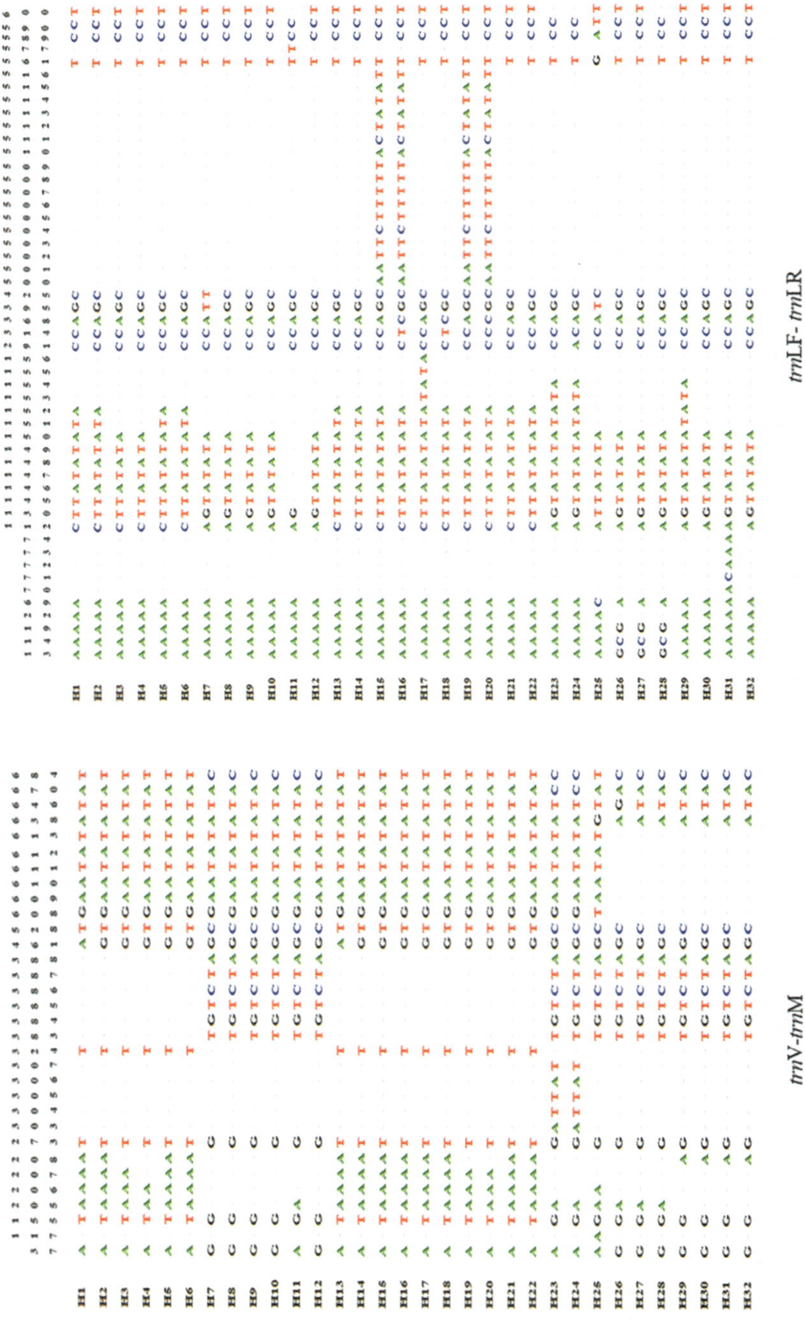

Fig. 16.2 SNP polymorphic sites based on the sequences of *trn*V-*trn*M and *trn*LF-*trn*LR in *Dalbergia oliveri*

Table 16.2 Chloroplast DNA haplotypes based on the sequences of *trn*S-*trn*G, *trn*V-trnM, and *trn*LF-*trn*LR in *Dalbergia oliveri* in Thailand

Haplotype	Central		North Eastern							Northern				Eastern		Total
	RKH	SSL	HSL	OliPT	OliPJ	PLC	NYN	PUD	PSD	TPT	MAV	LNM	LND	KTR	NTP	
H1	7															7
H2	3															3
H3		1														1
H4		1														1
H5		1														1
H6		7														7
H7			1													1
H8			1													1
H9			1													1
H10			7		5											12
H11				10												10
H12					5											5
H13						10										10
H14							3									3
H15							1									1
H16							1									1
H17							1									1
H18							1									1
H19							2									2
H20							1									1
H21								8								8
H22								1								1
H23								1	8							9
H24									2							2
H25										10	10	10	10			40

(continued)

Table 16.2 (continued)

Haplotype	Central		North Eastern							Northern				Eastern		Total
	RKH	SSL	HSL	OliPT	OliPJ	PLC	NYN	PUD	PSD	TPT	MAV	LNM	LND	KTR	NTP	
H26														7		7
H27														1		1
H28														2		2
H29															2	2
H30															1	1
H31															1	1
H32															4	4
Total	10	10	10	10	10	10	10	10	10	10	10	10	10	10	8	148

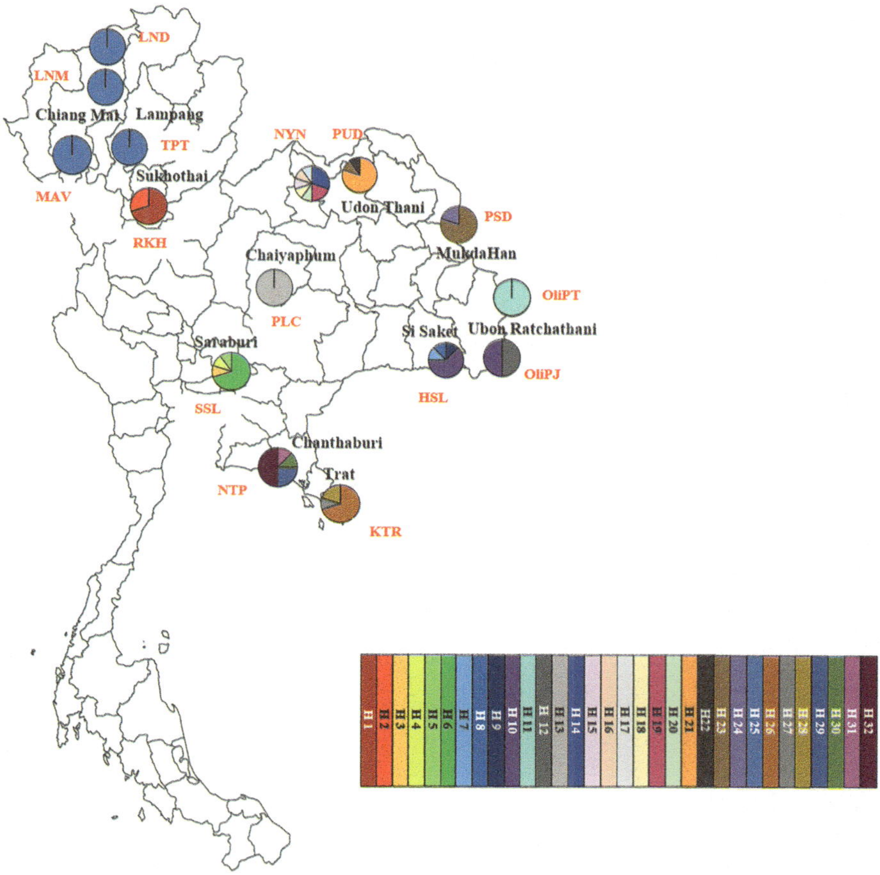

Fig. 16.3 Geographic distribution of cpDNA haplotypes in *Dalbergia oliveri* populations in Thailand

Chiang Mai Province, which were in line with only one haplotype found in those populations as shown in Table 16.3.

According to this study, it is suggested that *D. oliveri* populations in Thailand had unique chloroplast DNA haplotype specific to each population in each region except in the northern part of Thailand.

In contrast to previous study on chloroplast DNA haplotype diversity of *D. cochinchinensis* in Thailand that it could be distinguished only at regional level not at provincial level (Changtragoon et al. 2019).

The outcome of this study suggested that one population of *D. oliveri* in the northern part of Thailand might be selected as in situ gene conservation. However, in other regions, each population in each region is valued to be established as in situ gene conservation, since there is haplotype specific to each population. Regarding ex

Table 16.3 Molecular diversity indexes of *Dalbergia oliveri*

Populations	Haplotype (nh)	Haplotype diversity (h)	Nucleotide diversity (π)
RKH	2	0.467	0.00026
SSL	2	0.356	0.00020
HSL	3	0.511	0.00043
OliPT	1	0.000	0.00000
OliPJ	1	0.000	0.00000
PLC	1	0.000	0.00000
NYN	3	0.511	0.00046
PUD	2	0.356	0.00020
PSD	3	0.511	0.00031
TPT	1	0.000	0.00000
MAV	1	0.000	0.00000
LNM	1	0.000	0.00000
LND	1	0.000	0.00000
KTR	2	0.467	0.00027
NTP	2	0.250	0.00014
Total	32	0.857	0.00637

situ gene conservation and plantation, seed collection for plantation should be well planned and designed and should not be mixed as each population has unique haplotype.

16.4 Conclusion

A total of 148 samples of *D. oliveri* were studied for their diversity and genetic structure in 15 populations in four regions, examining differences in nucleotide sequences from noncoding regions of the chloroplast genome. Three sections with a total length of 1812 bp were *trn*S-*trn*G (504 bp), *trn*V-*trn*M (702 bp), and *trn*LF-*trn*LR (606 bp). A total of 182 variable positions were found, 37 of which were SNP positions and 145 indel positions.

The results revealed a total of 32 haplotypes of DNA, with each haplotype appearing in different proportions. The most common haplotype was H25, which was found in 4 of the 15 populations, representing 27.03%. The 29th haplotype was found only in one population. The number of haplotypes found in each population ranged from 1 to 7 haplotypes. Genetic diversity with a haplotype diversity (h) of 0.857 and a nucleotide diversity (π) of 0.00637 showed that haplotype diversity (π) was 0.00637. There was genetic diversity within 15 populations of *D. oliveri* with high haplotype diversity (h) and low nucleotide diversity (π).

The haplotypes of *D. oliveri* populations in four regions are completely different. Moreover, the specific haplotypes were found in each population group, allowing populations' provinces and regions to be distinguished, except for four populations in the northern region, namely Lampang Province and Chiang Mai Province.

Therefore, haplotypes in other locations and populations should be further studied in order to distinguish the populations in each province separately. The results of this study are expected to be used to identify illegally logged wood as well as in situ *and* ex situ *gene conservation design of D. oliveri* in Thailand. These data can be used to guide the development of DNA markers for future identification of the origin of *D. oliveri* genetic resources.

Acknowledgements The Forest and Plant Conservation Research Office, Department of National Park, Wildlife and Plant Conservation, is highly appreciated on supporting the budget for this research project. Mrs. Panida Roongrattanakul and the head of the mentioned national parks and wildlife sanctuary are appreciated for helping on collecting some samples in the national parks and wildlife sanctuary. Special thanks are given to Ms. Matiya Changjalern and Mr. Sanguan Wongsa for helping on the work in the DNA laboratory.

References

Aerts R, Volkaert H, Roongruangsree N, Roongruangsree UT, Swennen R, Muys B (2009) Site requirements of the endangered rosewood *Dalbergia oliveri* in a tropical deciduous forest in northern Thailand. For Ecol Manag 259:117–123

Changtragoon S, Roongrattanakul P, Singthong K (2019) Development of DNA markers specific to *Dalbergia cochinchinensis* populations in Thailand. In: Forest conference 2018 (FORCONS 2019), the Thai elephant conservation center Lampang, Thailand, 19–21 August 2018 pp 349–350

Changtragoon S, Roongrattanakul P, Thongboonsong N (2018) Evaluation of genetic diversity of Siamese rosewood (*Dalbergia cochinchinensis* Pierre) in Thailand using microsatellite markers. In: Forest conference 2018 (FORCONS 2018), the Kasetsart University, Bang Khen campus, Bangkok, Thailand, 22–24 august 2018 pp 129–130

CTSP (2004) Cambodian tree species. Monographs. Cambodia: Cambodia tree seed project, FA, and DANIDA, Denmark

Deguilloux MF, Pemonge MH, Bertel L, Kremer A, Petit RJ (2003) Checking the geographical origin of oak wood: molecular and statistical tools. Mol Ecol 12:1629–1636

Deguilloux MF, Pemonge MH, Petit RJ (2004) DNA-based control of oak wood geographic origin in the context of the cooperage industry. Ann For Sci 61:97–104

Demesure B, Sodzi N, Petit RJ (1995) A set of universal primers for amplification of polymorphic non-coding regions of mitochondrial and chloroplast DNA in plants. Mol Ecol 4:129–131

Doyle JJ, Doyle JL (1990) Isolation of plant DNA from fresh tissue. Focus 12:13–15

Hall TA (1999) BioEdit: a user-friendly biological sequence alignment editor and analysis program for Windows 95/98/NT. Nucleic Acids Symp Ser 41:95–98

Hamilton MB (1990) Four primer pairs for the amplification of chloroplast intergenic regions with intraspecific variation. Mol Ecol 8(3):521–523

Hartvig I, So T, Changtragoon S, Tran HT, Bouamanivong S, Theilade I, Kjaer ED, Nielsen LR (2017) Population genetic structure of the endemic rosewoods *Dalbergia cochinchinensis* and *D. oliveri* at a regional scale reflects the Indochinese landscape and life-history traits. Ecol Evol 8(1):1–16

Kelchner SA (2000) The evolution of non-coding chloroplast DNA and its application in plant systematics. Ann Mo Bot Gard 87:482–498

Kress WJ, Wurdack KJ, Zimmer EA, Weigt LA, Janzen DH (2005) Use of DNA barcodes to identify flowering plants. Proc Natl Acad Sci U S A 102(23):8369–8374

Librado P, Rozas J (2009) DnaSP v5: a software for comprehensive analysis of DNA polymorphism data. Bioinformatics 25:1451–1452

Ng CH, Lee SL, Tnah LH, Ng KKS, Lee CT, Diway B, Khoo E (2017) Geographic origin and individual assignment of *Shorea platyclados* (Dipterocarpaceae) for forensic identification. PLoS One 12(4):1–18

Ng CH, Ng KKS, Lee SL, Tnah LH, Lee CT, Zakaria N-F (2020) A geographical traceability system for Merbau (*Intsia palembanica* Miq.), an important timber species from peninsular Malaysia. Forensic Sci Int Genet 44:102188

Ng KKS, Lee SL, Tnah LH, Nurul-Farhanah Z, Ng CH, Lee CT, Tani N, Diway B, Lai PS, Khoo E (2016) Forensic timber identification: a case study of a CITES listed species, *Gonystylus bancanus* (Thymelaeaceae). Forensic Sci Int Genet 23:197–209

Petit RJ, Csaikl UM, Bordács S, Burg K, Coart E, Cottrell J, Van Dam B, Deans JD, Dumolin-Lapègue S, Fineschi S, Finkeldey R, Gillies A, Glaz I, Goicoechea PG, Jensen JS, König AO, Lowe AJ, Madsen SF, Mátyás G, Munro RC, Olalde M, Pemonge M-H, Popescu F, Slade D, Tabbener H, Taurchini D, de Vries SGM, Ziegenhagen B, Kremer A (2002) Chloroplast DNA variation in European white oaks phylogeography and patterns of diversity based on data from over 2600 populations. For Ecol Manag 156:5–26

Tnah LH, Lee SL, Ng KKS, Tani N, Bhassu S, Othman RY (2009) Geographical traceability of an important tropical timber (*Neobalanocarpus heimii*) inferred from chloroplast DNA. For Ecol Manag 258:1918–1923

USNH (2003) A checklist of the trees, shrubs, herbs, and climbers of Myanmar. Contributions from the United States National Herbarium. http://botany.si.edu/myanmar/checklistNames.cfm

Van Sam H, Nanthavong K, Kessler PJA (2004) Trees of Laos and Vietnam: a field guide to 100 economically or ecologically important species. Blumea 49:201–349

Viglietti G, Galla G, Porceddu A, Barcaccia G, Curk F, Luro F, Scarpa GM (2019) Karyological analysis and DNA barcoding of Pompia citron: a first step toward the identification of its relatives. Plan Theory 8(83):1–16

Status, Conservation, and Sustainability on Medicinal Plant Resources of India

17

Kanagam Nachiappan, Nagaraj Nallakaruppan, Mariadoss Alphonse, Manoj Sekaran, Chandra Veluchamy, Siva Ramamoorthy, Kalaivani Thaigarajan, and Rajasekaran Chandrasekaran

Abstract

India is one among the top eight mega diversity countries in the globe enriched with bioresources, and has four biodiversity hotspots. India is a biologically diverse nation varied with climatic conditions, topography, flora, and fauna. Around 70% of medicinal and aromatic plants are found in the forest area, while the remaining 30% are present in the non-forest area or cultivable lands. Using wild plant varieties to cure ailments is not new to the world. Chinese have been using about 5000 plant species as drugs for the past 4000 years. In the Indian traditional system of medicine, about 8000 plants were used, whereas the local health workers or practitioners only depend on the locally available medicinal plants. In South Asia, around 1.4 billion people are unaware of modern medications and they are also dependent on native species for traditional medicines. Due to high phytomedicinal market demand, many national and international companies are involved in the biopiracy of medicinal plant resources and their indigenous knowledge creating a serious threat to the biological resources. On proper understanding and implementation of the traditional ethnobotanical knowledge from the local inhabitants by the biotech, pharma, and agri sectors, it is possible to conserve the threatened and endangered

K. Nachiappan
Department of Biotechnology, School of Bio Sciences and Technology, Vellore Institute of Technology, Vellore, Tamilnadu, India

Department of Biotechnology, Sri Venkateswara College of Engineering, Sriperumbudur, Kanchipuram, Tamilnadu, India

N. Nallakaruppan · M. Alphonse · M. Sekaran · C. Veluchamy · S. Ramamoorthy · K. Thaigarajan · R. Chandrasekaran (✉)
Department of Biotechnology, School of Bio Sciences and Technology, Vellore Institute of Technology, Vellore, Tamilnadu, India

S. Ramamoorthy et al. (eds.), *Plant Genetic Resources, Inventory, Collection and Conservation*, https://doi.org/10.1007/978-981-16-7699-4_17

species. As an alternative approach, it is advisable to switch over to large-scale cultivation of unexplored plants using low-cost biotechnological techniques, thereby creating self-employment opportunities and economic upliftment to local tribes. Alternative medicinal plants with similar healing properties act as a boon to the pharma industry and indirectly lead to conservation of threatened species. India also must reach for such agreements or at least create its own regulations to conserve the diverse biological resources with high market value, thereby benefitting the tribal population.

Keywords

Threatened species · MAPs · Unexplored · Alternative medicine · Conservation · Sustainable ecosystem

17.1 Introduction

The World Health Organization (WHO) declared that 80% of the total global populace rely on traditional medicines for their major health issues (Bannerman et al. 1983). Currently, the wild flora of medicinal and aromatic plants (MAPs) are infusing excess pressure on the plant resources (Nauityal et al. 2015). There is a vast international market for MAPs in drug development. With 40% of global diversity, India and China contribute majorly for MAPs cultivation and product export catering the global healthcare (Rajasekaran 2000). However, this increased market demand for MAPs simultaneously leads to overexploitation intentionally or un-intentionally, resulting in habitat destruction in the natural pockets. Nearly 150 wild species are at the threshold of extinction (Nair and Shastry 1990; Bhadula and Purohit 1994). There are still more unexplored MAPs, but used on a small scale for domestic use or treatment among the local tribes. So, collection, identification, documentation, exploration, and indigenous practices of MAPs cultivation may contribute more to the modern healthcare system and at the same time help to protect the overexploited threatened species (Polunin and Stainton 1984).

India has around 6 lakh licensed traditional medicinal practitioners (Ayurveda, Unani, Siddha, homeopathy, and allopathy), who depend on herbal drugs for their treatment either directly or indirectly. The awareness and demand of herbal plants not only contribute to the growth of international trade but also act detrimental to the natural population of the bioresources. At this juncture, it is mandatory to revive the socioeconomic status of local tribes and non-tribes and also conserve the diversity on the other hand. So different cultivation strategies and collection methods must be developed in order to conserve the bioresources. Ethnobotanical knowledge, research studies, and surveys often document new varieties of medicinal plants with indigenous knowledge of tribes that are yet to be explored. Focusing on these unexplored resources, we can conserve the threatened/endangered species and lead to discovering novel pharmacophore with therapeutic potential against new target drugs (Chakraborty et al. 2012).

17.2 Significant Biodiversity of India

India has been recognized as the home of the treasury of plant genetic resources due to its diverse climatic conditions and terrain. Therefore, India is considered as one among the twelve mega diversity countries in the world (Vijayalatha 2004). India's numerous biogeographical regions, climate, and altitude changes are linked to the country's vast ecological environment, resulting in the growth of colorful flora with a unique array of medicinal plants. India is also an important source of medicinal raw materials used in traditional medicine preparation and pharmaceutical industries (Chandra 2016). According to the biogeographical regions, Indian sub continent is divided into "seven" zone-wise distribution: Western Himalayas, Eastern Himalayas, North-east India, Gangetic plains, semiarid regions, Western Ghats, and Eastern Ghats (Table 17.1) (Bishop and Chatterjee 2019). Each of the above

Table 17.1 Zone-wise topographical and climatic provenance of India

Zone	Area covered	Topography	Climate	Reference
Zone 1—Western Himalayas	Jammu and Kashmir to Himachal Pradesh	Soaring heights, steep-sided jagged peaks, valley, and alpine glaciers	Subtropical highland climate	Bishop and Chatterjee (2021)
Zone 2—Eastern Himalayas	West Bengal, Sikkim, and Arunachal Pradesh	Tropical rainforest	Tropical montane ecosystem	Bishop and Chatterjee (2021)
Zone 3—North-east India	Arunachal Pradesh, Assam, Manipur, Meghalaya, Mizoram, Nagaland, Sikkim, and Tripura	Plains, plateaus, mountains, and valleys	Humid sub-tropical climate	Chakraborty et al. (2012)
Zone 4—Gangetic plains	Uttar Pradesh and Bihar	Foothills	Sub-humid to humid climate	Saini (2008)
Zone 5—Semi-arid regions	Rajasthan, the Punjab, and North Gujarat	Desert landscape	Semi-arid climate	
Zone 6—Western Ghats	Gujarat, Tamil Nadu, Maharashtra, Goa, Karnataka, and Kerala	Mountainous faulted and eroded edge of the Deccan Plateau	Humid and tropical	https://indianestates.co.in/climatewesternghats/#:~:text=The%20climate%20is%20humid%20and,C%20(60%20%C2%B0F)
Zone 7—Eastern Ghats	West Bengal, Odisha, Andhra Pradesh, and Tamil Nadu	Uneven broken hills	Tropical monsoon climate	http://www.ilearncana.com/details/Eastern-Ghats/701

zone-wise distributions exhibits different range of variations in altitude and topography that regulates the occurrence and distribution of medicinal plant biodiversity (Shankar and Rawat 2013).

17.3 Promising Medicinal Plants Among Ethnic Communities

Indian tribal medicine, in particularly traditional medicinal systems such as Ayurveda, Siddha, and Unani, uses a wide range of plant products to prepare curative, protective, and preventive medicines. In India, there are 550 communities comprising 227 ethnic groupings based on the classifications provided by anthropologists on a linguistic basis (Pushpangadan et al. 2014). Based on the ethnic and sociocultural affinities, the tribal communities living in different regions use various formulations of medicinal plants for curing diseases (Table 17.2).

17.3.1 Utilization of Medicinal Plants Among Traditional Medicinal Systems

17.3.1.1 Ayurveda

Ayurveda is one of the oldest traditional medicine system in India. The word "Ayurveda" is derived from Sanskrit from the word Ayur (life) and Veda (knowledge), which means balancing mind, body, and spirit equally (Parasuraman et al. 2014). Theories of Ayurveda are a pseudoscientific report that claims to have healing knowledge derived spiritually from the divine power of healing known as the "Science of Life and Longevity" in Hinduism (Jaiswal et al. 2016). Nowadays, this traditional system of medicine has become an alternative to conventional medicine. In Ayurveda, certain principles govern medication activity, such as taste (Rasa), characteristics (Guna), potency (Verya), post-digestion, metabolism effect (Vipaka), and synergistic activities (Prabhava) (Nair et al. 2020a, b). According to National Medicinal Plants Board (NMPB)-Government of India, more than 18,000 higher plants are grown throughout the country's 15 agroclimatic zones, where 8000 species are identified to be medicinally beneficial (Jaiswal et al. 2016). Astavarga is a group of eight rare and threatened species that form the essential ingredients in classical formulations in ayurvedic medicine. Kakoli, Jeevak, Kshirkakoli, Mahameda, Meda, Ridhhi, Rishbhak, and Vridhii are the most notable members of this category, which are extensively utilized in Himalayan regions. Around 560 plants in this group are currently enlisted under the threatened and endangered category (Singh et al. 2018).

17.3.1.2 Siddha

Siddha is one of the traditional system of medicine followed even before 2000 years by ancient Tamizhs/Dravidians inhabiting the southern peninsular of India. The name Siddha means entrenched truth and the person practicing this are called siddhars. They formulated the medicine by following yoga and nadi pariksha (pulse) diagnosis. They do not fit into any race, religion, and faith (Thas 2008).

Table 17.2 Zone-wise distribution pattern of threatened medicinal plants, tribal formulations, and its applications

Zone	Tribes	Botanical name	Vernacular name	Formulation	Disease treated	Reference
Zone 1	Bhotia Gaddi Gujjars Jaunsari Kanaura	*Aconitum naviculare*	Aconite	Powder	Cold, fever, and jaundice	Shyaula (2011)
		Allium wallichii	Doona	Decoction	Cholera and dysentery	https://www.flowersofindia.net/catalog/slides/Himalayan%20Onion.html
		Adhatoda zeylanica	Adusa	Decoction	Respiratory disorders	Tandel et al. (2018)
		Colchicum luteum	Hirantutiya	Paste	Rheumatic swelling	https://www.flowersofindia.net/catalog/slides/Yellow%20Colchicum.html
		Orchis latifolia	Salep	Powder	Dysentery and diarrhea	https://www.bimbima.com/ayurveda/orchis-latifolia-medicinal-uses/217/
Zone 2	Adi, Anal Angami Apatani Baugni Ho, Khond Saora	*Chlorophytum arundinaceum*	Safed musli	Tonic	Diabetes	Kaushik (2005)
		Coptisteeta sp.	Mamira	Powder	Cough and cold	http://www.ccras.nic.in/content/mamira-coptis-teeta-wall
		Gentiana kurroo	Karu	Tonic	Digestive disorders	Skinder et al. (2017)
		Nardostachys jatamansi	Nard	Powder	Epilepsy and hysteria	Pandey et al. (2013)
		Podophyllum hexandrum	Ban kakri	Powder	Cold and bacterial infection	Qazi et al. (2011)
Zone 3	Bodo Chakama Deori Kachari Miri, Mizo Naga, Nissi Santhals	*Alpinia galangal*	Kulanjan	Powder	Flatulence, dyspepsia, and vomiting	Chudiwal et al. (2010)
		Aquilaria agallocha	Agar	Oil and powder	Headache and arthritis	Janey and Mujahid Mohd Badr (2015)
		Hydnocarpus kurzii	Chaulmoogra	Oil	Leprosy	Sahoo et al. (2014)

(continued)

Table 17.2 (continued)

Zone	Tribes	Botanical name	Vernacular name	Formulation	Disease treated	Reference
		Mucuna pruriens	Kapikacchu	Powder	Male infertility and nervous disorders	Lampariello et al. (2012)
		Rauvolfia serpentine	Sarpagandha	Decoction	Snake and insect bites	Lobay (2015)
Zone 4	Agaria, Asur, Bagua, Bedia, Gadaba Jaunsari Tharu	*Aegle marmelos*	Bael	Powder	Cancer	Rahman and Parvin (2014)
		Crataeva nurvala	Varuna	Decoction and powder	Kidney stone	Rahman and Parvin (2014)
		Pluchea lanceolata	Rasna	Decoction and oil	Rheumatic and muscular pains	Bopana and Saxena (2008)
		Psoralea corylifolia	Babachi	Oil	Leukoderma and skin rashes	Khushboo et al. (2010)
Zone 5	Bhil Choudhri Dhodia Garasisa Kol, Mina	*Balanites aegyptiaca*	Ingudi	Decoction and powder	Skin diseases	Sanjeev and Nayan (2016)
		Boswellia serrate	Sallaki	Oil	Chronic inflammation	Siddiqui (2011)
		Commiphora wightii	Guggul	Gum resin	Arthritis	Sarup et al. (2015)
		Diospyros melanoxylon	Kendu	Powder	Urinary, skin and blood diseases	Hmar et al. (2017)
Zone 6	Irular Kadukurumba Kondareddy Malayali Marati, Naikda Naikpod Todas and Yeruvas	*Eranthemum roseum*	Kaaduneeli	Juice	Diabetes	Kamble et al. (2010)
		Clerodendrum infortunatum	Thaggi	Decoction	Fever	Korpenwar (2012)
		Syzygium caryophyllatum	Kunneralehannu	Decoction	Diabetes	Devi Prasad and Shyma (2013)
		Bridelia stipularis	Akshatheballi	Powder and decoction	Fever and cough	Devi Prasad et al. (2014)

Zone 7	Bhil, Bhumji Gond, Koya Kulas, Munda Oran, Saora Sugali, Yenadi Yeni	*Acalypha alnifolia*	Chinnikeerai	Paste	Dysentery	Kottaimuthu and Vasudevan (2014)
		Andrographis ovate	Periyanangai	Paste and juice	Diarrhea	
		Curcuma pseudomontana	AdaviPasupu	Powder	Antiseptic	Pattanaik et al. (2009)
		Entada pursaetha	Gillacheetu	Juice	Ulcers	
		Hildegardia populifolia	Galibudda	Powder	Snake bite	

Siddha medicine is based on five elements (Mann—soil, Thee—fire, Neer—water, Akayam—sky, and Kaatru—air) and six tastes (sweet, sour, astringent, pungent, bitter, and salt) and practiced by the siddhars in the concept of "Unavae Marunthu, Marunthae Unavu"—food is medicine, medicine is food (Malayappan 2017). The siddhars divide sickness into three categories: Vata, Pitta, and Kapha that refer to air, fire, and water, respectively. According to these theories, the misbalance of these three components will cause various diseases which can be treated mainly with plants and their derivatives (https://www.britannica.com/science/Siddha-medicine). Due to overexploitation of medicinally valuable plants, many species were getting endangered, according to the IUCN. Around 15,000 flowering plant species are endangered to extinction due to habitat destruction and overexploitation, among the 50,000–80,000 flowering plant species utilized for therapeutic reasons globally (Pimm et al. 1995). Some medicinal plants in Siddha like *Aconitum heterophyllum, Aquilaria malaccensis, Adhatoda beddomei,* and *Nardostachys jatamansi* are critically endangered species due to continuous exploitation from the wild (Gopi et al. 2018).

17.3.1.3 Unani

Unani is a historic alternative method that started in ancient Greece and evolved into a defined medical discipline by Arabs. Blood, yellow bile, phlegm, and black bile are the four humors, and the four major qualities of states in the living human body are hot, cold, wet, and dry (Husain et al. 2010). The Unani medicinal system considers that the human body is made of seven components Arkan, Mizaj, Akhlaat, Arwaah, Aaza, Quwa, and Afaal (elements, temperament, humors, life force, organs, faculties, functions) (Ahmad and Akhtar 2007). The Unani system of medicine (USM) has been recognized by the World Health Organization (WHO) as an alternative system for meeting the healthcare needs of the human population. Unani is a well-known traditional medicine system that incorporates the old traditional medical systems practiced in nations such as China, Egypt, India, Iraq, Persia, and Syria. In the modern world, Unani is being practiced, studied, and investigated under various regional names in over 20 countries, including Afghanistan, India, China, Canada, Germany, Denmark, Finland, Norway, the Netherlands, Korea, Poland, Japan, Sweden, Saudi Arabia, Switzerland, the United Kingdom, Turkey, and the United States. It has been accepted as one of the most powerful alternative systems of medicine in India. Like Ayurvedic medicine, Unani medicine also prescribes everyday usage of therapeutic methods like chewing sticks to brush the teeth in the morning as well as after taking a meal every time to prevent and maintain oral health (http://www.indiahomeclub.com/botanical_garden/medicinal_plants_uesd_in_ays.html).

17.3.1.4 Homeopathic

Homeopathic is a pseudoscientific medicinal system formulated in 1796 by the German physician Samuel Hahnemann. The person practicing homeopathic formulations is called a homeopathic doctor or homeopath (https://www.nhs.uk/conditions/homeopathy/). Another main principal practice believed is the dilution and shaking of drugs. The more the substance is diluted, the more it will increase the power to treat symptoms. The doctors believed that lowering the dose would significantly increase the power of the medicine. So, they will dilute the substance to a very minimal amount and this process is called potentization. They employ liquid drops, sugar pellets, creams, tablets, and gels for the majority of their medications (Frye 2003). In India, England, and other European nations, homoeopathy is commonly used to treat a variety of long-term ailments such as allergies, irritable bowel syndrome, and rheumatoid arthritis, as well as other minor injuries such as scrapes, cuts, and muscular strains or sprains (Rhee et al. 2017). There are 1500 large- and small-scale drug manufacturing industries in India, producing pharmaceuticals such as Potencies, Mother Tinctures, and Bio-chemic. Over 80% of medications are sourced from botanical resources, with 60% of them being unique to India (http://hplism.nic.in/about-us/sections/herbal-garden).

17.3.1.5 Western

Medical doctors, pharmacists, nurses, and other healthcare professionals are licensed to practice and heal symptoms and diseases in the Western system of medicine, often known as allopathic medicine. The word allopathic was coined by German physician Samuel Hahnemann in the 1800s where "allos" means "opposite" and "pathos" means "to suffer" which refers to the treatment of symptoms with its opposites. For example, constipation is treated with laxatives (https://www.healthline.com/health/allopathic-medicine#controversial-term). The Western medicinal system is one of the modern methods of treatment that have more significance in healthcare that it reduces 60% of death rates in conditions like heart disease and HIV/AIDS and 16% of death rate in cancer (https://www.news-medical.net/health/Alternative-and-Western-Medicine.aspx). However, 33.9% of the population depend on allopathic medicine because of the quicker response of medicine than other traditional methods of medicine (Phondani et al. 2014). In Western medicine, the plants are not directly utilized. Instead, particular potential compounds are isolated from the plant and used in drug development (Table 17.3). For example, ephedrine is a drug that was isolated from *Ephedra sinica* which is used to regulate blood pressure.

17.4 Various Threats to Traditional Medicinal Plants

17.4.1 Habitat Destruction

Habitat destruction is one of the major threats to medicinal plants due to rapid economic imbalance, environmental changes, and lack of knowledge in medicinal plants that results in the exhaustion of their populations and habitations in nature (Kala 2000). Destruction of wild plants from their original habitat also takes place by cultivating desired crops on a large scale which adversely affects the availability of the plants in natural habitat (Haq et al. 2011). Besides this, environmental changes like landslides, soil erosion, and forest fires also lead to deterioration or habitat loss, thereby overexploiting medicinal plants (Ganie et al. 2019). Previous reports

Table 17.3 Active component and medicinal values of MAPs

Medicinal plants	Active principles	Medicinal values	Reference
Azadirachta indica	Limonoids (nimbidinin), Di- and triterpenoids	Inhibitor of carcinoma, chemopreventive, inhibits colon cancer, antiallergic, blood purifier	Biswas et al. (2002)
Piper nigrum	Piperidine, dehydropipernonaline	Anticarcinogenic, antihyperlipidemic, epilepsy	Ahmad et al. (2012)
Tinospora cordifolia	Geloy Diterpenoid furanolactones (tinosporin), isoquinoline alkaloids	Immunomodulator, chemopreventive, cardioprotective, antidiabetic	Mittal et al. (2014)
Ocimum sanctum	Tulsi Apigenin, Taxol, Ursolic acid, Citral	Antidiabetic, hepatoprotective antibacterial, antifungal, antipyretic, and anticancer properties	Jain et al. (2015)
Phyllanthus emblica	Amla Emblicanin A, emblicanin B, punigluconin, pedunculagin	Antiviral, antimicrobial, anticancer, hepatoprotective, and antidiabetic	Khan (2009)

Table 17.4 Banned MAPs by Ministry of Environment, Forests and Climate Change, GOI

Botanical name	Family	Vernacular name
Aconitum heterophyllum Wall.	Ranunculaceae	Atis
Aquilaria agallocha Roxb.	Thymelaeaceae	Agar
Gentiana kurroo Royle.	Gentianaceae	Chirata
Nardostachys grandiflora DC.	Caprifoliaceae	Jatamansi
Picrorhiza kurroa Royle ex Benth.	Plantaginaceae	Kutki
Pterocarpus santalinus L.f	Fabaceae	Lal chandan
Rauwolfia serpentina [L.] Benth ex Kurz.	Apocynaceae	Barachandrika
Swertia chirayita [Roxb. ex Fleming] H. Karst.	Gentianaceae	Chirayata

prevails that majority of the medicinal plants are flowering plants and in that 33% are trees, with herbs, shrubs, and climbers following them. Due to habit fragmentation, many medicinal plants lead to severe and irreplaceable loss of genetic diversity. The Ministry of Environment and Forest has banned the export of 29 highly medicinal species (Table 17.4). The most significant conservation approach is the protection of habitats, leading to the development of the system amicably (Nishteswar 2014).

17.4.2 Bioprospecting

Bioprospecting is the exploration of natural sources and commercializing the natural product into different aspects. Molecules extracted from the natural products such as microbes and plants have the efficiency to produce novel and significant chemical structural compounds to develop new pharmaceutical products (Arunachalam and

Gayathri 2010). Bioprospecting is a major threat to medicinal plants because the discovery of the biological value of the natural resource makes it an inevitable integral component of modern industries (Ngo et al. 2013). Turmeric (*Curcuma longa*) is a common medicinal plant in India and Southeast Asian countries used in many forms for curing a variety of ailments. Das and Kohly of India who worked in the University of Mississippi Medical Center received the US patent in 1995 to prove the wound-healing property of turmeric (Nambisan 2017). Likewise, genus Moringa which belongs to Moringaceae family with 33 species exhibits various bioactive properties with numerous potential activities such as cardiac and circulatory stimulants, anticancer, antipyretic, antiepileptic, anti-inflammatory, antiulcer, diuretic, antidiabetic, antibacterial, and antifungal agent. But recently, many species from this family have been placed in endangering categories due to the bioprospecting of plant compounds (Arora et al. 2013). Kanis are the tribes living in the jungles of Agasthya Hills, Thiruvananthapuram, which is considered as a treasure trove of medicinal plants. The Kanis make their living by cultivating millets, areca nut, and rubber and collecting honey, etc.. Padmashree recipient Dr. Pushpangadan, Former Director, Tropical Botanical Garden and Research Institute (TBGRI)-Thiruvananthapuram proposed a unique and novel sustainable model called the Kani model using access-benefit-sharing method. The outcome of this method was the discovery of a novel drug called "Jeevani" that treated fatigue and stress by boosting immune protection to the liver after 8 years of research in 1995. They also shared half of their profit with the Kani trust. These types of community-based researches must be initiated by the scientists and governments to benefit more tribal inhabitants, thereby leading to novel drug development benefitting the society.

17.4.3 Overharvesting by Local Tribes

The commercial plea and massive biotic burden urge the local tribes to harvest most of the medicinal plants available in the forest. If this present condition continues, the medically important species will vanish soon from the natural forests (Shah and Kapoor 1978). The local peoples (tribes, non-tribes), ayurvedic manufacturers, and local traders mostly depend on the forest products leading to the population decline of natural resources. Some medicinal plants like *Chlorophytum borivilianum, R. serpentina, C. caesia, Celastrus paniculatus,* and *Dioscorea hispida* gradually diminish due to overharvesting (Manish 2013). According to the previous studies, overharvesting is the major hazard to medicinal plants when proceeded without any legalization, where the local herbal sellers collect medicinal plants vigorously (Shah et al. 2014). The IUCN status of few highly demanded medicinal plants is enlisted in Table 17.5.

Table 17.5 IUCN status of threatened medicinal plants and their applications

Plant species	Family name	Vernacular name	Medicinal use	IUCN status
Rauwolfia serpentina	Apocynaceae	Indian snakeroot	Hypertension	Endangered
Pterocarpus santalinus	Fabaceae	Sandalwood	Antipyretic, anti-inflammatory, anthelmintic, tonic, hemorrhage, dysentery, aphrodisiac, and diaphoretic activities	Endangered
Aquilaria agallocha	Thymelaeaceae	Eaglewood	Asthma, gout, and rheumatoid arthritis	Vulnerable
Picrorhiza kurroa	Plantaginaceae	Katuka	Treat liver disorders, reduce fevers, dyspepsia, chronic diarrhea, and scorpion sting	Endangered
Gentiana kurroo	Gentianaceae	Himalayan Gentian	Leucoderma and syphilis	Endangered
Swertia chirata	Gentianaceae	Chiretta	Fever, constipation, upset stomach, loss of appetite, intestinal worms, skin diseases, and cancer	Not extinct
Nordostachys grandiflora	Caprifoliaceae	Spikenard	Hysteria, syncope epilepsy, and mental weakness	Endangered
Aconitum heterophyllum	Ranunculales	Indian Atees	Anthelmintic, antidiarrheal, antiemetic, and anti-inflammatory	Not extinct
Moringa oleifera	Moringaceae	Drumstick tree	Cardiac and circulatory stimulants possess antitumor, antipyretic, antiepileptic, anti-inflammatory, antiulcer, antispasmodic, diuretic, antihypertensive, cholesterol-lowering, antioxidant, antidiabetic	Not extinct
Chlorophytum borivilianum	Asparagales	Musli	Sex tonic, diabetes, arthritis and increasing body immunity, curative for natal and postnatal problems, for rheumatism and joint pains	Endangered
Curcuma caesia	Zingiberaceae	Black tumeric	Leprosy, cancer, wounds, impotency, fertility, toothache, vomiting, allergies, leucoderma, asthma, tumors, piles, bronchitis, enlargement of the spleen, epileptic, menstrual disorder	Endangered

(continued)

Table 17.5 (continued)

Plant species	Family name	Vernacular name	Medicinal use	IUCN status
Dioscorea hispida	Dioscoreaceae	Asiatic Bitter Yam	Kill worms in wounds, Tubers used in birth control	Not extinct
Celastrus paniculatus	Celastraceae	Black oil plant	Anti-arthritic wound healing, hypolipidemic, and antioxidant	Vulnerable

17.5 Endangered Medicinal Plants and Their Therapeutic Applications

In the past 20 years, a greater extent of medicinally important plants have been identified because of their activity in boosting the monetary status of local communities who market these plants across the world (Phillips and Meelleur 1998; Ticktin et al. 2002). Current trend on globalization and commercialization has ensured in the overexploitation of some economically crucial medicinal herbs, many of them have turned into threatened medicinal plants (TMPs). It has become the spotlight of world consideration since they show disappearing and declining flora due to their fundamental applications in healthcare system (Gustafsson 2002). Reports on TMPs from Jhabua district, Madhya Pradesh, India, indicated about 6 critically endangered, 25 endangered, and 22 vulnerable species in different areas of the district. Few important TMPs and their medcinal uses are describe as follows: (1) gum and resin from *Boswellia serrata* have been utilized against leukorrhea and also used in lac bangle making; (2) *Butea monosperma var. lutea* seeds are applied to treat snakebite; (3) tuber of *Ceropegia bulbosa* is used as a curative in urine retention; (4) corm of *Amorphophallus paeoniifolius* is applied as a remedy of intestinal ulcers; (5) leaf extracted from *Andrographis paniculata* is used against malaria; (6) tuber of *Chlorophytum borivilianum* is consumed raw for muscular weakness; (7) rhizome of *Costus speciosus* is highly effective in curing bronchial disorders; (8) tuber of *Gloriosa superba* is employed for treating piles; (9) *Mucuna pruriens* seeds are used as a medicine for spermatorrhea; and (10) bark decoction of *Oroxylum indicum* is effectively used in healing pneumonia and cancer (Wagh and Jain 2013; Kalaivani and Mathew 2009; Mohan et al. 2016; Shalini et al. 2017). The medicinal applications of few threatened and endangered species are enlisted in Table 17.6.

Table 17.6 Distribution pattern of threatened MAPs and their applications

Plant species	Distribution	Medicinal uses	Reference
Angiopteris evecta	Arunachal Pradesh, Manipur, Meghalaya, Nagaland, Sikkim, and Tripura	Dysentery	Shankar and Rawat (2013)
Cissampelos pareira	Arunachal Pradesh, Manipur, Meghalaya, Mizoram, Nagaland	Fever, purgative, cough urinary troubles	
Desmotrichum fimbriatum	Arunachal Pradesh, Meghalaya, Sikkim	Whole plant stimulant, Tonic	
Gloriosa superba	Himachal Pradesh	To remove intestinal worms	Sajem et al. (2008)
Messua ferrea	Darjeeling, Himalaya	General medicine	
Rauvolfia serpentina	Himachal Pradesh	Stomach ache	
Boswellia serrata	Jhabua District	Leucorrhoea	Wagh and Jain (2013)
Ceropegia bulbosa		Urine retention	
Costus speciosus		Bronchial disorders	
Gloriosa superba		Bleeding piles	
Dactylorhiza hatagirea	Gurez Valley	Astringent, controls kidney, skin disorders, and increases vitality	Dad and Khan (2011)
Fritillaria roylei		Tonic, asthma, and ayurvedic formations	
Podophyllum hexandrum		Used to treat septic wounds, gastric problems; veterinary uses	
Podophyllum hexandrum	Jammu and Kashmir	Cancer, cough, cuts, wounds, ulcer, diarrhea, fever, skin disease, tumor	Pant and Pant (2011)
Swertia chirayita	Uttarakhand	Cough, antidiabetic, blood purifier and urinary discharge, malaria fever	

17.6 Current Biotechnological Approaches in Conservation of Endangered Medicinal Plants

Plants used medicinally are emerging as precious sources of novel medicines due to the expanding necessity of herbal medicine, natural well-being products, and bioactive compounds of medicinal herbs; the medicinal plants' usage has been increasing expeditiously all over the world (Balunas and Kinghorn 2005; Cole et al. 2007). The aim of conservation is to improve the medical sector by keeping and applying biological resources using various methods in order not to dwindle the world's diversity of species or devastate crucial habitats and ecosystems as well. Conservation of plant genes has long been recognized as highly essential in protecting biodiversity (Kasagana and Karumuri 2011). From the past two centuries, anthropogenic developments in land use have been transforming agricultural and natural areas into the man-made surface, thereby driving the medicinal plants to be looked at amidst the highly threatened species worldwide. In a few countries, beyond two-thirds of the actual habitation forms are found to be endangered. Various human activities are the major reason of risk for 83% of plants which are under the status of endangered. Destroying the plant habitat and losing it as well are also an issue since they drive to the fragmentation of the leftover habitat ending in further seclusion of the plant populace (Tasheva and Kosturkova 2013). Owing to the fact of countless advantages of biotechnology in various areas such as agriculture, pharmaceuticals, forestry, food industry, and other sectors, the field of biotechnology has developed into a center of attraction for conservation and also in a continuous supply of plants containing medicinal values. In order to deliver medicinal plants or medicinal plant-based raw materials in a feasible form, the biotechnological approaches, namely in vitro micropropagation, cryopreservation, genetic manipulations, and acclimatization, are playing an extraordinary role.

17.6.1 In Vitro Propagation

Plant tissue culture studies were done to conserve medicinal plant resources and adapt production of pharmaceutically potent secondary metabolites. Micropropagation methods were found to facilitate much better potential propagation of plants than the common methods like seedling or layering. In vitro protocol has been developed by the TBGRI, for expeditious revival and establishment of nearly 40 medicinally significant intermittent and threatened plant species located in the Western Ghats. Following this, Verma et al. (2012) gave a similar try to conserve the medicinal plants. Their studies covered in vitro cultivation of about 23 overutilized medicinal plants that belong to India (Verma et al. 2012). Using the micropropagation technique, amplification of the enormous number of plants from explants such as leaves, seeds, ovary, and tubers is achieved. Thus, propagated plantlets can be utilized for producing multiple genetically similar plantlets. Besides the conservation of medicinal plants that are rare and becoming extinct, micropropagation is also known as the most ancient profit-making biotechnology method based on clonal propagation (Rai 2010). *In vitro* propagation has some significant advantages, including (1) higher multiplication rate, (2) controlled environment to reach specific requirement of the plant, (3) availability of plant throughout the year despite seasonal variations, and (4) collection and conservation of endangered plant species by in vitro cultivation (Sidhu 2011).

17.6.2 Cryopreservation

Plant cryopreservation includes storing plant parts (usually seed or shoot tips) in liquid nitrogen (LN_2) at a temperature of −196 °C or −135 °C in the vapor phase to retain viability of stored materials effectively after rewarming (Day et al. 2008; Hamilton et al. 2009). The major reason to apply cryostorage is to conserve plants which are being endangered, especially where seeds are highly deficient or of problematic quality and the species is in immediate extinction (Mallon et al. 2008; Paunescu 2009; Sen-Rong and Ming-Hua 2009). In the past three decades, many diverse cryopreservation protocols, namely classical vitrification, slow cooling, encapsulation/dehydration, and encapsulation/vitrification protocols, have been used to store germplasm (Reed 2008). The following plant species, such as *Atropa belladonna*, *Digitalis lanata*, *Hyoscyamus* sp., and *Rauwolfia serpentina*, are few examples of successful conservation using the cryopreservation technique (Jain et al., 2012). Above 110 accessions of plant species found as rare and threatened were preserved by cryopreservation at the Kings Park and Botanic Garden in Perth, Australia (Touchell and Dixon 1994). Furthermore, the National Bureau for Plant Genetic Resources (NBPGR) located in New Delhi, India, has stored and maintained over 1200 accession out of 50 different plant species (Mandal 2000; Mandal et al. 2009). Cryopreservation of embryogenic cultures of *Dioscorea bulbifera* was done with the aid of the procedure called encapsulation/dehydration. This method of subculturing of embryogenic culture showed a growth recovery rate of 53.3%. These research methods of cryopreservation apparently bring insights about the efficiency of cryo-techniques to conserve important medicinal plants (Tashdeva and Kosturkova 2012).

17.6.3 Genetic Transformation

Genetic manipulations make better yield and also better quality of medicinal herbs, including gene introduction or modification that boosts the production of secondary metabolites in medicinal plants. Gene transfer technology has grown into a comprehensive area in both characteristic enhancement and genetic analysis activity in plant species. Manipulating genome aims to transfer genes by promoting certain techniques in order to transfer the desired gene into a plant to achieve better biosynthetic activity of the bioactive compounds. The necessary action in this view is finding the appropriate marker genes to transfer genes, as it facilitates figuring out the transformed cell. Many research workers have been focusing on the transferring and integration mechanism of the marker as well as reporter genes (Rai 2010). Gene transfer promotes the growth of medicinal herbs having long-lasting resistance against insects and infections. Similarly, transgenes may help to improve the resistance in plants against insects, salinity, and drought which increases the demand to achieve the desire of the world and manage land in order to conserve plant biodiversity in natural environments. In nearly 120 species among the 35 families, genetic transformation has been done flourishingly using

Agrobacterium and few other transformation methods (Birch 1997; Yun et al. 1992; Cucu and Gabriela 2002). Plant transformation technology has not only performed a crucial task in introducing genes in order to fight against insects into appropriate plants but also developed into a resourceful platform to improve cultivar and gene function studies as well in plants.

17.6.4 Acclimatization and Adaptation to Open Field

Acclimatization is the process of adjusting of an organism to the climatic conditions, specifically plants brought into different types of environment. It happens under the effective surveillance of human beings (Conover and Poole 1984). Acclimatization not only is particular to micropropagation but has also been applied for many years in conventional propagation. After leafy cuttings are formed, roots in a propagation bench are frequently acclimatized (hardened off) before carrying the entire plant to the greenhouse setup. The major method of acclimatization is to gently decrease the relative humidity or quantity of mist (Hartmann et al. 1990) to maintain greenhouse shade level at 50–70% shades before transplanting the crops into the field (Welch 1970). Acclimatization and adaptation effectiveness differ with the species. Bach and Pawłowska (2003) reported 65% durability of rooted plantlets of *Gentiana pneumonanthe* after potting in the soil in a greenhouse. Additionally, the plants are well planted outdoors under field circumstances. Compared to the natural habitat, these in vitro cultivars yield a higher count of both flowers and stems (Butiuc-Keul et al. 2005).

17.7 Alternative Medicines to Overcome Exploitation of Medicinal Plants

Medicinal plants have become an integral part of human society's efforts to cure the disease at the advent of humanity. The vast resources of medicinal plants have long been used to cure a variety of ailments (Mallikarjuna Gowda et al. 2019). India has a diverse range of natural resources. The Western Ghats stretch down the west coast, and the whole area is noted for its biodiversity, abundance, and endemism of various organisms. Human population and per capita intake increase have culminated excessive exploitation of Earth's biological diversity, thereby intensifying ocean acidification, climatic alterations, and other anthropogenic environmental impacts (Rands et al. 2010). While species protection is most successfully accomplished by managing wild populations and natural environments (in situ conservation), the majority of therapeutic plants either do not grow seeds or produce seeds that are too tiny to germinate in soils. For medicinal plants, it is understood that populations of a few genera have historically been preferred from specific locations (Karuppusamy 2018) by the local healer, which makes it more liable to loss of few selective medicinal plants. Besides many conservations that intervene, the use of alternate medicinal plants with similar therapeutic properties instead of endangered plants will safely guard the extinction of these traditionally used medicinal plants. Here, we suggest a few alternate medicinal plants against the IUCN red-listed plants in Table 17.7.

17.8 Future Prospective and Challenges

The major reasons for the depletion of forest medicinal plants are unscientific mining, quarrying, overgrazing, biotic pressure, forest fire, irregular collection of samples, etc.. Also, the scientific community and pharma industries are currently exploiting the wild plant population for commercial drug development. Petty contractors or middlemen interplay by engaging tribal people for collecting bulk herbal plants from forests and paying them very low fare. This great biological loss of plant species insists upon the need for conservation to attain sustainable ecosystem.

Therefore, this increased demand leads to species extinction due to overharvesting of traditional medicinal plants. In the near future to improve healthcare, collaboration of research and technology with traditional healers is a promising approach. The most feasible measure to save threatened plant species is to switch over to alternative medicines derived from unexplored or underutilized plant species with similar healing properties.

Providing opportunities for scientists to discover new compounds with medicinal properties, creating awareness on cultivation employment of unexplored plants to local inhabitants, and discovering novel drugs for pharma industries are big challenges to the future technocrats and environmentalists.

17.9 Conclusion

To minimize the loss of medicinal plant resources, it is necessary to adopt sustainable agriculture techniques and modern conservation methods. Besides various conservation techniques, the intervention of state and central governments to protect the environment and habitat restoration is necessary. This can be done by establishing protected area covering and regulating the traditional healing practiced by local communities to enhance the medicinal plant population in the wild. If we are careless to take appropriate measures, we are sure to lose numerous lifesaving treasures, even without a trace of their existence. Intensive studies on the hidden plant resources and their potency must be carried over by the scientific community to the global community through appropriate modern tools and techniques to achieve ecological sustainability. The utilization of alternative medicinal plants instead of the traditionally used endangered plants is considered as an efficient method, and their circumstances must be more systematically defined and such results are reproduced and scaled up in intellectual and evidence-based ways.

Table 17.7 Threatened MAPs and its alternative plants in usage

Threatened species	IUCN status	Alternative species	Application	Reference
Terminalia chebula	Least concern	*Abrus precatorius* *Andrographis paniculata* *Bauhinia variegate* *Ficus arnottiana*	Ulcers, cancer, Leprosy	Mallikarjuna Gowda et al. (2019)
Gentiana kurroo	Critically endangered	*Abies pindrow* *Barleria buxifolia* *Phyllanthus niruri* *Picrorhiza kurroo* *Thapsia garganica*	Acute bronchitis, pneumonia, anti-inflammatory, hepatoprotective	Harish and Shivanandappa (2006), Bhat et al. (2013), Sharma et al. (2014)
Swertia chirayta	Endangered	*Aloe vera* *Caesalpinia bonducella* *Cajanus cajan* *Coccinia indica* *Momordica charantia* *Ocimum sanctum* *Tinospora cordifolia*	Antidiabetic	Grover et al. (2002), Kumar and Van Staden (2016)
Gymnema sylvestre	Vulnerable	*Aloe vera* *Caesalpinia bonducella* *Cajanus cajan* *Coccinia indica* *Momordica charantia* *Ocimum sanctum*	Antidiabetic	Grover et al. (2002), Mukherjee et al. (2007)

(continued)

Table 17.7 (continued)

Threatened species	IUCN status	Alternative species	Application	Reference
Acorus calamus	Least concern	*Abies pindrow* *Adhatoda vasica* *Terminalia arjuna*	Anticancer, anti-inflammatory, bronchial asthma	Mukherjee et al. (2007), Pattanaik et al. (2009), Bhat et al. (2013)
Albizia thompsonii	Near threatened	*Impatiens balsamina* *Rauvolfia serpentina*	Respiratory problems, wounds and pain, skin diseases, sexually transmitted infections	Pattanaik et al. (2009), Maroyi (2018)
ycas beddomei	Endangered	*Alangium salvifolium* *Aloe vera* *Tinospora gulancha*	Rheumatoid arthritis and muscle pain	Kumar et al. (2015), Shravya et al. (2017), Solomon Raju et al. (2019)
Deodar cedar	Least concern	*Alangium salvifolium* *Aloe vera* *Tinospora gulancha*	Rheumatoid arthritis	Kumar et al. (2015), Shravya et al. (2017)
Psoralea corylifolia	Least concern	*Catharanthus roseus* *Leptadenia reticulata* *Podophyllum hexandrum* *Saussurea involucrate*	Anticancer	Ryu et al. (1992), Satdive et al. (2003)

Gloriosa superba	Threatened	*Caralluma bicolor* *Catharanthus roseus* *Leptadenia reticulata* *Ophiorrhiza rugosa* *Podophyllum hexandrum*	Anticancer	Anilkumar et al. (2013), Vanitha et al. (2019)
Pterocarpus marsupium	Near threatened	*Syzigium cumini* *Tinospora cordifolia* *Trigonella foenum*	Antidiabetic	Grover et al. (2002)
Saussurea involucrate	Endangered	*Achyranthes aspera* *Caralluma bicolor* *Catharanthus roseus* *Podophyllum hexandrum* *Ophiorrhiza rugose*	Anticancer	Vanitha et al. (2019)
Aegle marmelos	Near threatened	*Acacia Arabica* *Albizzia lebbeck* *Catharanthus roseus* *Ficus arnottiana* *Juniperus phoenica* *Moringa oleifera* *Ophiorrhiza rugosa* *Podophyllum hexandrum*	Anticancer, antipyretic, ulcer healing, antigenotoxic, diuretic, antifertility, and anti-inflammatory	Gadekar et al. (2010), Vimala and Gricilda Shoba (2014), Daniyal and Akram (2015)
Albizia amara	Least concern	*Andrographis paniculata* *Bauhinia variegate* *Ficus arnottiana*	Diarrhea, gonorrhea, skin diseases, poisonous bites, leprosy, anticancer, antihyperlipidimic, anti-inflammatory	Gadekar et al. (2010), Balasubramani et al. (2017)

(continued)

Table 17.7 (continued)

Threatened species	IUCN status	Alternative species	Application	Reference
Ginkgo biloba	Endangered	*Catharanthus roseus* *Commiphora myrrha* *Ophiorrhiza rugosa* *Podophyllum hexandrum* *Withania somnifera*	Neuroprotective, anticancer, cardioprotective, stress alleviating, and memory enhancing	Mahadevan and Park (2008), Gadekar et al. (2010), Elufioye et al. (2017)
Salvia officinalis	Least count	*Celastrus paniculatus* *Ficus arnottiana*	Ulcers, gout, rheumatism, inflammation, dizziness, paralysis, diarrhea, and hyperglycemia	Vimala and Gricilda Shoba (2014), Ghorbani and Esmaeilizadeh (2017)
Bacopa monnieri	Least count	*Celastrus paniculatus* *Withania somnifera*	Anti-inflammatory, cardio tonic, sedative, neuro-muscular blocking activities, rejuvenate intellect and memory	Joshua Allan et al. (2007), Patil et al. (2010), Vollala et al. (2011)
Centella asiatica	Least count	*Celastrus paniculatus* *Ficus arnottiana* *Withania somnifera*	Skin conditions such as leprosy, lupus, varicose ulcers, eczema, psoriasis, diarrhea, fever, amenorrhea, brain tonic	Gohil et al. (2010), Subbaiyan et al. (2014)
Taxus globosa	Endangered	*Catharanthus roseus* *Podophyllum hexandrum*	Anticancer	Kamble et al. (2011)
Taxus baccata	Endangered	*Catharanthus roseus* *Ophiorrhiza rugose* *Podophyllum hexandrum*	Anticancer	Giri and Lakshmi Narasu (2000), Das et al. (2020)

Taxus cuspidata	Endangered	*Catharanthus roseus* *Ophiorrhiza rugose* *Podophyllum hexandrum*	Anticancer	Giri and Lakshmi Narasu (2000), Nammi et al. (2003)
Pterocarpus marsupium	Near threatened	*Adhatoda vasica* *Ficus arnottiana* *Ricnus communis* *Terminalia arjuna*	Cholera, dysentery, urinary complaints, toothache, Antidiabetics, leprosy, skin diseases, diarrhea, asthma, bronchitis, grayness of hairs	Chaudhari and Mengi (2006), Jelastin et al. (2012), Ahmad and Rajagopal (2015), Hossain and Hoq (2016)
Santalum album	Vulnerable	*Adhatoda vasica* *Ricnus communis* *Terminalia arjuna*	Cough, bronchitis, fever, and sore mouth and throat. It is also used to treat urinary tract infections (UTIs), liver disease, gallbladder problems, heatstroke, gonorrhea, Hepatitis B	Pattanaik et al. (2009, b), Gupta and Chaphalkar (2016)
Taxus wallichiana	Endangered	*Adhatoda vasica* *Rauvolfia serpentina* *Terminalia arjuna*	Cough, fever, and pain relive, anticancer	Juyal et al. (2014)
Saussurea costus	Critically endangered	*Adhatodavasica* *Crocus sativus* *Ficus arnottiana* *Phillanthus niruri* *Terminalia arjuna*	Asthma, anti-inflammatory, antiulcer, anticancer, and hepatoprotective activities	Chaudhari and Mengi (2006), Pandey et al. (2007), Hossain and Hoq (2016), Javadi et al. (2017)
Aconitum heterophyllum	Endangered	*Adhatoda vasica* *Crocus sativus* *Glycyrrhiza glabra* *Hyssopus officinalis* *Ruta graveolens* *Terminalia arjuna*	Bronchitis	Ukani et al. (1996), Chaudhari and Mengi (2006), Beigh et al. (2006), Hossain and Hoq (2016), Javadi et al. (2017)

(continued)

Table 17.7 (continued)

Threatened species	IUCN status	Alternative species	Application	Reference
Coptis teeta	Endangered	*Aloe vera* *Caesalpinia bonducella* *Cajanus cajan* *Coccinia indica* *Momordica charantia* *Ocimum sanctum*	Antidiabetic, bitter tonic, carminative, expectorant, febrifuge, ophthalmic, and stomachic	Mukherjee and Chakraborty (2019)
Commiphora wightii	Rare	*Aloe vera* *Ricnus communis*	Arthritis, lowering high cholesterol, (atherosclerosis), acne and other skin diseases, and weigh loss	Sarup et al. (2015)
Mundulea sericea	Least count	*Catharanthus roseus* *Ophiorrhiza rugosa* *Podophyllum hexandrum*	Anticancer, antimicrobial, antioxidant, and antiplasmodial	Giri and Lakshmi Narasu (2000), Chepkirui et al. (2021)
Ophioglossum reticulatum	Least count	*Cassia occidentalis* *Rosmarinus officinalis* *Borago officinalis* *Sida cordifolia* *Zingiber officinale*	Anti-inflammatory medicine; leaves boiled in oil are applied to wounds	Ghasemian et al. (2016)
Sonneratia caseolaris	Least count	*Alpinia galanga* *Rosmarinus officinalis* *Borago officinalis*	Astringent and antiseptic, sprain poultices, in treating piles, and arresting hemorrhage	Varghese et al. (2010), Singh et al. (2014)
Berberis aristata	Least count	*Abutilon indicum* *Acalypha indica*	Antibacterial, antiperiodic, antidiarrheal, and anticancer	Mazumder et al. (2011), Potdar et al. (2012), Rathi et al. (2013)

Nypa fruticans	Least count	*Adhatoda vasica* *Andrographis paniculate* *Bauhinia variegate* *Ficus arnottiana*	Asthma, leprosy, tuberculosis, sore throat, liver disease, snake bite, as a pain reliever	Gadekar et al. (2010), Prasad et al. (2013), Hossain and Hoq (2016)
Toona ciliata	Least count	*Myrtus communis* *Polygonum aviculare* *Pistacia lentiscus*	Diarrhea, dysentery, wounds, skin diseases, bleeding disorders, menstrual disorders	Kavitha and Satish (2013), Mobli et al. (2015)
Xylocarpus granatum	Least count	*Andrographis paniculata* *Bauhinia variegate* *Ficus arnottiana*	Diarrhea, peptic ulcer	Rouf et al. (2007), Lakshmi et al. (2010), Gadekar et al. (2010)
Woodfordia fruticosa	Least count	*Andrographis paniculata* *Bauhinia variegate* *Ficus arnottiana*	Hepatoprotective, dysentery, leprosy, erysipelas, blood diseases, leucorrhoea, menorrhagia, toothache	Das et al. (2007), Gadekar et al. (2010), Baravalia et al. (2011)
Utleria salicifolia	Critically endangered	*Ficus arnottiana* *Bauhinia variegate* *Andrographis paniculata*	Hepatoprotective, antiulcer	Rao et al. (2004), Gadekar et al. (2010)
Syzygium travancoricum	Critically endangered	*Alangium salvifolium* *Barleria acuminate*	Diabetes and arthritis	Mobli et al. (2015)
Saraca asoca	Vulnerable	*Myrtus communis* *Pistacia lentiscus* *Polygonum aviculare*	Bleeding, hemorrhoids, infertility, insomnia, psoriasis, anticancer	Pradhan et al. (2009), Singh et al. (2015)
Pterocarpus santalinus	Endangered	*Ficus arnottiana* *Bauhinia variegate* *Andrographis paniculata*	Antioxidative, antidiabetic, antimicrobial, anticancer, anti-inflammatory properties, ulcers	Gadekar et al. (2010), Walpola et al. (2011), Bulle et al. (2016)

(continued)

Table 17.7 (continued)

Threatened species	IUCN status	Alternative species	Application	Reference
Ochreinauclea missionis	Vulnerable	*Ficus arnottiana* *Bauhinia variegate* *Andrographis paniculata*	Constipation, ulcers, antirheumatic, skin diseases, eye diseases, dropsy, piles, fever, jaundice	Gadekar et al. (2010)
Aegle marmelos	Not threatened	*Andrographis paniculata* *Bauhinia variegate* *Ficus arnottiana*	Antimicrobial, antiviral, radioprotective, anticancer, chemopreventive, antipyretic, ulcer healing, antigenotoxic, diuretic, antifertility and anti-inflammatory	Rahman and Parvin (2014), Pathirana et al. (2020)
Aphanamixis polystachya	Least count	*Andrographis paniculate* *Bauhinia variegate* *Ficus arnottiana*	Anthelmintic and ulcer curing	Wang et al. (2013)
Artocarpus hirsutus	Least count	–	Hydrocele, pimple, heal sores, cracks in the skin anti-acne	Nayak et al. (2017a, b)
Balanites aegyptiaca	Least count	*Adhatoda vasica* *Terminalia arjuna*	Wounds, malaria, syphilis, epilepsy, dysentery, constipation, diarrhea, hemorrhoid, stomach aches, asthma, fever	Chothani and Vaghasiya (2011), Hossain and Hoq (2016)
Calophyllum apetalum	Least count	*Adhatoda vasica* *Bauhinia variegate* *Boswellia serrate* *Ficus arnottiana*	Rheumatism and leprosy	Choudhary et al. (2015), Chandrappa et al. (2016)
Cayratia pedata var. glabra	Critically Endangered	*Andrographis paniculata* *Bauhinia variegate* *Ficus arnottiana*	Diarrhea, refrigerant, hysteria, astringent, and ulcers	Gadekar et al. (2010), Karuppusamy (2018)
Cinnamomum macrocarpum	Least count	*Borago officinalis* *Rosmarinus officinalis*	Blood purifier, digestive, antiseptic, antifungal, antiviral, antibacterial, antioxidant, anti-inflammatory, and immunomodulatory	Ghasemian et al. (2016), Kumar et al. (2019)

Curcuma pseudomontana	Vulnerable	*Adhatoda vasica* *Andrographis paniculata*	Leprosy, dysentery, and cardiac diseases	Kanase and Singh (2018)
Decalepis hamiltonii	Endangered	*Achyranthes aspera* *Adhatoda vasica* *Boswellia serrate*	Dysentery, uterine hemorrhages, skin diseases, chronic rheumatism, anemia, coughs, bronchitis, antiulcer	Naik et al. (2007), Ch et al. (2013), Choudhary et al. (2015)
Dipterocarpus indicus	Endangered	*Acacia arabica* *Adansonia digitata* *Amaranthus spinosus* *Ocimumtenuiflorum* *Pongamia pinnata* *Sida cordifolia*	Gonorrhoea, ulcer, ringworm, and skin diseases	Mollik et al. (2009), Vimala and Gricilda Shoba (2014), Aslam et al. (2015)
Dysoxylum malabaricum	Endangered	*Bauhinia variegata* *Ficus arnottiana*	Insecticide, arthritis, anorexia, cardiac debility, expelling intestinal worms, inflammation, leprosy, and rheumatism	Senthil Nathan et al. (2006), Gadekar et al. (2010), Bodare et al. (2013)
Humboldtia vahliana	Endangered	*Andrographis paniculata* *Bauhinia variegate* *Ficus arnottiana*	Biliousness, leprosy, ulcers, and epilepsy	Gadekar et al. (2010), Sreelekshmi and Deepa (2020)
Humboldtia unijuga	Endangered	*Andrographis paniculate*	Chicken pox, head ache, snake bite, anticancer agent	Nair et al. (2020a, b)
Kingiodendron pinnatum	Endangered	*Andrographis paniculate*	Gonorrhea, catarrhal conditions of genito-urinary, and respiratory tracts	Shahid et al. (2018)
Michelia champaca	Least count	*Adhatoda vasica* *Andrographis paniculate*	Diarrhea, cough, bronchitis, hypertension, dyspepsia, fever, dysmenorrhea	Saqib et al. (2018)
Alstonia scholaris	Least count	*Adhatoda vasica* *Boswellia serrate*	Anti-arthritic	Choudhary et al. (2015)

(continued)

Table 17.7 (continued)

Threatened species	IUCN status	Alternative species	Application	Reference
Michelia nilagirica	Vulnerable	*Andrographis paniculata* *Bauhinia variegate* *Butea frondose* *Ficus arnottiana*	Leprosy, fever, antiulcer, anti-inflammatory, anticancer	Wei et al. (2011), Venkatadri et al. (2017)
Nilgirianthus ciliatus	Vulnerable	*Acacia Arabica* *Andrographis paniculata* *Annona squamosa* *Ficus arnottiana*	Neurological disorders, ulcers, glandular swellings, poisonous infections, itching, leprosy, and other skin diseases	Gadekar et al. (2010), Vimala and Gricilda Shoba (2014), Rameshkumar et al. (2017)
Vateria indica	Vulnerable	*Acacia arabica* *Adhatoda vasica* *Andrographis paniculate* *Annona squamosa* *Ficus arnottiana* *Terminalia arjuna*	Chronic bronchitis and throat troubles, cough, asthma, leprosy, skin eruptions, crack infection, wounds, and ulcer	Gadekar et al. (2010), Vimala and Gricilda Shoba (2014), Hossain and Hoq (2016)

References

Ahmad B, Akhtar J (2007) Unani system of medicine. Pharmacogn Rev 1(2):210

Ahmad H, Rajagopal K (2015) Pharmacology of *Pterocarpus marsupium* Roxb. Med Plant Res 5

Ahmad N, Fazal H, Abbasi BH, Farooq S, Ali M, Khan MA (2012) Biological role of Piper nigrum L. (Black pepper): A review. Asian Pacific J Trop Biomed 2(3):S1945–S1953

Bach A, Pawłowska B (2003) Somatic embryogenesis in Gentiana pneumonanthe L. Acta Biol Crac Ser Bot 45(2):79–86

Bhadula SK, Purohit AN (1994) Adaptational strategies of plants at high altitudes and future prospects for the conservation of biodiversity. Adv Plant Sci Res 1:1–24

Bishop BC, Chatterjee SP (2019) “Himalayas”. Encyclopedia britannica. https://www.britannica.com/place/Himalayas

Bishop BC, Chatterjee SP (2021) “Himalayas”. Encyclopedia britannica. https://www.britannica.com/place/Himalayas

Anilkumar KA, Kumar KMP, Udayan PS (2013) Caralluma bicolor Ramach. et al., (Apocynaceae: Asclepiadoideae) - a rare and little known endemic plant as a new record from Palakkad District, Kerala State, India. J Threat Taxa 5:5007–5009

Arora DS, Onsare JG, Kaur H (2013) Bioprospecting of Moringa (Moringaceae): microbiological perspective. J Pharmacogn Phytochem 1(6):193–215

Arunachalam C, Gayathri P (2010) Studies on bioprospecting of endophytic bacteria from the medicinal plant of *Andrographis paniculata* for their antimicrobial activity and antibiotic susceptibility pattern. Int J Curr Pharm Res 2(4):63–68

Aslam MS, Ahmad MS, Mamat AS (2015) A phytochemical, ethnomedicinal and pharmacological review of genus Dipterocarpus. Int J Pharm Pharm Sci 7:27–38

Bach A, Pawłowska B (2003) Somatic embryogenesis in *Gentiana pneumonanthe* L. Acta Biol Cracov Ser Bot 45(2):79–86

Balasubramani G, Ramkumar R, Raja RK et al (2017) *Albizia amara* Roxb. Mediated Gold Nanoparticles and Evaluation of Their Antioxidant, Antibacterial and Cytotoxic Properties. J Clust Sci 28:259–275. https://doi.org/10.1007/s10876-016-1085-9

Balunas MJ, Kinghorn AD (2005) Drug discovery from medicinal plants. Life Sci 78(5):431–441

Bannerman RHO, Burton J, Chen WC (1983) Traditional medicine and Health care coverage: A Reader for Health Administrators and practitioners. World Health Organization, Geneva

Baravalia Y, Vaghasiya Y, Chanda S (2011) Hepatoprotective effect of *Woodfordia fruticosa* Kurz flowers on diclofenac sodium induced liver toxicity in rats. Asian Pac J Trop Med 4:342–346. https://doi.org/10.1016/S1995-7645(11)60100-4

Beigh SY, Nawchoo IA, Iqbal M (2006) Cultivation and Conservation of Aconitum heterophyllum: A Critically Endangered Medicinal Herb of the Northwest Himalayas. Int J Geogr Inf Syst 11: 47–56. https://doi.org/10.1300/J044v11n04_06

Bhat JA, Kumar M, Bussmann RW (2013) Ecological status and traditional knowledge of medicinal plants in Kedarnath Wildlife Sanctuary of Garhwal Himalaya, India. J Ethnobiol Ethnomed 9:1. https://doi.org/10.1186/1746-4269-9-1

Birch RG (1997) Plant transformation: problems and strategies for practical application. Annu Rev Plant Biol 48(1):297–326

Biswas K, Chattopadhyay I, Banerjee RK, Bandyopadhyay U (2002) Biological activities and medicinal properties of neem (*Azadirachta indica*). Curr Sci 2002:1336–1345

Bodare S, Tsuda Y, Ravikanth G et al (2013) Genetic structure and demographic history of the endangered tree species *Dysoxylum malabaricum* (Meliaceae) in Western Ghats, India: implications for conservation in a biodiversity hotspot. Ecol Evol 3:3233–3248. https://doi.org/10.1002/ece3.669

Bopana N, Saxena S (2008) *Crataeva nurvala*: a valuable medicinal plant. J Herbs Spices Med Plants 14(1–2):107–127

Bulle S, Reddyvari H, Nallanchakravarthula V, Vaddi DR (2016) Therapeutic Potential of *Pterocarpus santalinus* L.: An Update. Pharmacogn Rev 10:43–49. https://doi.org/10.4103/0973-7847.176575

Butiuc-Keul A, Suteu A, Deliu C (2005) In vitro organogenesis of *Gentiana punctata*. Notulae Botanicae HortiAgrobotanici Cluj-Napoca 33:38

Ch M, Reddy R, Sri K, Murthy R (2013) A review on *Decalepis hamiltonii* Wight & Arn. J Med Plant Res 7(41):3014–3029

Chakraborty R, De B, Devanna N, Sen S (2012) North-East India an ethnic storehouse of unexplored medicinal plants. J Nat Prod Plant Resour 2(1):143–152

Chandra LD (2016) Bio-diversity and conservation of medicinal and aromatic plants. Adv Plants Agric Res 5(4):00186

Chandrappa CP, Govindappa M, Chandrasekar N et al (2016) Endophytic synthesis of silver chloride nanoparticles from *Penicillium* sp. of *Calophyllum apetalum*. Adv Nat Sci Nanosci Nanotechnol 7:025016. https://doi.org/10.1088/2043-6262/7/2/025016

Chaudhari M, Mengi S (2006) Evaluation of phytoconstituents of *Terminalia arjuna* for wound healing activity in rats. Phytother Res 20:799–805. https://doi.org/10.1002/ptr.1857

Chepkirui C, Ochieng PJ, Sarkar B et al (2021) Antiplasmodial and antileishmanial flavonoids from *Mundulea sericea*. Fitoterapia 149:104796. https://doi.org/10.1016/j.fitote.2020.104796

Chothani DL, Vaghasiya HU (2011) A review on *Balanites aegyptiaca* Del (desert date): phytochemical constituents, traditional uses, and pharmacological activity. Pharmacogn Rev 5:55–62. https://doi.org/10.4103/0973-7847.79100

Choudhary M, Kumar V, Malhotra H, Singh S (2015) Medicinal plants with potential anti-arthritic activity. J Intercult Ethnopharmacol 4:147. https://doi.org/10.5455/jice.20150313021918

Chudiwal AK, Jain DP, Somani RS (2010) *Alpinia galanga* Willd. – An overview on phytopharmacological properties.

Cole IB, Saxena PK, Murch SJ (2007) Medicinal biotechnology in the genus Scutellaria. Vitro Cell Dev Biol Plant 43(4):318–327

Conover CA, Poole RT (1984) Acclimatization of indoor foliage plants. Horticultural Rev 6:120–124

Cucu N, Gabriela GL (2002) Genetically modified medicinal plants. II. Transfer and expression of a marker kanamycin resistance gene in Atropa belladonna plants. Roumanian Biotech Lett 7:869–874

Dad JM, Khan AB (2011) Threatened medicinal plants of Gurez valley, Kashmir Himalayas: distribution pattern and current conservation status. Int J Biodivers Sci Ecosyst Serv Manage 7(1):20–26

Daniyal M, Akram M (2015) Antifertility activity of medicinal plants. J Chin Med Assoc 78:382–388. https://doi.org/10.1016/j.jcma.2015.03.008

Das A, Sarkar S, Bhattacharyya S, Gantait S (2020) Biotechnological advancements in *Catharanthus roseus* (L.) G. Don. Appl Microbiol Biotechnol 104:4811–4835. https://doi.org/10.1007/s00253-020-10592-1

Das PK, Goswami S, Chinniah A et al (2007) *Woodfordia fruticosa*: Traditional uses and recent findings. J Ethnopharmacol 11:189–199

Day JG, Harding KC, Nadarajan J, Benson EE (2008) Cryopreservation. In: Molecular Biomethods Handbook. Humana Press, Totowa, NJ, pp 917–947

Devi Prasad AG, Shyma TB (2013) Medicinal plants used by the tribes of Vythiri taluk, Wayanad district (Kerala state) for the treatment of human and domestic animal ailments. J Med Plant Res 7(20):1439–1451

Devi Prasad AG, Shyma TB, Raghavendra MP (2014) Traditional herbal remedies used for management of Reproductive disorders in Wayanad district, Kerala. Int J Res Pharma Chem 4 (2):333–341

Elufioye TO, Berida TI, Habtemariam S (2017) Plants-Derived Neuroprotective Agents: Cutting the Cycle of Cell Death through Multiple Mechanisms. Evid Based Complement Alternat Med 2017:e3574012. https://doi.org/10.1155/2017/3574012

Frye JC (2003) Herbal and homeopathic medicine: understanding the difference. Semin Integrative Med 1(3):158–166

Gadekar R, Singour PK, Chaurasiya PK et al (2010) A potential of some medicinal plants as an antiulcer agents. Pharmacogn Rev 4:136. https://doi.org/10.4103/0973-7847.70906

Ganie AH, Tali BA, Khuroo AA, Reshi ZA, Nawchoo IA (2019) Impact assessment of anthropogenic threats to high-valued medicinal plants of Kashmir Himalaya, India. J Nat Conserv 50: 125715

Ghasemian M, Owlia S, Owlia MB (2016) Review of Anti-Inflammatory Herbal Medicines. Adv Pharmacol Sci 2016:e9130979. https://doi.org/10.1155/2016/9130979

Ghorbani A, Esmaeilizadeh M (2017) Pharmacological properties of Salvia officinalis and its components. J Tradit Complement Med 7:433–440. https://doi.org/10.1016/j.jtcme.2016.12.014

Giri A, Lakshmi Narasu M (2000) Production of podophyllotoxin from *Podophyllum hexandrum*: a potential natural product for clinically useful anticancer drugs. Cytotechnology 34:17–26. https://doi.org/10.1023/A:1008138230896

Gohil KJ, Patel JA, Gajjar AK (2010) Pharmacological Review on *Centella asiatica*: A Potential Herbal Cure-all. Indian J Pharm Sci 72:546–556. https://doi.org/10.4103/0250-474X.78519

Gopi DK, Mattummal R, Narayana SKK, Parameswaran S (2018) IUCN red listed medicinal plants of Siddha. J Med Res 1(1):15–22

Grover JK, Yadav S, Vats V (2002) Medicinal plants of India with anti-diabetic potential. J Ethnopharmacol 81:81–100. https://doi.org/10.1016/S0378-8741(02)00059-4

Gupta A, Chaphalkar S (2016) Immunopharmacological screening of aqueous root extract of Santalum album. J Herbmed Pharmacol 5:7–11

Gustafsson L (2002) Presence and abundance of red-listed plant species in Swedish forests. Conserv Biol 16(2):377–388

Hamilton KN, Turner SR, Ashmore SE (2009) Cryopreservation. In: Offord CA, Meagher PF (eds) Plant germplasm conservation in Australia: strategies and guidelines for developing, managing and utilising ex situ collections. Australian Network for Plant Conservation, Albury, Australia, p 129

Haq F, Ahmad H, Alam M (2011) Traditional uses of medicinal plants of Nandiar Khuwarr catchment (District Battagram), Pakistan. J Med Plants Res 5(1):39–48

Harish R, Shivanandappa T (2006) Antioxidant activity and hepatoprotective potential of *Phyllanthus niruri*. Food Chem 95:180–185. https://doi.org/10.1016/j.foodchem.2004.11.049

Hartmann HT, Kester DE, Davies FT Jr (1990) Plant Propagation Principles and Practices, 5th edn. Prentice Hall, Englewood Cliffs, NJ

Hayta S, Akgun IH, Ganzera M, Bedir E, Gurel A (2011) Shoot proliferation and HPLC-determination of iridoid glycosides in clones of *Gentiana cruciata* L. Plant Cell Tissue Organ Culture (PCTOC) 107(1):175–180

Hmar BZ, Mishra S, Pradhan RC (2017) Physico-chemical, mechanical and antioxidant properties of Kendu (*Diospyros melanoxylon* Roxb.). Curr Res Nutr Food Sci J 5(3):214–222

Hossain MT, Hoq MO (2016) Therapeutic use of *Adhatoda vasica*. Asian J Med Biol Res 2:156–163. https://doi.org/10.3329/ajmbr.v2i2.29005

Husain A, Sofi GD, Tajuddin T, Dang R, Kumar N (2010) Unani system of medicine-introduction and challenges. Med J Islam World Acad Sci 18(1):27–30

Jain M, Johnson TS, Krishnan P (2012) Biotechnological approaches to conserve the wealth of nature: endangered and rare medicinal plant species, a review. J Nat Remedies 12(2):93–102

Jain S, Mohan R, Rai R (2015) Ocimum sanctum as an herbal medicine: A Review. Int J Maxillofacial Res 1(1):1–12

Jaiswal Y, Liang Z, Zhao Z (2016) Botanical drugs in Ayurveda and traditional Chinese medicine. J Ethnopharmacol 194:245–259

Janey A, Mujahid Mohd Badr B (2015) An insight of pharmacognostic study and phytopharmacology of *Aquilaria agallocha*. J Appl Pharmaceut Sci 5(8):173–181

Javadi B, Sahebkar A, Emami SA (2017) Medicinal Plants for the Treatment of Asthma: A Traditional Persian Medicine Perspective. Curr Pharm Des 23:1623–1632. https://doi.org/10.2174/1381612822666161021143332

Jelastin KSM, Tresina PS, Mohan VR (2012) Antioxidant, antihyperlipidaemic and antidiabetic activity of *Eugenia floccosa* Bedd leaves in alloxan induced diabetic rats. J Basic Clin Pharm 3: 235–240

Joshua Allan J, Damodaran A, Deshmukh NS et al (2007) Safety evaluation of a standardized phytochemical composition extracted from *Bacopa monnieri* in Sprague–Dawley rats. Food Chem Toxicol 45:1928–1937. https://doi.org/10.1016/j.fct.2007.04.010

Juyal D, Thawani V, Thaledi S, Joshi M (2014) Ethnomedical Properties of Taxus wallichiana Zucc. (Himalayan Yew). J Tradit Complement Med 4:159–161. https://doi.org/10.4103/2225-4110.136544

Kala CP (2000) Status and conservation of rare and endangered medicinal plants in the Indian trans-Himalaya. Biol Conserv 93(3):371–379

Kalaivani T, Mathew L (2009) Phytochemistry and free radical scavenging activities of *Oroxylum indicum*. Environ We Int J Sci Tech 4:45–52

Kamble S, Gopalakrishnan R, Eapen S (2011) Production of camptothecin by hairy roots and regenerated transformed shoots of *Ophiorrhiza rugosa* var. decumbens. Nat Prod Res 25:1762–1765. https://doi.org/10.1080/14786419.2011.559469

Kamble SY, Patil SR, Sawant PS, Sawant S, Pawar SG, Singh EA (2010) Studies on plants used in traditional medicine by Bhilla tribe of Maharashtra. IJTK 9(3):591–598

Kanase V, Singh JD (2018) A Pharmacognostic and pharmacological review on *Curcuma pseudomontana* J.Graham. Asian J Pharm Clin Res:17–21. https://doi.org/10.22159/ajpcr.2018.v11i12.28131

Karuppusamy S (2018) Diversity and conservation status of red-listed medicinal plants in Tamil Nadu. Kongunadu Res J 5:41–49. https://doi.org/10.26524/krj269

Kasagana VN, Karumuri SS (2011) Conservation of medicinal plants (past, present & future trends). J Pharm Sci Res 3(8):1378

Kaushik N (2005) Saponins of Chlorophytum species. Phytochem Rev 4(2–3):191–196

Kavitha KS, Satish S (2013) Evaluation of antimicrobial and antioxidant activities from *Toona ciliata* Roemer. J Anal Sci Technol 4:23. https://doi.org/10.1186/2093-3371-4-23

Khan KH (2009) Roles of *Emblica officinalis* in medicine - A review. Bot Res Int 2(4):218–228

Khushboo PS, Jadhav VM, Kadam VJ, Sathe NS (2010) *Psoralea corylifolia* Linn.—"Kushtanashini". Pharmacogn Rev 4(7):69

Korpenwar AN (2012) Ethnomedicinal plant *Clerodendrum infortunatum* Linn used by the traditional practitioners for the treatment of cuts, wounds and burns. IJIBS 2(1):55–58

Kottaimuthu R, Vasudevan N (2014) Endemic and Red-listed Medicinal Plants used by the Valaiyars of Karandamalai, Southern Eastern Ghats, Tamil Nadu, India, Munich, GRIN Verlag. https://www.grin.com/document/269170

Kumar S, Kumari R, Mishra S (2019) Pharmacological properties and their medicinal uses of Cinnamomum: a review. J Pharm Pharmacol 71:1735–1761. https://doi.org/10.1111/jphp.13173

Kumar SS, Bhosle D, Janghel A et al (2015) Indian Medicinal Plants Used for Treatment of Rheumatoid Arthritis. Res J Pharm Technol 8:597–610. https://doi.org/10.5958/0974-360X.2015.00099.2

Kumar V, Van Staden J (2016) A review of *Swertia chirayita* (Gentianaceae) as a traditional medicinal plant. Front Pharmacol 6:308. https://doi.org/10.3389/fphar.2015.00308

Lakshmi V, Singh N, Shrivastva S et al (2010) Gedunin and Photogedunin of *Xylocarpus granatum* show significant anti-secretory effects and protect the gastric mucosa of peptic ulcer in rats. Phytomedicine 17:569–574. https://doi.org/10.1016/j.phymed.2009.10.016

Lampariello LR, Cortelazzo A, Guerranti R, Sticozzi C, Valacchi G (2012) The magic velvet bean of *Mucuna pruriens*. J Tradit Complement Med 2(4):331–339

Lobay D (2015) Rauwolfia in the treatment of hypertension. Integrative Med 14(3):40

Mahadevan S, Park Y (2008) Multifaceted therapeutic benefits of ginkgo biloba L.: chemistry, efficacy, safety, and uses. J Food Sci 73:R14–R19. https://doi.org/10.1111/j.1750-3841.2007.00597.x

Malayappan MS (2017) Paediatric management in Siddha system of medicine. https://www.britannica.com/science/Siddha-medicine

Mallikarjuna Gowda AP, Ranjini TN, Peethambar SK, Praneeth YS (2019) Underutilized, yet potential medicinal plants of Western Ghats. Acta Hortic 15–22. https://doi.org/10.17660/ActaHortic.2019.1241.3

Mallón R, Bunn E, Turner SR, González ML (2008) Cryopreservation of *Centaurea ultreiae* (Compositae) a critically endangered species from Galicia (Spain). Cryo-Letters 29(5):363–370

Mandal BB (2000) Cryopreservation research in India: present status and future perspectives. In Cryopreservation of tropical plant germplasm: current research progress and application. Proceedings of an international workshop, Tsukuba, Japan, October, 1998 (pp. 282-286). International Plant Genetic Resources Institute (IPGRI).

Mandal BB, Dixit-Sharma S, Srivastava PS (2009) Cryopreservation of embryogenic cultures of *Dioscorea bulbifera* L. by encapsulation-dehydration. Cryo-Letters 30(6):440–448

Manish M (2013) Current status of endangered medicinal plant *Hedychium coronarium* and causes of population decline in the natural forests of Anuppur and Dindori districts of Madhya Pradesh, India. Int Res J Biol Sci 2(3):1–6

Maroyi A (2018) *Albizia adianthifolia*: Botany, Medicinal Uses, Phytochemistry, and Pharmacological Properties. Sci World J 2018:e7463584. https://doi.org/10.1155/2018/7463584

Mazumder PM, Das S, Das S, Das MK (2011) Phyto-pharmacology of *Berberis aristata* DC: A review. J Drug Deliv Ther 1:46–50

Mittal J, Sharma MM, Batra A (2014) *Tinospora cordifolia*: a multipurpose medicinal plant-A. J Med Plants 2(2):32–47

Mobli M, Qaraaty M, Amin G et al (2015) Scientific evaluation of medicinal plants used for the treatment of abnormal uterine bleeding by Avicenna. Arch Gynecol Obstet 292:21–35. https://doi.org/10.1007/s00404-015-3629-x

Mohan S, Thiagarajan K, Sundaramoorthy B et al (2016) Alleviation of 4-nitroquinoline 1-oxide induced oxidative stress by *Oroxylum indicum* (L.) leaf extract in albino Wistar rats. BMC Complement Altern Med 16:229. https://doi.org/10.1186/s12906-016-1186-x

Mollik AH, Islam T, Khatun A et al (2009) Medicinal plants used against syphilis and gonorrhea by traditional medicinal practitioners of Bangladesh. Planta Med 75:PH40. https://doi.org/10.1055/s-0029-1234757

Mukherjee D, Chakraborty S (2019) *Coptis teeta*: Conservation and Cultivation Practice - A Rare Medicinal Plant on Earth. Curr Investig AgricCurr Res 6:845–851. https://doi.org/10.32474/CIACR.2019.06.000244

Mukherjee PK, Kumar V, Mal M, Houghton PJ (2007) *Acorus calamus*: Scientific Validation of Ayurvedic Tradition from Natural Resources. Pharm Biol 45:651–666. https://doi.org/10.1080/13880200701538724

Naik Y, Jayaram S, Harish Nayaka MA et al (2007) Gastroprotective effect of swallow root (*Decalepis hamiltonii*) extract: Possible involvement of H+-K+-ATPase inhibition and antioxidative mechanism. J Ethnopharmacol 112:173–179. https://doi.org/10.1016/j.jep.2007.02.021

Nair MP, Shastry ARK (1990) Red data book of Indian Plants, Vol-I, 1987, Vol-II, 1988, Vol-III, 1990. Botanical Survey of India, Calcutta, India

Nair RVR, Jayasree DV, Biju PG, Baby S (2020b) Anti-inflammatory and anticancer activities of erythrodiol-3-acetate and 2, 4-di-tert-butylphenol isolated from *Humboldtia unijuga*. Nat Prod Res 34:2319–2322. https://doi.org/10.1080/14786419.2018.1531406

Nair SV, Ved DK, Ravikumar K, Tabassum IF, Sureshchandra ST, Somasekhar BS, Sathya S, Barve V, Naveen S, Payyappalimana U, Shankar D (2020a) Indian medicinal plants database (IMPLAD) and threatened medicinal plants of India. In: Conservation and utilization of threatened medicinal plants. Springer, Cham, pp 63–92

Nambisan P (2017) An introduction to ethical, safety and intellectual property rights issues in biotechnology. Academic Press, Cambridge, MA

Nammi S, Boini MK, Lodagala SD, Behara RBS (2003) The juice of fresh leaves of *Catharanthus roseus* Linn. reduces blood glucose in normal and alloxan diabetic rabbits. BMC Complement Altern Med 3:4. https://doi.org/10.1186/1472-6882-3-4

Nautiyal S, Rajasekaran C, Varsha NP, Mannam S (2015) Traditional ecological knowledge documentation and conservation of medicinal plants resources of BR Hills through scientific and technological interventions. Institute for Social and Economic Changes Monograph Series No. 41. ISBN: 81-7791-140-6.

Nayak M, Nagarajan A, Majeed M, Majeed M (2017b) Pharmacognostic Evaluation of Leaf and Stem Wood Extracts of *Artocarpus hirsutus* Lam. Pharm J 9:887–894. https://doi.org/10.5530/pj.2017.6.139

Nayak M, Nagarajan A, Majeed M et al (2017a) Flavonoids from *Artocarpus hirsutus* Lam: Synthesis of new prenyl ethers, acetates and their anti-acne activity. Cogent Chem 3:1416557. https://doi.org/10.1080/23312009.2017.1416557

Ngo LT, Okogun JI, Folk WR (2013) 21st century natural product research and drug development and traditional medicines. Nat Prod Rep 30(4):584–592

Nishteswar K (2014) Depleting medicinal plant resources: A threat for survival of Ayurveda. Ayu 35(4):349

Pandey MM, Katara A, Pandey G, Rastogi S, Rawat AKS (2013) An important Indian traditional drug of Ayurveda jatamansi and its substitute Bhootkeshi: chemical profiling and antioxidant activity. Evid Based Compl Altern Med 2013:142517

Pandey MM, Rastogi S, Rawat AKS (2007) *Saussurea costus*: Botanical, chemical and pharmacological review of an ayurvedic medicinal plant. J Ethnopharmacol 12:379–390

Pant S, Pant VS (2011) Status and conservation management strategies for threatened plants of Jammu and Kashmir. J Phytol 3(7)

Parasuraman S, Thing GS, Dhanaraj SA (2014) Polyherbal formulation: Concept of Ayurveda. Pharmacogn Rev 8(16):73

Pathirana CK, Madhujith T, Eeswara J (2020) Bael (*Aegle marmelos* L. Corrêa), a medicinal tree with immense economic potentials. Adv Agric 2020:e8814018. https://doi.org/10.1155/2020/8814018

Pattanaik C, Reddy CS, Reddy KN (2009) Ethno-medicinal Survey of Threatened Plants in Eastern Ghats, India. Our Nat 7:122–128 https://doi.org/10.3126/on.v7i1.2559

Paunescu A (2009) Biotechnology for endangered plant conservation: a critical overview. Rom Biotechnol Lett 14(1):4095–4103

Patil RH, Prakash K, Maheshwari VL (2010) Hypolipidemic Effect of *Celastrus paniculatus* in Experimentally Induced Hypercholesterolemic Wistar Rats. Indian J Clin Biochem 25:405–410. https://doi.org/10.1007/s12291-010-0050-x

Phillips OL, Meelleur BA (1998) Usefulness and economic potential of the rare plants of the United States: a statistical survey. Econ Bot 52(1):57–67

Phondani PC, Maikhuri RK, Saxena KG (2014) The efficacy of herbal system of medicine in the context of allopathic system in Indian Central Himalaya. J Herbal Med 4(3):147–158

Pimm SL, Russell GJ, Gittleman JL, Brooks TM (1995) The future of biodiversity. Science 269(5222):347–350

Polunin O, Stainton A (1984) Flowers of the Himalaya. Oxford University Press, New Delhi, India, pp 413–416

Potdar D, Hirwani RR, Dhulap S (2012) Phyto-chemical and pharmacological applications of *Berberis aristata*. Fitoterapia 83:817–830. https://doi.org/10.1016/j.fitote.2012.04.012

Pradhan P, Joseph L, Gupta V et al (2009) *Saraca asoca* (Ashoka): A Review. J Chem Pharm Res:62–67

Prasad AD, Shyma TB (2013) Medicinal plants used by the tribes of Vythiri taluk, Wayanad district (Kerala state) for the treatment of human and domestic animal ailments. J Med Plant Res 7 (20):1439–1451

Prasad N, Yang B, Kong KW et al (2013) Phytochemicals and Antioxidant Capacity from *Nypa fruticans* Wurmb. Fruit Evid Based Complement Alternat Med 2013:e154606. https://doi.org/10.1155/2013/154606

Pushpangadan P, George V, Sreedevi P, Ijinu TP, Ninawe A (2014) Ethnozoological knowledge of Indian scheduled tribe, scheduled caste and rural communities. Indian J Trad Knowl 13(4): 735–741

Qazi PH, Rashid A, Shawl SA (2011) *Podophyllum hexandrum*: a versatile medicinal plant. Int J Pharm Pharm Sci 3(5):261–268

Rahman S, Parvin R (2014) Therapeutic potential of *Aegle marmelos* (L.) - An overview. Asian Pac J Trop Dis 4:71–77. https://doi.org/10.1016/S2222-1808(14)60318-2

Rai MK (2010) Biotechnological strategies for conservation of rare and endangered medicinal plants. Biodiversitas 11(3):157–166

Rajasekaran C (2000) Himalayan Plant Wealth, *Nirjharni,* pp. 180, H.N.B.G.U.

Rameshkumar R, Largia MJV, Satish L et al (2017) In vitro mass propagation and conservation of *Nilgirianthus ciliatus* through nodal explants: A globally endangered, high trade medicinal plant of Western Ghats. Plant Biosyst-Int J Deal Asp Plant Biol 151:204–211. https://doi.org/10.1080/11263504.2016.1149120

Rands MRW, Adams WM, Bennun L et al (2010) Biodiversity Conservation: Challenges Beyond 2010. Science 329:1298–1303. https://doi.org/10.1126/science.1189138

Rao CV, Ojha SK, Radhakrishnan K et al (2004) Antiulcer activity of *Utleria salicifolia* rhizome extract. J Ethnopharmacol 91:243–249. https://doi.org/10.1016/j.jep.2003.12.020

Rathi B, Sahu J, Koul S, Kosha RL (2013) Detailed pharmacognostical studies on *Berberis aristata* DC plant. Anc Sci Life 32:234–240. https://doi.org/10.4103/0257-7941.131981

Reed BM (ed) (2008) Plant cryopreservation: a practical guide. Springer, New York, p 513

Rhee TG, Ng JY, Dusek JA (2017) Utilization and perceived benefits of homeopathy and herbal therapies in US adults: Implications of patient-centered care. Complement Ther Clin Pract 29:9–15

Rouf R, Uddin SJ, Shilpi JA, Alamgir M (2007) Assessment of antidiarrhoeal activity of the methanol extract of *Xylocarpus granatum* bark in mice model. J Ethnopharmacol 109:539–542. https://doi.org/10.1016/j.jep.2006.08.015

Ryu SY, Choi SU, Lee CO, Zee OP (1992) Antitumor activity of *Psoralea corylifolia*. Arch Pharm Res 15:356–359. https://doi.org/10.1007/BF02974112

Sahoo MR, Dhanabal SP, Jadhav AN, Reddy V, Muguli G, Babu UV, Rangesh P (2014) Hydnocarpus: An ethnopharmacological, phytochemical and pharmacological review. J Ethnopharmacol 154(1):17–25

Saini HS (2008) Climate change and its future impact on the Indo-Gangetic Plain (IGP). Earth Science India 1:138–147

Sajem AL, Rout J, Nath M (2008) Traditional tribal knowledge and status of some rare and endemic medicinal plants of North Cachar Hills district of Assam, Northeast India Ethnobotanical Leaflets, Northeast India. Ethnobotanical Leaflets 2008:31

Sanjeev K, Nayan DK (2016) A review on historical aspect of *Balanites aegyptiaca* (L.) Del (INGUDI). Int J Appl Ayur Res 2:2347–6362

Saqib F, Mushtaq Z, Janbaz KH et al (2018) Pharmacological basis for the medicinal use of *Michelia champaca* in gut, airways and cardiovascular disorders. Asian Pac J Trop Med 11: 292. https://doi.org/10.4103/1995-7645.231470

Sarup P, Bala S, Kamboj S (2015) Pharmacology and Phytochemistry of Oleo-Gum Resin of *Commiphora wightii* (Guggulu). Scientifica 2015:e138039. https://doi.org/10.1155/2015/138039

Satdive RK, Fulzele DP, Eapen S (2003) Studies on Production of Ajmalicine in Shake Flasks by Multiple Shoot Cultures of *Catharanthus roseus*. Biotechnol Prog 19:1071–1075. https://doi.org/10.1021/bp020138g

Sen-Rong H, Ming-Hua Y (2009) High-efficiency vitrification protocols for cryopreservation of in vitro grown shoot tips of rare and endangered plant *Emmenopterys henryi* Oliv. Plant Cell Tissue Organ Culture (PCTOC) 99(2):217–226

Senthil Nathan S, Kalaivani K, Sehoon K (2006) Effects of *Dysoxylum malabaricum* Bedd. (Meliaceae) extract on the malarial vector *Anopheles stephensi* Liston (Diptera: Culicidae). Bioresour Technol 97:2077–2083. https://doi.org/10.1016/j.biortech.2005.09.034

Shalini M, Kalaivani T, Rajasekaran C (2017) Evaluation of ethyl gallate for its antioxidant and anticancer property against chemical induced tongue carcinogenesis in mice. Biochem J 474:3011–3025. https://doi.org/10.1042/BCJ20170316

Shah GM, Abbasi AM, Khan N, Guo X, Khan MA, Hussain M, Bibi S, Nazir A, Tahir AA (2014) Traditional uses of medicinal plants against malarial disease by the tribal communities of Lesser Himalayas – Pakistan. J Ethnopharmacol 155(1):450–462

Shah NC, Kapoor LD (1978) Depletive medicinal plants of Kumaon Himalayas. J Res Ind Med Yoga Homeo 13(3):38–43

Shahid AP, Sasidharan N, Salini S et al (2018) *Kingiodendron pinnatum*, a pharmacologically effective alternative for *Saraca asoca* in an Ayurvedic preparation, Asokarishta. J Tradit Complement Med 8:244–250. https://doi.org/10.1016/j.jtcme.2017.06.005

Shankar R, Rawat MS (2013) Conservation and cultivation of threatened and high valued medicinal plants in North East India. Int J Biodivers Conserv 5(9):584–591

Sharma A, Kaur R, Sharma N (2014) *In vitro* morphogenic response of different explants of *Gentiana kurroo* Royle from Western Himalayas—an endangered medicinal plant. Physiol Mol Biol Plants 20:249–256. https://doi.org/10.1007/s12298-013-0220-4

Shravya S, Vinod BN, Sunil C (2017) Pharmacological and phytochemical studies of *Alangium salvifolium* Wang. – A review. Bull Fac Pharm Cairo Univ 55:217–222. https://doi.org/10.1016/j.bfopcu.2017.07.001

Shyaula SL (2011) Phytochemicals, traditional uses and processing of Aconitum species in Nepal. Nepal J Sci Technol 12:171–178

Siddiqui MZ (2011) *Boswellia serrata*, a potential anti-inflammatory agent: an overview. Indian J Pharm Sci 73(3):255

Sidhu Y (2011) *In vitro* micropropagation of medicinal plants by tissue culture. The Plymouth Student Scientist, 2010 4(1):432–449

Singh N, Singh AK, Nilofer Absar VRS, Singh P (2018) Importance of endangered/rare, Astavarga medicinal plants in traditional system of medicine in Ayurveda. Int J Agric Sci 14(1):258–265

Singh S, Krishna THA, Kamalraj S et al (2015) Phytomedicinal importance of *Saraca asoca* (Ashoka): an exciting past, an emerging present and a promising future. Curr Sci 109:1790–1801

Singh TT, Sharma HM, Devi AR, Sharma HR (2014) Plants used in the treatment of piles by the scheduled caste community of Andro village in Imphal East District, Manipur (India). J Plant Sci 2:113–119

Skinder BM, Ganai BA, Wani AH (2017) Scientific study of *Gentiana kurroo* Royle. Medicines 4(4):74

Solomon Raju AJ, Venkata Ramana K, Suvarna Raju P, Dileepu Kumar B , Lakshminarayana G, Sravan Kumar S (2019) Traditional food and medicinal uses of Cycas sphaerica Roxb and Cycas beddomei Dyer (Cycadaceae). Species 20:24–27

Sreelekshmi R, Deepa MS (2020) Chronic Anticonvulsant Effect of Ghee Prepared With Humboldtia Vahliana (Attuvanchi Bark) – An Ethnomedicinal Practice In Kerala. Int J Ayurveda Pharma Res 2020:61–65. https://doi.org/10.47070/ijapr.v8iSupply2.1654

Subbaiyan B, Samydurai P, Prabu MK et al (2014) Inventory of Rare, Endangered and Threatened (RET) Plant Species in Maruthamalai Hills, Western Ghats of Tamil Nadu, South India. Our Nat 12:37–43. https://doi.org/10.3126/on.v12i1.12255

Tandel MH, Animasaun DA, Krishnamurthy R (2018) Growth and phytochemical composition of *Adhatoda zeylanica* in response to foliar application of growth hormones and urea. J Soil Sci Plant Nutr 18(3):881–892

Tasheva K, Kosturkova G (2012) Towards agrobacterium-mediated transformation of the endangered medicinal plant golden root. Agro Life Sci. Aust Dent J 1(1):132–138

Tasheva K, Kosturkova G (2013) Role of biotechnology for protection of endangered medicinal plants. In: Environmental biotechnology-New approaches and prospective applications. IntechOpen, pp 235–238

Thas JJ (2008) Siddha medicine—background and principles and the application for skin diseases. Clin Dermatol 26(1):62–78

Ticktin T, Nantel P, Ramirez F, Johns T (2002) Effects of variation on harvest limits for nontimber forest species in Mexico. Conserv Biol 16(3):691–705

Touchell DH, Dixon KW (1994) Cryopreservation for seed banking of Australian species. Ann Bot 40:541–546

Ukani MD, Mehta NK, Nanavati DD (1996) Aconitum heterophyllum (Ativisha) In Ayurveda. Anc Sci Life 16:166–171

Vanitha A, Kalimuthu K, Chinnadurai V, Nisha KMJ (2019) Phytochemical screening, FTIR and GCMS analysis of aqueous extract of Caralluma bicolor – An endangered plant. Asian J Pharm Pharmacol 5:1122–1130. https://doi.org/10.31024/ajpp.2019.5.6.7

Varghese KJ, Belzik N, Nisha AR et al (2010) Pharmacognostical and phytochemical studies of a mangrove (*Sonneratia caseolaris*) from Kochi of Kerala state in India. J Pharm Res 3:2625–2627

Venkatadri B, Khusro A, Aarti C et al (2017) In vitro assessment on medicinal properties and chemical composition of *Michelia nilagirica* bark. Asian Pac J Trop Biomed 7:782–790. https://doi.org/10.1016/j.apjtb.2017.08.003

Verma P, Mathur AK, Jain SP, Mathur A (2012) In vitro conservation of twenty-three overexploited medicinal plants belonging to the Indian subcontinent. Sci World J 2012:929650

Vijayalatha SJ (2004) An Ornamental garden with medicinal plants an indirect approach for conservation of medicinal plants. Indian J Arecanut Spices Med Plants 6(3):98–107

Vimala G, Gricilda Shoba F (2014) A Review on Antiulcer Activity of Few Indian Medicinal Plants. Int J Microbiol 2014:e519590. https://doi.org/10.1155/2014/519590

Vollala VR, Upadhya S, Nayak S (2011) Enhancement of basolateral amygdaloid neuronal dendritic arborization following *Bacopa monniera* extract treatment in adult rats. Clinics 66: 663–671. https://doi.org/10.1590/S1807-59322011000400023

Wagh VV, Jain AK (2013) Status of threatened medicinal plants of Jhabua district, Madhya Pradesh, India. Ann Plant Sci 2(10):395–400

Walpola BC, Subasinghe S, Yoon M-H, IndikaUpaliArunakumara KK (2011) *Pterocarpus santalinus* Linn. f. (Rathhandun): A review of its botany, uses, phytochemistry and pharmacology. J Korean Soc Appl BiolChem 54:495–500. https://doi.org/10.3839/jksabc.2011.076

Wang G-W, Jin H-Z, Zhang W-D (2013) Constituents from Aphanamixis species and their biological activities. Phytochem Rev 12:915–942. https://doi.org/10.1007/s11101-013-9317-1

Wei LS, Wee W, Siong JYF, Syamsumir DF (2011) Characterization of Antimicrobial, Antioxidant, Anticancer Property and Chemical Composition of *Michelia champaca* Seed and Flower Extracts. Stamford J Pharm Sci 4:19–24. https://doi.org/10.3329/sjps.v4i1.8862

Welch HJ (1970) Mist propagation and automatic watering. Faber and Faber, London

Yun DJ, Hashimoto T, Yamada Y (1992) Metabolic engineering of medicinal plants: transgenic Atropa belladonna with an improved alkaloid composition. Proc Natl Acad Sci 89(24): 11799–11803

Plant Genetic Resources Management in the Framework of Policy Developments 18

N. Sivaraj, S. R. Pandravada, Kamala Venkateswaran, Siva Ramamoorthy, and K. Anitha

Abstract

Plant genetic resources constitute our invaluable assets to meet the growing needs to increase crop production and productivity. Plant genetic resources constitute a unique global heritage and their conservation and utilization are of immediate concern. Over 7.4 million accessions of crop germplasm are currently being conserved worldwide in gene banks. Despite its overall advantage and promotion by the international community, in situ conservation is still inadequate. Various international conventions have addressed the topic of genetic erosion and declining use of agrobiodiversity in modern agriculture. Concern about the future vulnerability of agricultural production, food security and environmental stability has moved the plant genetic resource management including conservation and sustainable use to the top of the international development agenda. Implications and impact of global conventions/treaties/agreements, such as the Convention on Biological Diversity, Trade-Related Intellectual Property Rights and Sanitary and Phytosanitary Agreements under WTO, International Plant Protection Convention, Global Plan of Action, International Treaty on Plant Genetic Resources for Food and Agriculture, Convention on International Trade in Endangered Species of Wild Fauna and Flora, Intergovernmental Committee on Intellectual Property and Genetic Resources, Traditional Knowledge and Folklore (ICGTK), International Union for Protection of Plant Varieties, Nagoya Protocol on Access and Benefit Sharing, Global Crop Diversity Trust and Cartagena Protocol on Biosafety, on plant genetic resource activities are highlighted. The changing national

N. Sivaraj (✉) · S. R. Pandravada · K. Venkateswaran · K. Anitha
ICAR-National Bureau of Plant Genetic Resources, Regional Station, Hyderabad, India

S. Ramamoorthy
Department of Biotechnology, School of Bio Sciences and Technology, Vellore Institute of Technology, Vellore, India

S. Ramamoorthy et al. (eds.), *Plant Genetic Resources, Inventory, Collection and Conservation*, https://doi.org/10.1007/978-981-16-7699-4_18

scenario in the framework of the above global developments, particularly the provisions under Biological Diversity Act and Protection of Plant Varieties and Farmers' Rights Act, is discussed. The status of PGR conservation and utilization at global and national levels is briefly discussed. Measures to promote PGR utilization including core development, pre-breeding, gene prospecting, allele mining, public–private partnerships and safe transboundary movement are discussed. Major issues and future thrust areas in the field of plant genetic resources are listed.

Keywords

Plant genetic resources · Conservation · Utilization · Global treaties

Plant genetic resources constitute a unique global heritage and our invaluable assets to meet the growing needs to increase crop production, productivity and capturing genes for posterity. Though about 30 plant species provide the major food and energy sources, some 3000 species of cultivated plants are grown worldwide for various purposes. Besides, there is a rich genetic wealth of other economic plants located in different centres of diversity. Unfortunately, in view of the rapid pace of economic development, genetic diversity in such areas is now highly threatened. All these genetic resources of potential value for current and future needs of humankind are increasingly being exploited and in danger of extinction due to habitat destruction. Thus, plant genetic resources constitute a unique global heritage and their conservation and utilization are of immediate concern. There have been tremendous efforts in ex situ conservation of plant genetic resources. Conserving plant genetic resources ex situ in gene banks has been the most significant step towards prevention of loss of diversity and genetic erosion in farmers' fields. Over 1700 gene banks and 2500 botanical gardens conserve a total of 7.4 million germplasm accessions and 80,000 species, respectively, around the world. Despite its overall advantage and promotion by the international community, in situ conservation is still inadequate. India is endowed with vast genetic resources of crop plants and is known as one of the mega centres of biodiversity as well as one of the centres of diversity of agriculture. Several indigenous farmers' varieties and landraces have been cultivated by the farmers since the ancient time in India. These landraces have been evolved, refined, conserved and used by local people for their specific agronomic, nutritional and other diverse economic traits. In the recent times, aggressive promotion of monoculture and intensive agriculture through the use of high-yielding varieties, even in the stressed and fragile agroecosystems to increase production and productivity of crops, has threatened the native plant genetic resources. Plant genetic resources (PGR) have been playing a significant role in crop improvement and agricultural development in terms of providing the indispensable raw material, i.e. germplasm. Plant genetic resources assumed critical importance with the legislation of CBD, WTO and TRIPS. These recent developments in conjunction with rapid advances in agricultural research and development have led to the emergence of new legal, political and technological regimes that dictate crop biodiversity management. The issues of ownership and access and benefit sharing of agri-biodiversity resulted in a paradigm shift from plant genetic resources being a 'common heritage of mankind' to 'sovereign rights of nations'. The National Bureau of Plant Genetic Resources (NBPGR) has been advocating the immense importance of these local landraces and undertaking all-out efforts for their long-term conservation and sustainable utilization.

Currently, a number of new approaches offer advances in crop improvement. These include gene transfer and use of marker-assisted selection as tools for managing 'desired' trait diversity and improving the efficiency and scope of both conventional plant breeding and genetic engineering. An increasing range of techniques is available to facilitate wide crosses in order to access germplasm in the secondary and tertiary gene pools. There is sizeable investment in many of these approaches including public–private partnership (PPP) mode, although such investments are heavily biased towards the agricultural systems of developed countries and more export-oriented crops.

Various international conventions have addressed the topic of genetic erosion and declining use of agrobiodiversity in modern agriculture. Concern about the future vulnerability of agricultural production, food security and environmental stability has moved the conservation and sustainable use of plant genetic resources to the top of the international development agenda. Opportunities and challenges in view of the recent policy developments in PGR conservation and utilization are being discussed.

18.1 Policy Developments and Their Implications on Germplasm Ownership and IPR

The major global issues impacting genetic resource management include the recent international treaties, conventions, agreements, global climate change, use of biotechnology and other technological advances in agriculture, biosecurity and biosafety issues, etc. Various international and national policy developments that impacted plant genetic resource conservation and use are briefly summarized below:

International

1. The CBD, adopted in 1992 at Rio de Janeiro, provides national sovereignty over genetic resources and access conditions for other sovereign parties.
2. The World Trade Organization (WTO) is the only global international organization dealing with the rules of trade between nations. At its heart are the WTO agreements, negotiated and signed by the bulk of the world's trading nations and ratified in their parliaments. The goal is to help producers of goods and services, exporters, and importers conduct their business. TRIPS, adopted in Marrakesh in 1994, provides a minimum IP protection standard for biological matter such as plant varieties, microorganisms and microbiological processes.
3. The International Plant Protection Convention (IPPC) is a multilateral treaty for international cooperation in plant protection. The Convention makes provision

for the application of measures by governments to protect their plant resources from harmful pests (phytosanitary measures) which may be introduced through international trade. The IPPC was first adopted in 1951 and has been amended twice, most recently in 1997. The revision of the IPPC agreed in 1997 and which entered into legal force on 2 October 2005 represents an updating of the Convention to reflect contemporary phytosanitary concepts and the role of the IPPC in relation to the Uruguay Round Agreements of the WTO, particularly the SPS Agreement.

4. Sanitary and phytosanitary or SPS measures are an agreement on how governments can apply food safety and animal and plant health measures that set out the basic rules in the WTO. The WTO's SPS Agreement states that 'to harmonize sanitary and phytosanitary measures on as wide a basis as possible, Members shall base their sanitary or phytosanitary measures on international standards, guidelines or recommendations'. The SPS Agreement identifies the IPPC as the reference organization developing international standards for plant health (phytosanitary) measures.
5. The Global Plan of Action (adopted in 1996) for the conservation and sustainable use of plant genetic resources for food and agriculture (PGRFA) includes, among its priority activities, 'increasing genetic enhancement and base-broadening efforts', 'promoting sustainable agriculture through diversification of crop production and broader diversity in crops' and 'supporting on farm management and improvement of PGRFA'.
6. International Treaty on Plant Genetic Resources for Food and Agriculture entered into force on 29.06.2004. The Treaty provides an internationally agreed framework for the conservation and sustainable use of all plant genetic resources for food and agriculture. Indeed, the Treaty's objectives are 'the conservation and sustainable use of plant genetic resources for food and agriculture and the fair and equitable sharing of the benefits arising out of their use, in harmony with the Convention on Biological Diversity, for sustainable agriculture and food security'. The Treaty recognizes the enormous contribution that farmers and their communities have made and continue to make to the conservation and development of plant genetic resources. This is the basis for Farmers' Rights, which include the protection of traditional knowledge and the right to participate equitably in benefit sharing and in national decision-making about plant genetic resources.
7. The Convention on International Trade in Endangered Species of Wild Fauna and Flora (CITES) is an international agreement between governments. CITES was drafted as a result of a resolution adopted in 1963 at a meeting of members of IUCN (The World Conservation Union) and on 1 July 1975 CITES entered in force. Its aim is to ensure that international trade in specimens of wild animals and plants does not threaten their survival.
8. ICGTK was set up in 2001 by WIPO to discuss IP issues relating to access to genetic resources and protection of traditional knowledge, including disclosure requirements in patent applications.

9. UPOV provides legal protection for plant varieties fulfilling the NDUS criteria (new, distinct, uniform and stable) while including a breeder's exemption and farmer's privilege.
10. The International Treaty on Plant Genetic Resources for Food and Agriculture, adopted in Rome in 2001, provides a multilateral system of access and benefit sharing under a revised material transfer agreement (MTA) in relation to some 35 defined crops.
11. The Global Crop Diversity Trust, set up in 2002, is an attempt by the Food and Agriculture Organization (FAO) of the United Nations and the World Bank to establish a trust fund for global ex situ collections of germplasm of relevance for food and agriculture.
12. The Cartagena protocol, adopted in Montreal in 2000, provides rules for the transfer of genetically modified living organisms across borders.
13. In 2002, the CBD adopted the Bonn Guidelines on Access to Genetic Resources and Fair and Equitable Sharing of Benefits Arising out of their Utilization. A voluntary supplement to the CBD, the Bonn Guidelines offer basic information about the rules on access and concrete procedures (or protocols) to follow. The objectives of the Bonn Guidelines in relation to academic research are:
    (a) To promote awareness of the implementation of relevant provisions of the CBD
    (b) To provide parties to the CBD and stakeholders with a transparent framework to facilitate access to genetic resources and ensure fair and equitable sharing of benefits
    (c) To provide information about the practices and approaches to be adopted by users and providers in the context of access and benefit sharing
    (d) To promote capacity building and transfer of appropriate technology to providing parties
14. The Nagoya Protocol was adopted by the Conference of the Parties to the Convention on Biological Diversity at its tenth meeting on 29th October 2010 in Nagoya, Japan. It entered into force on October 12th, 2014. It sought to ensure commercial and research utilization of genetic resources leading to sharing its benefits with the government and the community that conserved such resources. It deals with access to genetic resources and the fair and equitable sharing of benefits arising from their utilization to the Convention on Biological Diversity. The Nagoya Protocol sets out obligations for its contracting parties to take measures in relation to access to genetic resources, benefit sharing and compliance. The parties to the protocol are obliged to take actions with respect to access to genetic resources, benefit sharing and compliance.
    - Access measures should have legal certainty, transparency and clarity.
    - The rules and procedures thereof should be fair and non-arbitrary.
    - There should be clear rules for prior informed consent and mutually agreed terms.
    - The rules should have provisions for the issuance of a permit (or its equivalent) when granted access.

- Encourage research that will contribute to the conservation of biodiversity and its sustainable use.
- Cases of imminent emergencies that threaten plant, animal or human health should be considered.
- Take into consideration the importance of genetic resources for food and agriculture, to have food security.

18.1.1 Intellectual Property Rights

IP rights are temporary, exclusive ownership rights to the *application* of an idea. Such rights may be granted in the form of patents, trademarks, industrial designs, copyrights, geographical indications, or trade secrets. Given the breakthroughs in biotechnology and information technologies in the last few decades, intellectual property has expanded considerably into the area of biological matter. For example, in the area of agricultural research the following biological matter falls under various IP regimes:

1. Plant seeds or other propagative plant parts collected after 1994
2. Plant and animal cell lines
3. Plasmids
4. Other recombinant vectors
5. Gene promoters
6. Gene markers
7. Transformed bacteria
8. Isolated plant DNA
9. Plant cDNAs
10. Isolated animal DNA
11. Bacteria (other than the transformed bacteria)
12. Isolated/purified proteins (other than those obtained by purchase of laboratory reagents)
13. Equipment for specialized laboratory purposes
14. Information regarding laboratory methods
15. Genomic sequence database(s)
16. Other nucleotide sequence database(s) such as PCR primer databases and cDNA sequences

Traditional and farming knowledge is also protected under the CBD and the Treaty, subject to national legislation. In general, researchers in the public sector, using proprietary biological materials and related information owned by private sector companies, may have to sign agreements stipulating further use and confidentiality conditions. Furthermore, public research products using proprietary materials and methods may be required to sign license and royalty agreements with those who hold the relevant IP rights.

18.2 National

Recognizing the values of plant genetic resources for food and agriculture and in harmony with several international treaties, Indian Government has come up with new initiatives such as enactment and implementation of the Biodiversity Act and Plant Varieties Protection and Farmers' Rights Act (PVP&FRA).

1. The Biological Diversity Act (BDA), 2002, identifies the strategies and the actions that need to be taken for access to genetic resources, their conservation, sustainable use and fair and equitable sharing of benefits as per the provisions of the CBD.
2. Protection of Plant Varieties and Farmers' Rights Act (PPVFRA), 2001, is unique in providing equal rights to the farmers as breeder and conserver of genetic resources of local importance.
3. Plant, Fruits and Seeds (PFS) Order, 1989, governs the quarantine processing of planting material imported from abroad.

18.3 Registration of Genetic Resources

The Indian Council of Agricultural Research has instituted a mechanism for registration of experimentally developed germplasm of potential value. More than 2000 proposals have been screened by NBPGR for registration of germplasm. So far, 1733 potentially valuable germplasms belonging to over 115 crops have been registered. The registration of genetic resources would help in protecting the country's genetic resources in the wake of IPR regime.

18.3.1 Implications of PPVFRA

- Plant Genetic Savior Community Recognition Award has been instituted by the Authority. Selection of individual/community may cause disharmony.
- Breeding objectives now need to take care of DUS characters of the crop concerned.
- Evaluation of new varieties needs to be done along with probable reference varieties for DUS testing.
- Tests for DUS should be kept in view.
- Annual maintenance and registration fee should be provided by the breeder.
- Annual maintenance fee amount should be decided by the authority on quantum of sale or other parameters.
- For new varieties, DUS testing fee (Rs. 15,000–35,000) should be deposited along with seed.
- Passport data of parental lines must be recorded by breeders (NBPGR help).
- Lawful acquisition of genetic material (MTAs/authorization from owners).

- Maintenance of pure seed/propagating material by depositories (institutions/ breeders).
- Evaluation in farmers' fields (without commercialization) to counter litigation by the farmer.

18.3.2 Implications of CBD/BDA

- Overriding of sovereignty rights over the easy access.
- Slower rate of germplasm exchange and transboundary movement.
- Exemption for collaborative research through government-sponsored or government-approved institutions subject to overall policy guidelines and approval of the central government.
- India is also strong in agricultural research and development (R&D) and exchange of resources for such R&D has been a common practice in the past. In some cases, such practices may be required in future too.
- In case of the agriculture sector, benefit sharing may be more modest, but on high-end commercial products (breakfast cereals, oil, syrups, etc.), benefit sharing may be calculated differently than seed sales.
- Research and development work by the academic community would also require free access to resources even though results of their research may have to be subject to some regulation for the purposes of commercial exploitation, at least by people outside India.
- Prior approval of SBB for commercialization.
- Mechanisms for monitoring new taxa discovered and deposition of voucher specimens in repositories.

18.3.3 Implications of ITPGRFA

- Article 12.3 (d) states, 'Recipients shall not claim any intellectual property or other rights that limit the facilitated access to the PGRFA or their genetic parts or components, in the form received from the Multilateral system (The term genetic parts and components is not defined and would be subject to different interpretations. More problematic is the term "in the form received"'.
- Article 12.3 (e) on access.
- Article 12.3 (a) states that access shall be provided 'solely for the purpose of utilization and conservation for research, breeding and training for food and agriculture'. The current MTA used in the context of agreements between FAO and the centres allows for access for direct production. Plant breeders may acquire materials from the multilateral system that they would want to release without any further research or breeding. Likewise, on occasion, farmers, NGOs or small companies may also wish to acquire a landrace and use it without further research or breeding.

Table 18.1 Genetic diversity collection and utilization

Commodity	Landraces	% in collections	Wild species	In situ collections	Ex situ collections (Acc.)	Utilization distribution
Rice	140,000	90	20	Few	420,000	High
Maize	65,000	90	–	Few	277,000	High
Sorghum	45,000	80	20	Few	169,000	Low
Millets	30,000	80	–	None	90,000	Low
Soybean	30,000	60	–	None	174,000	Low-medium
Chickpea	22,230*	90	19	None	33,782*	High
Pigeon pea	8220*	80	57	Few	13,628*	Medium-high
Groundnut	6374*	90	45	None	15,419	Medium-high
Potato	30,000	95	65	Few	31,000	High

Modified after Evenson et al. 1998; * ICRISAT & ICARDA holdings

- Annex I list of crops was the subject of passionate debate and substantial input from experts.
- Centres are obliged to conserve and make available designated accessions. Materials of non-Annex I crop that have been collected prior to the entry into force of the Treaty and are held by the centres are to be made available (Table 18.1).

18.4 Conservation

Since the 1970s, work on the conservation of crop genetic resources has increasingly become a large-scale independent activity detached from crop improvement efforts. A substantial germplasm collecting effort was launched in the 1970s in response to concerns about genetic erosion and crop vulnerability. Over 1700 gene banks have been established, holding about 7.4 million accessions. The germplasm holdings in major international agricultural research centres are provided in Table 18.2. Much has been invested in the characterization, evaluation and documentation; yet, for various reasons, the use of many collections has actually been rather limited (FAO 1998). This situation has led to criticism by some, including both plant breeders and nongovernmental organizations, that the genetic resources maintained in gene banks could be more usefully deployed in both plant breeding and farmers' fields. The following are among the important international seed banking facilities:

1. Seed Banks of Global Network of Agricultural Research Institutions
   International agricultural research institutions, co-ordinated by the Consultative Group on International Agricultural Research (CGIAR), Washington, are focused on crops and have extensive seed collections for such crops as rice, maize, wheat, barley, millets, pulses, oil seeds, tuber crops, banana, tropical forage and fruits.

Table 18.2 PGR holdings at IARCs. Source: SINGER database

Genebank	Total number of Accessions
International Crops Research Institute for the Semi arid Tropics, India	126,830
International Rice Research Institute, The Philippines	131,862
International Centre for Agricultural Research in Dry Areas, Syria	147,118
International Institute of Tropical Agriculture, Nigeria	32,584
International Livestock Research Institute, Ethiopea	20,229
Centro Internacional de Agricultura Tropical, Columbia	67,770
Centro Internacional de Mejoramiento de Maíz y Trigo, Mexico	178,000
West African Rice Development Association, Ivory Coast	26,098
Musa international Transit Centre, Bioversity International, Belgium	1,529
Centro Internacional de la Papa, Peru	19,093
Information and Communication Division, International Centre for Research in Agroforestry, Kenya	2,005
World Vegetable Centre (Asian Vegetable Research and Development Centre), Taiwan	61,000
Total	8,14,118

The collections in these seed banks are well documented and the institutions are networked among themselves and with several other institutions.

2. The Millennium Seed Bank Project
The Millennium Seed Bank Project (MSBP) at the Royal Botanic Garden, Kew, England, is one of the largest conservation projects. MSBP's 47 partner organizations in 17 countries intend to store 25% of the world's plant species by 2020. The Seed Information Database (SID) at Kew is an ongoing compilation of seed characteristics and traits worldwide, targeted at >24,000 species.
3. The Svalbard Global Seed Vault
On February 26, 2008, the Svalbard Global Seed Vault (SGSV) opened near Longyearbyen (Norway), 600 miles from the North Pole. SGSV is designed to hold 4.5 billion batches of seeds of the world's main crops. The SGSV is a glazed cave-like structure, drilled 500 ft. below permafrost, in the middle of a frozen Arctic mountain topped with snow, with the goal to store and protect samples from every seed collection in the world, which will stay frozen. An automated digital monitoring system controls temperature and humidity and provides high security. The SGSV is an insurance against natural disasters such as earthquakes and tsunamis, or deliberate attacks like bomb blasts or human errors such as nuclear disasters or failure of refrigeration that may erase the seeds of any important species in the other seed banks or in the wild, in other countries. Such seeds can be re-established using seeds from SGSV.
4. Global Crop Diversity Trust (GCDT)
Important collections of crop diversity face urgent and chronic funding shortages. These shortages can lead to loss of diversity, the very building blocks on which adaptive and productive agriculture depends. The sole global response to this threat is the Global Crop Diversity Trust. The Trust is a unique public–private

partnership raising funds from individual, corporate and government donors to establish an endowment fund that will provide complete and continuous funding for key crop collections, in eternity.

In line with the International Treaty on Plant Genetic Resources and the Global Plan of Action for the Conservation and Sustainable Utilization of Plant Genetic Resources for Food and Agriculture, GCDT goal is to advance an efficient and sustainable global system of ex situ conservation by promoting the rescue, understanding, use and long-term conservation of valuable plant genetic resources.

The Trust aims to ensure the conservation of the diversity within all crops of importance to food security. However, the Trust will give priority to the crops that are included in Annex 1, or referred to in Article 15, of the International Treaty on Plant Genetic Resources for Food and Agriculture. These are the crops that the international community has judged to be the most important for food security and interdependence.

The crops listed in Annex 1 of the International Treaty include breadfruit, asparagus, oat, beet, brassicas (the cabbage family including broccoli and cauliflower), pigeon pea, chickpea, citrus, coconut, aroids (including taro and cocoyam), carrot, yams, finger millet, strawberry, sunflower, barley, sweet potato, grass pea, lentil, apple, cassava, banana/plantain, rice, pearl millet, beans, pea, rye, potato, eggplant, sorghum, triticale, wheat, faba bean, cowpea, maize and more than 80 forage species from 30 different genera.

18.5 Genomic Resource Conservation and Information

18.5.1 National Genomic Resources Repository

A recent initiative of the Indian Council of Agricultural Research (ICAR) is the establishment of the National Genomic Resources Repository in the premises of NBPGR as an institutional framework for methodical and centralized efforts to collect, generate, conserve and distribute genomic resources for agricultural research. Current research (including underutilized legumes) (both routine cloning experiments and genome sequencing projects) generates a lot of genomic resources. These genomic resources are indispensable tools for postgenomic research, be it physiological and morphological characterization of a species or functional analysis of genes or comparative genomics or plant breeding. Therefore, it is necessary to maintain an efficient system for the conservation and management of spin-off DNA materials. Due to the availability of techniques that help characterize and utilize DNA sequences (without the requirement of whole organism), value-added products of gene banks can attract new clients involved in *allele mining* and *cisgenesis*, such as molecular biologists and geneticists alongside the traditional plant breeders.

18.5.2 Promoting Genetic Resource Utilization

Precise evaluation and documentation of plant genetic resources are a prerequisite for their utilization. The following areas of research need to be paid more attention for promoting effective utilization of PGR gene resources:

- Development of core collections
- Focused Identification of Germplasm Strategy (FIGS)
- Pre-breeding
- Gene prospecting and allele mining for a trait of interest from genetic resources

The development of a core set of collection particularly in the crops having large germplasm collection can be a powerful tool for promoting the utilization of germplasm. This cost-effective approach needs to be implemented for major plant genetic resources. An alternative approach to core collection is the Focused Identification of Germplasm Strategy (FIGS). The underlying concept of FIGS is the 'distribution of genetic variation as a function of the relationship between genotype, environment and conscious or un-conscious selection - evolutionary processes'. Pre-breeding of plant genetic resources aims 'to introduce new desirable traits/ genes into an adapted genetic background. It will broaden the genetic base in a breeding material in pace with environmental changes'. It is a vital contribution to increasing the total genetic diversity in crops and finding specific genes and traits and a link between conservation of PGR in gene bank collections and utilization of these resources in agriculture and horticulture. The most efficient way to capture specific traits—often rare alleles to enhance the utilization of existing genetic variation through biotechnological interventions—is gene prospecting and allele mining. These procedures are some of the promising ways to promote effective utilization of plant genetic resources in the coming years.

18.6 Molecular Approaches

Conservation and sustainable use of genetic resources are essential to meet the demand for future food security. Advances in biotechnology have generated new opportunities for genetic resource conservation and utilization. Techniques like in vitro culture and cryopreservation have made it easy to collect and conserve genetic resources, especially of species that are difficult to conserve as seeds. While technologies like enzyme-linked immunosorbent assay (ELISA) and polymerase chain reaction (PCR) have provided tools that are more sensitive and pathogen specific for seed health testing, tissue culture methods are now widely applied for the elimination of systemic diseases such as viruses for safe exchange of germplasm. Molecular markers are increasingly used for screening of germplasm to study genetic diversity, identify redundancies in the collections, test accession stability and integrity and resolve taxonomic relationships. The technology is also expanding the scope of genetic resource utilization.

18.7 PGR Conservation in India

The Indian subcontinent is immensely rich in plant genetic resources of both crop species and their wild relatives. PGR is fundamental to crop improvement programme and the key to establishing future food and nutritional security. The importance of plant genetic resources has increased significantly in the recent years with the changing global scenario in material ownership and the legal regimes with respect to access to PGR under the international agreements. The National Bureau of Plant Genetic Resources is the nodal organization in the country for acquisition and management of indigenous and exotic plant genetic resources (PGR) for food and agriculture and to carry out related research and human resource development for sustainable growth of agriculture.

India has a rich and varied heritage of biodiversity, encompassing a wide spectrum of habitats from tropical rainforests to alpine vegetation and from temperate forests to coastal wetlands. It is one of the 8 centres of origin and is one of the 12 mega gene centres of the world, possessing 11.9% of the world's flora. About 33% of the recorded floras of the country are endemic and distributed mainly in the North-East, Western Ghats, North-West Himalayas, Eastern Ghats and the Andaman and Nicobar Islands. Of the 49,219 higher plant species, 5725 are endemic and belong to 141 genera under 47 families (Nayar 1980). India also has two biodiversity hotspots out of 25 listed by Myers et al. (2000) {the Western Ghats/Sri Lanka and the Indo-Burma (covering the Eastern Himalayas)} and they are included among the top eight most important hotspots. In addition, India has 26 recognized endemic centres that are home to nearly one-third of all the flowering plants described to date (Gautam 2004). India is a homeland of 167 cultivated species and 329 wild relatives of crop plants (Arora 1991). It has about 30,000–50,000 landraces of rice, pigeon pea, mango, turmeric ginger, sugarcane, gooseberries, etc. and ranks seventh in terms of contributions to world agriculture. More than 1000 wild edible plant species are exploited by native tribal communities. These include 145 species of roots and tubers, 521 species of leafy vegetables/greens, 101 species of buds and flowers, 647 species of fruits and 118 species of seeds and nuts (Arora and Pandey 1996). In addition, nearly 9500 plant species of ethnobotanical uses have been reported, of which around 7500 are of ethnomedicinal importance and 3900 are multiple-purpose edible species (Arora and Pandey 1996). The endemic plant wealth of the country has also been supplemented with the species/forms that had been introduced from abroad. These species got naturalized over time and have undergone the process of domestication on being isolated climatically and spatially. Important among these are apple, pear, peach, apricot, grape, almond, date palm, maize, potato, sweet potato, tomato, bean, onion, garlic, chilli, lentil, clove, coriander, cumin, fennel, coffee, cocoa, cashew, litchi, strawberry, blueberry, tea, rubber and pine apple.

The National Herbarium of Cultivated Plants (NHCP) holds a total of 18,969 herbarium specimens, 2503 seed samples and 501 economic products, which represents 260 families, 1401 genera and 3618 species. The collection and assemblage of accessions from various sources have resulted in conservation of 452,651 seed samples representing 1762 species in the long-term storage (LTS) of the

Table 18.3 PGR holdings in the National Gene Bank, India (as on 31.05.2021)

Crop group	No. of accessions
Cereals	169,268
Millets	59,798
Pseudo-cereals	7873
Grain legumes	67,530
Oilseeds	61,176
Fibre crops	16,249
Vegetables	27,819
Fruits and nuts	295
Ornamental	678
Medicinal and aromatic	8446
Spices and condiments	3458
Agroforestry	1670
Safety duplicates	10,235
Trial material (wheat, barley)	10,771
Total	452,651

Source: www.nbpgr.ernet.in

National Gene Bank (Table 18.3). This includes more than 50,000 accessions restored from the international agricultural research centres like ICRISAT, IRRI and CIAT. A total of 5034 proposed/released varieties, hybrids, parental lines of different fields and horticultural crops have been assembled and conserved in the gene bank. A total of 1927 samples belonging to tropical fruits, temperate and minor tropical fruits, bulbous crops, medicinal and aromatic plants, spices and industrial crops have been conserved for over 6–18 years using in vitro methods. Further, a total of 11,961 sample species have been cryopreserved using embryos/embryonic axes and seeds. In addition, the crop-based National Active Germplasm Sites (NAGS—59 centres) linked up with the bureau also holds thousands of accessions of germplasm in the country.

The conservation of plant genetic resources involves two basic strategies: (1) in situ and (2) ex situ. In situ conservation of PGR has to be carried on farm, where landraces and locally improved/adapted materials are cultivated, evaluated, utilized and conserved as part of traditional farming systems. These farming systems are of particular importance in maintaining local genetic diversity and providing food for local consumption and local markets. Ex situ conservation requires collection and systematic storage of seeds/propagules outside the natural habitats of species for short, medium and long term after proper characterization and evaluation. Huge amount of germplasms already under ex situ conservation in the country are to be characterized and evaluated in a phased manner. A total of 187,439 germplasm accessions of various crops were characterized and preliminarily evaluated at the NBPGR and its regional stations. These have been documented and published as several crop catalogues for the utilization by breeders in various crop improvement programmes in the country.

18.8 Utilization

Plant breeding and production of new cultivars are widely regarded as underpinning agriculture and the development of society. Yet crop failures and risks associated with genetic uniformity, yield stagnation (below potentially attainable levels) and persistent failures to achieve sustainable production increases in important local cropping systems are widespread problems. The continuing success requires a long-term and sustainable commitment to the effective utilization of plant genetic resources by enhancing and expanding the genetic base from which future cultivars will be generated. Selections from diverse landrace populations can reveal high-yielding genotypes, which can be used directly or in breeding programmes.

The genetic diversity and collection by commodity for major crops are presented in Table 18.1. The world's naturally occurring plant genetic resources are threatened by loss of habitat, climate change or disasters caused by nature or activities of mankind such as development and overgrazing. Conservation of plant genetic resources, coupled with collection of information that helps us understand their nature, helps assure their availability to benefit society in the future.

Plant genetic resource conservation utilization has been the source of dramatic scientific changes over the course of the last few decades. Indirectly, it has been the source of value differentiation within the agricultural sector. Such value differentiations characterize complementarities across activities that aid breeders/ seed companies in identifying desired products and delivering those products to consumers at the lowest possible cost. The changes in genetic resources and their uses have motivated new research inquiries in attribute detection, preference determination, attribute production and coordination. New discoveries in the development of transgenic seeds as well as their conservation and utilization will continue to depend on both public and private research and development.

18.9 Safe Transboundary Movement

Unregulated germplasm exchange activity may result in inadvertent introduction of insect-pests, diseases and weeds into the country. There are examples of enormous crop losses caused by the introduced pests including weeds in our country. Large economic losses occurred in grain yield of *Cicer arietinum* in the states of Haryana, MP and Punjab and adjoining areas due to infection by a virulent biotype of *Ascochyta* introduced from the Middle East. The introduction of exotic weeds such as *Lantana camara* in the early nineteenth century from Central America, *Parthenium hysterophorus* from Central and South America and *Phalaris minor* from Mexico in the mid-twentieth century into India has become a threat to our crop production and environment. Plant quarantine through a set of legislative measures is undertaken to prevent the introduction of pests. As per Plants, Fruits and Seeds (PFS) Order, 1989, the NBPGR is working as the nodal agency for quarantine processing of planting material imported for research purposes. To undertake it, elaborate laboratory and greenhouses have been developed in New Delhi and

Hyderabad. There is a need to carry out extensive surveys to collect endemic pest data and comprehensive epidemiological data on pests of national and international importance in order to fix tolerance limits, conduct pest risk analysis and develop diagnostic protocols using molecular techniques for detection of exotic pests. A total of 2,796,117 samples of various crops comprising seeds, vegetative propagules, in vitro and transgenic material were processed for quarantine clearance of which 2,137,579 samples were imported and 658,538 were meant for export. Detailed quarantine examination revealed infestation/infection/contamination in 148,959 samples [insects and mites—98,645; pathogens—30,256; nematodes—16,719; and weeds—3339], of which 148,872 were salvaged and released to the indenters by the NBPGR.

The emergence of transgenic crops has raised concerns among the public at large regarding their effects on safety, health and environment and these issues have been discussed at various national and international forums. The regulations for release of transgenics are well established in most of the developed countries. In our country, the Ministry of Environment and Forests, Department of Biotechnology and the Ministry of Science and Technology have developed regulatory mechanisms. As per these mechanisms and the PFS Order, the NBPGR issues the import permit for introducing transgenic materials for research purpose and undertaking quarantine processing of the introduced transgenic materials. A National Containment Facility has been established at the NBPGR under a collaborative project of ICAR and DBT. It has the objectives of quarantine processing of imported transgenic, developing molecular probes for the detection of transgenes (promoter or terminator sequences) and developing human resources in the field of biosafety.

The introduction of valuable exotic germplasm has enriched the Indian agriculture in the past. Introduction of short-statured, lodging-resistant, input-responsive, high-yielding wheat and rice ushered the Green Revolution in the country. Introduction of cytoplasmic-nuclear male sterility and fertility restoration genes brought an era of hybrid breeding in crops like sorghum, pearl millet and rice. Further, soybean and sunflower became the major field crops. Over 21 lakh samples of promising genetic stocks, varieties, transgenic materials and wild species of different agricultural and horticultural crops were imported from 113 countries and 8 IARCs. A total of 9298 germplasm consignments were processed for export, for which phytosanitary certificates were issued by the NBPGR.

18.10 Public–Private Partnerships

Public–private partnership is broadly described as any joint effort between public and private entities in which each contributes to planning, commits resources, shares risks and benefits and conducts plant genetic resource activities to accomplish a mutual objective of increased food and ecological security. Partnerships between public research organizations and private firms offer a means of tapping the strengths of diverse actors and channelling knowledge and resources into areas where they can address complex development problems that are relevant to the needs of resource-poor farmers and food-insecure consumers. By exploiting the potential for research synergies, complementariness, scale economies and knowledge-sharing among stakeholders, there are greater chances of success at lower costs that public or private stakeholders might otherwise expect when acting alone. Most importantly, PPP are valuable because they bring private sector resources and expertise to bear on public research priorities in developing nations like India.

The emergence of plant breeding programmes in the twentieth century created a demand for germplasm exchange among breeders. This initiated collecting missions and explorations to satisfy the growing need for such crop plant characters as pest/disease resistance, earliness, stiff stalks and grain quality. All over the globe, the balance between public and private industry roles in varietal research and development has shifted in the past three decades. The private sector expanded considerably and is focused on commodities with major markets enhancing their share for the development of agriculture. Public institutions are involved in the various germplasm-related activities, viz., collection, characterization, conservation and pro-poor crop improvement still playing a major role in the development of agricultural research.

18.10.1 International Scenario

Partnerships are particularly relevant in PGR activities including Agbiotech R&D, where the technical requirements include access to cutting-edge research tools, intellectual property, advanced scientific expertise, expensive and sophisticated equipment and solid infrastructure (Byerlee and Fischer 2002). Public–private partnerships have become the order of the day in several CGIAR centres to address specific areas such as apomixis in maize (CIMMYT-PHI, Syngenta, Limagrain), Bt genes for rice transformation (IRRI-Novartis, Plantech), golden rice (IRRI-Syngenta), potato/sweet potato transformation (CIP-Plant Genetic Systems, Axis Genetics, Monsanto) and cassava transformation (CIAT-Novartis). Interesting collaborations are also being successfully implemented such as harvest plus (CIAT and IFPRI-Syngenta) and unlocking crop genetic diversity for poor people (CIMMYT, BI, IRRI-MAHYCO, Bayer Crop Science, PHI). A recent study on Agbiotech research partnerships found that critical assets and competencies from the private sector are not being adequately brought to bear on the research challenge in collaboration with public research (Spielman 2006). IRRI-supported Hybrid Rice Development Consortium and SOL genomics network are excellent models for sharing information, avoiding duplication, prioritizing and distributing work with low focus on common external funding.

Millennium Seed Bank and Kew Botanical Gardens have active support and collaboration with the private sector even though the trustees are nominated by the Government of the UK.

China also has operational models for agricultural research having joint ventures at provincial level with first refusal privilege to the participating private partners followed by non-exclusive sale of technology/products developed in such ventures.

Similarly, several industrialized nations took lead in implementing collaborative research projects with the private sector.

18.10.2 National Scenario

Some of the successful models of PPP in India are described below:

- Hybrid Rice (IARI)
  The Indian Agricultural Research Institute (IARI), a flagship research institute of the Indian Council of Agricultural Research (ICAR), has developed a very effective public–private partnership model for the promotion of hybrid rice in India. IARI, Indian Foundation Seeds and Services Association (IFSSA) and Barwale Foundation had earlier signed a memorandum of agreement (MoA) for seed multiplication of parental lines of Pusa RH 10, the first superfine grain aromatic rice hybrid developed by the IARI. In addition to IFSSA, the IARI also signed a memorandum of understanding with 18 other seed companies to produce hybrid seed of Pusa RH 10. The partnership with IFSSA helped the area under Pusa RH 10 to reach nearly 0.5 million ha during *Kharif* (crop) season.
- Collaboration on Insect Management for Brassicas in Asia and Africa (CIMBAA)
  The project showcases the joint investment and collaborative research involving ICAR, India; AVRDC, Taiwan; University of Melbourne, Australia; Natural Resources Institute, University of Greenwich, UK; University of Cornell, USA; and Nunhems, India. It relies on the public sector for socio-economic and ethical evaluation of the project, facilitating regulatory approval within the established norms and infrastructure facilities, and management of stewardship training in taking the product to the public.
- Agricultural Biotechnology Support Project Model (ABSP II Model)
  In ABSP II model, Mahyco; IIVR, Varanasi; UAS, Dharwad; and Tamil Nadu Agricultural University, Coimbatore, are involved in the development of transgenic brinjal varieties resistant to fruit and shoot borer. In this project, ABSP provided the funding, DBT offered the regulatory support, Mahyco provided the cry gene and IIVR has the responsibility to develop the resistant varieties.
- Consortium Model (ICRISAT)
  The international Crops Research Institute for Semi-Arid Tropics (ICRISAT), India, formed a consortium with private seed companies and others for collaborative research (local/regional programmes) for sorghum and millet research. Also, ICRISAT has established Biotech incubator involving private sector biotechnology companies.

Consortium approach even within the private sector for Bt cotton is a success story. Current level of interactions between private and public sectors in Indian agricultural research is limited to few consultancies, outsourcing for mandatory tests and sale of public technologies/parental lines on non-exclusive basis in addition to public sector facilitating protection of plant varieties, germplasm and import/export of germplasm and commercial consignments in a limited way. ICAR has provided mechanisms for patenting, sharing and commercializing technologies and products (ICAR, 2006). Leading institutes such as IARI and IIHR have taken initiatives for taking public research technologies and products to the needy farming communities through commercialization on a non-exclusive basis with the private sector. NBPGR has standardized several procedures and coped with the changed scenario in facilitating mechanisms for import/export of germplasm exchange. Unfortunately, there is hardly any example to quote a joint venture/research project between public and private sectors in agricultural research in India comparable to the projects in CG centres, industrialized countries or provincial level joint ventures as in China. However, there are ample opportunities for joint projects and ventures in PGR activities.

The public–private partnership networks for plant genetic resource activities need to expand among the entire National Agricultural Research System (NARS), comprising ICAR institutes, agricultural universities and other public sector institutions. It is a challenge that the PPP model should take into their ambit the whole chain from innovation and product development to marketing. Varied operating systems which are prevailing within public and private firms need to be recognized and harmonious working relationships are to be established between stakeholders. Germplasm management is a highly complex enterprise with respect to developing workable partnerships. PPP models in PGR activities have to be need based, case based and region based and should achieve through teamwork goals of equal partners (APAARI, 2007). Some of the key issues that promote PPP mode in PGR activities, which are relevant to the Indian context, are listed below:

- Cataloguing landraces with fingerprints, enlisting all valuable national germplasm collections with identity, facilitating designation of proprietary lines developed by the private sector
- Developing mechanisms across the sectors for exchange of scientists for short term (1 year) and long term (3 years)
- Joint evaluation in identified areas and crops (for example, exploring diverse male sterile systems for hybrids in jowar, bajra, cotton and rice)
- Collaborations for hybrid development using diverse public domain germplasm in minor millets and pulses with market support
- Sharing mechanisms including sale of germplasm, elite lines, inbred lines and discarded lines prior to release of varieties
- Balancing the proposal with justification for amendments of BDA, PPVFRA and PQO with due discussions by both the sectors
- Long-term collaborative projects for allele mining, functional genomics, climate change (heat, drought, flood and salinity), chronic unsolved problems of biotic stresses, search for alternative to Bt, etc.
- Developing mechanisms for dual-lock system in conservation at the developer's location with provisions for use in national emergencies

- ITPGRFA and implications of commercialization—public and private sector responsibilities
- Empowering one clearing house for exchange of germplasm

18.11 Major Issues and Future Thrusts

Approaches to better management of plant genetic resources need to focus on the following major elements: (i) looking for more species and genes to provide bio-alternatives and (ii) using both traditional breeding approaches and modern technologies. It has to be addressed in order to match the increasing human and animal feed in the country. Following are the major issues and future thrusts for the effective management of PGR on sustainable terms:

- Harnessing of the eco-regional potentials, more particularly in rainfed, coastal and hill eco-regions and hot spots to meet the climate change
- Acquiring of diverse germplasm at the extreme ends of the genetic diversity to bolster resistance to increasing biotic and abiotic stresses
- Need to have geo-referenced data and time series on PGR to take informed decisions
- Need to develop methodologies and tools to make the dynamic conservation of the genetic diversity of multiple agricultural species compatible with poverty alleviation and increased well-being for its keepers
- National policy or long-term action plan for promoting on-farm conservation
- Development of core collections and validation
- Focused Identification of Germplasm Strategy (FIGS)
- Dissecting of the physiological/biochemical/molecular basis of resistance
- Base broadening/genetic enhancement to enhance utilization particularly the CWR
- Indicators and criteria to identify crops in need of genetic base broadening
- Stronger linkages between PGR conservation institutions, genetic enhancement and plant breeding programmes
- Development of genomic resources from genetic resources and their conservation
- Efficient utilization of genomic resources
- Discovery of novel traits such as genes for biotic and abiotic stresses including climate change and quality attributes
- Phenotyping and phenomics to add value to PGR
- Augmentation of introductions: Import of germplasm for the crops which have narrow genetic base (arid, temperate and tropical fruits, oilseeds, pulses and ornamentals)
- Introduction of resistant sources identified from different countries for biotic and abiotic stresses
- Introduction of commercial hybrids/clones/varieties in different crops and CMS lines in pearl millet, rice and sorghum

- Introduction of new crop species, promoting underutilized crop species and novel germplasm to enhance the food and nutritional security of the country in the wake of global change in climate
- Development/use of advanced biotechnological tools for detection
- Preparedness for risk analysis of pests under WTO regime
- Characterization for highly heritable traits, joint evaluation for quantitative traits/ agronomic performance
- Pre-breeding/germplasm enhancement
- Low-energy-based conservation
- Documentation of trait/gene-specific germplasm
- In situ/on-farm conservation
- Development of cost-effective in vitro multiplication and conservation protocols
- Development of widely applicable cryopreservation protocols
- DNA banking of clonally propagated/non-orthodox seeded/threatened species
- Development of improved marker systems for genomic applications in PGR
- Exploitation of genomic resources for improved utilization of PGR
- Mapping of genetic diversity for quality conservation
- Addressing of biosafety issues related to transgenics
- Development of repository for genomic resources in agriculture and bioinformatic
- Enhancement of public, private and civil society partnership investment in research on genetic resource handling and use and confidence building to increase the share of private sector in research and development, exchange and collaborative use of genetic resources for food and agriculture (CSO + national bureaux + public institute + private industry)
- New initiatives in the area of processing, product development and value addition of suitable crops and varieties by appropriate exploitation of their genetic resource
- Encouraging of networking for exchange, evaluation and sustainable utilization of genetic resources (ARIS, databases, MOUs)
- Development of ability, appropriate institutional arrangements and policy framework for handling intellectual property rights-related issues
- Need for safeguarding the national interests in order to meet the challenges of the new legal regimes of the CBD, the WTO and TRIPS, UPOV and ITPGRFA

References

APAARI (2007). http://www.apaari.org/wp-content/uploads/2009/05/2007-apcoab-models-of-ppp-2007.pdf

Arora RK (1991) Plant diversity in Indian Gene Centre. In: Paroda RS, Arora RK (eds) Plant genetic resources–conservation and management. IPGRI, Regional Office for South Asia, New Delhi, India, pp 25–54

Arora RK, Pandey A (1996) Wild edible plants of India: diversity, Conservation and use. National Bureau of Plant Genetic Resources, New Delhi, India

Byerlee D, Fischer K (2002) Accessing modern science: policy and institutional options for agricultural biotechnology in developing countries. World Dev 30(6):931–948

Evenson RE, Gollin P, Santaniello V (eds) (1998) Agricultural values of plant genetic resources. CABI Publishing, Wallingford, UK

FAO (1998) The state of the world's plant genetic resources for food and agriculture. FAO, Rome, Italy

Gautam, P.L. 2004. Trends in plant genetic resource management. In: Dhillon BS, RK Tyagi, Arjun Lal and S. Saxena. Plant genetic resource management. Narosa Publishing House, New Delhi, India pp. 18-30

Myers N, Mittermeier RA, Mittermeier CG, da Fonseca GAB, Kent J (2000) Biodiversity hotspots for conservation priorities. Nature 403:853–858

Nayar MP (1980) Endemism and pattern of distribution of endemic genera (angiosperm) in India. J Econ Tax Bot 1:99–110

Spielman DJ (2006) A critique of innovation systems perspectives on agricultural research in developing countries. Innovation Strategy Today 2:25–38

Dynamics of Agricultural Livelihoods in Peripheral Villages of a Protected Area in South India

19

Sunil Nautiyal and Mrinalini Goswami

Abstract

Agriculture and protected areas are generally perceived as poles apart; however, agriculture has been primary livelihoods for forest dwelling and forest fringe communities of several protected areas of India. Population growth has been exerting tremendous pressure on the agricultural sector to increase productivity with new crops. On the other hand, protected area and resource rights conflicts are also seen as barriers in both effective conservation and acquiring livelihoods. The state of Karnataka, situated in South India, has a number of protected areas and a large portion of its population depends upon agriculture for its livelihood. The current research was conducted in MM Hills Wildlife Sanctuary, Karnataka, and aimed to develop an understanding about the role played by agriculture in supporting local livelihoods, different types of crops cultivated and changes in the pattern of crop cultivation. For this purpose, four points (1990, 2000, 2010 and 2020) of time were considered. Data were collected by undertaking questionnaire survey in villages located in and around the Wildlife Sanctuary, during which randomly selected household heads were interviewed. The main analysis was done by taking into account certain demographic aspects and linking them with the agricultural sector. Additionally, an index was also created that took into account of crops cultivated, agricultural land area and production. It was found that agriculture was the chief means of livelihood and 29 different crops were cultivated, out of which 2 (maize and ragi) were predominant. Choice of crops, production and area cultivated rapidly fluctuated, whereas the application of chemical fertilizers has increased. A shift from subsistent agriculture to diversified market-oriented cultivation was observed. As a whole, the findings

S. Nautiyal (✉) · M. Goswami
Centre for Ecological Economics and Natural Resources (CEENR), Institute for Social and Economic Change (ISEC), Nagarabhavi, Bengaluru, India

S. Ramamoorthy et al. (eds.), *Plant Genetic Resources, Inventory, Collection and Conservation*, https://doi.org/10.1007/978-981-16-7699-4_19

of the study reveal the importance of agriculture in providing rural livelihoods in fringe villages of protected areas. The study suggests that further actions and planning for protected areas must consider this important aspect of local livelihoods for socio-economic development of indigenous people.

Keywords

Cropping choice · Crop production · Forest periphery · Wildlife Sanctuary

19.1 Introduction

Protected areas cover more than 15% of the earth's surface. Association of people with protected area in terms of their subsistence and acquiring livelihoods is evident worldwide, and more prominent in India. Protected areas are known for maintaining ecosystem processes and also as a source of ecosystem services such as pollination and pest control, clean water and resilience to disaster risks essential for sustaining agriculture (FAO 2014; Duran et al. 2013). Agriculture and allied sector serve as the largest livelihood providers in India where maintaining those ecosystem services are crucial for both food and livelihood security (FAO 2020). A major challenge for India is to meet the ever-increasing and diversified food demand for ensuring food and nutritional security with degrading land and agricultural land conversion to other land uses (Kumar et al. 2019; Kumar and Mittal 2006). The conversion of wild lands to agriculture and other land uses occurs due to population growth and related development including urbanization (Goswami et al. 2020; Slaughter 2011). Consequently, a large proportion of ecological hotspots and protected areas are directly threatened by agricultural expansion (Veach et al. 2017). The pressure of increasing production and diversified food crops has also led to several biotic and abiotic stresses on food production systems (Birthal et al. 2014). In addition, impacts of climate change with erratic rainfall have affected agriculture, more importantly subsistent agriculture in mountainous region (Azhoni and Goyal 2018). In the current times, the issues of arable land depletion, soil fertility loss and environmental degradation arising from agricultural activities and excessive human pressure have received much attention (Singha et al. 2014).

An additional 100 million hectares of land under agriculture will be required to meet the projected food demand which is double of the current food supply by 2050 (Pastor et al. 2019). A severe food crisis can be foreseen if the agricultural production continues to grow at the current rate, and at this rate, India would only be able to fulfil 59% of its total food demand by 2030 (Jitendra 2015). There is a severe pressure in the country to ensure food security. Consequently, sustainable management of agricultural land resource is urgently needed in India because there is likely to be greater demand for food to satisfy the dietary requirements for the entire growing population (Nath et al. 2015). In the South Indian state of Karnataka, 17% of the forests have been designated as protected areas (https://aranya.gov.in/new/StaticPages/WildLife.aspx). In many instances, isolationist approach of conservation and traditional agricultural systems closely associated with protected area ecosystems, both inside and outside the boundaries, are disvalued, which affects the livelihoods of indigenous communities living in the periphery. The state has diverse agroclimatic conditions and varying natural resource features and cultivation is undertaken in land areas stretching across low-rainfall eastern region to high-rainfall Western Ghats and coastal regions (Kannan 2011). There are 10.80 million hectares of cropped area in Karnataka, out of which 21.50% is irrigated (Chandrakanth et al. 2013). However, stagnant crop yields and greater input costs have caused agrarian distress in the state, leading to the urgent need for studies to improve the sustainability of farming systems in the state (Patil et al. 2014).

Considering all the above and the fact that the economy of the communities in peripheries of forests has traditionally been dependent on subsistence agriculture (Shithi and Pati 2012), this research was designed to study the agricultural practices in and around MM Hills Wildlife Sanctuary, Karnataka, in South India. This is because the role of fringe communities is of paramount importance for sustainable development of forests and their socio-economic condition is interlinked in this context (Singha 2013). The study is designed with the objective to understand the role of agriculture in terms of providing local livelihoods. Furthermore, the study also documents the crops cultivated in the fringe villages and temporal changes in cultivation.

19.2 Materials and Methods

19.2.1 Study Area

The Male Mahadeshwara Wildlife Sanctuary (MM Hills, Latitude: 11° 55′ and 12° 13' N and Longitude: 77° 30′ to 77° 47′ E) covers an area of 911.80 km^2 and has a highly undulating terrain. This is one of the oldest forest divisions and is located in Kollegal Taluk of Chamarajanagar District in southeast Karnataka, bordering Tamil Nadu along the junction of the Eastern and Western Ghats, and acts as an important elephant corridor between Biligiri Ranganatha Swamy Temple Wildlife Sanctuary and Cauvery Wildlife Sanctuary. It is a part of Nilgiri Biosphere Reserve and was notified as a Wildlife Sanctuary in 2013. The area is predominantly covered by dry and deciduous forests, where 64% is dry deciduous, 20.5% is scrub forests and 2.45% is with moist deciduous and riparian vegetation. The buffer zone of the Wildlife Sanctuary has 26 villages. The most dominant communities residing in and around MM Hills are Soligas and Lingayats. Soligas are hunters and gatherers who practised shifting cultivation till settled in hamlets, on lands allotted by the government, whereas Lingayats are temple priests. Non-timber forest products (NTFP) are an important source of income for both the communities (https://www.atree.org/content/atree-male-mahadeshwara-hills, Uma Shaanker et al. 2004). The current study was conducted in four villages (Pacchedoddi, Kanchally, Gundapura and Boodhipadaga) located in and around MM Hills as depicted in Fig. 19.1. Agriculture and wage labour are the chief means of livelihood in the villages.

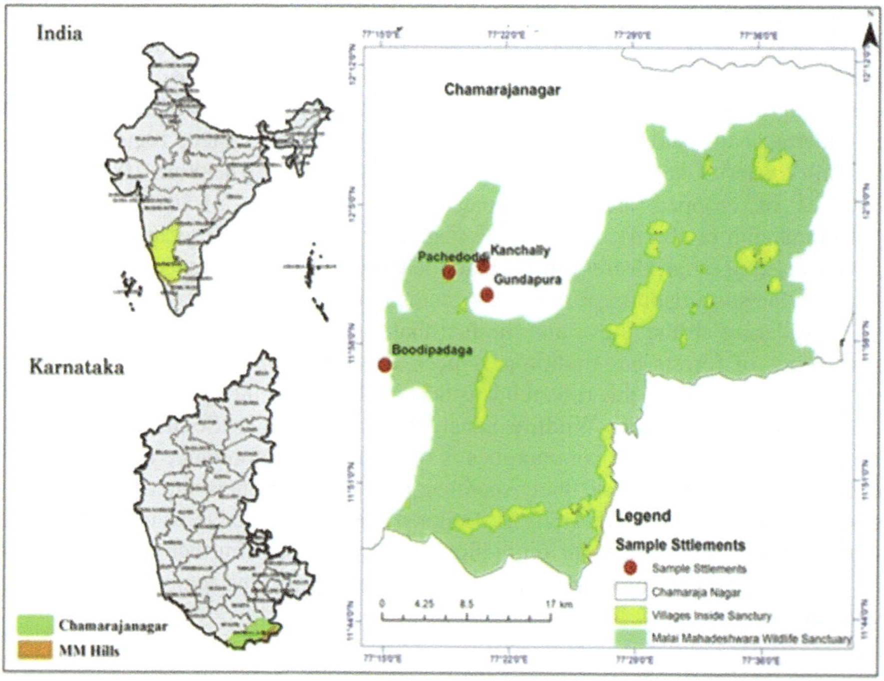

Fig. 19.1 Location of study area

19.2.2 Data Collection and Analyses

The study is based on primary data collected from households in fringe villages. Data were collected by conducting a questionnaire survey (Fanning 2005) in four villages (Pacchedoddi, Kanchally, Gundapura and Boodhipadaga) located in and around MM hills. A sample of 200 households was drawn from the study villages with 50 randomly selected households from each village. The questionnaire used to conduct the survey included groups of questions under (a) general household profile, (b) socio-economic information, (c) landholding and land use, (d) livelihoods and income, (e) crops cultivated and (f) cropping practices and production. Past information based on the memory of the household head/elderly was collected for four points of time (as decades), namely: 1990, 2000, 2010 and 2020. In case a household derived incomes from more than one source, then the source mentioned by the household head was considered as the primary occupation.

The data were analysed in MS Excel 2007 Past 3.07. The data were first tested for normality and suitable statistical tests were done. In order to analyse the data obtained from the questionnaire survey, we proposed the Crop Cultivation Index (CCI) based on three variables, namely: proportion of households cultivating a particular crop (H), proportion of the overall cultivated area in which a particular crop was grown (A) and proportion of the overall production contributed by a particular crop (P). The index has been proposed to compare the changes in the trends of cropping among four points of time (1990, 2000, 2010 and 2020) and is described below:

$$A = \frac{\text{Area in which a particular crop is cultivated}}{\text{Total area under cultivation of all crops}}$$

$$H = \frac{\text{Total number of households that cultivate a particular crop}}{\text{Total number of households that owned cultivable land (102)}}$$

$$P = \frac{\text{Annual production of a particular crop}}{\text{Total annual production of all crops}}$$

Now, CCI $= A + H + P$.

All the crops grown in the surveyed households for the different points of time considered were listed and the CCI was calculated for every crop. The highest possible value of CCI for a particular crop for a particular point of time was 03 indicating that only that particular crop was cultivated at that particular point of time. CCI is an indicator of the performance of a crop relative to other crops cultivated simultaneously. An increase in the CCI of a crop does not necessarily mean that its cultivation or production has increased and vice versa. Monetary values have been expressed as Indian National Rupees (INR; 1 INR = 0.014 Dollars as on 13 January 2019).

19.3 Results

19.3.1 Significance of Agriculture as a Means of Livelihood

Agriculture and wage labour had been the chief means of primary livelihood in the studied villages. In 1990, 51.93% of the surveyed households depended on agriculture, which increased to 58.01% in 2020. On the contrary, the dependency on wage labour decreased from 31.49% to 23.76% of households during the same period (Table 19.1). In 2020, the contribution of agriculture to annual income had been at least 40%, even in households in which it was not considered as a primary source of livelihood. The highest contribution of this sector in this regard was about 85%. Government service was the most lucrative employment in terms of annual income and the highest proportion of the households engaged in this profession (50%) resided in pucca houses. Agriculture, private sector and wage labour were the sources of primary livelihood for the highest section of households belonging to general category (49.53%), scheduled castes (73.33%) and scheduled tribes (48.84%), respectively (Table 19.2). Different proportions of households changed their original means of primary income during 1990–2020, except those depending on other sources. In case of all sectors, except government service, the maximum percentage of households transitioned to agriculture. On the other hand, the highest

Table 19.1 Sources of primary livelihood in the surveyed fringe villages of MM Hills Wildlife Sanctuary, Karnataka, over four decades

	Points of time			
Source of primary livelihoods	1990	2000	2010	2020
Government service	2.21	2.76	2.21	1.10
Private sector	8.29	8.29	7.18	8.29
Agriculture	51.93	54.14	58.56	58.01
Daily wage	31.49	27.07	25.97	23.76
Animal husbandry	3.31	4.97	3.87	6.08
Others	2.76	2.76	2.21	2.76

Table 19.2 Socio-economic and demographic information of households having different primary income sources in 2020

Type of occupation	Percentage of total annual income derived from agriculture and allied activity	Average annual household income	Percentage of households			Percentage of households	
			Gen	SC	ST	Concrete house	Mud house
Government service	40.48	105,000	50	50.00	0	50.00	50.00
Private sector	79.33	34,667	6.67	73.33	20	6.67	93.33
Agriculture	84.93	54,952	49.53	33.33	17.14	33.33	66.67
Daily wage	65.80	43,690	23.25	27.91	48.84	34.88	65.12
Animal husbandry	56.99	26,500	36.37	45.45	18.18	9.09	90.91
Others	55.38	65,000	20	40.00	40	40.00	60.00

proportion of households (84%) which continued to retain their original means of primary income since 1990 was found in case of agriculture (Table 19.3).

19.3.2 Cultivation in General

Among the surveyed households, 51% owned cultivable land. The ownership of such land among these households ranged from 0.1 to 20 acres and the average landholding was 3.52 ± 2.95 acres. A significant majority of the landholders owned 2–8 acres of land ($\chi^2 = 15.57$, df $= 1, p < 0.05$). Overall, there is a total of 358.95 acres of cultivable land in the four villages at present, out of which 43.97% is irrigated. In 21.57% of the households, the entire cultivated land area was irrigated, whereas there was no irrigation in 41.18%. There was no significant difference between the irrigated and non-irrigated areas ($z = 1.92$, $n_1 = 102$, $n_2 = 102$, $p > 0.05$). The residents of the surveyed villages have been growing 16 types of agricultural crops (cotton, cowpea, foxtail millet, groundnut, horse gram, little

Table 19.3 Transition matrix for percentage of households shifting of primary livelihood options (from 1990 to 2020)

Primary occupation	Government service	Private sector	Agriculture	Daily wage	Animal husbandry	Others
Government job	50	50				
Private job		33.33	20.00	20.00	20.00	6.67
Agriculture		5.32	84.04	5.32	4.26	1.06
Daily wage		5.26	36.84	56.14		1.75
Animal husbandry			16.67	33.33	50	
Others						100

millet, maize, niger seeds, paddy, pearl millet, ragi, red gram, sugarcane, sunn hemp, sunflower and wheat) and 13 types of horticultural crops (beetroot, brinjal, carrot, castor, cauliflower, chilli, coconut, French bean, hyacinth bean, onion, potato, tomato and turmeric) since 1990 (Table 19.4). Maize, ragi and groundnut were the three most common agricultural crops and were being cultivated currently or were cultivated in the past by 85.29%, 85.29% and 32.35% of the households, respectively. Hyacinth bean, onion and brinjal were the three most common horticultural crops and were being cultivated currently or were cultivated in the past by 25.49%, 17.65% and 16.67% of the households, respectively.

19.3.3 Trends in Crop Cultivation

During 1990, 2000, 2010 and 2020 farmers were cultivating 20, 21, 19 and 28 types of crops respectively in the study region. There was no significant difference in the cultivated areas under different crops per household in 1990 ($F = 0.66$, df $= 19, 206$, df $= p > 0.05$), 2000 ($F = 1.42$, df $= 20, 191$, df $= p > 0.05$) and 2010 ($F = 1.61$, df $= 18, 191$, df $= p > 0.05$). However, in 2020 a significant difference was found in this regard ($F = 4.60$, df $= 27, 227$, df $= p < 0.05$). On the other hand, there was a significant difference in the number of households that cultivated different crops in 1990 ($\chi^2 = 541.36$, df $= 19$, $p < 0.01$), 2000 ($\chi^2 = 717.45$, df $= 20$, $p < 0.01$), 2010 ($\chi^2 = 716.66$, df $= 18$, $p < 0.01$) and 2020 ($\chi^2 = 742.62$, df $= 27$, $p < 0.01$).

Out of the 29 crops, 18 (castor, chilli, coconut, cotton, cowpea, groundnut, horse gram, hyacinth bean, little millet, maize, paddy, pearl millet, ragi, sugarcane, sunn hemp, tomato, turmeric, wheat) were grown continuously from 1990 to 2020. The cultivation of six crops (beetroot, carrot, cauliflower, French bean, potato and sunflower) was started in 2020, whereas the villagers have discontinued the cultivation of niger seeds since 1990. The percentage of crop fields in which both manure and fertilizer were applied increased from 57.59% in 1990 to 73.70% in 2020. On the other hand, the percentage of crop fields in which only manure and only fertilizers

Table 19.4 The CCI indices of the crop grown in the fringes of MM Hills, Karnataka, at four points of time (1990, 2000, 2010 and 2020)

Name of crop	Point of time			
	1990	2000	2010	2020
Beetroot	0	0	0	0.181
Brinjal	0	0	0.016	0.236
Carrot	0	0	0	0.037
Castor	0.057	0.046	0.046	0.044
Cauliflower	0	0	0	0.035
Chilli	0.026	0.026	0.041	0.047
Coconut	0.016	0.017	0.020	0.015
Cotton	0.077	0.146	0.042	0.059
Cowpea	0.048	0.049	0.062	0.133
Foxtail millet	0.077	0.011	0	0.011
French bean	0	0	0	0.090
Groundnut	0.491	0.431	0.329	0.318
Horse gram	0.084	0.044	0.041	0.111
Hyacinth bean	0.225	0.228	0.174	0.255
Little millet	0.107	0.029	0.013	0.016
Maize	1.103	1.326	1.579	1.366
Niger seeds	0.028	0	0	0
Onion	0	0.018	0	0.282
Paddy	0.127	0.110	0.121	0.119
Pearl millet	0.100	0.095	0.060	0.084
Potato	0	0	0	0.076
Ragi	1.332	1.244	1.202	1.01
Sugarcane	0.094	0.037	0.056	0.032
Sunn hemp	0.056	0.045	0.046	0.058
Sunflower	0	0	0	0.057
Red gram	0	0.011	0	0.018
Tomato	0.015	0.018	0.016	0.171
Turmeric	0.109	0.105	0.162	0.131
Wheat	0.041	0.047	0.032	0.039

Table 19.5 Percentage of crop fields according to manure and fertilizer used

Period	Manure and fertilizer	Manure	Fertilizer
1990	57.59	32.14	10.27
2000	56.54	25.70	17.76
2010	71.23	9.43	19.34
2020	73.70	5.84	20.45

were used decreased and increased, respectively, from 32.14% and 10.27% in 1990 and 2020 to 5.84% and 20.45% (Table 19.5).

The costs of cultivation per acre were the highest for cotton (25,000 INR), turmeric (24,107.14 INR) and pearl millet (13,777.78 INR) in 1990, whereas in 2000, sugarcane, (30,000 INR), cotton (25,556 INR) and turmeric (23,833 INR) were the top three crops in this regard. In 2010, cotton (33,333.33 INR), coconut

Table 19.6 Cost of cultivation per acre of different crops grown in the surveyed fringe villages of MM Hills, Karnataka

	Point of time			
Crop	1990	2000	2010	2020
Beetroot	0	0	0	27,384.62
Brinjal	0	0	10,000	33,103.45
Carrot	0	0	0	16,666.67
Castor	9714.286	3200	8000	6400
Chilli	3000	7333	6808.511	26,216.22
Coconut	10,000	10,000	30,000	10,000
Cotton	25,000	25,556	33,333.33	49,090.91
Cauliflower	0	0	0	100
Cowpea	1730.769	1731	2916.667	8952.381
Foxtail millet	1714.286	800	0	4000
French bean	0	0	0	26,666.67
Groundnut	4119.403	6615	10925.93	16,913.32
Horse gram	1304.348	1429	4400	6666.667
Hyacinth bean	8471.698	4399	10838.1	18,733.33
Little millet	3615.385	0	0	0
Maize	5904.04	8311	10393.46	14326.62
Niger seeds	575	0	0	0
Onion	0	3000	0	67941.18
Paddy	10,350	10,286	6083.333	10,000
Pearl millet	13,777.78	12,000	20,000	24,000
Potato	0	0	0	49,230.77
Ragi	4639.175	7119	9724.564	14,780.14
Sugarcane	2500	30,000	8000	16,000
Sunn hemp	2286	3200	14,000	8888.889
Sunflower	0	0	0	8666.667
Red gram	0	1000	0	20,000
Tomato	2000	1000	1000	29,242.42
Turmeric	24,107.14	23,833	10,790.7	37,435.9
Wheat	4250	5550	15,000	45,000

(30,000 INR) and pearl millet (20,000 INR) and, in 2020, onion (67,941.18 INR), potato (49,230.77 INR) and cotton (49,090.91 INR) were found to be the crops with the highest costs of cultivation (Table 19.6).

19.3.4 Dominant Crops

Ragi (1.33) had the highest value of CCI in 1990, whereas maize had the highest values of CCI in 2000 (1.32), 2010 (1.56) and 2020 (1.37), respectively (Table 19.4). These were the only crops that scored CC indices exceeding 01 during the different points of time considered. In 1990, ragi and maize were cultivated by 68.63% and 50.98% of the households over 126.1 and 98 acres of land, respectively. In spite of this, the contribution of maize (31.52%) was higher than ragi (29.85%) in the overall annual crop production. In 2000, ragi (68.63%) was cultivated by a greater proportion of households than maize (50.81%). However, the total area under maize cultivation (106.1 acres) and its contribution to the overall crop production (37.87%) were higher than the total area under ragi cultivation (93.9 acres) and its contribution to the overall crop production (24.86%). In 2010, the contribution of maize to the overall annual production further increased to 50.90%, and that of ragi fell down to 22.45%. The former was grown in 113.10 acres of land and the latter was grown in 88.95 acres of land at this point of time. In 2020, it was found that ragi and maize were cultivated by 65.69% and 69.61% of the households over 74.5 and 111.75 acres of land yielding 13.05% and 33.28% of the overall crop production, respectively.

19.4 Discussion

Rural livelihoods are constituted by variations in gender and class, mixed up with complicated and multiple negotiations and processes (Arun 2012). In the rural areas of most of the developing economies, agriculture is a crucial sector that provides a livelihood to a major proportion of the population, especially in rural areas (Mathur et al. 2006). In order to integrate nature conservation and agriculture, social and economic aspects need to be coordinated (Kremen 2015, Scariot 2013). Agriculture and wage were the major means of livelihood in the surveyed villages. However, in due course of time, the contribution and relevance of agriculture further increased with respect to income generation. It, in fact, even contributed immensely to households, in which it was not considered as the primary source of income. The declaration of Male Mahadeshwara Hills as a Wildlife Sanctuary and construction of a dam were important factors that affected different aspects of the agricultural sector in the area. In Gundapura, several villagers lost their agricultural lands when a dam was constructed during the 1990s. The affected farmers received monetary compensation from the government authority. On the other hand, agriculture was comparatively well developed in Kaanchali and many of its farmers had been shifting towards the cultivation of horticultural and other commercial crops in lieu of traditional crops. Such shifting towards irrigated crops was due to the availability of water from a dam constructed in Gundapura. Boodhipadaga was predominantly inhabited by people from the Soliga community which mainly depended upon forest products for their livelihoods. However, when Male Mahadeshwara Hills was declared a Wildlife Sanctuary in 2013, the entry into the forest for the collection of resources was restricted and so the villagers shifted to agriculture and other sectors for livelihood generation. Consequently, the area under cultivation increased. However, some people of this village collected forest resources with due permission from the authorities. The landless people often migrated to nearby areas for daily-wage labour. The few villagers, who had adequate sources of water (e.g. borewells) to irrigate their agricultural fields, cultivated profitable crops. The others practiced rainfed agriculture. Greater dependency on agriculture was also evident in Pacchedoddi, from where a number of families had emigrated to other towns and nearby areas because of social and economic reasons. As a whole, it was observed that in due course of time, more and more people quit their original occupations to eventually adopt a livelihood based on agriculture. On the contrary, most of the people who earlier depended on agriculture retained it as the source of income. In other words, agriculture emerged as the most preferred means of supporting livelihoods. However, the livestock population greatly decreased due to the imposition of restrictions on grazing after the area was declared a Wildlife Sanctuary. The condition in MM Hills can be compared with the situation when Bwindi, Uganda, was declared as a National Park, and a formidable pressure was imposed on its adjacent communities, especially the poor who were mostly dependent on the forest resources. As a result, the household income and the level of food security were reduced (Blomley et al. 2010). Among all the different professions in the studied villages, government service yielded the highest financial security, as reflected in household incomes, but the lowest proportion of villagers had been involved in this profession. This reflects the unavailability of government jobs as well as inability of people to attain such jobs.

The farmers cultivated a wide variety of crops instead of being strictly selective towards the monocropping of one or two. Maize and ragi were the most dominant crops, but the farmers also grew crops that even required higher costs of cultivation. The number of households cultivating different crops tended to change with time and so did the areas in which different crops were cultivated. At times, the number of households cultivating a specific crop would decrease, whereas the respective area cultivated would decrease and vice versa. Fluctuations were also evident with respect to production. At times, farmers did not obtain any production from certain crops even after spending excessively on their cultivation due to incidental factors such as unavailability of water at the peak stage of requirement. On the other hand, some others were able to obtain greater profits at even lower input costs because of the optimum availability of all factors at the required stages of the crop. Under such circumstances, it was frequently observed that sometimes, crop production from smaller fields exceeded that of larger fields. Therefore, the overall performance of different crops as reflected by the CCI values rapidly fluctuated. A number of fields were artificially irrigated, and the extent and efficiency of irrigation were also important factors in this context. The impacts of droughts on productivity are reduced when farmers are equipped with supplemental irrigation (Sharma et al. 2010). This is crucially important because water scarcity has shaped up as a major issue in case of agricultural production in arid and semi-arid areas of India (Dhanya and Ramachandran 2015).

The cropping preference of farmers was dynamic that frequently changed and accordingly there were changes in crops cultivated and areas in which different crops are grown. Cost of cultivation was an important factor in this regard, but it was not the only determinant. This was evident from the fact that even though maize and ragi had the highest values of CCI since 1990, these crops never had the lowest costs of cultivation. Rapid variations were also evident in the costs of cultivation of different crops. During the study, it was found that presently the farmers were showing an inclination towards horticultural crops. This was evident from the fact that they started to cultivate a number of vegetables, which had not been grown since 1990. In fact, with the technological development, modernization and alterations in consumption pattern, the agricultural sector of India is shifting towards the horticultural crops (Singha et al. 2014). Overall, the total number of crops grown increased near MM Hills, even though the farmers discontinued the cultivation of a few. Crop diversification could play an important role as it is a strategy for the overall agricultural development in the country. The development of basic infrastructural facilities is an essential requirement for enabling conditions to facilitate agricultural development and crop diversification (Acharya et al. 2011).

Organic farming is an important adopted strategy to stabilize agriculture (Patil et al. 2014). The villagers prepared organic manure for their crop fields. For this, livestock excreta, especially cow dung, was mixed with leftover plant material from crop harvest such as straw. The mixture was gathered in heaps or put in a ditch dug in the ground and left to decompose in exposure to different weather conditions. After decomposition for about 2 months to 1 year, the resultant material was applied in crop fields as manure. In case the production of manure was more than the requirement, the excess amount was sold. The combined application of this organic manure and chemical fertilizers was the most common practice. However, overall application of chemical fertilizers increased and that of organic manure decreased. The excessive use of fertilizers can have serious environmental impacts. This is evident from the fact that the Green Revolution has resulted in continuous environmental degradation, particularly of soil, vegetation and water resources in Haryana. Already 82% of the geographic area of the state is under cultivation, and further increase in productivity is crucially dependent on more energy-intensive inputs (Singh 2000). On the other hand, in Puducherry, South India, it has been found that agricultural workers are not adequately aware about pesticides. Consequently, the improper disposal of pesticides and their containers can harm the environment (Mohanty et al. 2013). Thus, there is the necessity to ensure that farmers are aware of the different aspects of pesticide use. Simultaneously, for the protection and restoration of biodiversity in agricultural landscapes, the quantity and quality of habitat on and around farms should be improved, and this also includes the establishment of protected areas (Syngenta International 2019).

19.5 Conclusion

Agriculture is the backbone of the economy and is the major source of livelihood generation in the fringes of MM Hills. Choice of crops cultivated has changed in the course of time and it is likely that it would continue to change in the future. However, the trend is towards cultivating greater number of crops, where the new crops are predominantly vegetable crops. This also reflects the readiness of farmers to attempt the cultivation of new types of market-oriented crops. The application of chemicals has also increased and hence attempts should be made to limit their application to reasonable levels. This would enable MM Hills to emerge as a site of biodiversity conservation that also sustainably contributes to achieving food security.

Acknowledgements The authors are grateful for the support received from the local people and forest department. We extend our sincere thanks to the Science for Equity, Empowerment & Development (SEED) Division,Department of Science & Technology (DST), Technology, New Delhi for financial support. The views expressed in this research article are those of the authors.

Declaration There is no conflict of interest.

References

Acharya SA, Basavaraja H, Kunnal LB, Mahajanashetti SB, Bhat ARS (2011) Crop diversification in Karnataka: an economic analysis. Agric Econ Res Rev 24:351–357

Arun S (2012) 'We are farmers too': agrarian change and gendered livelihoods in Kerala, South India. J Gend Stud 21(3):271–284

Azhoni A, Goyal MK (2018) Diagnosing climate change impacts and identifying adaptation strategies by involving key stakeholder organisations and farmers in Sikkim, India: challenges and opportunities. Sci Total Environ 626:468–477

Birthal PS, Md KT, Negi DS, Agarwal S (2014) Impact of climate change on yields of major food crops in India: implications for food security. Agric Econ Res Rev 27(2):145–155

Blomley T, Namara A, McNeilage A, Franks P, Rainer H, Donaldson A, Malpas R, Olupot W, Baker J, Sandbrook C, Bitariho R, Infield M (2010) Development and gorillas? Assessing fifteen years of integrated conservation and development in South-Western Uganda, natural resource issues, vol 23. IIED, London

Chandrakanth MG, Priyanka CN, Mamatha PM, Patil KK (2013) Economic benefits from micro irrigation for dry land crops in Karnataka. Indian J Agric Econ 68(3):326–338

Dhanya P, Ramachandran A (2015) Farmers' perceptions of climate change and the proposed agriculture adaptation strategies in a semi-arid region of South India. J Integr Environ Sci 13:1–18. https://doi.org/10.1080/1943815X.2015.1062031

Durán AP, Casalegno S, Marquet PA, Gaston KJ (2013) Representation of ecosystem services by terrestrial protected areas: Chile as a case study. PLoS One 8(12):e82643

Fanning E (2005) Formatting a paper-based survey questionnaire: best practices. Practical Assessment Research and Evaluation 10(12):1–14

FAO (2014) Protected areas, people and food security. An FAO contribution to the World Parks Congress, Sydney, 12–19 November 2014

FAO (2020) FAO in India. http://www.fao.org/india/fao-in-india/india-at-aglance/en/#:~:text=Agriculture%2C%20with%20its%20allied%20sectors,275%20million%20tonnes%20(MT)

Goswami M, Nautiyal S, Manasi S (2020) Drivers and consequences of biophysical landscape change in a peri-urban–rural interface of Guwahati, Assam. Environ Dev Sustain 22(2):791–811

Jitendra (2015). https://www.downtoearth.org.in/news/current-agricultural-growth-rate-wont-meet-world-food-demand-in-2030-47269

Kannan E (2011) Relationship between agricultural credit policy, credit disbursements and crop productivity: a study in Karnataka. Indian J Agric Econ 66(3):444–456

Kremen C (2015) Reframing the land-sparing/land-sharing debate for biodiversity conservation. Ann N Y Acad Sci 1355:52–76

Kumar P, Mittal S (2006) Agricultural productivity trends in India: sustainability issues. Agric Econ Res Rev 19:71–88

Kumar M, Kumar R, Rangnamei KL, Das A, Meena KL, Rajkhowa DJ (2019) Crop diversification for enhancing the productivity for food and nutritional security under the Eastern Himalayas. Indian J Agric Sci 89(7):1157–1161

Mathur AS, Das S, Sircar S (2006) Status of agriculture in India: trends and prospects. Econ Polit Wkly 41(52):5327–5336

Mohanty MDMK, Behera BK, Jena MDSK, Srikanth MDS, Mogane C, Samal S, Behera AA (2013) Knowledge attitude and practice of pesticide use among agricultural workers in Puducherry, South India. J Forensic Legal Med 20(8):1028–1031

Nath R, Luan Y, Yang W, Yang C, Chen W, Li Q, Cui X (2015) Changes in arable land demand for food in India and China: a potential threat to food security. Sustainability 7:5371–5397

Pastor AV, Palazzo A, Havlik P, Biemans H, Wada Y, Obersteiner M et al (2019) The global nexus of food–trade–water sustaining environmental flows by 2050. Nat Sustain 2(6):499–507

Patil S, Reidsma P, Shah P, Purushothaman S, Wolf J (2014) Comparing conventional and organic agriculture in Karnataka, India: where and when can organic farming be sustainable? Land Use Policy 37:40–51

Scariot A (2013) Land sparing or land sharing: the missing link. Front Ecol Environ 11:177–178

Sharma BR, Rao KV, Vittal KPR, Ramakrishna YS, Amarasinghe U (2010) Estimating the potential of rainfed agriculture in India: prospects for water productivity improvements. Agric Water Manag 97(1):23–30

Shithi PK, Pati CK (2012) Non-timber forest products for livelihood security of tribal communities: a case study in Paschim Medinipur District, West Bengal. J Human Ecol 40(2):149–156

Singh RB (2000) Environmental consequences of agricultural development: a case study from the Green Revolution state of Haryana, India. Agric Ecosyst Environ 82(1–3):97–103

Singha AK (2013) Assessing forest villagers' role perception and role performance behaviour towards social development with forest resources in Assam. J Hum Ecol 41(1):77–82

Singha K, Choudhary R, Vishnu K (2014) Growth and diversification of horticulture crops in Karnataka: an inter-district analysis. SAGE Open 2014:1–12

Slaughter RA (2011) Welcome to the Anthropocene. Futures 44(2):19–26

Syngenta International (2019) Our policy position: Syngenta and enhancing agricultural biodiversity. Syngenta International, Cham, Switzerland

Uma Shaanker R, Ganeshaiah KN, Krishnan S, Ramya R, Meera C, Aravind NA, Kumar A, Rao D, Vanaraj G, Ramachandra J, Gathier R, Ghazoul J, Poole N, Reddy Chinnapa BV (2004) Livelihood gains and ecological costs of NTFP dependence: assessing the roles of dependence, ecological knowledge and market structure in three contrasting human and ecological setting in South India. Environ Conserv 31:242–253

Veach V, Moilanen A, Di Minin E (2017) Threats from urban expansion, agricultural transformation and forest loss on global conservation priority areas. PLoS One 12(11):e0188397

Conservation of Plant Genetic Resources: A Special Reference to Dye-Yielding Plants

20

Ressin Varghese, C. George Priya Doss, Chandrasekaran Rajasekaran, R. Seenivasan, T. Senthilkumar, and Siva Ramamoorthy

Abstract

Plant genetic resources are the potential sources for farming, experimenting, and conserving plants. The crisis imparted by genetic erosion instigated global conservation schemes to protect plants. Dye-yielding plants are gaining much importance due to severe toxic effects of synthetic dyes. Biotechnology-based approaches like in vitro propagation, metabolic engineering, gene markers, and genome sequencing have substantially contributed to the conservation of dye-yielding plants. Cryopreservation of plants is also part of this conservation policy. Indigenous communities, their traditional knowledge, and community-based approach must be considered for an efficient conservation of dye-yielding plants.

Keywords

Conservation · Genetic resources · Dye yielding plants · Genetic markers · In vitro propagation

R. Varghese · C. G. P. Doss · C. Rajasekaran · S. Ramamoorthy (✉)
School of Bio Sciences and Technology, Vellore Institute of Technology, Vellore, India

R. Seenivasan
Center for Nanotechnology, Vellore Institute of Technology, Vellore, India

T. Senthilkumar
Department of Plant Science, Bharathidasan University, Tiruchirappalli, India

S. Ramamoorthy et al. (eds.), *Plant Genetic Resources, Inventory, Collection and Conservation*, https://doi.org/10.1007/978-981-16-7699-4_20

20.1 Introduction

The primitive forms of plants first emerged on the earth's surface during the Ordovician period around 470 million years back. Approximately, 374,000 different plant species have been identified to date and still more are waiting to be explored (Christenhusz and Byng 2016). Plants are the major resources in the biosphere providing different commodities ranging from food to fuel. Unfortunately, it is a shocking and distressing fact that around three seed-bearing species are disappearing from the earth every year (Ledford 2019). One of the huge challenges confronted by the twenty-first century is to utilize this treasured resource without exploiting it. Yet the global and national level efforts for conservation are giving us hope.

The conservation strategies are giving particular emphasis to food and agriculture crops while some categories are getting exempted. Dye-yielding plants are one such category, a multihued treasure to uphold. The attraction of men towards color is a centuries-old phenomenon and the testimony of this has been discovered in cave arts, colored garments in the Indus valley, and Egyptian pyramids. The barks, seeds, flowers, and leaves of plants have been identified as sources of a variety of colors. Seeds of *Bixa orellana* (orange-reddish dye), leaves of *Indigofera tinctoria L.* (blue dye), and barks of *Alnus glutinosa* (L.) Gaertn. (black dye) are some notable examples. Around 450 taxa of dye-producing plants have been identified in India itself. These dyes have been used as colorants in foods, garments, and cosmetics. Besides the coloring property, many of the plants are rich in medicinal properties. For instance, yellow-colored turmeric is a well-known ayurvedic herb (Siva 2007). The rise of synthetic dyes reduced the demand for natural dyes considerably. But later studies proved that synthetic dyes are posing a serious threat to health (Manzoor and Sharma 2019). Thereafter, natural sources of dyes have been considerably investigated. The research studies are still going on to elicit pigment production in plants via novel biotechnology tools. However, the recent "*organic*" marketing strategy has led to the uncontrolled exploitation of dye-yielding plants for extracting organic natural dyes. This book chapter deals with the conservation strategies of genetic resources of dye-yielding plants. Particular significance has been given to recent biotechnological approaches used for assessing as well as conserving the genetic diversity of dye-yielding plants. Reports on indigenous groups farming dye-synthesizing plants are also subjected for review.

20.2 Plant Genetic Resources and Their Conservation

The mother earth is opulent with a vast variety of living beings ranging from microscopic microbes to gigantic whales. These variants of living beings constitute the biodiversity of nature. Out of the 374,000 reported plants, 308,312 are vascular plants, with 295,383 flowering plants. Rest is liverworts, mosses, ferns, etc. (Christenhusz and Byng 2016). Even in the midst of global conservation policies, around two in five plant species are facing the threat of extinction (Nic Lughadha et al. 2020). Rampant deforestation, capricious climate changes, emerging variants of pathogens, and growing industrialization are all adding fuel to fire.

It is indeed necessary to conserve the biodiversity for a sustainable development and balanced endurance of all living beings. Genetic resources are described as the genetic material of potential value. These resources are of plant, animal, and microbial origin which act as the functional unit to develop the whole organism. The conservation and effective utilization of plant genetic resources (PGRs) have been documented centuries ago. The first reported incident was in AD 795 when Charlemagne (Capitulare de villis) ordered to cultivate 77 different food crops in his farm (Sonnino 2017). The genetic resources were considered as a common commodity and were exchanged without restrictions throughout the world. Later, the world witnessed the development of different crop breeds during the era of Green Revolution. The introduction of improved plant breeds led to the vanishing of wild genes and gene combinations. This phenomenon was named as gene erosion in the 1960s. The loss of wild varieties in this form pinpoints the inevitability of gathering and protecting plant genomic resources. Trailing this, new policies like the International Union for the Protection of New Varieties of Plants (UPOV) Convention came into existence (Roa et al. 2016). The major incidents in the history of conservation of plant genomic resources are illustrated in Fig. 20.1.

20.2.1 Framework of PGR Conservation Systems

Rights over PGRs is a hot disputable topic among developing and developed countries. The various policies and protocols have been written in such a way as to avoid disputes over IPR rights. Tropical countries are bestowed with its megadiversity harboring thousands of plant species. Developed countries which are financially competent exploit these resources and often attain more profits through their unparallel monopoly. Various treaties have been implemented to secure the rights of farmers over PGRs, mainly the International Treaty on Plant Genetic Resources for Food and Agriculture, 2001. The International Plant Genetic Resources Institute (IPGRI) holds the authority over collection and conservation of plant germplasms and has been espoused by more than 40 countries. The Global Crop Diversity Trust is another organization managing funds to securely safeguard plant genomic resources in gene banks across the world (Sonnino 2017). In addition to this, large databases like EURISCO are available to access the information regarding millions of PGRs (Weise et al. 2017). Morals in plant genomic resources are a multidimensional fact which requires the collaboration of farmers, researchers, multinational companies, and governing authorities (Deplazes-Zemp 2019). Sadly, no considerable efforts to safeguard the treasure of dye-yielding plants are observed. The efforts at both organizational and individual levels are required for the conservation, utilization, and documentation of thousands of dye-yielding plants across the world.

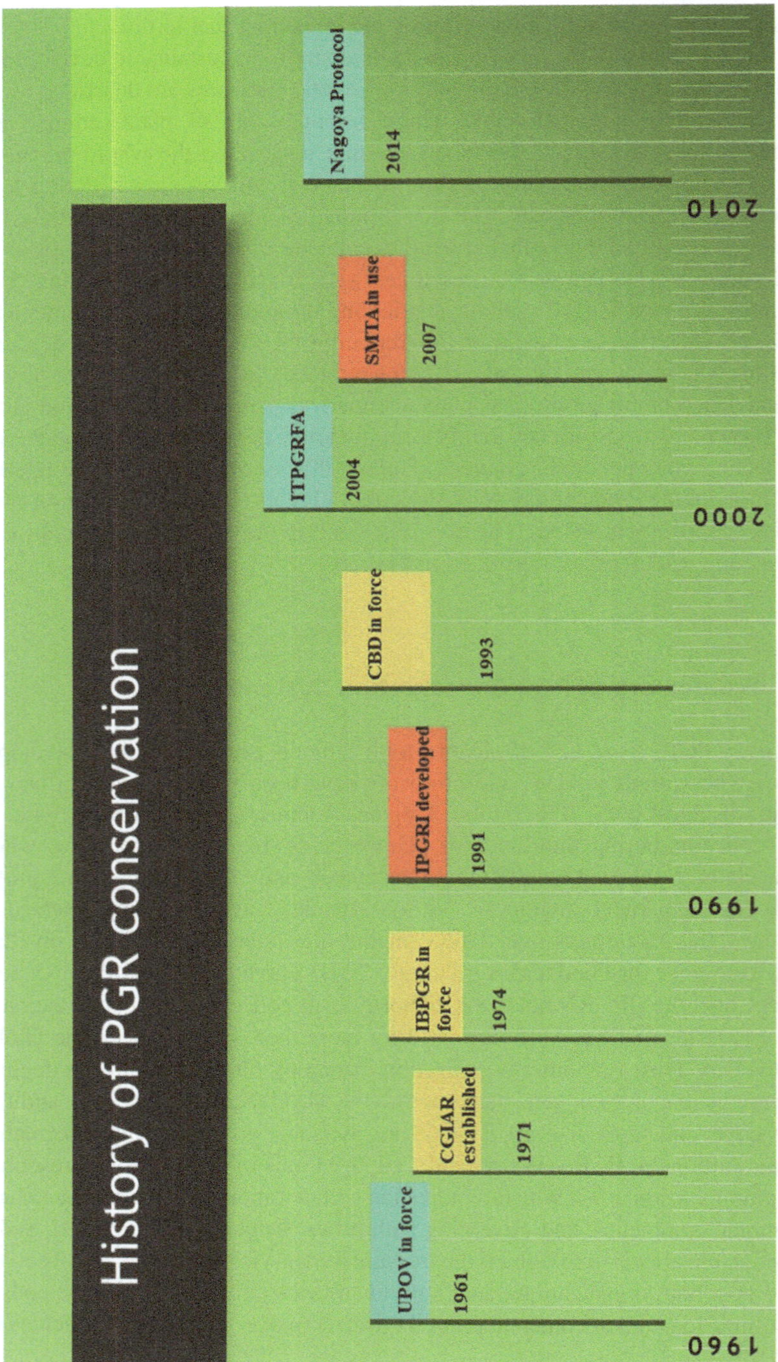

Fig. 20.1 Major incidents in the history of PGR conservation

20.2.2 Different Approaches in Conservation of PGRs

After the increased incidents of genetic erosion, researchers have formulated various strategies to conserve PGR. In 1960s, cold storage systems were developed to collect and keep germplasm collections. Gene banks are one of the most explored ex situ conservation strategies. Large cryogenic systems are available throughout the world for storing seeds and pollens. These systems are the major resources of research for ameliorating the productivity and also for mounting new breeds. Various invitro conservation strategies have been employed to protect and investigate the mysteries of pigments in dye-yielding plants. However, some researchers came with the opinion that seeds should be cultivated for natural genetic variation rather than long-term storage. In situ conservation where species are protected in their natural habitat and are allowed to complete the growth cycle by undergoing evolutionary changes was later popularized. On-farm management and protected area approaches have gained much acceptance in many countries. In the context of dye-yielding plants, they are often the integral part of botanic gardens because of their appealing flowers. Additionally, highly protected areas like Western Ghats harbor many varieties of these plants. A considerable leap in the area of protected lands up to 17 million km^2 evidenced the efforts made by various countries in protecting their biodiversity (FAO 2010). Nevertheless, adversities have been encountered in maintaining protected areas because of the anthropogenic pressures and fluctuating climatic variations (Zegeye 2017).

The efficient implementation of ex situ and in situ strategies is a requisite to conserve the rare plant species like dye-yielding plants. But in situ conservation strategies often have an upper hand over ex situ strategy. The reason is the huge capital required to maintain long-term storage of germplasm (Cohen et al. 1991). Community-based approaches structured and implemented by M.S. Swaminathan are another strategy to safeguard and utilize landrace cultivars with the cooperation of local people, for instance Potato Park situated in Peru (Sonnino 2017). Developing countries like India and Mexico have made attempts also to document the traditional ethnobotanical knowledge possessed by farmers for the proper cultivation coupled with the conservation of PGRs. Ethnobotanical data of many dye-yielding plants have been collected and documented from the tribal areas of India. The proper incorporation of in situ, ex situ, and community-based schemes should be exercised globally and at national level for ensuring conservation of PGRs. The questions that pop up before thinking about the conservation of dye-yielding PGR are what is the need of conserving and how should it be conserved. As depicted in Fig. 20.2, the enormous benefits of pigments of these plants are the reason for the uncontrolled exploitation.

Fig. 20.2 Benefits of dye-yielding plants. (**a**) Textiles, (**b**) cosmetics, (**c**) perfumes, (**d**) food coloring, (**e**) drug coloring, (**f**) antibacterial, (**g**) anticancer, (**h**) antidiabetic

20.3 Biotechnology-Oriented Conservation Strategies

The integration of biotechnology into conservation strategies imparted beneficial effects in the gathering, characterizing, maintaining, exchanging, utilizing, and depositing of germplasms (Rao 2004). Developments in in vitro propagation strategies and bioinformatic tools considerably fasten up the process of collection of plant germplasm (Cruz-Cruz et al. 2013). It is very crucial to determine the genetic diversity of species for efficient conservation in case of genetically vulnerable species. Advent of molecular markers was a revolutionary milestone in the history of plant genomics. Tissue culture techniques are the most reliable, cost-effective option to maintain the rare, endangered dye-yielding plants. The exchange of germplasm is a relatively sensible procedure where chances of contamination are extremely high. Biotechnological tools like ELISA and PCR are efficient means to detect the microbial attacks in germplasm deposits (Rao 2004). Moreover, a substantial hike has been observed in the improvement of plant varieties via metabolic engineering. A proper optimized amalgamation of biotechnology with orthodox conservation approaches will indubitably help in securing many rare, vulnerable dye-yielding plants.

20.3.1 Markers in Assessing Genetic Diversity

Analyzing genetic variation is a preliminary step in the conservation of germplasm and is important in determining which variant should be conserved and how to be conserved. Moreover in an economical viewpoint, diversity data will help in developing cultivars with much-needed traits like high yield and stress resistance. Sound knowledge in genetic diversity will help to understand the evolutionary relationships and taxonomy of less explored dye-yielding plants. These data will help to form core collections of pigment-producing plant species with least number of likeness (Rao and Hodgkin 2002). It is imperative to maintain diversity at species, genus, and ecosystem levels to achieve the real goals of conservation. Different techniques have been practiced to illustrate the diversity in species. Morphological markers are the oldest tool to differentiate plants based on parameters like color, size, and shape of flowers and leaves. Despite the required cost and time, this tool is very apposite in perceiving pathogenic attack in plants (Govindaraj et al. 2015). The biochemical markers are another category where isoenzymes and proteins are the criteria explored to study the diversity (Bretting and Widrlechner 1995). For instance, native breeds of *Bixa orellana,* a renowned pigment-producing plant, were primarily categorized on the basis of fruit shape. Later biochemical markers like water-soluble polysaccharide content became the parameter to classify the varieties (Parimalan et al. 2014). Molecular markers have been well utilized to study genetic variations within many dye plants and thus pivotal in framing conservation strategies. These genetic markers principally contributed to constructing QTL mapping, map-based cloning, and marker-assisted selection in plant breeding programs (Moose and Mumm 2008). The use of markers helps in computing the polymorphism within genomic DNA and the resultant data will be processed via similarity indexes like Nei's similarity. Subsequent tree construction by means of UPGMA gives a systematic knowledge of genetic relatedness in plant species. Markers with more PIC values are considered to be more beneficial. It entails the level of polymorphism or allelic diversity (Singh et al. 2015). Genetic diversity can be studied by using a single marker or a combination of markers for more meticulous results. In case of *Curcuma longa* in the northeastern part of India, RAPD, ISSR, and AFLP markers were prolifically employed and evolutionary lineage of curcuma was well understood (Das et al. 2011). Similar investigations were performed in *Carthamus tinctorius* (safflower) (Sehgal and Raina 2005) and *Justicia adhatoda* (Kumar et al. 2014). Table 20.1 represents the list of markers used in dye-yielding plants and their results with references.

20.3.1.1 RAPD (Random Amplified Polymorphic DNA) Markers

RAPD markers are a rapid tool for illustrating genomic map and are used for conservation as well as breeding purposes. In fact, RAPD is an amended version of PCR where genomic DNA is amplified with random nucleotide sequence primers (Senthilkumar and Gurusubramanian 2011). The RAPD markers have been used for the germplasm characterization as part of conservation methodologies (Elyasi Zarringhabaie et al. 2012). RAPD markers were used to investigate the genetic likeness in Acacia trees. The bark of Acacia tress like *A. catechu* and *A. dealbata* is a good source of black and brown dyes (Siva 2007). The output evidenced how the species are clustered and also in-between similarity (Nanda et al. 2004). *B. orellana* is one of the commercially important dye-yielding plants which easily acclimate

Table 20.1 List of markers used in some of the dye-yielding plants

Plant name	Marker used	Remarks	Reference
Acacia leucophloea	RAPD	36 Provenance-specific markers and 162 species-specific markers generated	Mutharaian et al. (2018)
Alnus sp.	ITS, *rbcL*, *matK* and *trnH-psbA*	Around 23sp. exhibited their own molecular integrity.	Ren et al. (2010)
Barleria sp.	rbcL, matK, and RAPD	Distinguished at subgenera level	Meenakshi et al. (2016)
Betula utilis D. Don.	RAPD and ISSR	Investigated diversity of accessions in North-western Himalaya	Wani et al. (2018a)
	EST-SSR	Twenty accessions were clustered into two	Wani et al. (2018b)
Butea monosperma	ISSR	Polymorphism was about 85.25%	Kandasamy et al. (2013)
	RAPD	Polymorphism was about 66.3%	Vashishtha et al. (2014)
	RAPD	36 RAPD primers were used	Thamilarasi et al. 2016
	RAPD, ISSR, and SRAP	High polymorphism identified	Vashishtha et al. (2013)
Cassia auriculata	RAPD, DAMD, and ISSR	Markers discriminates the plant from *C. surrattensis*	Panapitiya and Welikala (2018)
Citrus medica L.	SSR	Monophyletic nature identified	Ramadugu et al. (2015)
	SSR, EST-SSR	Greater diversity	Luro et al. (2012)
	RAPD,ISSR	Samples from different parts of India. Elite genotypes detected.	Uchoi et al. (2017)
Clitoria ternatea	RAPD	Polymorphism detected	Yeotkar et al. (2011)
	RAPD and ISSR	Polymorphism observed was ISSR (29.52%) and RAPD (27.73%)	Bishoyi et al. (2014)
Cordia myxa	RAPD	High genetic diversity	Sivalingam et al. (2012)
Coscinium fenestratum	ITS	For distinguishing from adulterants	Venkatasubramanian (2011)
Curcuma longa	matK, rbcL, ITS-LSU D1/D3	Categorize the species within commercial breeds	Duan et al. (2017)
Crocus sativus	SCAR	Used as screening tool to detect saffron	Torelli et al. (2014)
Dioscorea bulbifera	SSR	Studied diversity of variants in Brazil	Silva et al. (2016)
	SSR	Studied diversity of variants in west Africa	Osuagwu and UL (2020)

(continued)

Table 20.1 (continued)

Plant name	Marker used	Remarks	Reference
	LFY	Distinguish between various discorea sp.	Techen et al. (2017)
Dipterocarpus turbinatus	SSR	–	Cvetković et al. (2019)
Euphorbia sp.	*matK*, *rbcL*, ITS, *psbA-trnH*	Evaluating the efficacy of markers for sps differentiation	Aubriot et al. (2013)
Galium sp.	ISSR and RAPD	Greater level of genetic diversity	Abdel Khalik et al. (2014)
Garcinia mangostana	ISSR	Distinguish between various *Garcinia* sp.	Sobir et al. (2011)
	SSR	Developing SSR markers via ISSR suppression-PCR	Qosim et al. (2011)
	ISSR	Alikeness in accessions from the same region	Mansyah Ellina et al. (2010)
Gardenia jasminoides	AFLP	Diversity at sp. level identified	Han et al. (2007)
	EST-SSR	Higher polymorphism	Deng et al. (2015)
Jatropha curcas L.	EST-ISSR	Developed and described these markers	Kumar Yadav et al. (2011)
	RAPD and AFLP	To access genetic fidelity	Sharma et al. (2011)
	SSR, EST-SSR	Intergroup diversity was found to be more	Wen et al. (2010)
	ISSR	Analyzed efficiency of ISSR markers	Grativol et al. (2011)
	RAPD	Higher diversity	Rafii et al. (2012)
	RAPD and ISSR	Moderate level of diversity in accessions from Rajasthan	Khurana-Kaul et al. (2012)
	RAPD, AFLP, and SSR	Comparatively lower diversity among accessions	Mastan et al. (2011)
	SSR	Heterozygosity was low	Maurya and Yadav (2016)
Malpighia glabra L.	SRAP	To analyze genetic stability of regenerated plants	Yamamoto et al. (2017)
Toona ciliata Roem.	SRAP	Diversity among populations were identified	Li et al. (2015)
Ziziphus mauritiana	RAPD	Polymorphism was about 98.6%	Vashishtha et al. (2014)

with different ecogeographical conditions. RAPD markers were successfully utilized to discover the high-yielding variants in cultivars of Kerala (Joseph and Siril 2014). Similar region-specific genome studies were also carried out in case of *C. longa* accessions in Pakistan (Jan et al. 2011) and India (Ashraf et al. 2017) and safflower in Asia (Hacioglu et al. 2013). Genetic diversity of accessions from various sites across the world was assessed in dye-producing plants; RAPD markers were also employed to evaluate the genetic clonality of in vitro-propagated plants of *B. orellana* (Siril and Joseph 2013), *C. longa* (Tyagi et al. 2007), and *Butea monosperma* (Yarra et al. 2018). This aided in assuring that the genetic nature has been maintained from the wild to in vitro plants.

20.3.1.2 AFLP (Amplified Fragment Length Polymorphism) Markers

AFLP is another dominant DNA fingerprinting means that selectively amplifies restricted portions of DNA (Vos et al. 1995). These markers were used to research the genetic diversity of *Carthamus tinctorius* accessions from more than 40 countries. Interestingly, corresponding diversities differing with geographical sites were discovered. *Carthamus tinctorius* is the major source of carthamin, a red dye. The Near East and Iran–Afghanistan regions exhibited maximum diversity par with the Indian subcontinent (Kumar et al. 2015).

20.3.1.3 Microsatellite Markers

Microsatellite markers (SSR) are short sequence repeats of 2–4 bp length found universally in every organism. SSRs have been widely utilized for genome mapping and to evaluate the evolutionary relationship among species. SSR is also founded on PCR and polymorphic DNA is identified by designing primers flanking the short sequence repeats. Inter SSR (ISSR) is a prompt technique for estimating the genetic diversity (Godwin et al. 1997). Interestingly, SSR markers were used to study the genetic diversity within *Bixa orellana* populations of Brazil. A higher heterozygosity was observed and divergence was found more in "Rondonia" groups by employing 16 SSR markers (Dequigiovani et al. 2017). Analysis of genomic variance via eight SSR markers in *C. longa* (turmeric) cultivars in Rapa Nui region led to the identification of different alleles and interrelatedness between three major clusters of native breeds (Balada et al. 2021). SSR markers have also been used to study the region-specific germplasm collections of turmeric like in Brazil (Sigrist et al. 2011) and India (Siju et al. 2010a; Senan et al. 2013). Similarly, ISSR and DAMAD markers were used to explore the genetic diversity in ethnic turmeric collections of India. High polymorphism and the resultant dendrogram accentuate the level of divergence in germplasm (Verma et al. 2015). Moreover, *C. longa* was appraised as the high curcumin-containing species via molecular studies using matK sequences (Hayakawa et al. 2011). A collective approach of RAPD and ISSR has been accomplished in *Acacia senegal* in four Kenyan populations. It was elucidated that variation has occurred in populations (Chiveu et al. 2008). ISSR marker-based diversity studies of germplasm accessions were also reported in various other dye-yielding plants like safflower (Panahi and Ghorbanzadeh Neghab 2013; Houmanat et al. 2016).

20.3.1.4 EST-SSR (Expressed Sequence Tags-Simple sequence repeats) Markers

EST-SSR markers are more robust and functionally potential markers developed by rectifying the demerits of SSR markers. SSR markers stemmed from EST databases are more transferable and economical. These markers have been employed to illustrate genomic mapping, diversity, and phylogenetic and population genetic studies (Varshney et al. 2005). EST-SSR markers successfully estimated high polymorphism in curcuma cultivars collected from different sites of India. Consequently, the transferability rate among species was also subjected for investigation (Siju et al. 2010b). EST-SSR markers used in *C. longa* (Lakadong and Suvarna varieties) were found to be more expedient than conventional biochemical markers in assessing diversity. Phylogenetic tree-built methodology also evidenced the genetic variance in these two breeds (Sahoo et al. 2017). Similarly, EST-SSR markers were used to study the genetic relatedness in native breeds of *Curcuma* in Chhattisgarh and above 80% of polymorphism was found between different species (Jain and Parihar 2019). These markers have been successfully utilized to evaluate genetic diversity of many other dye-producing varieties like *Carthamus tinctorius* (Chapman et al. 2009; Derakhshan et al. 2014). The resultant data could act as the foundation for standardizing breeding programs and in situ conservation strategies. Remarkably, genetic pureness of crossbreeds can also be evaluated by EST-ISSR markers, for example safflower (Naresh et al. 2009).

20.3.1.5 SRAP (Sequence related amplified Polymorphism) Markers

SRAP marker, a very recent dominant marker, is developed for amplifying DNA binding to primers aiming predominantly ORFs. This tool has been primarily used in horticulture technique and subsequently for assessing germplasm diversity (Robarts and Wolfe 2014). High genetic diversity was identified in annatto varieties using SRAP markers. Variation was exhibited on the basis of difference in gathering locations (Valdez-Ojeda et al. 2008). A genetic linkage map was constructed for *B. orellana* using SRAP markers along with β-cyclase marker. The resultant map covers around 879.054 cM, encompassing 103 markers (Romero-López et al. 2019). This map is worthwhile in conservational policies and also for mounting high-breed variants.

20.3.1.6 Plastid-Coding Sequence Markers

Plastid-coding sequences are widely used markers to research genetic diversity as well as to generate DNA barcodes. A number of plastid-coding sequences like atpF–atpH spacer, rpoB gene, and rpoC1 gene are used for species distinction. Identification in taxonomical level is a very foremost step in conservation policies (CBOL plant working group 2009). DNA barcodes of *Acacia* were constructed using matK and rbcL markers and a dendrogram was generated depicting the evolutionary relationships within *Acacia* species. Scannable QR code was also produced by analyzing the data obtained (Ismail et al. 2020). Similarly, matK and rbcL together with ITS-LSU D1/D3 were used to categorize the species within commercial breeds of *C. longa*. Interestingly, among the three ITS-LSU D1/D3 was found to be the most apt marker for species classification (Duan et al. 2017).

20.3.1.7 Other Markers

Diversity and population studies were worked out in germplasm accessions of safflower from 28 countries using iPBS-retrotransposon markers. Most diversity was observed in Asian countries. iPBS-retrotransposon is a robust retrotransposon marker and works on the presence of tRNA as a primer-binding site (Ali et al. 2019).

20.3.2 Role of Bioinformatics in Accelerating Plant Genomic Conservation

Computational approaches were always a groundwork tool in inspecting the genetic diversity of species. Geographic information systems (GIS) can be defined as a database management system which instantaneously manages digital spatial information together with nonspatial, attached information. GIS gave the answers to the questions of a conservationist like where to protect (Jarvis et al. 2005). The phase of computational approaches in germplasm conservation underwent a transition by the advent of genomics. The dawn of transcriptomic techniques began in the nineteenth century intending to understand the biosynthetic pathways and functions via genomic knowledge (Lowe et al. 2017). Different strategies like expression sequence tags (EST) were practiced earlier. But these were later replaced by sophisticated techniques like NGS (Hawkins et al. 2010; McGettigan 2013) and microarrays (Auburn et al. 2005). Genotyping via sequencing upgraded the understanding of polymorphisms and thus genetic variation present within the plant family.

NGS, which is far more advantageous than conventional sequence protocols, is commercially available as PacBio (Pacific Bioscience), Oxford nanopore, and Illumina (Lu et al. 2016). Opportunely, de novo sequencing had been accomplished in commercially important dye-producing plants. A gene map was constructed in *C. tinctorius* via whole-genome shotgun sequencing, and 2,008,196 genetically located SNPs in 1178 unique positions were discovered. Genome-wide SNP genotyping is a potent means for association mapping, phylogenetic studies, and comparative genomics (Bowers et al. 2016). Whole-chloroplast genome sequencing was accomplished in *C. tinctorius* (Wu et al. 2019), *Pterocarpus santalinus* (produces red santalin) (Jiao et al. 2019), and *Beta vulgaris* (Li et al. 2014). Transcriptomics approach was utilized in dye-producing plants to expound biosynthetic gene expressions. This in turn will help the researchers to enhance the pigment production, to implement microbial reactors, and ultimately to lessen the maltreatment of plant for dyes: for instance, to elucidate indigo biosynthetic pathways of *Indigofera tinctoria* and *Polygonum tinctorium* (Sarangi et al. 2015) and to study apocarotenoid synthesis of *Crocus sativus* (Jain et al. 2016). Illumina HiSeq 2500 platform was utilized to sequence *B. orellana* to elucidate the biosynthetic pathways of bixin and assembled de novo using CAP3 software, Velvet, CLC Genomics Workbench (Cárdenas-Conejo et al. 2015). Similarly, probable genes for *C. tinctorius* yellow pigment production were explicated via Roche's 454 sequencing (Liu et al. 2015). A combined metabolomics and transcriptomics approach was carried out in *C. tinctorius* to elucidate the role of methyl jasmonate in the flavonoid biosynthesis (Chen et al. 2020). A comparative genomics study on four cultivars of *C. longa* was performed via Illumina sequencing, followed by gene assembly, KEGG analysis, and molecular studies, and the resultant data revealed their genetic disparities (Sahoo et al. 2019). Parallel studies using genomics will help in apprais-ing the diversity and in turn empower the conservation strategies. A better knowl-edge of plants at genomic level will amply help to produce breeds with desirable traits like high yield and stress resistance (Dhanapal 2012).

20.3.2.1 Gene Banks

Gene banks are built to safeguard and sustain rare species without losing their genetic integrity. Gene banks require an efficient infrastructure to maintain as well as to regenerate the germplasm. Above seven million plant germplasm accessions are presently well maintained in 1750 gene banks around the world, and interestingly two million accessions among them are exceptional. However, poor funding, loss of genetic integrity, political scams, and inadequate research are some of the hurdles in gene bank management (Fu 2017). Gene bank systems generally give much prefer-ence to agricultural crops. However, dye-synthesizing plants with nutritional values are stored in many gene banks like safflower (Australian Grain Gene Bank) (Zhao et al. 2021).

Field gene bank, a variant of gene bank, is attaining much popularity now. It is a living collection of germplasm for maintenance and storage. For instance, dye-yielding plants like *Acacia* sp., *B. orellana*, *Albizia odoratissima* (L.f.) Benth., *Aloe barbadensis*, *Artocarpus heterophylla*, *Basella alba* L., *Cassia* sp., *Crescentia cujete* L., *Delonix regia* Raf., *Dioscorea bulbifera* L., *Tectona grandis* Linn.f. *Momordica charantia*, *Morinda* sp., *Mangifera indica* Linn., *Lawsonia inermis* L., *Jatropha curcas* L., *Hibiscus rosa-sinensis* L., and *Dioscorea bulbifera* are conserved in the field gene bank of the National Centre for Genetic Resources and Biotechnology, Ibadan, Nigeria (Borokini et al. 2010). Plant-specific field gene banks were established for *B. orellana* (Poornima and Ambika 2012).

The integration of genomics into management and utilization of gene bank resources will considerably transform the face of conventional gene bank into a digital resource. The genetic integrity and less redundancies in gene banks can be achieved via genotyping using sequences and markers (Wambugu et al. 2018; Mascher et al. 2019). Besides, online databases are also now available to access, deposit, and manage germplasm collections in gene banks like GBIS (Gene Bank Information System), which smoothened the working of IPK gene bank in Europe (Schmutzer et al. 2017).

20.3.3 In Vitro Propagation for Conservation of PGRs

Growing cells, tissues, and organs of plants under artificial synthetic media and appropriate environmental conditions is termed as tissue culture (Van Eck and Smith 2018). Tissue culture is an appropriate choice for the mass propagation of extinction facing dye-yielding plants. Several attempts and excellent protocols have been made in this arena. The biotechnology-integrated approaches stand out because of the ease of cultivation, reduced space, lowered time, substantial number of clones, and most importantly reduced biotic and abiotic stress in culture. Most importantly, tissue culture is an elementary part of all gene transformation techniques. Choosing the explant is the most preliminary, vital phase in commencing micropropagation. Nodal and internodal segments, axillary meristems, shoot, and shoot tip culture are the common explants chosen (Yarra et al. 2018). Contrarily, propagating rhizomes are used as explants for plants like *Curcuma longa* where root is the source of dye (Nasirujjaman et al. 2005). Seeds are another option in propagating and conserving plants like *B. orellana* (Teixeira da Silva et al. 2019). Interestingly, leaves were employed in the micropropagation of *Rubia cordifolia*, another red dye-yielding plant (Mariadoss et al. 2020). In vitro-generated microcorms are another explored source for micropropagation in case of *Crocus sativus* (Sharma and Piqueras 2010).

Apposite sterilization is a major step ensuring the success of tissue culture. Preliminary sterilization is generally performed at the gathering site itself followed by efficient treatments in aseptic environment. Proper sterilization will remove pathogen contamination in tissues to a major extent. Utmost attention should be paid to steer clear of diseased and damaged organs. 70% (v/v) ethanol (EtOH), 0.5–4.5% (v/v) sodium hypochlorite (NaOCl), and 0.05–0.2% (w/v) mercuric chloride are conventional disinfectants in the sterilization course (Nikhil et al. 2014; Teixeira da Silva et al. 2019). Tween-20 is another regular detergent used in washing explants (Kashyap and Kale 2019). Concentrations, combinations, and duration of sterilization procedures vary for different plants. Murashige and Skoog (MS) medium is generally preferred for explant growth. 3% sucrose and 0.8% agarose are usual parameters considered for the media preparation for every dye-yielding plant (Vijayakumar et al. 2017; Yarra et al. 2018). Other media like Gamborg B5 have also been utilized for the in vitro cultivation of *Pterocarpus santalinus* L. (red sandalwood tree) (Warakagoda and Subasinghe 2013). On top, in vitro propagation of *Semecarpus anacardium* L. (produces black dye) was done using woody plant medium (Panda and Hazra 2012). Several additives and pretreatments have also been employed to facilitate the overall growth and metabolite production of particularly dyes of plants. Helium-neon laser treatment upraised the overall growth of *C. longa* in terms of height, rhizome growth, and pigment production (El-Sherif 2019). The media along with growth factors, optimal conditions, and outcome observed for the micropropagation of various dye-yielding plants are listed in Table 20.2.

In vitro clonal propagation by means of somatic embryogenesis and organogenesis is widely experimented in pigment-producing plants and is progressively more fundamental in ex situ conservation of germplasm. For instance, embryogenic callus was induced in *Oldenlandia umbellata* (roots produce red dye) via culturing MS media with appropriate concentrations of NAA, BA, and coconut milk (Siva et al. 2009). In vitro organogenesis was performed in various dye-yielding plants like *B. orellana* (Da Cruz et al. 2014) and *Adhatoda vasica* Nees (Mandal and Laxminarayana 2014), whereas somatic embryogenesis was done in *C. longa* (He and Gang 2014) and *B. orellana* (Parimalan et al. 2011b). By the same token,

Table 20.2 The explants, media with PGRs and results observed in micropropagation of some dye-yielding plants

Plant name	Explant	Media with supplements	Outcome	Reference
Acacia nilotica	Seeds	B5 with BAP (3.0) + NAA (0.5)	MSt	Abbas et al. (2010)
		MS with BAP (2.0) + NAA (0.5)	MSt	
Acacia sinuata	Hypocotyls, cotyledons and leaves	MS with 2,4-D (2.26–13.56 μM), NAA (2.69–16.14 μM) and IAA (2.85–17.10 μM) and the cytokinins, BAP (2.22 and 4.44 μM) and kinetin (2.32 and 4.65 μM)	Or	Vengadesan et al. (2000)
Adhatoda vasica	Leaf	MS with BAP (2) + BAP (3)	C	Kashyap and Kale (2019)
	Nodal segments	MS with NA (0.6–1.4)	S	Panigrahi et al. (2017)
		MS with IBA (1) + NAA (0.25)	R	
		MS with CW + BAP (5)	S	Bimal and Shahnawaz (2012)
		MS with IBA (1)	R	
	Petiole and leaf segment	MS/SH with TDZ + NAA	C	Mandal and Laxminarayana (2014)
Althaea rosea	Shoot tip	MS with BAP (2.5 μM) + NAA (1.0 μM)	S	Tyub et al. (2016)
		MS with IBA (5–10 μM)	R	
	Cotyledon	MS with BAP (4) + NAA (3) or MS with BAP (4) + 2,4-D (0.03)	C	Munir et al. (2012)
Arnebia Benthamii	Shoot tip	MS with BA (5 μM)	S	Quadri et al. (2012)
		MS with IBA (4 μM)	R	
Arnebia hispidissima	Shoot tip, young leaves, nodal and internodal segments	MS with IAA + KIN + BAP + CH	Pl	Pal and Chaudhury (2010)
Azadirachta indica	Immature flower buds	Semisolid MS with 2,4-D (9.05 μM)	C	Gehlot et al. (2017)
		MS with BA (8.88 μM)	S	
		MS with IBA (4.92 μM)	R	

(continued)

Table 20.2 (continued)

Plant name	Explant	Media with supplements	Outcome	Reference
	Root	MS with BAP (8.88 μM) + 2iP (9.84 μM) + IAA (5.71 μM) + adenine hemisulfate (81.43 μM) and putrescine (2.27 μM)	S	Arora et al. (2011)
		MS with BAP (1.11 μM), IAA (1.43 μM), and adenine hemisulfate (135.72 μM)	R	
	Leaf and petiole	MS with TDZ (0.2–0.6) and 2,4-D (0.2–0.6)	C	Ashokhan et al. (2020)
Barleria prionitis	Apical segments, leaf, seeds, and nodal segments	MS with TDZ + NAA + BAP	S	Lone et al. (2012)
		MS with IBA + IAA	R	
	leaf	MS with IAA (2)+ BAP (1.5)	C	Kumari et al. (2013)
	Shoot tips and leaves	MS with 2,4-D and kinetin	C	Shukla et. al. (2011)
Bauhinia variegate	Shoot tip, single nodes	MS with BA	S	Papafotiou et al. (2010)
Bauhinia Tomentosa	Coteledonary node and stem nodes	MS with TDZ (0.8 μM)	MSt	Naz et al. (2012)
Bixa Orellana	Root	MS liquid with ZEA + TDZ + BA	Abu	da Cruz et al. (2014)
	Leaf petiolar	MS with 2iP (9.8 μM) + NAA (10.7 μM)	S	Mohammed et al. (2015)
		MS with IAA (1.7 μM) + 2iP 14.7 μM	R	
Bridelia stipularis	Leaf and internode	MS with NAA (21.8 μM)	C	Sreenivas et al. (2011)
Carthamus tinctorius	Cotyledonary leaf segments	MS with TDZ (2.0) + NAA (0.5)	C	Nikhil et al. (2014)
		MS with TDZ (2.0 mg/L) + NAA (0.1 mg/L)	S	
		MS with NAA	R	
Caralluma stalagmifera var. longipetala	Nodes	MS with BAP (2) + Kn (0.5) + IAA (0.3)	MSt	Ramachandra et al. (2014)

(continued)

Table 20.2 (continued)

Plant name	Explant	Media with supplements	Outcome	Reference
Clitoria ternatea	Shoot tip, node, and cotyledonary node	MS with BAP (2)	MSt	Pandeya et al. 2010
	Nodal segments	MS with TDZ (250 μM)	MSt	Ismail et al. (2011)
		MS with IBA (2 μM)	R	
	Leaf	DKW with BAP (1)	S	Mohamed and Taha (2011)
		DKW with NAA (2)	R	
		DKW with BAP (2) + NAA (1)	C	
	Coteledonary node	MS with BAP (1.5)	MSt	Singh and Tiwari (2010)
		MS with IBA (0.2)	R	
Crocus sativus	Apical, axillary buds, and young true leaves	MS with TDZ + piclrocam	SE	Devi et al. (2014)
Curcuma longa	Rhizome buds	MS with BAP, Kin, NAA	S	Pal et al. (2020)
		MS with BAP, NAA, Kin, IBA	R	
	Shoot tips	Market sugar + Isabgol (gelling agent)	P	Tyagi et al. (2007)
	Leaf base	MS with 2,4-D (4.49 μM) in dark to MS with BA (1.32 μM)	SEm	Soundar Raju et al. (2015)
		MS with GA_3	Pl	
Coscinium fenestratum	Epicotyl	MS with Kinetin (1 μM) + 2,4-D (0.25 μM)	MSt	Senarath (2010)
		MS with IBA (2.5 μM)	R	
	Nodes	WPM with BAP (2) + TDZ (1) + 2,4-D (0.4)	MSt	Warakagoda et al. (2017)
Gardenia jasminoides	Node	MS with NAA + BA	S	Gaber and Barakat (2012)
		MS with IBA + NAA	R	
	Leaves at different stages	WPM with 2,4-D (2) + NAA (3)	C	Farzinebrahimi et al. (2014)
Garcinia mangostana	Shoots	MS with BA	S	Rostika et al. (2008)
		MS/WPM with IBA	R	
Gynochthodes umbellata	Single nodes and shoot tips	MS with BA (2)	MSt	Anjusha and Gangaprasad (2016)
		MS with IAA/NAA	R	
Indigofera tinctoria	Nodes	MS with BA (1) and IAA (0.1)	S	Nair et al. (2016)
		MS with IAA (1.5)	R	

(continued)

Table 20.2 (continued)

Plant name	Explant	Media with supplements	Outcome	Reference
Jatropha curcas	Embryo and nodes	MS with BAP (3.0) and IBA (1.0)	S	Ram et al. (2012)
		MS with IBA (3)	R	
	Nodes	MS with BAP (2.0) and IAA (0.1)	S	Rathore et al. (2015)
		MS with AC (2) and IBA (3)	R	
	Leaf and shoot tips	MS with 2, 4-D (0.5) + BAP (5)	SEm	Medipally et al. (2014)
Lawsonia inermis.	Nodes	MS with BA (0.5)	MSt	Shiji and Siril (2018)
Malpighia glabra	Shoot tip	MS with BA + Zeatin	S	Yamamoto et al. (2017)
Melastoma malabathricum	Internodes	MS with TDZ + NAA	C	Karimah et al. (2020)
Nyctanthes arbor-tristis	Cotyledonary node	MS with BAP (22.19 μM)	S	Mishra et al. (2019)
Oxalis corniculata	Nodes	MS with BAP (1) + NAA (0.5)	S	Prasuna and Srinivas (2015)
		MS with IBA + NAA	R	
Plumbago rosea	Leaf	Semisolid MS with IAA (1.5) + IBA (1)	R	Silja and Satheeshkumar (2015)
Pterocarpus santalinus	Cotyledonary nodal, mesocotyl segments, in vitro-derived shoot tips, immature and semi-hard wood cuttings detached from one-year-old plants and in vitro germinated seedlings	B5 with BAP (8 μM)+ NAA (2 μM)	S	Warakagoda and Subasinghe (2013)
		MS with IBA (1.5)	R	
Semecarpus anacardium	Nodal buds	WPM with TDZ	S	Panda and Hazra (2012)
Shorea robusta	Nodal segments	WPM with BAP (1.0) + NAA (0.5)	S	Singh et al. (2014)
		WPM with IBA (0.5)	R	
Strobilanthes crispus	Shoot tips	MS with BAP, Kn and TDZ	S	Ahmad Asnawi et al. (2017)
Toddalia asiatica	Shoot and nodes	MS with BA	MSt	Praveena and Veeresham (2014)
		MS with IBA	R	

(continued)

Table 20.2 (continued)

Plant name	Explant	Media with supplements	Outcome	Reference
Toona ciliate	Cotyledons and cotyledonary node	MS with BA (0.5), KT (0.05), and NAA (0.05)	C	Song et al. (2021)
		MS BA (0.3) and NAA (0.2)	S	
Wrightia tinctoria	Nodes	MS with BAP, NAA, activated charcoal	Shoots with roots	Mridula and Nair (2018)

Media with supplements: *MS* Murashige Skoog media, *WPM* woody plant media, *BA* benzyl adenine, *IAA* indole acetic acid, *NAA* naphthalene acetic acid, *IBA* indole 3-butyric acid, *TDZ* thidizuron, *Kin* kinetin, *2,4-D* 2,4-dichlorophenoxyacetic acid
Outcomes: *S* shoot, *MSt* multiple shoot, *R* root, *C* callus, *SEm* somatic embryo, *Or* organogenesis, *Abu* axillary bud formation, *Pl* plantlet

dyes were also successfully extricated from in vitro cultures, for instance, anthraquinones from *Oldenlandia umbellata* L. (Siva et al. 2012).

20.3.3.1 Elicitation for Pigment Production

Dye-yielding plants are blindly hunted for getting dyes for industrial purposes. In vitro elicitation for pigment is the most steadfast way to avoid the "production does not meet requirement" fate. Both abiotic and biotic elicitors have been efficaciously exercised for augmented pigment production (Namedo 2007). Abiotic factors are mainly chemicals and radiation while fungal extracts constitute the biotic factors. Abiotic elicitors like gamma radiation have increased the anthraquinone contents of *Rubia cordifolia* suspension cultures. 13-fold increase was observed in alizarin content, while around ninefold in purpurin with the treatment of grama irradiation (8 gyr) (Mariadoss et al. 2020). In addition, abiotic elicitors like NaCl and MgSO4 enhanced red and yellow pigment synthesis. Similarly, abiotic elicitors like salicylic acid, paclobutrazol, and methyl jasmonate enhanced annatto production in *B. orellana* (Parimalan et al. 2011a). Environmentally friendly compounds like chitosan were also reported as elicitors for augmented curcumin production (Sathiyabama et al. 2016). Use of biotic factors like *Nostoc linckia, Anabaena cylindrica,* and *Anabaena variabilis* had considerably increased red pigment synthesis in *Carthamus tinctorius.* Fungal parts like mycelium and polysaccharide content were also testified to be good elicitors in this plant (Hanagata et al. 1994). Similarly, *Aspergillus niger* extracts have augmented 3.5 times dye production in *Indigofera* (Dutta et al. 2019).

20.4 Metabolic Engineering Strategies in Germplasm Conservation

Increased demand for pigments for various industrial purposes caused an extensive upsurge in metabolic engineering approaches. Metabolic engineering is defined as the manipulation of metabolic pathway in a selected organism with the intention of producing economically valuable metabolites (Wuest et al. 2019). Gene manipulating and transformation strategies have a prominent role in producing elite stress-resistant, disease-resistant breeds. Above all, it will substantially trim down the exploitation of plants for dyes and thus enhance their safeguarding in a socioeconomic way. Standard agrobacterium-mediated gene transformation protocols have been existing for dye-synthesizing plants like *C. longa* (He and Gang 2014) and *B. orellana* (Parimalan et al. 2011b). Metabolic and transformation strategies employed in dye-yielding plants have many aims like to elucidate biosynthetic pathway, to increase productivity, and to generate stress-resistant plants. Bixin biosynthetic pathway was elucidated by expressing bixin biosynthetic genes like lycopene cleavage dioxygenase, bixin aldehyde dehydrogenase, and norbixin carboxy methyltransferase in *E. coli* and this resulted in in vitro synthesis of bright bixin from lycopene (Bouvier et al. 2003). Correspondingly, the biosynthetic role of CCD4 and CCD1 in bixin production was evidenced via *E. coli* system (Carballo-Uicab et al. 2019). Secondly, overexpression of AtCPK1 in *Rubia cordifolia* elevated the anthraquinone biosynthesis (Shkryl et al. 2016). Many productive endeavors have been made by researchers to transform microbes as pigment bioreactors by integrating biosynthetic genes. Pigments like crocin or crocetins were fruitfully produced in many microbial systems like *E. coli*, *Saccharomyces cerevisiae*, and *Chlorella vulgaris* (Liu et al. 2020). Curcumin production was achieved in bacterial cells by genetically expressing curcumin synthase gene, and membrane engineering strategy was also experimented for boosted production (Wu et al. 2020).

Not only microbes but model plants like *Nicotiana benthamiana* were also illustrated as factories for pigment production. Incorporation of viral system carrying CCD genes from *Crocus sativus* and *Buddleja davidii* produced crocin and picrocrocin in *Nicotiana benthamiana* within a limited time span of 2 weeks (Martí et al. 2020). Rice endosperm was subjected to genetic engineering to synthesize betanin, one of the commercially important red pigments, and this resulted in the production of betanin-enriched rice (Tian et al. 2020). It is a declared fact that roots are the major source of some dye-yielding plants. Transformation strategies via *Agrobacterium rhizogenes* are an obvious choice for elevated pigment production for that genre of plants. Augmented synthesis of anthraquinones was accomplished in *Oldenlandia umbellate* L. by co-cultivating leaf explants with *rhizogenes* (Saranya Krishnan and Siril 2016). Interestingly, curcumin was synthesized in *A. belladonna* roots via rhizogenes-mediated transformation of biosynthetic genes. Intriguingly, the high spot of all these transformation experiments is the elevated production of pigments in the model plants. These plant systems are promising candidates in reducing the uncontrolled use of dye-generating plants and thus in its germplasm conservation. Unfortunately, metabolic engineering studies are performed only in very limited plants that produce commercially creditable pigments.

20.4.1 Synthetic Seed for Germplasm Storage

The artificial seed technology was a breakthrough in the field of in vitro propagation as well as in germplasm conservation. It can be defined as an encapsulated somatic embryo or any meristematic tissue like shoot tips that can be cultivated as seeds and which are capable to regenerate the whole plant (Chandrasekhara 2012). Artificial seed production is considered as the most apposite, reasonable means for exchange and short-term conservation of rare, extinction-facing plants. Explants are soaked in sodium alginate and round beads are produced on contact with calcium chloride. Polyox and agarose are some other gelling agents used for synthetic seed preparation. Above all, somatic embryos are deemed as the best candidates for artificial seed production (Chandrasekhara 2012). Shoot tips and axillary buds of dye-synthesizing plants like *Acacia* (Li et al. 2014) and somatic embryos of *C. longa* (Babu et al. 2016) and *Ziziphus* (leaf produces pink-red color) (Sudhersan et al. 2013) are some other examples. Intriguingly, most of the studies successfully regenerated plantlets from these synthetic seeds.

20.5 Other Conventional Approaches

Cryopreservation is an in situ conservation approach for long-term storage in extreme cold conditions. Widely used cryopreservation practices are vitrification and encapsulation. In vitrification, highly concentrated solutions like PVS2 are used for dehydration while in encapsulation air-drying is mostly exercised. Desiccated samples are preserved customarily in liquid nitrogen (Matsumoto 2017). Efficacious encapsulation technique was practiced in *Indigofera tinctoria* and regeneration rate attained was 62%. Withal, genetic fidelity was confirmed by means of RAPD markers. Beads were prepared by introducing axillary bud-sodium alginate aliquots into calcium chloride supplemented with 0.29 M sucrose and MS media (Nair et al. 2016).

Protected areas constitute around 13% of earth's surface. They are highly diverse ecosystems housing rare, endangered species. Approximately 173,461 protected areas and 2059 proposed sites are present in the world. There are 104 national parks, 551 sanctuaries, 88 conservation reserves, and 127 community reserves in India (Bhargav 2020). Rajiv Gandhi National Park is one of the most important biodiversity spots in India forming a fraction of the Western Ghats. This area harbors many dye-synthesizing plants like *Semecarpus anacardium* L. f., *Barleria prionitis*, *Desmos chinensis Lour*, *Bixa orellana L.*, *Indigofera* species, *Toddalia asiatica*, *Tectona grandis* L., *Terminalia catappa* L., *Urena lobata* L., *Tamarindus indica* L., *Rubia cordifolia* L., and *Mallotus philippensis* Muell. Most of the protected areas in

Table 20.3 Elicitors and outcome observed in some dye-yielding plants

Plant species	Elicitor used	Response	Reference
Arnebia euchroma	L-Phenylalanine	Shikonin ↑	Sykłowska-Baranek et al. (2012)
	Rhizoctonia solani	Sevenfold Shikalkin ↑	Arghavani et al. (2015)
	MeJA	Napthaquinones ↑	Hao et al. (2014)
Bougainvillea	Calcium chloride, FeEDTA and copper sulfate (abiotic), MeJA, β-glucan (biotic)	Betacyanin and betaxanthin ↑	Lakhotia et al. (2014)
Cassia alata	Salicylic acid and Ethrel	12- to 13-fold anthraquinone ↑	Shah and George (2019)
Cassia tora	Salicylic acid	Twofold anthraquinone ↑	Shah and George (2020)
Celosia cristata	*Fusarium oxysporum*	Betalain ↑	Warhade and Badere (2018)
Curcurma longa	He-Ne laser	Pigment yield ↑	El-Sherif (2019)
Curcuma zedoaria	Salicylic acid and yeast extract	1.61- to 2.53-fold curcumin	Lan et al. (2019)
Hypericum perforatum	Acetic acid	550% Xanthonine ↑	Valletta et al. (2016)
Impatiens balsamina	MeJA	Napthaquinones ↑	Sakunphueak and Panichayupakaranant (2010)
Oldenlandia umbellata	Pectin, yeast extract and xylan	Anthraquinone ↑	Krishnan and Siril (2018)
Rosa hybrida	MeJA	Anthocyanin ↑	Ram et al. (2013)
Plumbago rosea	Naphthalene, acetic acid	5.59-fold Plumbagin ↑	Silja et al. (2014)
	Jasmonic acid, yeast extract, and sodium salicylate	Plumbagin ↑	Silja and Satheeshkumar (2015)
Tinospora cordifolia	Tyrosine	3.19-fold Berberin ↑	Pillai and Siril (2019)
Vaccinium corymbosum	Naphthalene acetic acid	Anthocyanin ↑	Lucioli et al. (2018)
Vitis vinifera	MeJA, phenylalanine	4.6- and 3.4-fold anthocyanin ↑	Qu et al. (2011)

India are plenteous in plants producing dyes. Around 195 dye-yielding plants reported in the state of Maharashtra itself indicates the copiousness (Patil et al. 2019). Table 20.3 lists out the main dye-yielding plants commonly found in Indian forests and the list is accessible in the Environmental Information (ENVIS) website of the Indian Government.

Botanic gardens are living plant genomics resource conservation systems. They are also channels for educating the importance of biodiversity conservation to the public. Botanic gardens are also a study material for many researchers for their biological investigations. But still many possibilities are to be explored in botanic gardens (Donaldson 2009). Many dye-synthesizing plants are conserved in botanic gardens. For instance, *Tectona grandis*, *Albizia lebbekoides,* and *Terminalia catappa* are part of Singapore Botanic Gardens (Website Singapore Government 2021); *Prunus* species is seen in Victoria Garden, Bangalore (Elder et al. 2021); and flowering plants like *Crocus sativus* are found in Chicago Botanic Garden. Acacia trees are well protected in the Jawaharlal Nehru Tropical Botanic Garden and Research Institute, Thiruvananthapuram (Mohanan et al. 2016).

20.6 Case Studies

Many rural groups are upholding their own traditional knowledge of dyeing and subsequently conserving rare dye-yielding plants. Nonetheless, these ethnic knowledges are vanishing as years pass. Hasty industrialization and more waged daily jobs are magnetizing people to abandon the traditional dyeing practices. However, a community-based approach including the local people is a most effective in situ conservation strategy to look after the plant species exclusive to that vicinity. Community-based management is a conservation approach targeted to build community institutions to protect PGRs. A local governing agency or an NGO will guide and involve in decision-making in these communities which wholly rest on agriculture. Community-based strategies were framed to benefit the farmers with a sustainable development as well as to safeguard natural biodiversity (de Boef et al. 2013). Most importantly, unparallel knowledge of indigenous rural groups should be properly documented. Some of the ethnic groups with natural dye resources are subjected to review, and possibilities of community-based approaches in these groups are discussed.

20.6.1 Bai People and Their Traditional Dyeing Knowledge

Bai people residing in the Northwest Yunnan, China, are well known for the traditional technique of dyeing. Interestingly, 23 species belonging to 19 plant taxonomic families have been used for dye synthesizing by Bai people. *Strobilanthes cusia* (Nees) O. Kuntze is the most widely used plant for traditional coloring of textiles. Moreover, they have developed different dyeing techniques unique to a community. Apart from the traditionally used plants like *Rubia cordifolia* L., *Punica granatum*, and *Hibiscus sabdariffa* Linn., Bai people discovered the coloring property of some invasive plants. *Eupatorium adenophora* Spreng is an exotic species with unpleasant effects on the farming sector of Bai. Accidently, the local people identified the immense dyeing property and now it is an industrially used dye. Similarly, the dyeing properties of *Coriaria nepalensis* Wall., *Cyclobalanopsis delavayi* (Franch.) Schott., *Alnus ferdinandi-coburgii* Schneid., and *Quercus acutissima* Carruth. were primarily stated by Bai. The unique dyeing technique of Bai called tie and dye has been industrialized in small factories to enhance the production of these valuable dyes (Fan et al. 2018). High production cost and requirement of large amount of raw material are some of the limiting factors in industrial production. However, proper financial funding and guidance from government agencies will help to establish a socioeconomic conservation approach for these rare dyeing plants. A proper committee involving community leaders and government representatives should be developed to teach people about the importance of diversity conservation and the strategies to manage and benefit the biodiversity. Furthermore, these strategies also safeguard the ethnobotanical knowledge of Bai people for generations.

20.6.2 Dong Community and Dye-Synthesizing Plants

Dong is an orthodox, linguistic group living in the highlands of Central China. Around 13 plants from 9 botanical families were reported with coloring property. Dong people extract dyes from roots, leaves, flowers, stems, bark, and tubers of plants to color their foods as well as textiles. *Gardenia jasminoides*, *Sophora japonica*, *Phellodendron amurense*, *Curcuma longa*, *Myrica rubra*, and *Mahonia fortunei* are food coloring plants of Dong society. The medicinal properties of these plants are also reported which lifts the importance of conservation. *Polygonum tinctorium*, *Strobilanthes cusia*, and *Dioscorea cirrhosa* are the plants used for textile dyeing via the traditional dyeing method (Liu et al. 2014). However, these cultural practices got considerably reduced by the intervention of modernization in China. Dye-synthesizing plants should be properly documented and conserved via community-based approaches. Setting up of small-scale units with dye production on industrial scale will elate the living of rural people and importantly the upkeeping of this variety of plants.

20.6.3 Case of Central Himachal Pradesh

Himachal Pradesh is an Indian state affluent with diversified flora and fauna. The rural community depends on plants for coloring their textiles and other traditional purposes. 17 species belonging to 12 families have been identified as dye sources in this area. *Acacia catechu, Curcuma domestica, Punica granatum,* and *Impatiens balsamina* are some of the plants used for dyeing purposes. The natural dyes are the heart of their religious ceremonies too. Yellow dyes from curcuma are used in marriage ceremonies. The extraction and dyeing procedures are exceptional to each community similar to aforementioned cases. Sadly, the indigenous knowledge is performed and esteemed by the old generation (Kumari et al. 2019). As part of the community-based approach, local community should be educated about the importance of the conservation of these plants. A conjoined approach with authorities should be commenced to endorse the dye production at an industrial level. In situ and ex situ conservation policies should be executed with the support from indigenous people.

20.7 Conclusion and Perspectives

Conservation of plants in a socioeconomic way is the necessity of the decade. In situ and ex situ conservation strategies have been followed worldwide to protect this treasure. Crop and medicinal plants are given a special significance in conservation policies. In the midst of rapidly increasing negative reports of artificial dyes, dye-yielding plants must be properly conserved and utilized. Issues in the conservation of dye-yielding plants have discrete aspects. Plants like curcuma are highly exploited and expansively experimented. Contrarily, some plants are not even in vitro cultured, while some other classes of plants are not hitherto documented. In this scenario, conservation strategies of dye yielding should be implemented with the below goals:

- To document and protect rare dye-yielding plants
- To safeguard plants from exploitation via biotechnology-based approaches
- To increase the raw material for augmented industrial production in a community-based approach

Many biotechnology-oriented approaches like micropropagation, gene mapping, and metabolic engineering have been well documented as conservation means for dye-synthesizing plants. Efforts should be taken to extend biotechnology-oriented studies to a greater number of plants. Many rural communities are protecting a variety of dye-producing plants and efficiently utilizing them for their textiles and food coloring. A properly structured community-based approach will safeguard these plants as well as the knowledge of indigenous people.

References

Abbas H, Qaiser M, Naqvi B (2010) Rapid in vitro multiplication of Acacia nilotica subsp. hemispherica, a critically endangered endemic taxon. Pak J Bot 42:4087–4093

Abdel Khalik K, Abd El-Twab M, Galal R (2014) Genetic diversity and relationships among Egyptian Galium (Rubiaceae) and related species using ISSR and RAPD markers. Biol 69:300–310. https://doi.org/10.2478/s11756-013-0314-z

Ahmad Asnawi M, Hartinie M, Jualang AG (2017) Effect of cytokinins on shoot induction from nodal explant of Pecah Beling (Strobilanthes crispus). Transactions Of The Malaysian Society Of Plant Physiology, pp 8–12

Ali F, Yılmaz A, Nadeem MA et al (2019) Mobile genomic element diversity in world collection of safflower (*Carthamus tinctorius* L.) panel using iPBS-retrotransposon markers. PLoS One 14:1–19. https://doi.org/10.1371/journal.pone.0211985

Anjusha S, Gangaprasad A (2016) In vitro propagation and anthraquinone quantification in Gynochthodes umbellata (L.) Razafim. & B. Bremer (Rubiaceae)—a dye yielding plant. Ind Crop Prod 81:83–90. https://doi.org/10.1016/j.indcrop.2015.11.043

Arghavani P, Haghbeen K, Mousavi A (2015) Enhancement of shikalkin production in Arnebia euchroma callus by a fungal elicitor, Rhizoctonia solani. Iran J Biotechnol 13:10–16. https://doi.org/10.15171/ijb.1058

Arora K, Sharma M, Srivastava J et al (2011) In vitro cloning of Azadirachta indica from root explants. Biol Plant 55:164–168. https://doi.org/10.1007/s10535-011-0023-9

Ashokhan S, Othman R, Hafiz M et al (2020) Effect of plant growth regulators on coloured callus formation and accumulation of azadirachtin, an essential biopesticide in Azadirachta indica. Plants 9:352

Ashraf K, Ahmad A, Shah SAA, Mujeeb M (2017) Genetic diversity in accessions of Indian turmeric (*Curcuma longa* L.) using RAPD markers. Int J Pharm Pharm Sci 9:288–291

Aubriot X, Lowry PP, Cruaud C et al (2013) DNA barcoding in a biodiversity hot spot: potential value for the identification of Malagasy Euphorbia L. listed in CITES Appendices I and II. Mol Ecol Resour 13:57–65. https://doi.org/10.1111/1755-0998.12028

Auburn RP, Kreil DP, Meadows LA et al (2005) Robotic spotting of cDNA and oligonucleotide microarrays. Trends Biotechnol 23:374–379. https://doi.org/10.1016/j.tibtech.2005.04.002

Babu KN, Divakaran M, Pillai GS, Kusumanchi P (2016) Protocols for in vitro propagation, conservation, synthetic seed production, microrhizome production, and molecular profiling in turmeric (*Curcuma longa* L.). In: Protocols for in vitro cultures and secondary metabolite analysis of aromatic and medicinal plant. Springer, Cham, pp 387–399

Balada C, Castro M, Fassio C et al (2021) Genetic diversity and biological activity of *Curcuma longa* ecotypes from Rapa Nui using molecular markers. Saudi J Biol Sci 28:707–716. https://doi.org/10.1016/j.sjbs.2020.10.062

Bhargav P (2020) India's protected area (PA) network. In: Conserv. India

Bimal R, Shahnawaz M (2012) Plant regeneration from nodal explants of Adhatoda vasica Nees. J Med Plant Res 6:1229–1233. https://doi.org/10.5897/jmpr11.1055

Bishoyi AK, Pillai VV, Geetha KA, Maiti S (2014) Assessment of genetic diversity in Clitoria ternatea populations from different parts of India by RAPD and ISSR markers. Genet Resour Crop Evol 61:1597–1609. https://doi.org/10.1007/s10722-014-0145-y

Borokini TI, Okere AU, Giwa AO et al (2010) Biodiversity and conservation of plant genetic resources in field gene bank of the National Centre for Genetic Resources and Biotechnology, Ibadan, Nigeria. Int J Biodivers Conserv 2:037–050

Bouvier F, Dogbo O, Camara B (2003) Biosynthesis of the food and cosmetic plant pigment bixin (annatto), vol 300. Science, pp 2089–2091. https://doi.org/10.1126/science.1085162

Bowers JE, Pearl SA, Burke JM (2016) Genetic mapping of millions of SNPs in safflower (*Carthamus tinctorius* L.) via whole-genome resequencing. G3 Genes Genomes Genet 6: 2203–2211. https://doi.org/10.1534/g3.115.026690

Bretting PK, Widrlechner MP (1995) Genetic markers and plant genetic resource management. Plant Breeding Reviews, In, pp 11–86

Carballo-Uicab VM, Cárdenas-Conejo Y, Vallejo-Cardona AA et al (2019) Isolation and functional characterization of two dioxygenases putatively involved in bixin biosynthesis in annatto (Bixa orellana L.). Peer J 2019:1–24. https://doi.org/10.7717/peerj.7064

Cárdenas-Conejo Y, Carballo-Uicab V, Lieberman M et al (2015) De novo transcriptome sequencing in *Bixa orellana* to identify genes involved in methylerythritol phosphate, carotenoid and bixin biosynthesis. BMC Genomics 16. https://doi.org/10.1186/s12864-015-2065-4

CBOL plant working group (2009) A DNA barcode for land plants. Proc Natl Acad Sci U S A 106: 12794–12797

Chandrasekhara MR (2012) Synthetic seeds: a review in agriculture and forestry. African J Biotechnol 11:14254–14275. https://doi.org/10.5897/ajb12.770

Chapman MA, Hvala J, Strever J et al (2009) Development, polymorphism, and cross-taxon utility of EST-SSR markers from safflower (*Carthamus tinctorius* L.). Theor Appl Genet 120:85–91. https://doi.org/10.1007/s00122-009-1161-8

Chen J, Wang J, Wang R et al (2020) Integrated metabolomics and transcriptome analysis on flavonoid biosynthesis in safflower (*Carthamus tinctorius* L.) under MeJA treatment. BMC Plant Biol 20:1–12. https://doi.org/10.1186/s12870-020-02554-6

Chiveu CJ, Dangasuk OG, Omunyin ME, Wachira FN (2008) Genetic diversity in Kenyan populations of *Acacia senegal* (L.) Willd. based on ISSR markers. African J Biotechnol 7: 2333–2340. https://doi.org/10.1515/sg-2009-0003

Christenhusz MJM, Byng JW (2016) The number of known plants species in the world and its annual increase. Phytotaxa 261:201–217

Cohen JI, Williams JT, Plucknett DL, Shands H (1991) Ex situ conservation of plant genetic resources: global development and environmental concerns, vol 253. Science, pp 866–872. https://doi.org/10.1126/science.253.5022.866

Cruz-Cruz CA, González-Arnao MT, Engelmann F (2013) Biotechnology and conservation of plant biodiversity. Resources 2:73–95. https://doi.org/10.3390/resources2020073

Cvetković T, Hinsinger DD, Strijk JS (2019) Exploring evolution and diversity of Chinese Dipterocarpaceae using next-generation sequencing. Sci Rep 9:1–11. https://doi.org/10.1038/s41598-019-48240-y

da Cruz ACF, Rocha DI, Iarema L et al (2014) In vitro organogenesis from root culture segments of *Bixa orellana* L. (Bixaceae). Vitr Cell Dev Biol-Plant 50:76–83. https://doi.org/10.1007/s11627-013-9580-2

Das A, Kesari V, Satyanarayana VM et al (2011) Genetic relationship of Curcuma species from Northeast India using PCR-based markers. Mol Biotechnol 49:65–76. https://doi.org/10.1007/s12033-011-9379-5

de Boef WS, Thijssen M, Peroni N, Subedi A (2013) Community biodiversity management and in situ conservation. In: Community biodiversity management: promoting resilience and the conservation of plant genetic resources. Springer, Cham, pp 378–386

Deng SY, Wang XR, Zhu PL et al (2015) Development of polymorphic microsatellite markers in the medicinal plant Gardenia jasminoides (Rubiaceae). Biochem Syst Ecol 58:149–155. https://doi.org/10.1016/j.bse.2014.11.009

Deplazes-Zemp A (2019) Challenges of justice in the context of plant genetic resources. Front Plant Sci 10:1–12. https://doi.org/10.3389/fpls.2019.01266

Dequigiovani G, Ramos SLF, Alves-Pereira A et al (2017) Genetic diversity and structure in a major Brazilian annatto (*Bixa orellana*) germplasm bank revealed by microsatellites and phytochemical compounds. Genet Resour Crop Evol 64:1775–1788. https://doi.org/10.1007/s10722-017-0535-z

Derakhshan E, Majidi MM, Sharafi Y, Mirlohi A (2014) Discrimination and genetic diversity of cultivated and wild safflowers (Carthamus spp.) using EST-microsatellites markers. Biochem Syst Ecol 54:130–136. https://doi.org/10.1016/j.bse.2014.01.003

Devi K, Sharma M, Ahuja PS (2014) Direct somatic embryogenesis with high frequency plantlet regeneration and successive cormlet production in saffron (Crocus sativus L.). South Afr J Bot 93:207–216. https://doi.org/10.1016/j.sajb.2014.04.006

Dhanapal AP (2012) Genomics of crop plant genetic resources. Adv Biosci Biotechnol 03:378–385. https://doi.org/10.4236/abb.2012.34054

Donaldson JS (2009) Botanic gardens science for conservation and global change. Trends Plant Sci 14:608–613. https://doi.org/10.1016/j.tplants.2009.08.008

Duan Z, Song W, Chen K et al (2017) Assessment of genetic and chemical variability in curcumae longae rhizoma (*Curcuma longa*) based on DNA barcoding markers and HPLC fingerprints. Biol Pharm Bull 40:1638–1645. https://doi.org/10.1248/bpb.b17-00020

Dutta S, Mutaz E, Farag N, Sarangi BK (2019) Enhancement of indigo dye yield from Indigofera plant biomass by using fungal elicitors during fermentation. Indian J Exp Biol 57:619–624

Elder CB, Maple J, Maple J, et al (2021) List of trees in Victoria Garden

Ellina M, Sobir SE, Roedhy P (2010) Assessment of inter simple sequence repeat (ISSR) in technique in mangosteen (Garcinia mangostana L.) grown in different Sumatra region. J Hortic For 2:127–134

El-Sherif F (2019) Laser irradiation induces DNA polymorphism and alters phytochemicals compositions as well as growth and yield of *Curcuma longa*. J Dis Med Plants 5:29–38

Elyasi Zarringhabaie G, Javanmard A, Pirahary O (2012) Random amplified polymorphic markers as indicator for genetic conservation program in Iranian pheasant (*Phasianus colchicus*). ScientificWorldJournal 2012:1–5. https://doi.org/10.1100/2012/640381

Fan Y, Zhao Y, Liu A et al (2018) Indigenous knowledge of dye-yielding plants among Bai communities in Dali, Northwest Yunnan, China. J Ethnobiol Ethnomed 14:1–11. https://doi.org/10.1186/s13002-018-0274-z

FAO (2010) The state of ex situ conservation

Farzinebrahimi R, Mat Taha R, Rashid K, Syafawati Yaacob J (2014) The effect of various media and hormones via suspension culture on secondary metabolic activities of (Cape Jasmine) Gardenia jasminoides Ellis. Sci World J 2014:1–7. https://doi.org/10.1155/2014/407284

Fu Y (2017) The vulnerability of plant genetic resources conserved ex situ. Crop Sci 57:2314–2328. https://doi.org/10.2135/cropsci2017.01.0014

Gaber MK, Barakat AA (2012) Micropropagation and somatic embryogenesis induction of Gardenia jasminoides plants. Exchange J 40(1):190–202. https://doi.org/10.21608/ASEJAIQJSAE.2019.29981

Gehlot A, Dev I, Sarita A (2017) Regeneration of complete plantlets from callus culture of Azadirachta indica A. Juss using immature flower buds. Adv For Sci 4:71–76

Godwin I, Aitken E, Smith L (1997) Application of inter simple sequence repeat markers to plant genetics. Electrophoresis 18:1524–1528

Govindaraj M, Vetriventhan M, Srinivasan M (2015) Importance of genetic diversity assessment in crop plants and its recent advances: an overview of its analytical perspectives. Genet Res Int 2015:1–14. https://doi.org/10.1155/2015/431487

Grativol C, Da Fonseca L-MC, Hemerly AS, Ferreira PCG (2011) High efficiency and reliability of inter-simple sequence repeats (ISSR) markers for evaluation of genetic diversity in Brazilian cultivated Jatropha curcas L. accessions. Mol Biol Rep 38:4245–4256. https://doi.org/10.1007/s11033-010-0547-7

Hacioglu BT, Yaman H, Arslan Y, Katar D (2013) Investigation of molecular diversity of Asian safflower (*Carthamus tinctorius* L.) accessions by RAPD markers for using in hybridization program. Res Crop 14:169–174

Han J, Zhang W, Cao H et al (2007) Genetic diversity and biogeography of the traditional Chinese medicine, Gardenia jasminoides, based on AFLP markers. Biochem Syst Ecol 35:138–145. https://doi.org/10.1016/j.bse.2006.05.021

Hanagata N, Uehara H, Ito A et al (1994) Elicitor for red pigment formation in Carthamus tinctorius cultured cells. J Biotechnol 34:71–77. https://doi.org/10.1016/0168-1656(94)90167-8

Hao H, Lei C, Dong Q et al (2014) Effects of exogenous methyl jasmonate on the biosynthesis of shikonin derivatives in callus tissues of Arnebia euchroma. Appl Biochem Biotechnol 173:2198–2210. https://doi.org/10.1007/s12010-014-1025-9

Hawkins RD, Hon GC, Ren B (2010) Next-generation genomics: an integrative approach. Nat Rev Genet 476–486. https://doi.org/10.1038/nrg2795

Hayakawa H, Kobayashi T, Minaniya Y et al (2011) Development of a molecular marker to identify a candidate line of turmeric (*Curcuma longa* L.) with a high curcumin content. Am J Plant Sci 02:15–26. https://doi.org/10.4236/ajps.2011.21002

He R, Gang DR (2014) Somatic embryogenesis and agrobacterium-mediated transformation of turmeric (*Curcuma longa*). Plant Cell Tissue Organ Cult 116:333–342. https://doi.org/10.1007/s11240-013-0407-y

Houmanat K, Charafi J, Mazouz H et al (2016) Genetic diversity analysis of safflower (*Carthamus tinctorius*) accessions from different geographic origins using ISSR markers. Int J Agric Biol 18:1081–1087

Ismail M, Ahmad A, Nadeem M et al (2020) Development of DNA barcodes for selected Acacia species by using rbc L and matK DNA markers. Saudi J Biol Sci 27:3735–3742. https://doi.org/10.1016/j.sjbs.2020.08.020

Ismail N, Rani U, Batra A (2011) Crucial role of TDZ in the quick regeneration of multiple shoots of Clitoria ternatea L. Int J Pharm Sci Rev Res 6:23–26

Jain A, Parihar DK (2019) Molecular marker based genetic diversity study of wild, cultivated and endangered species of Curcuma from Chhattisgarh region for in situ conservation. Biocatal Agric Biotechnol 18:101033. https://doi.org/10.1016/j.bcab.2019.101033

Jain M, Srivastava PL, Verma M et al (2016) De novo transcriptome assembly and comprehensive expression profiling in Crocus sativus to gain insights into apocarotenoid biosynthesis. Sci Rep 6:1–13. https://doi.org/10.1038/srep22456

Jan HU, Rabbani MA, Shinwari ZK (2011) Assessment of genetic diversity of indigenous turmeric (*Curcuma longa* L.) germplasm from Pakistan using RAPD markers. J Med Plants Res 5:823–830

Jarvis A, Yeaman S, Guarino L, Tohme J (2005) The role of geographic analysis in locating, understanding, and using plant genetic diversity. In: Methods in enzymology. Elsevier, Amsterdam, pp 279–298

Jiao L, Lu Y, He T et al (2019) A strategy for developing high-resolution DNA barcodes for species discrimination of wood specimens using the complete chloroplast genome of three Pterocarpus species. Planta 250:95–104. https://doi.org/10.1007/s00425-019-03150-1

Joseph N, Siril EA (2014) Evaluation and selection of elite annatto (*Bixa orellana* L.) and identification of RAPD markers associated with yield traits. Rev Bras Bot 37:1–8. https://doi.org/10.1007/s40415-013-0039-9

Kandasamy T, Kumari K, Kaprakkaden A et al (2013) Molecular diversity analysis of flower colour variants of Butea monosperma (Lam.) Taub. using inter simple sequence repeats. Bioscan 8: 969–974

Karimah K, Yuniati R, Handayani W (2020) In vitro culture from internodes of Melastoma malabathricum L. on Murashige and Skoog (1962) modified medium with thidiazuron and 1-naphthaleneacetic acid. IOP Conf Ser Earth Environ Sci 481(1):012007

Kashyap S, Kale RD (2019) Development of callus and cell suspension culture from the leaf of *Adhatoda vasica* Nees using economical growth media. Indian J Exp Biol 57:195–200

Khurana-Kaul V, Kachhwaha S, Kothari SL (2012) Characterization of genetic diversity in Jatropha curcas L. Indian J Biotechnol 11:54–61

Krishnan SRS, Siril EA (2018) Elicitor mediated adventitious root culture for the large-scale production of anthraquinones from Oldenlandia umbellata L. Ind Crop Prod 114:173–179. https://doi.org/10.1016/j.indcrop.2018.01.069

Kumar A, Mishra P, Singh SC, Sundaresan V (2014) Efficiency of ISSR and RAPD markers in genetic divergence analysis and conservation management of *Justicia adhatoda* L., a medicinal plant. Plant Syst Evol 300:1409–1420. https://doi.org/10.1007/s00606-013-0970-z

Kumar S, Ambreen H, Murali TV et al (2015) Assessment of genetic diversity and population structure in a global reference collection of 531 accessions of *Carthamus tinctorius* L. (safflower) using AFLP markers. Plant Mol Biol Report 33:1299–1313. https://doi.org/10.1007/s11105-014-0828-8

Kumar Yadav H, Ranjan A, Asif MH et al (2011) EST-derived SSR markers in Jatropha curcas L.: development, characterization, polymorphism, and transferability across the species/genera. Tree Genet Genomes 7:207–219. https://doi.org/10.1007/s11295-010-0326-6

Kumari P, Samant SS, Puri S et al (2019) Diversity, indigenous uses and traditional practices of dye yielding plants in Central Himachal Pradesh, north western Himalaya. Indian J Tradit Knowl 18: 560–564

Kumari P, Yadav P, Arya A, Kumar S (2013) In vitro callus production and anti-bacterial activity of Barleria prionitis Linn. against dental caries pathogens. Int J Bot Res 3:1–6

Lakhotia P, Singh KP, Singh SK et al (2014) Influence of biotic and abiotic elicitors on production of betalain pigments in bougainvillea callus cultures. Indian J Hortic 71:373–378

Lan TTP, Huy ND, Luong NN et al (2019) Effect of salicylic acid and yeast extract on curcuminoids biosynthesis gene expression and curcumin accumulation in cells of Curcuma zedoaria. J Plant Biotechnol 46:172–179. https://doi.org/10.5010/JPB.2019.46.3.172

Ledford H (2019) Global plant extinctions mapped. Nat Commun 570:1–2

Li H, Cao H, Cai YF et al (2014) The complete chloroplast genome sequence of sugar beet (*Beta vulgaris* ssp. vulgaris). Mitochondrial DNA 25:209–211. https://doi.org/10.3109/19401736.2014.883611

Li P, Zhan X, Que Q et al (2015) Genetic diversity and population structure of Toona ciliata Roem: based on sequence-related amplified polymorphism (SRAP) markers. Forests 6:1094–1106. https://doi.org/10.3390/f6041094

Liu T, Yu S, Xu Z et al (2020) Prospects and progress on crocin biosynthetic pathway and metabolic engineering. Comput Struct Biotechnol J 18:3278–3286. https://doi.org/10.1016/j.csbj.2020.10.019

Liu X, Dong Y, Yao N et al (2015) De novo sequencing and analysis of the safflower transcriptome to discover putative genes associated with safflower yellow in *Carthamus tinctorius* L. Int J Mol Sci 16:25657–25677. https://doi.org/10.3390/ijms161025657

Liu Y, Ahmed S, Liu B et al (2014) Ethnobotany of dye plants in Dong communities of China. J Ethnobiol Ethnomed 10:1–8. https://doi.org/10.1186/1746-4269-10-23

Lone SA, Yadav AS, Bajaj A et al (2012) Conservation strategies for threatened medicinal plant—Barleria prionitis L.—using in vitro and ex vitro propagation techniques. Arch Phytopathol Plant Protect 45:1327–1340. https://doi.org/10.1080/03235408.2012.673301

Lowe R, Shirley N, Bleackley M et al (2017) Transcriptomics technologies. PLoS Comput Biol 13:1–23. https://doi.org/10.1371/journal.pcbi.1005457

Lu H, Giordano F, Ning Z (2016) Oxford Nanopore min ION sequencing and genome assembly. Genomics Proteomics Bioinforma 14:265–279. https://doi.org/10.1016/j.gpb.2016.05.004

Lucioli S, Di Bari C, Forni C et al (2018) Anthocyanic pigments from elicited in vitro grown shoot cultures of Vaccinium corymbosum L., cv. Brigitta Blue, as photosensitizer in natural dye-sensitized solar cells (NDSSC). J Photochem Photobiol B 188:69–76. https://doi.org/10.1016/j.jphotobiol.2018.09.002

Luro F, Venturini N, Costantino G et al (2012) Genetic and chemical diversity of citron (Citrus medica L.) based on nuclear and cytoplasmic markers and leaf essential oil composition. Phytochemistry 77:186–196. https://doi.org/10.1016/j.phytochem.2011.12.013

Mandal J, Laxminarayana U (2014) Indirect shoot organogenesis from leaf explants of *Adhatoda vasica* Nees. Springerplus 3:1–8. https://doi.org/10.1186/2193-1801-3-648

Manzoor J, Sharma M (2019) Impact of textile dyes on human. Health Environ 2019:162–169. https://doi.org/10.4018/978-1-7998-0311-9.ch008

Mariadoss A, Satdive R, Fulzele DP et al (2020) Enhanced production of anthraquinones by gamma-irradiated cell cultures of Rubia cordifolia in a bioreactor. Ind Crop Prod 145:111987. https://doi.org/10.1016/j.indcrop.2019.111987

Martí M, Diretto G, Aragonés V et al (2020) Efficient production of saffron crocins and picrocrocin in Nicotiana benthamiana using a virus-driven system. Metab Eng 61:238–250. https://doi.org/10.1016/j.ymben.2020.06.009

Mascher M, Schreiber M, Scholz U et al (2019) Genebank genomics bridges the gap between the conservation of crop diversity and plant breeding. Nat Genet 51:1076–1081. https://doi.org/10.1038/s41588-019-0443-6

Mastan SG, Sudheer PDVN, Rahman H et al (2011) Molecular characterization of intra-population variability of Jatropha curcas L. using DNA based molecular markers. Mol Biol Rep 39:4383–4390. https://doi.org/10.1007/s11033-011-1226-z

Matsumoto T (2017) Cryopreservation of plant genetic resources: conventional and new methods. Rev Agric Sci 5:13–20. https://doi.org/10.7831/ras.5.13

Maurya R, Yadav HK (2016) Microsatellite markers based heterozygosity assessment in Jatropha curcas L.: a potential bioenergy crop. Trop Plant Res 3:191–198

McGettigan PA (2013) Transcriptomics in the RNA-seq era. Curr Opin Chem Biol 17:4–11. https://doi.org/10.1016/j.cbpa.2012.12.008

Medipally SR, Naresh B, Manoj Kumar S et al (2014) Somatic embryogenesis from leaf and shoot tip explants of Jatropha curcas L. Indian J Sci Technol 7:1842–1846. https://doi.org/10.17485/ijst/2014/v7i11.3

Meenakshi K, Jamkhedkar S, George IA (2016) Molecular profiling of some Barleria species using rbcL, matK gene sequences and RAPD markers. J Appl Hortic 18:203–206. https://doi.org/10.37855/jah.2016.v18i03.35

Mishra AK, Tiwari KN, Mishra P et al (2019) Effect of cytokinin and MS medium composition on efficient shoot proliferation of Nyctanthes arbor-tristis L. through cotyledonary node explant and evaluation of genetic fidelity and antioxidant capacity of regenerants. South Afr J Bot 127: 284–292. https://doi.org/10.1016/j.sajb.2019.09.008

Mohamed N, Taha RM (2011) Plant regeneration of Clitoria ternatea from leaf explants cultured in vitro. J Food, Agric Environ 9:268–270

Mohammed A, Chiruvella KK, Namsa ND, Ghanta RG (2015) An efficient in vitro shoot regeneration from leaf petiolar explants and ex vitro rooting of Bixa orellana L.—a dye yielding plant. Physiol Mol Biol Plants 21:417–424. https://doi.org/10.1007/s12298-015-0297-z

Mohanan NN, Nehru J, Botanic T, Mathew PJ (2016) Live plants of JNTBGRI

Moose SP, Mumm RH (2008) Molecular plant breeding as the foundation for 21st century crop improvement. Plant Physiol 147:969–977. https://doi.org/10.1104/pp.108.118232

Mridula MR, Nair AS (2018) Rapid micro-propagation of Wrightia tinctoria (ROXB.) R Br: a medicinal tree. Int J Bot Stud 3:126–131

Munir M, Hussain A, Ul-Haq I et al (2012) Callogenesis potential of cotyledonary explants of Althaea rosea L. from Pakistan. Pakistan J Bot 44:271–275

Mutharaian VN, Kamalakannan R, Mayavel A et al (2018) DNA polymorphisms and genetic relationship among populations of Acacia leucophloea using RAPD markers. J For Res 29: 1013–1020. https://doi.org/10.1007/s11676-017-0574-5

Nair DS et al (2016) In vitro regeneration and conservation of indigo (*Indigofera tinctoria* L.) by slow growth induction. Int J Environ Agric Biotechnol 1:873–884

Namedo GA (2007) Plant cell elicitation for production of secondary metabolites: a review. Rev Lit Arts Am 1:69–79

Nanda RM, Nayak S, Rout GR (2004) Studies on genetic relatedness of Acacia tree species using RAPD markers. Biologia (Bratisl) 59:115–120

Naresh V, Yamini KN, Rajendrakumar P, Kumar VD (2009) EST-SSR marker-based assay for the genetic purity assessment of safflower hybrids. Euphytica 170:347–353. https://doi.org/10.1007/s10681-009-9995-3

Nasirujjaman K, Uddin S, Zaman S, Md R abu (2005) Micropropagation of turmeric (*Curcuma longa* Linn.) through in vitro rhizome bud culture. J Biol Sci 5: 490–492. doi:https://doi.org/10.3923/jbs.2005.490.492

Naz R, Anis M, Aref I (2012) Assessment of the potentiality of TDZ on multiple shoot induction in Bauhinia tomentosa L., a woody legume. Acta Biol Hung 63:474–482. https://doi.org/10.1556/ABiol.63.2012.4.6

Nic Lughadha E, Bachman SP, Leão TCC et al (2020) Extinction risk and threats to plants and fungi. Plants People Planet 2:389–408. https://doi.org/10.1002/ppp3.10146

Nikhil M, Dudhare MS, Jadhav PV et al (2014) In vitro shoot regeneration and plantlet in vitro shoot regeneration and plantlet development in safflower (*Carthamus tinctorius* L.). Bioscan 9: 551–555

Osuagwu AN, Edem UL (2020) Evaluation of genetic diversity in aerial yam (Dioscorea bulbifera L) using simple sequence repeats (SSR) markers agrotechnology. Agrotechnology 9:1–7. https://doi.org/10.35248/2168-9881.20.9.202

Pal M, Chaudhury A (2010) High frequency direct plant regeneration, micropropagation and Shikonin induction in Arnebia hispidissima. J Crop Sci Biotechnol 13:13–19. https://doi.org/10.1007/s12892-009-0127-3

Pal K, Chowdhury S, Dutta SK et al (2020) Analysis of rhizome colour content, bioactive compound profiling and ex-situ conservation of turmeric genotypes (Curcuma longa L.) from sub-Himalayan Terai region of India. Ind Crops Prod 150:112401. https://doi.org/10.1016/j.indcrop.2020.112401

Panahi B, Ghorbanzadeh Neghab M (2013) Genetic characterization of Iranian safflower (*Carthamus tinctorius*) using inter simple sequence repeats (ISSR) markers. Physiol Mol Biol Plants 19:239–243. https://doi.org/10.1007/s12298-012-0155-1

Panapitiya PDDM, Welikala N (2018) Development of RAPD, DAMD and ISSR markers for authentication of medicinal plant Cassia auriculata and its adulterant Cassia surattensis. Int J Environ Agric Biotechnol 3:1906–1912. https://doi.org/10.22161/ijeab/3.5.43

Panda BM, Hazra S (2012) Micropropagation of *Semecarpus anacardium* L.: a medicinally important tree species. Plant Biosyst 146:61–68. https://doi.org/10.1080/11263504.2012.727877

Pandeya K, Tiwari KN, Singh J et al (2010) In vitro propagation of Clitoria ternatea L.: a rare medicinal plant. J Med Plant Res 4:664–668. https://doi.org/10.5897/ajb2014.13852

Panigrahi J, Gantait S, Patel IC (2017) An efficient in vitro approach for direct regeneration and callogenesis of Adhatoda vasica Nees, a potential source of quinazoline alkaloids. Natl Acad Sci Lett 40:319–324. https://doi.org/10.1007/s40009-017-0596-8

Papafotiou M, Antoniou I, Akoumianaki-Ioannidou I (2010) Studies on in vitro propagation of Bauhinia variegata. Acta Hortic 885:263–266

Parimalan R, Mahendranath G, Giridhar P (2014) Analysis of water soluble polysaccharides as a potential chemotaxonomic marker for landraces in Bixa orellana. Indian J Biochem Biophys 51: 81–86

Parimalan R, Mahendranath G, Giridhar P, Ravishankar G a (2011a) Abiotic elicitor mediated augmentation of annatto pigment production in standing crop of *Bixa orellana* L. Indian J Fundam Appl Life Sci 1: 229–236

Parimalan R, Venugopalan A, Giridhar P, Ravishankar GA (2011b) Somatic embryogenesis and Agrobacterium-mediated transformation in *Bixa orellana* L. Plant Cell Tissue Organ Cult 105: 317–328. https://doi.org/10.1007/s11240-010-9870-x

Patil SH, Kurlapkar DD, Gaikwad DK (2019) Short communication: dye-yielding plant resources of Maharashtra, India: a checklist. Biodiversitas 20:250–266

Pillai SK, Siril EA (2019) Enhanced production of berberine through callus culture of Tinospora cordifolia (Willd.) Miers ex Hook F. and Thoms. Proc Natl Acad Sci India Sect B Biol Sci 323–331. doi:https://doi.org/10.1007/s40011-019-01106-9

Poornima S, Ambika S (2012) Intraspecific diversity and conservation of an important natural food dye yielding plant *Bixa orellana* L. Indian J Fundam Appl Life Sci 2:114–125

Prasuna S, Srinivas B (2015) In vitro regeneration studies of Oxalis corniculata from nodal explants. Int J Recent Sci Res 6:6216–6220

Praveena C, Veeresham C (2014) Multiple shoot regeneration and effect of sugars on growth and nitidine accumulation in shoot cultures of Toddalia asiatica. Pharmacogn Mag 10:S480–S486. https://doi.org/10.4103/0973-1296.139777

Qosim WA, Patarapuwadol S, Watanabe KN (2011) Development of SSR markers of mangosteen (Garcinia mangostana L.). Int Res J Biotechnol 2:1–8

Qu J, Zhang W, Yu X (2011) A combination of elicitation and precursor feeding leads to increased anthocyanin synthesis in cell suspension cultures of Vitis vinifera. Plant Cell Tiss Org Cult 107: 261–269. https://doi.org/10.1007/s11240-011-9977-8

Quadri RR, Kamili AN, Teixeira JA (2012) In vitro multiplication of Arnebia benthamii Wall., a critically endangered medicinal herb of the western Himalayas. Med Aromat Plant Sci Biotchnol 6:54–57

Rafii MY, Shabanimofrad M, Puteri Edaroyati MW, Latif MA (2012) Analysis of the genetic diversity of physic nut, Jatropha curcas L. accessions using RAPD markers. Mol Biol Rep 39: 6505–6511. https://doi.org/10.1007/s11033-012-1478-2

Ram GD, Chauhan SS, Verma DK (2012) In vitro propagation of Jatropha curcas from embryo and nodal explants. An Int Quart J Life Sci 7:251–254

Ram M, Prasad KV, Singh SK et al (2013) Influence of salicylic acid and methyl jasmonate elicitation on anthocyanin production in callus cultures of Rosa hybrida L. Plant Cell Tiss Org Cult 113:459–467. https://doi.org/10.1007/s11240-013-0287-1

Ramachandra MN, Karuppusamy S, Krishnamurthy YL (2014) Micropropagation of Caralluma stalagmifera var. longipetala: a rare succulent medicinal plant from Karnataka. India Afr J Biotechnol 13:3553–3559. https://doi.org/10.5897/ajb2014.13852

Ramadugu C, Keremane ML, Hu X et al (2015) Genetic analysis of citron (Citrus medica L.) using simple sequence repeats and single nucleotide polymorphisms. Sci Hortic 195:124–137. https://doi.org/10.1016/j.scienta.2015.09.004

Rao K (2004) Plant genetic resources: advancing conservation and use through biotechnology. Afr J Biotechnol 2004:136–145

Rao RV, Hodgkin T (2002) Genetic diversity and conservation and utilization of plant genetic resource. Plant Cell Tissue Organ Cult 68:1–19

Rathore MS, Yadav S, Yadav P et al (2015) Micropropagation of elite genotype of Jatropha curcas L. through enhanced axillary bud proliferation and ex vitro rooting. Biomass Bioenergy 83:501–510. https://doi.org/10.1016/j.biombioe.2015.10.023

Ren BQ, Xiang XG, Chen ZD (2010) Species identification of Alnus (Betulaceae) using nrDNA and cpDNA genetic markers. Mol Ecol Resour 10:594–605. https://doi.org/10.1111/j.1755-0998.2009.02815.x

Roa C, Hamilton RS, Wenzl P, Powell W (2016) Plant genetic resources: needs, rights, and opportunities. Trends Plant Sci 21:633–636

Robarts DWH, Wolfe AD (2014) Sequence-related amplified polymorphism (SRAP) markers: a potential resource for studies in plant molecular biology. Appl Plant Sci 2:1400017. https://doi.org/10.3732/apps.1400017

Romero-López N, Luna-Martínez F, Gallegos-Brito C et al (2019) An integrated genetic linkage map of *Bixa orellana* L. Tree Genet Genomes 15:1–11. https://doi.org/10.1007/s11295-019-1371-4

Rostika I, Sunarlim N, Mariska I (2008) Micropropagation of mangosteen (Garcinia mangostana). Indones J Agric 1:28–33. https://doi.org/10.5070/s5231022264

Sahoo A, Jena S, Kar B et al (2017) EST-SSR marker revealed effective over biochemical and morphological scepticism towards identification of specific turmeric (*Curcuma longa* L.) cultivars. 3. Biotech 7:1–12. https://doi.org/10.1007/s13205-017-0701-1

Sahoo A, Kar B, Sahoo S et al (2019) De novo transcriptome sequencing explored cultivar specific sequence variation and differential expression of pigment synthesis genes in turmeric (*Curcuma longa* L.). Ind Crop Prod 134:388–402. https://doi.org/10.1016/j.indcrop.2019.04.021

Sakunphueak A, Panichayupakaranant P (2010) Increased production of naphthoquinones in Impatiens balsamina root cultures by elicitation with methyl jasmonate. Bioresour Technol 101:8777–8783. https://doi.org/10.1016/j.biortech.2010.06.067

Sarangi BK, Minami Y, Thul ST (2015) RNA-Seq analysis for indigo biosynthesis pathway genes in Indigofera tinctoria and *Polygonum tinctorium*. Genomics Data 6:212–213. https://doi.org/10.1016/j.gdata.2015.09.021

Saranya Krishnan SR, Siril EA (2016) Induction of hairy roots and over production of anthraquinones in *Oldenlandia umbellata* L.: a dye yielding medicinal plant by using wild type *Agrobacterium rhizogenes* strain. Indian J Plant Physiol 21:271–278. https://doi.org/10.1007/s40502-016-0229-0

Sathiyabama M, Bernstein N, Anusuya S (2016) Chitosan elicitation for increased curcumin production and stimulation of defence response in turmeric (*Curcuma longa* L.). Ind Crop Prod 89:87–94. https://doi.org/10.1016/j.indcrop.2016.05.007

Schmutzer T, Bolger ME, Rudd S et al (2017) Bioinformatics in the plant genomic and phenomic domain: the German contribution to resources, services and perspectives. J Biotechnol 261:37–45

Sehgal D, Raina SN (2005) Genotyping safflower (*Carthamus tinctorius*) cultivars by DNA fingerprints. Euphytica 146:67–76. https://doi.org/10.1007/s10681-005-8496-2

Senan S, Kizhakayil D, Sheeja TE et al (2013) Novel polymorphic microsatellite markers from turmeric, *Curcuma longa* L. (Zingiberaceae). Acta Bot Croat 72:407–412. https://doi.org/10.2478/botcro-2013-0002

Senarath WTPSK (2010) In vitro propagation of Coscinium fenestratum (Gaertn.) Colebr. (Menispermaceae)—an endangered medicinal plant. J Natl Sci Found Sri Lanka 38:219–223. https://doi.org/10.4038/jnsfsr.v38i4.2648

Senthilkumar N, Gurusubramanian G (2011) Random amplified polymorphic DNA (RAPD) markers and its applications. In: Proceedings of the "National Level Workshop on Random Amplified Polymorphic DNA (RAPD) Markers and It's Applications" organized. pp 116–124

Shah MR, George IA (2019) Increased biomass and pigment production from Cassia alata L. callus cultures and their potential as a textile dye. Ind Crop Prod 128:346–353. https://doi.org/10.1016/j.indcrop.2018.11.018

Shah M, George IA (2020) Pigment elicitation and sun protection factor of callus induced from Cassia tora seedling explants. Plant Cell Tiss Org Cult 143:201–210. https://doi.org/10.1007/s11240-020-01913-3

Sharma KD, Piqueras A (2010) Saffron (*Crocus sativus* L.) tissue culture: micropropagation and secondary metabolite production. Funct Plant Sci Biotechnol 2010:64–73

Sharma S, Sudheer Pamidimarri DVN, Vijay Anand KG, Reddy MP (2011) Assessment of genetic stability in micropropagules of Jatropha curcas genotypes by RAPD and AFLP analysis. Ind Crop Prod 34:1003–1009. https://doi.org/10.1016/j.indcrop.2011.03.008

Shiji PC, Siril EA (2018) An improved micropropagation and ex vitro rooting of a commercially important crop Henna (Lawsonia inermis L.). Physiol Mol Biol Plants 24:1273–1284. https://doi.org/10.1007/s12298-018-0600-x

Shkryl YN, Veremeichik GN, Makhazen DS et al (2016) Increase of anthraquinone content in *Rubia cordifolia* cells transformed by native and constitutively active forms of the AtCPK1 gene. Plant Cell Rep 35:1907–1916. https://doi.org/10.1007/s00299-016-2005-z

Shukla P, Singh A, Gawri S, Alexander A, Sonwane S (2011) In Vitro Propagation Of Barleria Prionitis Linn And Its Antibacterial. Int J Pharma Prof Res 2:198–200

Sigrist MS, Pinheiro JB, Filho JAA, Zucchi MI (2011) Genetic diversity of turmeric germplasm (*Curcuma longa*; Zingiberaceae) identified by microsatellite markers. Genet Mol Res 10:419–428. https://doi.org/10.4238/vol10-1gmr1047

Siju S, Dhanya K, Syamkumar S et al (2010a) Development, characterization and utilization of genomic microsatellite markers in turmeric (*Curcuma longa* L.). Biochem Syst Ecol 38:641–646. https://doi.org/10.1016/j.bse.2010.08.006

Siju S, Dhanya K, Syamkumar S et al (2010b) Development, characterization and cross species amplification of polymorphic microsatellite markers from expressed sequence tags of turmeric (*Curcuma longa* L.). Mol Biotechnol 44:140–147. https://doi.org/10.1007/s12033-009-9222-4

Silja PK, Satheeshkumar K (2015) Establishment of adventitious root cultures from leaf explants of Plumbago rosea and enhanced plumbagin production through elicitation. Ind Crop Prod 76:479–486. https://doi.org/10.1016/j.indcrop.2015.07.021

Silja PK, Gisha GP, Satheeshkumar K (2014) Enhanced plumbagin accumulation in embryogenic cell suspension cultures of Plumbago rosea L. following elicitation. Plant Cell Tiss Org Cult 119:469–477. https://doi.org/10.1007/s11240-014-0547-8

Silva DM, Siqueira MVBM, Carrasco NF et al (2016) Genetic diversity among air yam (Dioscorea bulbifera) varieties based on single sequence repeat markers. Genet Mol Res 15:1–11. https://doi.org/10.4238/gmr.15027929

Singh J, Tiwari KN (2010) Evaluation of cotyledonary node of Clitoria ternatea L. for high frequency in vitro axillary shoot proliferation. Asian J Plant Sci 9:351–357. https://doi.org/10.3923/ajps.2010.351.357

Singh M, Sonkusale S, Niratker C, Shukla P (2014) Micropropagation of Shorea robusta: an economically important woody plant. J For Sci 60:70–74. https://doi.org/10.17221/80/2013-jfs

Singh AK, Nanda P, Singh A, Singh B (2015) Genetic diversity analysis in turmeric (*Curcuma longa* L.) based on SSR markers. J Biol Eng Res Rev 2:20–24

Siril EA, Joseph N (2013) Micropropagation of annatto (*Bixa orellana* L.) from mature tree and assessment of genetic fidelity of micropropagated plants with RAPD markers. Physiol Mol Biol Plants 19:147–155. https://doi.org/10.1007/s12298-012-0150-6

Siva R (2007) Status of natural dyes in India. Curr Sci 92:916–925

Siva R, Mayes S, Behera SK, Rajasekaran C (2012) Anthraquinones dye production using root cultures of *Oldenlandia umbellata* L. Ind Crop Prod 37:415–419. https://doi.org/10.1016/j.indcrop.2011.12.027

Siva R, Rajasekaran C, Mudgal G (2009) Induction of somatic embryogenesis and organogenesis in *Oldenlandia umbellata* L., a dye-yielding medicinal plant. Plant Cell Tissue Organ Cult 98:205–211. https://doi.org/10.1007/s11240-009-9553-7

Sivalingam PN, Singh D, Chauhan S (2012) Morphological and molecular diversity of an underutilized fruit crop—Cordia myxa L. germplasm from the arid region of Rajasthan, India. Genet Resour Crop Evol 59:305–316. https://doi.org/10.1007/s10722-011-9781-7

Sobir S, Poerwanto R, Santosa E et al (2011) Genetic variability in apomictic mangosteen (Garcinia mangostana) and its close relatives (Garcinia spp.) based on ISSR markers. Biodiversitas J Biol Divers 12:59–63. https://doi.org/10.13057/biodiv/d120201

Song H, Mao W, Shang Y et al (2021) A regeneration system using cotyledons and cotyledonary node explants of Toona ciliata. J For Res 32:967–974. https://doi.org/10.1007/s11676-020-01189-5

Sonnino A (2017) International instruments for conservation and sustainable use of plant genetic resources for food and agriculture: an historical appraisal. Diversity 9:1–19. https://doi.org/10.3390/d9040050

Soundar Raju C, Aslam A, Shajahan A (2015) High-efficiency direct somatic embryogenesis and plant regeneration from leaf base explants of turmeric (Curcuma longa L.). Plant Cell Tiss Org Cult 122:79–87. https://doi.org/10.1007/s11240-015-0751-1

Sreenivas VK, Jisha VN, Martin KP, Madhusoodanan PV (2011) Bridelia stipularis: a new source for anthocyanin production in vitro. Acta Physiol Plant 33:2051–2056. https://doi.org/10.1007/s11738-011-0735-7

Sudhersan C, Jibimanuel S, Ashkanani J (2013) Somatic embryogenesis and production of synthetic seeds in ziziphus. Acta Hortic 993:83–90. https://doi.org/10.17660/ActaHortic.2013.993.12

Sykłowska-Baranek K, Pietrosiuk A, Naliwajski MR et al (2012) Effect of l-phenylalanine on PAL activity and production of naphthoquinone pigments in suspension cultures of Arnebia euchroma (Royle) Johnst. In Vitro Cell Dev Biol Plant 48:555–564. https://doi.org/10.1007/s11627-012-9443-2

Techen N, Parveen I, Khan IA (2017) A single molecular marker to distinguish between species of Dioscorea. Genome 60:1–7. https://doi.org/10.1139/gen-2015-0105

Teixeira da Silva JA, Zeng S, Godoy-Hernández G et al (2019) *Bixa orellana* L. (achiote) tissue culture: a review. Vitr Cell Dev Biol-Plant 55:231–241. https://doi.org/10.1007/s11627-019-09969-3

Thamilarasi K, Ram MK, Lohot VD, et al (2016) Genetic diversity analysis of Palas [Butea monosperma (Lam.) Taub.] flower variants through random amplified polymorphic DNA. Indian J Plant Genet Resour 29187. doi:https://doi.org/10.5958/0976-1926.2016.00027.9

Tian YS, Fu XY, Yang ZQ et al (2020) Metabolic engineering of rice endosperm for betanin biosynthesis. New Phytol 225:1915–1922. https://doi.org/10.1111/nph.16323

Torelli A, Marieschi M, Bruni R (2014) Authentication of saffron (Crocus sativus L.) in different processed, retail products by means of SCAR markers. Food Control 36:126–131. https://doi.org/10.1016/j.foodcont.2013.08.001

Tyagi RK, Agrawal A, Mahalakshmi C et al (2007) Low-cost media for in vitro conservation of turmeric (*Curcuma longa* L.) and genetic stability assessment using RAPD markers. Vitr Cell Dev Biol Plant 43:51–58. https://doi.org/10.1007/s11627-006-9000-y

Tyub S, Kamili AN, Bhat MM (2016) In vitro propagation of Althaea rosea—a valuable medicinal plant of Kashmir Himalaya. J Nat Nat Sci 1(5):1–4. https://doi.org/10.26859/jnnsci.v1i01.9596

Uchoi A, Malik SK, Choudhary R et al (2017) Molecular markers in assessing genetic variation of Indian citron (Citrus medica L.) cultivars collected from different parts of India. Indian J Biotechnol 16:346–356

Valdez-Ojeda R, Hernández-Stefanoni JL, Aguilar-Espinosa M et al (2008) Assessing morphological and genetic variation in annatto (*Bixa orellana* L.) by sequence-related amplified polymorphism and cluster analysis. Hort Science 43:2013–2017. https://doi.org/10.21273/hortsci.43.7.2013

Valletta A, De Angelis G, Badiali C et al (2016) Acetic acid acts as an elicitor exerting a chitosan-like effect on xanthone biosynthesis in Hypericum perforatum L. root cultures. Plant Cell Rep 35:1009–1020. https://doi.org/10.1007/s00299-016-1934-x

Van Eck JM, Smith FD (2018) Tissue culture. In: Environmentally safe approaches to crop disease control. Routledge, Abingdon, UK, pp 317–331

Varshney RK, Graner A, Sorrells ME (2005) Genic microsatellite markers in plants: features and applications. Trends Biotechnol 23:48–55

Vashishtha A, Jehan T, Lakhanpaul S (2013) Genetic diversity and population structure of Butea monosperma (Lam.) Taub.—a potential medicinal legume tree. Physiol Mol Biol Plants 19: 389–397. https://doi.org/10.1007/s12298-013-0170-x

Vashishtha A, Sharma K, Lakhanpaul S (2014) DNA profiling of Ziziphus mauritiana (Lam.) and Butea monosperma (Lam.) taub. for Lac cultivation in India. Proc Natl Acad Sci India Sect B Biol Sci 84:593–601. https://doi.org/10.1007/s40011-014-0332-9

Vengadesan G, Ganapathi A, Anand RP, Anbazhagan VR (2000) In vitro organogenesis and plant formation in Acacia sinuata. Plant Cell Tiss Org Cult 61:23–28

Venkatasubramanian SPB (2011) Molecular identification and development of nuclear DNA its sequence based marker to distinguish Coscinium fenestratum Gaertn. (Menispermaceae) from its adulterants. Curr Trends Biotechnol Pharm 5:1163–1172

Verma S, Singh S, Sharma S et al (2015) Assessment of genetic diversity in indigenous turmeric (*Curcuma longa*) germplasm from India using molecular markers. Physiol Mol Biol Plants 21: 233–242. https://doi.org/10.1007/s12298-015-0286-2

Vijayakumar J, Ponmanickam P, Samuel P et al (2017) Influence of meta-topolin on efficient plant regeneration via micropropagation and organogenesis of safflower (*Carthamus tinctorius* L.) cv. NARI-H-15. Am J Plant Sci 08:688–705. https://doi.org/10.4236/ajps.2017.84048

Vos P, Hogers R, Bleeker M et al (1995) AFLP: a new technique for DNA fingerprinting. Nucleic Acids Res 23:4407–4414. https://doi.org/10.1093/nar/23.21.4407

Wambugu PW, Ndjiondjop MN, Henry RJ (2018) Role of genomics in promoting the utilization of plant genetic resources in genebanks. Brief Funct Genomics 17:198–206

Wani MS, Gupta RC, Munshi AH, Sharma V (2018a) Genetic diversity and structure of Betula utilis accessions of north-western Himalaya based on RAPD and ISSR markers. Nucleus 61: 145–152. https://doi.org/10.1007/s13237-018-0239-6

Wani MS, Sharma V, Gupta RC, Munshi AH (2018b) Development and characterization of SSR markers in Himalayan species Betula utilis. J For Res 31. https://doi.org/10.1007/s11676-019-00932-x

Warakagoda PS, Subasinghe S (2013) In vitro propagation of Pterocarpus santalinus L. (red sandalwood) through tissue culture. J Natl Sci Found Sri Lanka 41:53–63. https://doi.org/10.4038/jnsfsr.v41i1.5333

Warakagoda PS, Subasinghe S, Gunasekare MTK (2017) In vitro clonal propagation of Cosoinium fenestratum (Gertn.) Colebr. (Weniwel) through nodal explants. J Natl Sci Found Sri Lanka 45: 133–141. https://doi.org/10.4038/jnsfsr.v45i2.8179

Warhade MI, Badere RS (2018) Fusarium oxysporum cell elicitor enhances betalain content in the cell suspension culture of Celosia cristata. Physiol Mol Biol Plants 24:285–293. https://doi.org/10.1007/s12298-018-0511-x

Website Singapore Government (2021) Our gardens. Singapore Bot. Gard. http://www.hort.vt.edu/hhg/gardens.html

Weise S, Oppermann M, Maggioni L et al (2017) EURISCO: the European search catalogue for plant genetic resources. Nucleic Acids Res 45:D1003–D1008. https://doi.org/10.1093/nar/gkw755

Wen M, Wang H, Xia Z et al (2010) Development of EST-SSR and genomic-SSR markers to assess genetic diversity in Jatropha Curcas L. BMC Res Notes 3:1–8. https://doi.org/10.1186/1756-0500-3-42

Wu J, Chen W, Zhang Y et al (2020) Metabolic engineering for improved curcumin biosynthesis in *Escherichia coli*. J Agric Food Chem 68:10772–10779. https://doi.org/10.1021/acs.jafc.0c04276

Wu ZH, Liao R, Dong X et al (2019) Complete chloroplast genome sequence of *Carthamus tinctorius* L. from PacBio sequel platform. Mitochondrial DNA Part B Resour 4:2635–2636. https://doi.org/10.1080/23802359.2019.1643799

Wuest DM, Hou S, Lee KH (2019) Metabolic engineering. In: Comprehensive biotechnology. Elsevier, Amsterdam, pp 647–658

Yamamoto M, Ito A, Sogabe A et al (2017) Shoot-tip culture of acerola (Malpighia glabra L.). Trop Agric Dev 61:57–61. https://doi.org/10.11248/jsta.61.57

Yarra R, Mushke R, Velmala M (2018) In vitro approaches for conservation and sustainable utilization of *Butea monosperma* (Lam.) Taub. Var. Lutea (Witt.) Maheshwari: a highly valuable medicinal plant. In: Biotechnological approaches for medicinal and aromatic plants. Springer, Cham, pp 346–351

Yeotkar SD, Malode SN, Waghmare VN, Thakre P (2011) Genetic relationship and diversity analysis of Clitoria ternatea variants and Clitoria biflora using random amplified polymorphic DNA (RAPD) markers. Afr J Biotechnol 10:18065–18070. https://doi.org/10.5897/AJB11.1423

Zegeye H (2017) In situ and ex situ conservation: complementary approaches for maintaining biodiversity. Int J Res Environ stu 4:1–12

Zhao H, Li Y, Petkowski J et al (2021) Genomic prediction and genomic heritability of grain yield and its related traits in a safflower genebank collection. Plant Genome 14:1–15. https://doi.org/10.1002/tpg2.20064

Perception of Advanced Techniques in Conservation and Utilization of Wild Genetic Resources of Rice

21

Disna Ratnasekera, Asanka Tennakoon, Salinda Sandamal, and Parakkrama Wijerathna

Abstract

Breeders need access to unique genetic variability to meet the growing demand for food while maintaining sustainable agricultural production with the impacts of climate change for generating high-quality nutritional food. Changes in climate and anthropogenic activities and a multitude of environmental influences pose severe threats to food supply and preservation of natural diversity. For example, unpredictable droughts, elevated temperature, and new diseases and pests threaten crop production. Thus, breeding with crop wild relatives (CWR) gives significant resilience to modern agricultural systems and the ability to help sustainably boosting agricultural productivity. As a result, numerous genotype screenings are necessary for broad adaptability, producing a segregating material through fast breeding or rapid generation to shorten the breeding cycle and improving genetic gain. In addition, CWR genomics generates data that support CWR's usage to boost agricultural genetic diversity. QTL mapping, identifying of candidate genes by next-generation sequencing, gene-based marker development, or significant candidate gene pyramiding of stress-responsive loci in popular cultivar are required to maintain the sustainability of crop production.

D. Ratnasekera (✉) · A. Tennakoon · P. Wijerathna
Department of Agricultural Biology, Faculty of Agriculture, University of Ruhuna, Matara, Sri Lanka
e-mail: disnar@agbio.ruh.ac.lk

S. Sandamal
Department of Agricultural Biology, Faculty of Agriculture, University of Ruhuna, Matara, Sri Lanka

State Key Laboratory of Systematic and Evolutionary Botany, Institute of Botany, Chinese Academy of Sciences, Beijing, China

University of Chinese Academy of Sciences, Beijing, China

S. Ramamoorthy et al. (eds.), *Plant Genetic Resources, Inventory, Collection and Conservation*, https://doi.org/10.1007/978-981-16-7699-4_21

Thus, genomic data is useful for identifying and isolating novel and dominant alleles of genes from crop gene pools that are agronomically important, which can be used to generate improved crop cultivars. Hence, the natural allelic difference in candidate genes that influences major agronomic characteristics and crop development initiatives is being investigated via allele mining. Among the CWRs of economically important crops, the wild species of rice is essential to improve modern rice cultivars. The awareness of novel genetic and genomic approaches of rice genetic resources for efficient utilization is crucial. Further, their conservation status and availability have not been quantified globally. As a result, a joint effort is required to improve the conservation and accessibility of crop wild relatives for rice breeding.

Keywords

Genomics of CWR · Crop improvement · Rice genetic resources

21.1 Introduction

The world's population is predicted to expand by two billion people by 2050 (United Nations 2019), putting even undue demands on plant-based food (Godfray et al. 2010). These changes provide more effort for crop improvement programs to handle various challenges, and new cultivars need to meet growers' and producers' limitations on biophysical and socioeconomic restrictions, as well as consumer expectations and aspirations. Thus, modern plant breeders rarely go back to wild lineages to investigate crop wild relatives as a prominent source of genetic variation in crop improvement (Dempewolf et al. 2017). Crop wild relatives (CWRs) are taxonomically related to domesticated crop cultivars that survive in the wild and exist as a possible contributor to beneficial genes for crop improvement. The acquiring of desirable characteristics to modern cultivars was developed from CWRs by selection or hybridization. Even though CWRs are becoming more popular and essential, their contributions to new cultivar development are still lower than the expected level. Therefore, it is critical to characterize CWRs for desirable traits using emerging advanced techniques and upgraded procedures for intercrossing species from different lineages, molecular strategies for managing backcross initiatives, and continuing to expand the collection of wild species and accessions in ex situ gene banks. This is because introducing wild genetic material or exotic crop germplasm to existing breeding plants causes disruption. The genome is disturbed by the inflow of novel alleles when they are crossed to elite breeding lines. Therefore, it is required to carefully design gene complexes which are the source of particularly valuable characteristics in commercial cultivars (Dempewolf et al. 2012). However, when beneficial alleles are added, associated deleterious alleles that need an attempt to remove frequently accompany them. Sterility barriers can make it more challenging to produce fertile offspring, extending the cost and time of cultivar improvement (Zamir 2001). To address this issue, identifying CWR genes/

QTL by genome-wide association studies (GWAS), QTL mapping, and association mapping can be used for genomics-assisted breeding for genetic improvement (Jamwal and Bijral 2020). Because crop wild relatives (CWRs) are adapted to different environments than their domesticated relatives, genetic material from these wild species has the potential to serve a significant role in inbreeding for greater pest and disease and environmental stress tolerance. Positively, CWR has been used to combat potentially harmful pest and disease outbreaks, as well as climate change. Thus, their diversity is still underutilized, and global efforts to collect and preserve CWR diversity continue to be a top priority (Warschefsky and Rieseberg 2021).

Among the CWRs, wild rice has been recognized as the valuable genetic source for rice genetic improvement (Khush 1997; Vaughan et al. 2003; Sang and Ge 2007). Because of their significance in rice domestication and genetic modification, extensive research has been conducted to investigate their classification, population genetics, evolutionary history, adaptation potential, and agronomical traits (Morishma et al. 1984; Barbier 1989; Zhu and Ge 2005; Grillo et al. 2009; Huang and Schaal 2012; Liu et al. 2015). Furthermore, these rice genetic reserves contain a considerable amount of favorable genes that serve as one of the most significant raw materials for meeting the present and future demands of rice improvement and production programs. These are the primary resources for developing modern cultivars to address sustainable production and meet the demands for maintaining food security. Rice is the staple food that satisfies nearly half of the world's population (Bierschenk et al. 2020), supplying up to 80% of daily calorie consumption, especially in Asia (Sahebi et al. 2018). Although rice has been successfully domesticated, genetic erosion has occurred, and it is accompanied by domestication, which has resulted in decreased genetic diversity among cultivars and loss of certain germplasms (Brush 2000). Due to the introduction and continuous release of modern rice varieties, hybridization and selection for yield and yield components, especially quality characteristics, have led to the progressive loss of genes important for regulating biotic and abiotic stressors (Ullah et al. 2012). In addition, allele mining is essential to identify key alleles responsible for distinct stressors for a specific trait and their variability in other genotypes used for further trait improvement of crop plants.

Moreover, recent advances in genomics of wild relatives of rice provide insight to screen and incorporate genetic materials for modern cultivars. Meanwhile, genes responsible for biotic and abiotic stresses for growth and development must be investigated. Knowledge of gene pool classification of rice genetic resources is also required to select candidate parents and breeding techniques for developing modern agricultural plants to meet the growing demand for plant-based foods. Therefore, efficient conservation of prevalent wild resources is vital for present and future crop enhancement for food and nutrition security (Ferrer et al. 2020).

21.2 The Concept of the Gene Pool

Gene pools were initially described by Harlan and De Wet (1971). Throughout the era of crop improvement, gene pools have been identified as the most promising source for the usage of plant genetic resources. A gene pool is the entire set of unique alleles present in a species or population. The size of the gene pool determines the extent of genetic diversity available in a population. Further, the gene pool is classified into three groups based on interfertility between related species (Harlan and De Wet 1971). The reproductive isolation of two clusters of living organisms promotes speciation (Lowry et al. 2008). The development of reproductive barriers is the final element in inhibiting viable intercrosses among diverse species of plants.

Further, natural selection may have accelerated the emergence of interspecific reproductive barriers in sympatric environments (Turelli et al. 2001). In CWR's breeding efforts, there have been documented failures to develop viable and fertile hybrids, which may threaten the effectiveness of breeding programs (Kubota et al. 2012; Roth et al. 2018). Therefore, the gene pool classification system, described by Harlan and De Wet, is still one of the most useful and widely used in CWR breeding programs.

21.2.1 Primary Gene Pool

Members of closely related taxa, such as cultivated, landrace, weedy, and wild crop types, make up the primary gene pool. Crossing individuals from the gene pool is simple, resulting in fertile and vigorous offspring with minimal gene transfer problems. Hybrids exhibit perfect seed fertility due to regular meiotic chromosomal pairing and recombination. It is the most widely utilized gene pool, and the majority of breeding efforts rely on it. It was further divided into purposefully cultivated species and spontaneous species (i.e., wild or weedy).

21.2.2 Secondary Gene Pool

The secondary gene pool included several species that might be crossed with the target crop to enhance gene transfer; however, this is associated with many challenges due to reproductive barriers and genetic incompatibilities. Ploidy differences, chromosome alterations, and genetic barriers are some of the causes of sexual incompatibility. In comparison to the primary gene pool, individuals of this category are used less in breeding programs. It was conceivable to breed from such CWRs, but it was more laborious and more costly than breeding from the primary gene pool (Choudhary et al. 2017).

21.2.3 Tertiary Gene Pool

All remaining CWRs that could not be transferred to the crop were placed in the tertiary gene pool. When crossings were feasible, the offspring were sterile, deformed, or short-lived. In such circumstances, producing additional generation or even F1 hybrids was rare, and gene transfer into the crop might require a significant amount of attempt using methods like embryo culture, grafting, artificial polyploidization, or usage of bridging species. The tertiary gene pool effectively dictated the limitations of the possible gene that could be transferred into a crop species through the crossing. It is commonly used to describe all CWRs for which traditional crossbreeding methods have not produced fertile hybrids. These species representing the extreme outer limit of the potential germplasm are included in this type of gene pool (Dwivedi et al. 2008).

21.2.4 Gene Pool Classification of Genus *Oryza*

The genus *Oryza* contains 24 species (Gramineae family); two are cultivated, *O. sativa* L. and *O. glaberrima* Steud. Moreover, there are 22 wild relatives of *Oryza* (2n = 24 or 48) with a broad geographic range (Khush 1997; Vaughan 1989). They represent ten genome types (Sanchez et al. 2013) (Table 21.1). *O. sativa* is grown and consumed as a staple food in most Asian countries. *O. glaberrima* is cultivated as African rice and is grown in a few regions of West Africa. There are two subspecies of *O. sativa*: *indica* and *japonica*. Comparatively, *indica* and *japonica* subspecies have a broad and narrow genetic variation, respectively.

Previously, wild *Oryza* species was grouped into three major groups or complexes based on the easiness of gene transmission from wild to cultivated rice. These are the complexes of *O. sativa, O. officinalis, O. ridleyi,* and *O. meyeriana* (Morishima and Oka 1960). These complexes were later referred to as primary, secondary, and tertiary gene pools of *Oryza* (Khush 1997).

The *O. sativa* complex consists of two farmed species and six wild species with the AA genome (Table 21.1). These species have homologous chromosomal pairing and are diploid. The Asian cultivated rice *O. sativa* is derived from perennial *O. rufipogon*, whereas *O. glaberrima* is derived from perennial *O. barthii* as the African cultivated rice (Chang 1976; Oka 1988; Vaughan et al. 2008; Zhu and Ge 2005).

There are ten wild species in the *O. officinalis* complex with a wide geographic distribution. Six types of genomes are uncovered in the *O. officinalis* complex. These are *O. punctata, O. officinalis, O. rhizomatis, O. eichingeri, O. minuta, O. latifolia, O. alta, O. grandiglumis, O. australiensis,* and *O. brachyantha*, either diploid or tetraploid. These species have nonhomologous chromosomal pairing, and cross-incompatibility with *O. sativa* makes gene transfer difficult.

O. meyeriana and *O. granulata* are two diploid wild species in the *O. meyeriana* complex, both of which have the GG genome. A cross-incompatibility exists between these two species and *O. sativa*. The *O. ridleyi* complex comprises two

Table 21.1 Chromosome number, genomic composition, and gene pool classification in the wild species of genus *Oryza*

Wild species	Chromosome Number	Genome	Gene Pool
O. rufipogon Griff.	24	AA	Primary
O. nivara Sharma et Shastry	24	AA	Primary
O. longistaminata Chev. et Roehr	24	AA	Primary
O. barthii Chev. et Roehr	24	AA	Primary
O. meridionalis Ng	24	AA	Primary
O. glumaepatula Steud.	24	AA	Primary
O. punctata Kotschy ex Steud.	24, 48	BB, BBCC	Secondary
O. minuta J.S. Presl. ex C.B. Presl.	48	BBCC	Secondary
O. officinalis Wall ex. Watt	24	CC	Secondary
O. rhizomatis Vaughan	24	CC	Secondary
O. eichingeri Peter	24	CC	Secondary
O. latifolia Desv.	48	CCDD	Secondary
O. alta Swallen	48	CCDD	Secondary
O. grandiglumis Prod.	48	CCDD	Secondary
O. australiensis Domin.	24	EE	Secondary
O. brachyantha Chev. et Roehr	24	FF	Secondary
O. granulata Nees et Arn. ex. Watt	24	GG	Tertiary
O. meyeriana Baill	24	GG	Tertiary
O. longiglumis Jansen	48	HHJJ	Tertiary
O. ridleyi Hook	48	HHJJ	Tertiary
O. schlechteri Pilger	24	HHKK	Tertiary
O. coarctata Roxb.	48	HHKK	Tertiary

tetraploid wild species with HHJJ genomes (*O. longiglumis* and *O. ridleyi*); both are strongly cross-incompatible with *O. sativa* (Jena 2010). Moreover, wild species of rice are widely distributed throughout Asia, Africa, and Australia (Liu et al. 2015).

21.2.5 Rice Genetic Resources in Sri Lanka

Sri Lanka is one of the secondary diversity centers for rice genetic resources. Five wild species of *Oryza*, *viz. O. nivara* (AA genome), *O. rufipogon* (AA genome), *O. eichingeri* (CC genome), *O. rhizomatis* (CC genome), and *O. granulata* (GG genome), are found in Sri Lanka. *Oryza rhizomatis* is considered endemic to the country. Among these, *O. nivara* and *O. rufipogon* are closely related to Asian cultivated rice (Banaticla-Hilario 2012). Their distribution and habitats are well described under Sri Lankan conditions (Liyanage and Senanayake 2010; Sandamal et al. 2018a; Tennakoon et al. 2019). Recent studies reported that the genetic diversity of the two species in Sri Lanka, either at the population level or at the species level, was relatively high given the small area of the island, which is the essential primary resource for crop improvement (Tennakoon et al. 2018; Sandamal et al. 2018a, b).

Rather than gene pool classification, advanced genomic approaches were used to screen CWR for current crop improvement programs.

21.3 Genomics of Crop Wild Relatives of Rice

Genomics is used to characterize the wild relatives of cultivated plants to support their conservation and efficient utilization. Genome analysis in CWRs might facilitate the invention of new genetic resources (Brozynska et al. 2014). Ex situ germplasm collections are also easier to manage and utilization of CWR genomics. CWR sequencing (Henry 2014) can reveal the genetic diversity of some genomes with limited mutations due to genetic bottlenecks related to crop improvement or domestication (Waters and Henry 2014). This information is required to focus on the introduction of favorable genetic variation into specific crop genomes.

Edwards and Henry (2011) focused on plants and plant species at the whole-genome level because of frequent advancements in DNA sequencing techniques. Sequencing the genomes of wild relatives enabled the development of species-specific reference genome sequences. Data for resequencing compared to the domesticated genome is acquired in different settings, whereas transcriptome analysis allows the expression genome to be defined. Maternal genome sequencing, particularly chloroplast genome sequencing is an effective method for investigating genetic variation in naturally existing CWR populations and germplasm collections (Nock et al. 2011).

21.3.1 Model Plant Genome Sequencing and Their Crop Wild Relatives

Certain CWRs may serve as a model plant genome sequence that assists in the characterization and awareness of more complex domesticated species. Because of its tiny and compact genome compared to other main cereal crops, *Oryza sativa* has been used as a model organism for studying the genetic composition, domestication, and modification of cereal crops (Brozynska et al. 2016).

The ultimate goal for the efficient use of a CWR genetic resource is reference genome sequences generated via de novo sequence data assembling. In addition, limited CWRs are sequenced, but main domesticated species are significantly sequenced; this becomes a key aim of plant genome sequencing (Chen et al. 2013; Zhang et al. 2014). The high amount of heterozygosity in CWR makes assembling of a reference genome sequence more difficult (Brozynska et al. 2016).

Shotgun sequencing by next-generation sequencing (NGS) generates a lot of quality sequence results in short sequences. Longer read technology is now available complementary to short-read technology. Although the quality was initially inadequate (sequencing precision), it has since improved, and read lengths of about

15,000 bp are now accessible. They must typically be linked to the genome's genetic or physical map to be placed along the chromosomes. Although alignment to close lineages (wild relatives) is frequently used, it carries the risk of biasing the assembly towards the reference genome's structure. Because substantial genomic rearrangements between the species may not have been discovered, it is best to utilize genetically close lineages. When it comes to rice (*Oryza sativa*), a systematic research effort was begun to produce reference genomes for each significant species, as well as all for A genome species (primary gene pool) and members of each of the remaining genome categories (Jacquemin et al. 2013; Zhang et al. 2014). This strategy could be effective in promoting CWR access for other main crops.

21.3.2 Nuclear Genome Resequencing of Crop Wild Relatives

Shotgun sequencing of individual genotypes reveals genetic differences between genotypes in the domesticated lineage. Furthermore, direct mapping is also practiced compared to the reference genome (Subbaiyan et al. 2012). In certain instances, shotgun CWR sequencing is utilized to obtain sequence reads for direct comparison with the reference genome of domesticated species (Huang et al. 2012).

This technique enables the selective reintroduction of genetic variation into the domesticated gene pool. This type of resequencing may not reveal significant genomic rearrangements or genetic identities that have been lost due to the domestication process. This technique is likely to be beneficial for plants that are very closely related. Further, it could be used to determine specific genomic sites where diversity has been lost owing to domestication selection (Subbaiyan et al. 2012).

21.3.3 Sequencing Crop Wild Relatives That Contribute to Polyploid Genomes

From one or even more diploid CWRs, polyploid genomes are generated, and they are found in many valuable crop plants. The first step in generating a reference genome sequence for these complicated polyploid genomes is to sequence the CWR responsible for their formation.

21.3.4 Chloroplast Sequencing of Crop Wild Relatives

The well-preserved genome of chloroplast was widely used to barcode plants and discover genetic relations in plant populations (Kane et al. 2012; McPherson et al. 2013). Because the chloroplasts are usually inherited uniparentally (Birky 2001; Hirosawa et al. 2004), they are not recombinant in flower-producing plants, so they may be recognized as a single phylogenetic unit comprising multiple genes. Whole-plant DNA sequencing was performed to obtain the whole-chloroplast genome sequence. This can be achieved by mapping the sequencing readings directly into the chloroplast genome reference (Nock et al. 2011) by first de novo assembling the data and then using homology search to isolate chloroplast contigs (McPherson et al. 2013). New populations of CWR in rice have been detected by whole-chloroplast genome sequencing (Waters et al. 2012). The introduction of whole-chloroplast genome sequencing facilitates a better understanding of the interrelationship between CWRs and domesticated lineages. Phylogenetic relations or divergence time between species within families was determined using whole-chloroplast genome sequences of CWR compared to domesticated relatives (Saski et al. 2007).

21.3.5 Sequencing the Transcriptomes of Crop Wild Relatives

CWR transcriptome sequencing may quickly reveal new alleles or expression variation for crop growth and development, even if the species' genome has yet to be sequenced (Bao et al. 2015). In addition, transcriptome sequencing also assists in the identification of newly discovered genomes. The advancement of RNA-Seq methods has been contributed by NGS, enabling concurrent identification of new alleles and genes in CWR with a broad range of expression levels. Transcriptome analysis (TA) is used to detect gene expression patterns and differentially expressed genes (DEGs) in CWR and their domesticates. Further, alterations in the allelic expression of genes encoding paralogues and orthologues in polyploids may also lead to discovering of causal SNPs. They can be utilized in crop improvement using molecular assays. Research into the transcriptome of the CWR may aid researchers in better understanding the genomic basis of the CWR's adaptability to various environments as well as the domestication and diversification of plant species. Apart from that, comparative genetic studies enable scientists to get a better knowledge of multidirectional gene flow between members of a domesticated wild weedy complex, evolving of weeds, and consequences of gene flow from GMO crops to their CWR (Ellstrand 2003). For example, the halophyte *Oryza coarctata* is remarkable for its ability to adapt to extreme salinity (20–40 dS/m) and its exceptional tolerance to prolonged submersion in saltwater. Jena (1994) generated a sexual hybrid between *O. sativa* and *O. coarctata*. Candidate genes have also been identified by analyzing uniquely expressed genes following treatments with varied salt and submergence levels. It is used to understand plant improvement processes of salt and submergence tolerances (Garg et al. 2014). Further, dissection of naturally existing variation at candidate genes/loci through allele mining provides clues to direct access to essential alleles responsible for distinct stressors for a specific trait and their variability in other genotypes.

21.4 Allele Mining in Rice

Introgressions of new alleles from wild species into modern cultivars (McCouch et al. 2007) have shown that particular alleles and allele combinations can cause alterations in trait expression by avoiding the genetic bottlenecks. Therefore, the substantial germplasm reserves are continuously re-examined for novel alleles to enhance crop varieties' genetic potential in various agronomic characteristics. New genes will be found and annotated quickly as sequence and expression data accumulates in multiple genomic databases, enabling the construction of allele-specific markers (Spooner 2005). Polymerase chain reaction (PCR) methods are developed based on gene and genome sequences to identify beneficial alleles of genes from various species (Latha et al. 2004). Easy access to beneficial alleles provides resistance to biotic or abiotic stressors and enhances nutrient-use efficiency, increased production, and quality improvements. Emerging genomic techniques are also used to discover identical alleles related to various characteristics and underlying variations in different genotypes. The process is known as "dissection of naturally occurring variation at candidate genes/loci" or "allele mining."

Early allele mining experiments concentrated on detecting SNPs/InDels at coding regions or exons of a gene until they were expected to alter the encoded protein function and/or structure. Recent research has discovered that nucleotide changes in noncoding regions (5′ UTR), such as the promoter region, introns, and 3′ UTR, significantly impact transcript synthesizing and accumulation, enabling alterations in trait expression. The significance of intronic mutations in gene regulation was shown by expressing specific rice genes, such as tubulin and rubi3 gene (Fiume et al. 2004; Samadder et al. 2008). In rice, a tenfold boost in gene expression was caused by a mutation over the first intron of the waxy (Wx) gene's 5′ splice site (Isshiki et al. 1998). Mining for sequence variation, especially in the regulatory part of genes, has become increasingly important in gene expression. Because promoter elements are well known for their importance in gene regulation, any alteration in their sequences will substantially affect gene expression, resulting in a wide range of trait expressions. Sequence diversity in the binding sites of cis-acting/trans-acting regulatory elements in the upstream region is the primary source of changes in gene expression.

Mutations in the promoter region of the candidate gene for recessive bacterial blight resistance (xa13) were important in generating disease resistance in rice (Chu et al. 2006). Similarly, mutations in the promoter regions of a gene involved with partial grain filling (GIF1) have been identified as the source of a restricted expression pattern in cultivars relative to a wide expression pattern in the wild species of *Oryza* (Wang et al. 2008). "True" allele mining also includes the study of noncoding and regulatory regions of the target genes in evaluating sequence variations in the coding sites of the agronomically important genes to cover the majority of functional variants of interest in the genes (Rangan et al. 1999; Latha et al. 2004). In order to better understand the significance of "mining" in terms of regulatory region differences, promoters of genes have been identified as a potential target for allele mining, also known as "promoter mining."

There are two major ways for detecting sequence polymorphisms in naturally occurring populations for a given gene. EcoTilling, a modified TILLING (Targeting Induced Local Lesions in Genomes) technique, and sequencing-based mining are the main approaches of allele mining.

21.4.1 EcoTilling

TILLING is a heteroduplex analysis technique for identifying polymorphisms in a target gene caused by induced mutations (Till et al. 2003). EcoTilling is a technique for assessing how much natural variation there is in particular genes in crops. The technique is similar to TILLING, except that mutations are assessed in naturally existing alleles in both primary and secondary crop gene pools rather than being artificially induced (Comai and Henikoff 2006; Comai et al. 2004). EcoTilling focuses on the enzymatic cleavage of heteroduplexed DNA (produced by a single-nucleotide variation in sequence between the reference and test genotype) using a single strand-specific nuclease under specified conditions, followed by detection with Li-Cor genotypes (Li-Cor, USA). The nuclease will cleave the product at point mutations, resulting in two cleaved products of equal size. The presence, location, and type of gene mutations or SNP will be identified by sequencing the amplicon from the mutation-carrying genotype tested.

21.4.2 Sequencing-Based Allele Mining

The method includes PCR amplification of alleles in different genotypes along with DNA sequencing to identify nucleotide variation. Individuals could be analyzed for haplotype structure and diversity using sequencing-based allele mining to infer genetic association experiments in plants. EcoTilling can be achieved at a lower cost than SNP/haplotyping methods, but the sophisticated instrument and expertise needed and the validation of sequencing necessary in the EcoTilling procedure restrict its broad application. Any technology has benefits and drawbacks that must be carefully considered in terms of the intended use.

A strong connection between effective genetic and genomic utilization in genomics-driven plant breeding may be observed in allele mining, which has tremendous potential and application. To cope effectively with the ever-increasing sequence of data in GenBank and the ever-expanding agricultural gene banks, new and efficient "mining" methods are required. Efforts should be made to develop tools and techniques for dealing with genetic and genomic resources. The field of comparative functional genomics in rice has not been actively pursued, given the fast-growing resources for mining novel alleles. As more germplasm resequencing data becomes available, phylogenetic footprinting should be used to conduct a meaningful comparative study of upstream regulatory sequence variation to discover novel alleles and networks with potential uses in rice breeding.

21.5 A Potential Avenue for Utilization of Genetic Resources of Rice

Plant genetic resources are described as the overall genetic diversity of cultivated varieties and their descendants, and they assist breeders in crop improvement (Jackson et al. 1990). Crop genetic resources are responsible for food and agriculture linked with human food production. Because human demands are constantly changing, genetic resources can offer critical characteristic features such as disease resistance and climatic adaptation for the crops. It takes a long time, much money, and many human resources to successfully introduce genetic variation from wild relatives into crops for food security. It is a lengthy procedure that begins in the wild species with genetic resource experts, botanists, and taxonomists searching for identifying, and collecting propagating material.

Furthermore, it is conserved and characterized by gene bank curators, geneticists, agronomists, entomologists, and pathologists. They are also involved in the evaluation of traits and gene discovery and validation. Finally, pre-breeders, breeders, or germplasm improvement programs develop new varieties, which are then fed into seed systems and eventually end up in farmers' fields. Moreover, plant genetic resources include breeding lines, genetic stocks, old cultivars, landraces, and traditional varieties.

Cross-compatibility, taxonomic relation between cultivars and their wild counterpart, fertility in F1 and subsequent offspring, existence or conservation of CWRs, and investigation and use of wild germplasms are potential causes for an efficient usage of CWRs in certain crops (Zamir 2001). Wild species genetic diversity can be employed in different ways; however, there are two basic approaches to consider: (i) "Choose first": wild materials that express a specific trait of interest are selected based on the phenotypic, genotypic, or collection local data and used in targeted crossing, after which the offspring are evaluated. (ii) "Cross first": a broad range of wild and domesticated species are crossed, and the resultant offspring is tested for characteristics of interest in the domesticated setting. In order to analyze agronomic performance data (measurable traits expressed directly by an individual), genotypic data (alleles that play a role in a specific trait of interest), or a combination of both, the first methodology can be carried out using statistical genetics methods. This concept is referred to as predictive characterization (Thormann et al. 2014). Once traits of interest in a wild accession or individual have been discovered, they must be introduced into crop backgrounds. Alternatively, first, make crosses between wild and cultivated lineages, and then examine the progeny for desirable features as F1 or subsequent materials. Subsequent work concentrated on discovering unexpected sources of variation that can only be seen when introgressed in a domesticated context. After transferring alleles of interest into a target background, a specific number of backcrosses are required to decrease the deleterious variation associated with CWR introgressions. Linkage drag occurs when undesirable traits are linked to a desirable trait (Zamir 2001). Many traits associated with inadequate agronomic performance in domesticated species have been deliberately chosen out of wild counterparts. Examples include poor yields, seed shattering, and low seed or fruit size (Salamini et al. 2002). These characteristics present various issues to breeders, including the inability to recover sufficient numbers of seeds, failure to use standard equipment, and failing to meet unusual agronomic or horticultural needs. However, comprehensive genetic mapping followed by marker-assisted selection (MAS) may be employed in some circumstances to overcome this obstacle. Thus, linkage drag appears to be a barrier to adopting wild lineages in many crop development initiatives. Adaptive farming is a possible solution to modifying the planting date or changing to alternating crop types with many benefits. However, the generation of novel crop types (Lobell et al. 2008) has been proposed as the biggest benefit that can be accomplished on time. It is a costly procedure requiring investments in farming, and new cultivars will need beneficial genes and gene resources. CWR is recognized as an important potential source for such genes that are used to incorporate, but only if efficient conservation ensures long-term survival.

Rice consumption on a global scale is expected to enhance from 723 million tons in 2015 to 852 million tons in 2035 (Brar and Khush 2018). Therefore, rice yield must be increased and maintained to feed the growing population. Improving yield potential and incorporating abiotic and biotic stress tolerance, particularly in agroecology, are two ways to accomplish this objective. Plant breeding has played a critical role in mating-induced genetic variability to boost rice yield by adding genetic material (new alleles or loci) from wild species to crop varieties. However, pre and post fertilization obstacles, hybrid sterility, karyotypic separation between cultivars and wild lineage chromosomes, limited recombination, hybrid breakdown, and linkage drag are obstacles to the precise transfer of genomic segment from wild species to the crop (Brar and Khush 2018). Recent progress in plant tissue culture, genetic modification, and rice genome sequence has created an opportunity to study the roles and usage in the reinforced germplasm of significant genes/QTL. Modern instruments of genetic engineering have made it far more accurate than in the past to make use of wild species.

The high-yielding *O. sativa* is found worldwide, while the poor-yielding African *O. glaberrima* is only found in West Africa. Rice genetic resources are rich in new genes/QTLs for traits such as pest and disease resistance (Sarao et al. 2016; Dong et al. 2020; Yang et al. 2020), tolerance to environmental stress (Lei et al. 2013; Luo et al. 2016; Zhang et al. 2017; Cao et al. 2020), and characteristics that are economically significant such as grain quality (Gaikwad et al. 2018; Qi et al. 2018), grain yield (Gaikwad et al. 2019; Guttikonda et al. 2020), fertility restoration (Gaikwad et al. 2014), cytoplasmic male sterility (Hu et al. 2016), and heterosis for yield attribute traits (Gaikwad et al. 2019; Guttikonda et al. 2020). For gene/QTL mapping, wild species are used to generate alien introgression lines (ILs), backcross inbred lines (BILs), chromosomal segment substitution lines (CSSLs), doubled haploids (DH), and near-isogenic lines (NILs). The genes and QTLs that regulate important agronomic characteristics have been molecularly identified and introduced into elite genetic backgrounds. Certain wild forms have unique traits that may be used in breeding. *O. rufipogon* has the potential to tolerate acidic soil (Mandal and Gupta 1997), *O. meyeriana* and *O. granulata* have shade or low light intensity tolerance (Vaughan et al. 2003), and *O. nivara* has high micronutrient content

(Swamy et al. 2014). To get distinct yield characteristics, researchers used *Oryza rufipogon* (Fu et al. 2010; Luo et al. 2011; Thalapati et al. 2012; Ngu et al. 2014) and *Oryza nivara* (Gaikwad et al. 2014; Swamy et al. 2014).

Large-scale genetic and genomic resources must be collected initially to use for developing these traits. Some worldwide initiatives were established, including the *Oryza* Map Alignment Project (OMAP), the largest genetic database of wild rice germplasm in the world. Its purpose is to align the *O. sativa* ssp. *japonica* cv. "Nipponbare" reference genome sequence to the BAC-based physical maps of 18 wild types. The goal of OMAP is to lay a solid foundation for comparing wild rice genomes to cultivated rice genomes and using that information to access the genetic potential of wild species of rice to identify new agriculturally useful genes and QTLs. It is vital to discuss the importance of rice genetic resources that contain genes/QTLs for agricultural production and their application in rice breeding programs (Wing et al. 2005).

21.6 Trait Improvement

Agricultural productivity is victimized by many causes, including water scarcity and irrigation system failures, declining soil fertility, irregular rainfall, rising air temperatures and heat waves, pest and disease, and their coevolution into virulent strains (Singh et al. 2020). Between 1964 and 2007, grain crop yields globally were reduced by 10% due to extreme heat and frequent droughts. The cost of the damage was higher in wealthy nations, with a loss of 8–11% more than in developing countries (Lesk et al. 2016). Using biotechnology tools and breeding techniques to develop crop improvement initiatives resulted in identifying characteristics and introducing improved cultivars that may help minimize the negative impacts (Table 21.2). Plant breeding is the process of selecting superior kinds among variations in terms of crop production and quality of edible portions, ease of cultivating, harvesting and processing, tolerance to environmental stressors, and pest and disease resistance. Plant breeding's challenge is to simultaneously enhance most of the characteristics of interest, a process made more difficult by genetic correlations between characteristics, which pleiotropic genes may produce, physical linkage among genes in chromosomes, or genetic structure of a particular population.

Moreover, selecting one trait impacts associated traits, sometimes positively and sometimes negatively (Breseghello and Coelho 2013). Traditional breeding has evolved into molecular breeding to solve agricultural issues, and genomics-assisted breeding must be adopted. RNAi, VIGS, and gene/genome editing have been utilized to create breeding lines with improved agronomic and climate-resilient characteristics by moving away from gene cloning and overexpression or knockout. Whole-genome high-throughput genomics methods, including genetic association mapping, map-based cloning, genomic selection, and rapid breeding, have also been proven to enhance genetic gains and accelerate crop development processes. These

Table 21.2 List of wild species-contributed QTLs that have a beneficial impact on crop production and grain quality characteristics

Traits	QTL/loci	References
Oryza rufipogon		
Grain yield	*yld1.1, yld2.1*	Xiao et al. (1996)
	yld3.2	Thomson et al. (2003)
	qyld2.1	Guttikonda et al. (2020)
Panicle length	*pl1*	Xiong et al. (1999)
	pl1.1	Septiningsih et al. (2003)
	qPL5–2	Ma et al. (2019)
Grain weight	*gw1.1*	Septiningsih et al. (2003), Thomson et al. (2003)
	qgw8.1	Kang et al. (2018)
	qgw5.1	Bhatia et al. (2018)
Grain length	*qgl12–1*	Qiao et al. (2016)
	qgl12.2	Qi et al. (2017)
	qgl7.1	Bhatia et al. (2018)
	qGL9–1	Ma et al. (2019)
	qGL3.1	Yuan et al. (2020)
	qGL3	Hussain et al. (2020)
Grain width	*qGW10*	Ma et al. (2019)
	qgw3–1	Qiao et al. (2016)
1000-grain weight	*qtgw5.1*	Bhatia et al. (2018)
	qtgw8	Qi et al. (2017)
	qGWT1.1	Yuan et al. (2020)
Grain setting density	*qMPW7*	Liang et al. (2019)
	qNPB7	Liang et al. (2019)
	qGSD7	Liang et al. (2019)
Plot yield	*gw9.2*	Marri et al. (2005)
Panicle number	*pnl12*	Lee et al. (2005)
Milled rice recovery	*mrr11*	Gaikwad et al. (2018)
Head rice recovery	*hrr1*	Gaikwad et al. (2018)
Stigma exsertion rate	*qSERr3–3*	Zou et al. (2020)
Stigma exsertion ratio	*qSER3, qSER8*	Bakti and Tanaka (2019)
Grain size	*qGS3*	Hussain et al. (2020)
Early heading date	*ehd8*	Sun et al. (2019)
Glossiness of grain	*qgcr9*	Yun et al. (2016)
Spikelets/panicle	*spp1.1*	Lee et al. (2005)
	qspp5	Tan et al. (2007)
	hsp11	Luo et al. (2011)
Panicles/plant	*ppl6.1*	Moncada et al. (2001)
	ppl7.1	Thomson et al. (2003)
	qpn1	Tian et al. (2006)
	ppl7.1	Fu et al. (2010)

(continued)

Table 21.2 (continued)

Traits	QTL/loci	References
Grains/panicle	*gpp3.1*	Thomson et al. (2003)
	qGP3	Ma et al. (2019)
	qNGPP12.1	Yuan et al. (2020)
LB ratio	*qlwr8*	Qi et al. (2017)
	lbr12	Gaikwad et al. (2018)
Primary branches/panicle	*qPBP8*	Ma et al. (2019)
Panicle length grains/panicle	*pl1.1 gpp1.1*	Xiao et al. (1998)
Yield/plant	*yld1.1*	Septiningsih et al. (2003)
	qgw8	Tian et al. (2006)
Grain length, breadth ratio	*qGLWR3.1*	Yuan et al. (2020)
Filled grains, main panicle weight	*qNFG6*	Liang et al. (2019)
Grain weight, grain yield	*gw8, yd8*	Jin et al. (2009)
Grains/plant, grain yield	*gpl1.1, yld1.1*	Moncada et al. (2001)
1000-grain weight, grain yield	*gw11.1, yld11.1*	Moncada et al. (2001)
Grain length, LB ratio	*gl1, lw1*	Lee et al. (2005)
Grain weight, yield/plant	*gnp5.1*	Marri et al. (2005)
Grains/panicle, grain weight	*qgpa8, qfg8*	Tian et al. (2006)
Spikelets/plant, grains/plant	*snp5.1*	Marri et al. (2005)
Grains/panicle, yield/plant	*qgpp3, qgwp3*	Tan et al. (2007)
	gpp1.1, yld1.1	Fu et al. (2010)
Primary branches, panicle length, spikelets/ panicle	*qbp2, qpl2* & *qspp2*	Ji et al. (2012)
Main panicle weight, number of primary branches/panicle	*qMPW6*	Liang et al. (2019)
Brown rice recovery, protein content, gel consistency	*br3, pc3, con3*	Yuan et al. (2010)
Milled rice recovery, head rice recovery, amylose content	*mr6, hr6* &, *ac6*	Yuan et al. (2010)
Spikelets/panicle, grains/panicle, grain yield	*spp1, gpp1, yld1*	Gaikwad et al. (2014)
Spikelets/panicle, grains/panicle, panicle length, yield/plant	*sn9.1, gn9.1, pl9.1, yd9.1*	Xie et al. (2008)
Oryza nivara		
Kernel length	*qkl1*	Eizenga et al. (2016)
Kernel width	*kw6.1*	Swamy et al. (2012)
Kernel length after cooking	*klac12.1*	Swamy et al. (2012)
Panicle length	*pl1*	Gaikwad et al. (2014)
100-seed weight	*qhsdwt10*	Eizenga et al. (2016)
Seed length	*qsdlg3*	Eizenga et al. (2017)
Seed width	*qsdwd2–2*	Eizenga et al. (2017)
Seedling vigor	*qsv6*	Haritha et al. (2018)
Grain yield	*yld8*	Gaikwad et al. (2014)
	qyld8.3	Addanki et al. (2019)

(continued)

Table 21.2 (continued)

Traits	QTL/loci	References
LB ratio	*lbr6*	Gaikwad et al. (2018)
	qklwr1	Eizenga et al. (2016)
	lwr6.1	Swamy et al. (2012)
	Qsdlgwdro 12–1	Eizenga et al. (2017)
Milled rice recovery	*mp1.2*	Swamy et al. (2012)
	mrr7	Gaikwad et al. (2018)
Milling percentage	*mp1.2*	Swamy et al. (2012)
Amylose content	*ac2.1*	Swamy et al. (2012)
	qaac6	Eizenga et al. (2016)
1000-grain weight	*gw10*	Gaikwad et al. (2014)
Tiller number	*nt12.1*	Swamy et al. (2014)
Biomass	*nfg1.1*	Swamy et al. (2014)
Brown rice recovery	*brr7*	Gaikwad et al. (2018)
Per day productivity	*qPDP-11–1*	Balakrishnan et al. (2020)
Total dry matter	*qTDM11–1*	Balakrishnan et al. (2020)
Panicle weight	*qpw9.1*	Surapaneni et al. (2017)
	qPW-11–1	Balakrishnan et al. (2020)
Panicles/plant	*qpnnb7*	Eizenga et al. (2016)
	qppl3	Ma et al. (2016)
Yield/plant	*yldp8.1*	Kaladhar et al. (2008)
	bm11.1	Swamy et al. (2014)
	qYLDP-11–1	Balakrishnan et al. (2020)
Productive tillers/plant	*qnpt1.1*	Surapaneni et al. (2017)
Bulk yield/square meter	*qBY-11–1*	Balakrishnan et al. (2020)
Grains/panicle, spikelet fertility	*gpp7, sf7*	Gaikwad et al. (2014)
Spikelets/plant and grains/plant	*nsp1.1*	Swamy et al. (2014)
Spikelets on primary branches, secondary branches, spikelets on secondary branches and spikelets/plant, grain length	*qsnpb7, qsbn7*	Ma et al. (2016)
	qsnsb7	Ma et al. (2016)
	qspp7	Ma et al. (2016)
	qgl3.1	Ma et al. (2016)
Oryza glaberrima		
Milled rice recovery	*qmr5*	Aluko et al. (2004)
Grain length	*kl1.1*	Li et al. (2004)
Grain width	*kw11.1*	Li et al. (2004)
LB ratio	*lwr12.2*	Li et al. (2004)
Grains/panicle	*qgnp7*	Hu et al. (2018)
Oryza barthii		
Grain weight	*qgw3*	Bessho-Uehara et al. (2017)
Seed setting rate	*qSSb1–1*	Zhao et al. (2019)
1000-grain weight	*qGWTb1–1*	Zhao et al. (2019)
Stigma exsertion rate	*qSER5–2*	Zou et al. (2020)
Grains/panicle	*qgpp8*	Bessho-Uehara et al. (2017)

(continued)

Table 21.2 (continued)

Traits	QTL/loci	References
Oryza glumaepatula		
1000-grain weight	*qtgw6.1*	Bhatia et al. (2018)
	qGWTgl–1	Bhatia et al. (2018)
Panicle number	*qpnr8*	Brondani et al. (2002)
	qpnr11	Brondani et al. (2002)
Grain yield	*qyld8*	Rangel et al. (2008)
Grain length	*qgl7.1*	Bhatia et al. (2018)
Seed setting rate	*qSSgl–1*	Zhao et al. (2019)
Stigma exsertion rate	*qSER1a*	Tan et al. (2020)
	qSER1b	Tan et al. (2020)
	qSER10	Tan et al. (2020)
Oryza longistaminata		
Panicle weight, spikelets/panicle, spikelet fertility	*qpw6.1*	Gichuhi et al. (2016)
	qnsp6.1	Gichuhi et al. (2016)
100-grain weight	*qsf6.1*	Gichuhi et al. (2016)
Grain width	*qwg5.1*	Gichuhi et al. (2016)
	qgw3.1	Gichuhi et al. (2016)
Stem breaking strength	*qLR8*	Long et al. (2020)
Grains/panicle	*gpp1*	Ramos et al. (2016)
Spikelets/panicle	*qSPP2.2*	Kaur et al. (2018)
Stem diameter, stem length, stem breaking strength	*qLR1*	Long et al. (2020)
Oryza grandiglumis		
Panicle length, spikelets/panicle, grain length	*gw2, tgw2*	Yoon et al. (2006)
Grain width, 1000-grain weight, LB ratio	*lw2*	Yoon et al. (2006)
Oryza minuta		
Tiller number	*tn3*	Rahman et al. (2007)
Panicle number	*pn4*	Rahman et al. (2007)
Panicle length	*pl6*	Rahman et al. (2007)
LB ratio	*glw2*	Rahman et al. (2007)
1000-grain weight	*qTGW1,9,12*	Wei et al. (2015)
Spikelets/panicle	*qspp7*	Balkunde et al. (2013)
Spikelets/panicle, yield	*qSPP1, qYGP12*	Wei et al. (2015)
Spikelets/panicle, fertility	*spp6, sf6*	Rahman et al. (2007)
Grain length, 1000-grain weight	*gl7, tgw7*	Rahman et al. (2007)
Spikelets/panicle, seeds/panicle, yield/plant	*spp7, sp7, yd7*	Linh et al. (2008)
O. meridionalis		
Grain length	*qgl4–4*	He et al. (2017)
LB ratio	qrlw12–1	He et al. (2017)
Stigma exsertion rate	qSERm3	Zou et al. (2020)
Oryza meyeriana		
Panicles/plant, spikelet/panicle, 1000-grain weight	*Qpnp, qspp6, qtgw12*	Chen et al. (2012)

Modified from Gaikwad et al. (2021)

techniques have improved the yields of primary crops, including maize, rice, and wheat (Bailey-Serres et al. 2019).

CSSLs, BILs, and ILs have been utilized in many studies to introgress QTLs of yield-related characteristics and grain quality traits from diverse *O. rufipogon* accessions (wild species) into elite *indica* and *japonica* lines (Gaikwad et al. 2019). Developing rice varieties that are tolerant of low-nitrogen conditions will be both cost effective and environmentally friendly. Root biomass, root length, number of root tips, and root volume are critical architectural components contributing to nitrogen-use efficiency (Ranjan et al. 2019). It is vital to enhance parental lines to generate high-yielding rice hybrids. Genes from wild species can be used to boost the production of elite parental lines. These species have delivered fertility restoration genes to cultivated rice CMS systems (Hu et al. 2016). Identifying crop production improvement genes/QTLs in wild rice and transfer to hybrid rice parents (Table 21.2) will dramatically increase heterosis in rice hybrids (Xiao et al. 1996; Wu 2009).

21.7 Utilization of CWRs in Crop Improvement for Plant Stresses

The climate change caused by global warming and deforestation in recent years has dramatically affected the global scenario and consequently increased the damage to the ecosystem (Bellard et al. 2012). Humans, animals, reptiles, birds, pests, pathogens, and plants are all affected. Yield reduction caused by unfavorable climate changes has been a significant problem in recent years to increase specific abiotic and biotic stressors. The emergence of novel pests and diseases and changes in the environment are significant challenges to sustainable crop production and biotic and abiotic stress resistance (Hasan et al. 2015). An integrated approach is required to address the main genes/QTL conferring resistance/tolerance to important abiotic and biotic stressors and select for high yields in the absence of stress and minimize the disease infestation and yield loss due to abiotic and biotic stresses (Pang et al. 2017; Jena et al. 2017; Wang et al. 2017; Das et al. 2018; Kumar et al. 2018a; Sandhu et al. 2019; Muthu et al. 2020).

21.7.1 Abiotic Resistance of Rice

In the near future, abiotic stressors like severe weather, high temperatures, and a rise in carbon dioxide (CO_2) are expected to become much more frequent. These impacts would be particularly noticeable due to climate change in low latitudes and tropical areas, where rice cultivation is prevalent and where most developing nations are situated. Crop production has suffered in many parts of the globe in recent years because of unfavorable climatic changes, owing to increased incidences of many abiotic stressors, including salinity, drought, submergence, high temperature, cold, and heavy metals (Lafitte et al. 2004). Further, abiotic stressors may reduce rice yields up to 70% by negatively impacting the survival, development, and grain filling of rice (Akram et al. 2019). Rice crops may adapt to mild environmental conditions, but the yield is frequently decreased or entirely hindered when various abiotic stresses frequently occur since crops have an insufficient capacity to adapt due to their low genetic diversity (Des Marais et al. 2013). In addition, wild relatives of rice can adapt to the same environmental changes as crops due to the high genetic diversity of diverse populations and higher phenotypic plasticity (Des Marais et al. 2013; Fahad et al. 2017; Cohen and Leach 2019; Fogliatto et al. 2019).

For instance, drought is a key limiting factor for rice production, with nearly 42 million hectares in Asia experiencing substantial yield loss due to different degrees of drought stress at different phases of crop development (Saikumar et al. 2016; Sandhu and Kumar 2017; Mukamuhirwa et al. 2019). Aside from drought, salinity stress and alkalinity are the other factors that have a negative impact on rice yields throughout the vegetative and reproductive stages of the plant. Salinity stress represents approximately 20% of cultivated land and 33% of irrigated agricultural regions of the world (Shrivastava and Kumar 2015). It is among the most damaging abiotic stressors on rice plants that limit their growth and development and lead to yield reduction for more than 50% (Molla et al. 2015). Submergence is another most significant abiotic stress in rice-growing areas vulnerable to flash floods (Iftekharuddaula et al. 2015). In lowland rain-fed conditions, submergence tolerance is essential for rice (*Oryza sativa*). Hence, identifying stress-tolerant rice cultivars is vital for sustaining production and meeting global food security requirements.

Recent advancements in molecular biology have resulted in quantitative trait loci (QTLs) linked with important abiotic stress, i.e., drought tolerance (qDTY1.1, qDTY2.1, qDTY3.1, qDTY6.1, and qDTY12.1 (Bernier et al. 2007; Kumar et al. 2007; Venuprasad et al. 2009; Vikram et al. 2011), submergence tolerance (Sub1) (Xu and Mackill 1996), and salinity tolerance (Saltol) (Thomson et al. 2010). Genetic enhancement of rice abiotic stress tolerance has limited success due to the complex nature of the mechanisms controlling these characteristics and the challenges associated with phenotyping the characteristics (Muthu et al. 2020). Therefore, continuous improvements in rice production need the development of resistant rice varieties adapted to marginal environments with significant yield differences caused by abiotic stressors (Table 21.3).

21.7.2 Biotic Resistance of Rice

During infestations, biotic stressors like diseases (fungi and bacteria), insects, pests, and weeds cause significant yield losses (Hasan et al. 2015). The most common biotic stressors are blast, bacterial leaf blight (BLB), brown planthopper (BPH), and gall midge (GM) that cause yield loss in rice. Rice cultivars with improved resistance to biotic stresses, good agronomic potential of yield, and high quality rice grain have been in demand for decades (Khush 2005). Hence, genetic improvement seems to be the most successful approach for reducing the negative impact on the environment

Table 21.3 List of key abiotic resistance genes and QTLs discovered for wild species of rice

Trait	Gene/ QTLs	Wild species	Genome	References
Abiotic resistance				
Tolerance to iron toxicity	Unknown	*O. rufipogon*	AA	Brar and Khush (2006), Brar and Khush (2002)
	Unknown	*O. glaberrima*	AA	Brar and Khush (2006), Brar and Khush (2002)
Drought-related traits	QTLs	*O. glaberrima*	AA	Brar and Khush (2006), Brar and Khush (2002)
Tolerance to aluminum toxicity	QTLs	*O. rufipogon*	AA	Nguyen et al. (2003), Sanchez et al. (2013)
Tolerance to acidic conditions	Unknown	*O. glaberrima*	AA	Sanchez et al. (2013), Brar and Khush (2006), Brar and Khush (2002)
	Unknown	*O. rufipogon*	AA	Brar and Khush (2006), Brar and Khush (2002)
Tolerance to phosphorus deficiency	QTLs	*O. glaberrima*	AA	Brar and Khush (2006), Brar and Khush (2002)
	QTLs	*O. rufipogon*	AA	Brar and Khush (2006), Brar and Khush (2002)
Weed competitive ability	QTLs	*O. glaberrima*	AA	Brar and Khush (2006), Brar and Khush (2002)
Salinity	Saltol	*O. coarctata*	KKLL	Bonilla et al. (2002)
Submergence	Sub1	*O. coarctata*	KKLL	Zhang and Xie (2014)
Cold tolerance	QTLs	*O. rufipogon*	AA	Koseki et al. (2010)
Drought and heat tolerance	QTLs	*O. glaberrima*	AA	Ndjiondjop et al. (2010), Sanchez et al. (2013)
	QTLs	*O. barthii*	AA	Ndjiondjop et al. (2010), Sanchez et al. (2013)
	QTLs	*O. meridionalis*	AA	Ndjiondjop et al. (2010), Sanchez et al. (2013)
	QTLs	*O. australiensis*	EE	Ndjiondjop et al. (2010), Sanchez et al. (2013)
	QTLs	*O. longistaminata*	AA	Ndjiondjop et al. (2010), Sanchez et al. (2013)

Modified from Brar and Kush (2018)

by reducing pesticide usages, which affects our biodiversity and ecosystem directly or indirectly. Rice disease and insect resistance have been developed using recent advances in QTL and genotype detection methods, marker-assisted selection (MAS), and sophisticated genomic approaches (Table 21.4) (Kumar et al. 2018b). This will provide valuable information to develop future climate-smart rice varieties.

Table 21.4 List of major genes and QTLs for biotic stress resistance identified from wild species of rice

Trait	Gene/ QTLs	Wild species	Genome	References
Biotic resistance				
Bacterial blight	Xa21	*O. longistaminata*	AA	Huang et al. (1997)
	Xa32	*O. australiensis*	EE	Zheng et al. (2009)
	Xa23	*O. rufipogon*	AA	Yinglun et al. (2006)
	Xa27	*O. minuta*	BBCC	Gu et al. (2004)
	Xa29	*O. officinalis*	CC	Tan et al. (2004)
	Xa30	*O. rufipogon*	AA	Xuwei et al. (2007)
	xa32	*O. australiensis*	EE	Zheng et al. (2009)
	xa32	*O. meyeriana*	GG	Ruan et al. (2008)
	Xa33	*O. nivara*	AA	Kumar et al. (2012)
	Xa35	*O. minuta*	BBCC	Guo et al. (2010)
	Xa38	*O. nivara*	AA	Bhasin et al. (2012)
	xa45(t)	*O. glaberrima*	AA	Neelam et al. (2020)
	qBBR1	*O. meyeriana*	GG	Han et al. (2014)
	qBBR3	*O. meyeriana*	GG	Han et al. (2014)
	qBBR5	*O. meyeriana*	GG	Chen et al. (2012)
Rice blast	Pi54rh	*O. rhizomatis*	CC	Das et al. (2012)
	Pi57	*O. longistaminata*	AA	Xu et al. (2015)
	Pi68	*O. glumaepatula*	AA	Devi et al. (2020)
	Pi69(t)	*O. glaberrima*	AA	Dong et al. (2020)
	qBLAST8	*O. nivara*	AA	Eizenga et al. (2013)
	qBLAST12	*O. nivara*	AA	Eizenga et al. (2013)
	qKI307–2	*O. minuta*	BBCC	Rahman et al. (2011)
Sheath blight	qShB6	*O. nivara*	AA	Eizenga et al. (2013)
Brown planthopper (BPH)	Bph10	*O. australiensis*	EE	Ishii et al. (1994), Zhao et al. (2016)
	Bph10	*O. officinalis*	CC	Thi Lang and Chi Buu (2003)
	bph11	*O. officinalis*	CC	Hirabayashi (1998)
	Bph12	*O. latifolia*	CCDD	Qiu et al. (2012)
	Bph13	*O. officinalis*	CC	Renganayaki et al. (2002)
	Bph14	*O. officinalis*	CC	Du et al. (2009)
	Bph15	*O. officinalis*	CC	Yang et al. (2004), Cheng et al.(2013)
	Bph16(t)	*O. minuta*	BBCC	Hirabayashi et al. (2004)
	Bph18	*O. australiensis*	EE	Ji et al. (2016b)
	Bph18(t)	*O. rufipogon*	AA	Li et al. (2010)
	Bph19(t)	*O. rufipogon*	AA	Li et al. (2010)
	Bph20(t)	*O. minuta*	BBCC	Rahman et al. (2009)
	bph20(t)	*O. rufipogon*	AA	Yang et al. (2012)

(continued)

Table 21.4 (continued)

Trait	Gene/ QTLs	Wild species	Genome	References
	Bph21(t)	*O. minuta*	BBCC	Rahman et al. (2009), Zhao et al. (2016)
	Bph21(t)	*O. rufipogon*	AA	Yang et al. (2012)
	Bph22(t)	*O. minuta*	BBCC	Harini et al. (2010)
	Bph22(t)	*O. glaberrima*	AA	Ram et al. (2010)
	bph22(t)	*O. rufipogon*	AA	Hou et al. (2011)
	Bph23(t)	*O. minuta*	BBCC	Ram et al. (2010)
	bph23(t)	*O. rufipogon*	AA	Hou et al. (2011)
	bph24(t)	*O. rufipogon*	AA	Deen et al. (2010)
	Bph27	*O. rufipogon*	AA	Huang et al. (2013)
	bph29	*O. rufipogon*	AA	Wang et al. (2015)
	bph30	*O. rufipogon*	AA	Yang et al. (2012)
	Bph34	*O. nivara*	AA	Kumar et al. (2018b)
	Bph36	*O. rufipogon*	AA	Li et al. (2019)
	Bph38	*O. rufipogon*	AA	Yang et al. (2020)
	qBph3	*O. officinalis*	CC	Hu et al. (2015)
	qBph4	*O. officinalis*	CC	Hu et al. (2015)
	qBph4.2	*O. australiensis*	EE	Hu et al. (2015)
White-backed planthopper (WBPH)	Wbph7	*O. officinalis*	CC	Tan et al. (2004)
	Wbph8	*O. officinalis*	CC	Tan et al. (2004)
	qWbph2	*O. rufipogon*	AA	Chen et al. (2010)
	qWbph5	*O. rufipogon*	AA	Chen et al. (2010)
	qWbph9	*O. rufipogon*	AA	Chen et al. (2010)
Small brown planthopper	qSbph3d	*O. officinalis*	CC	Zhang et al. (2014)
	qSbph7a	*O. officinalis*	CC	Zhang et al. (2014)
	qSbph12b	*O. officinalis*	CC	Zhang et al. (2014)
Green rice leafhopper	Grh4	*O. nivara*	AA	Fujita (2004)
	Grh5	*O. rufipogon*	AA	Fujita et al. (2006)
	Grh6	*O. nivara*	AA	Phi et al. (2019)
	qGrh9	*O. glaberrima*	AA	Fujita et al. (2010)
Grassy stunt resistance	Gs	*O. nivara*	AA	Khush and Ling (1974)
Tungro tolerance	Unknown	*O. rufipogon*	AA	Unknown
Green leafhopper (GLH)	Glh11	*O. glaberrima*	AA	Angeles and Khush (2000)
Gall midge	Gm1	*O. australiensis*	EE	Biradar et al. (2004)
	Gm4	*O. australiensis*	EE	Nair et al. (1996)

Modified from Gaikwad et al. (2021)

Rice blast, caused by the fungal pathogen *Magnaporthe oryzae*, is one of the most frequent and destructive rice diseases, posing a significant challenge to world food security (Khush and Jena 2009). Rice crops are infected by the pathogen at all stages of development, from seedling to grain filling. Lesions appear on plant sections above ground level as disease symptoms, leading to reduced yield up to 52% (Babujee and Gnanamanickham 2000; Van Oort and Zwart 2018). Host resistance to this disease is well documented, with over 100 blast resistance (R) genes and approximately 500 QTLs discovered so far (Ashkani et al. 2015).

Studies were done by Das et al. (2012), Wu et al. (2015), and Zheng et al. (2016) who cloned and described 28 genes. Among those characterized genes, Pi9 is a key resistant gene that has been identified in the wild species *Oryza minuta* (Sitch et al. 1989; Amante-Bordeos et al. 1992) and provides broad-spectrum resistance to various *M. oryzae* isolates (Qu et al. 2006).

Another serious disease in rice is bacterial leaf blight (BLB), caused by *Xanthomonas oryzae* pv. *Oryzae* results in 20–40% yield reductions at the maximum tillering stage and yields losses of about 50% at the initial stage (Yasmin et al. 2017). Until now, 43 genes have been found that confer BLB resistance (Busungu et al. 2016; Dilla-Ermita et al. 2017), as well as a few are being used to improve resistant lines using MAS. A long-lasting resistance to BLB was observed for pyramiding of xa5+ xa13+ Xa21 (Singh et al. 2001; Kottapalli et al. 2010; Pradhan et al. 2015; Dokku et al. 2013a, 2013b), xa13+ Xa21 (Arunakumari et al. 2016, Swathi et al. 2019, Joseph et al. 2004), Xa4+ xa5+ Xa21 (Jeung et al. 2006; Suh et al. 2015), and Xa4+ xa5+ xa13+ Xa21 (Das and Rao 2015).

The brown planthopper (BPH) is another damaging insect in rice cultivation. Numerous BPH-resistant genes have been discovered (Bph1, Bph3, Bph14, Bph15, Bph17, and Bph18), conferring resistance to this insect. Effective introgression of Bph3 (Qing et al. 2019), Bph14, and Bph15 (He et al. 2019; Wang et al. 2019) has been identified as providing greater resistance to the recurrent parent.

Another serious rice pest is gall midge, which causes significant yield reductions due to the silver shoot development (Bentur et al. 2003). Gall midge resistance genes Gm1, Gm2, Gm3, Gm4, Gm5, Gm6, Gm7, Gm8, Gm9, Gm10, and Gm11 have been discovered and used in rice breeding to combat various gall midge biotypes (Kumar et al. 2017).

The pyramiding of several resistance genes is an essential molecular breeding technique for achieving durable resistance to desired target diseases while also preventing or delaying resistance breakdown (Shinada et al. 2014; Feng et al. 2018; Kumar et al. 2018a; Liu et al. 2020). Ashkani et al. (2015) reviewed various effective molecular breeding strategies and gene pyramiding approaches to enhance biotic disease resistance genes in rice. Significantly, Ji et al. (2016a) utilized a gene pyramiding technique to develop restorer lines that were resistant to bacterial blight (Xa23 and Xa5), brown planthopper (Bph3), and blast (Pita, Pi1, and Pi2). Consequently, restorer lines are helpful in the production of hybrid rice.

21.8 Conservation of Wild Genetic Resources

Climate change is causing a significant threat to global agriculture. Therefore, it is essential to be aware of plant genetic resources, even though it has ignored conservation (Perrino and Perrino 2020). Furthermore, it is an essential element of functional genomics resources for food and agriculture (PGRFA) to produce crops for catering the food security with climate change. In situ and ex situ conservation measures should be used to ensure their accessibility for usage (Zair et al. 2021). As a result, international agreements are being formed to guarantee the security of genetic resources, and conservation efforts were already promoted to provide these resources more widely accessible (Heywood et al. 2007; Khoury et al. 2010; Maxted et al. 2010; Maxted et al. 2012). CWRs occupy around 2–6% of worldwide gene bank collections (Maxted and Kell 2009). Approximately 1.6% of those CWR species are the most urgently endangered species due to their significance in global food security (Maxted et al. 2012). The use of functional genomics in gene bank operations is becoming more widespread as sequencing and genotyping have transformed genomic resource conservation and utilization (McCouch et al. 2007). In contrast, other wild species have been investigated for their conservation and utilization (Song et al. 2005; Henry et al. 2010).

21.8.1 Ex Situ Conservation

Ex situ conservation, which may take seed banks, field gene banks, or botanic gardens, is the most frequent technique of germplasm conservation. Another alternative for genetic resource preservation is to conserve DNA in DNA banks (Henry et al. 2010). The size and scope of the gene bank collections show that the most genetic diversity has been collected. However, comprehensive gap analysis would assist in properly monitoring these gene banks and protecting existing resources, guiding future conservation strategies. This research type will support assessing whether a taxon's whole genetic variation is reflected in both in situ and ex situ conserving locations (Maxted et al. 2008a, b). Because using molecular data, ecogeographic surveys, and even more precise geographic referencing is essential in detecting limits and redundancy in current collections (Allender 2010), these techniques must also be used to assess and document genetic diversity and its conservation status more precisely and systematically. A comparison of ex situ diversity with in situ diversity will be highly informative in identifying potential collection gaps.

In contrast to other species, *Oryza* has a higher number of herbarium specimens than conserved germplasm accessions, which might indicate a limitation of ex situ conservation facilities. According to Hay et al. (2010), wild rice species have been seen collecting gaps in regions other than Asia and Africa. As a result, genotyping using molecular markers or sequencing will become the standard procedure before accession is deposited, ensuring that individuals employing unique alleles or allele sequences are entered into the collection. Aside from the ex situ collections, it is widely acknowledged that developing and maintaining in situ collections to provide additional protection to genetic resources are necessary.

21.8.2 In Situ Conservation

For decades, in situ conservation of wild species has been used as an alternative to ex situ conservation. It has been shown to have certain benefits, like enabling to continue the process of natural evolution. Despite the proven significance of in situ conservation, there are concerns about the alarming rates of genetic diversity loss among wild relatives of rice (Vaughan and Chang 1992; Song et al. 2005).

Multiple investigations have suggested programs in rice species (Song et al. 2005; Xie et al. 2010). According to Vaughan and Chang (1992), ex situ conservation of wild species with high genetic variation is costly and time consuming, highlighting the importance of in situ conservation. Extensive genetic analyses of wild rice species are essential for establishing in situ genetic reserves because they contribute to selecting priority sites for in situ conservation (Maxted and Kell 2009). Changing climate, overgrazing, floods, ecological shift, invasive species, and negative pollution impact *Oryza* populations, either in protected or in non-protected habitats. Climate change is anticipated to act as a significant challenge to wild species, demanding immediate actions to safeguard their genetic resources since the risk of extinction for endemic and narrowly adapted species is considerable. Some in situ populations that can survive the harsh effects of climate change may provide functional, promising characteristics that enable rice improvement (Allender 2010; Pettersson et al. 2009). Conversely, it is widely accepted that transferring genes from wild to cultivated species might have substantial and positive effects; gene flow, in the other way, can result in genetic diversity losses or even the extinction of tiny populations (Ellstrand et al. 1999).

21.8.3 On-Farm Conservation

During the cultivation, farmers discovered to preserve a substantial genetic diversity of wild rice within the farming system. This conservation method, also known as on-farm conservation, usually differs from the farmer/locally produced seed and small or no certified seed purchase from the official seed industry.

Furthermore, the system is strongly reliant on the usage of a wide range of traditional varieties, most of which have higher levels of genetic variation than developed forms. Although these landraces are low yielding, they can withstand a wide range of abiotic and biotic conditions, leading to yield stability. According to the investigation, farmers desire their yield stability to the highest possible yields as risk management (Wambugu et al. 2012). These landraces may contain valuable traits that help rice improvement. Thus, it is vital to protect them on-farm and ex situ and conduct genetic studies (Wambugu et al. 2013).

21.9 Future Perspectives

Crop yield is being limited due to global climate change, environmental constraints, and biotic factors. Fertile agricultural land is also decreasing due to fast urbanization and soil degradation. Furthermore, pandemic circumstances, such as the current outbreak of COVID-19, have caused significant food and nutritional shortages in the global population. As a consequence, ecologically friendly high-yielding cultivars must be introduced quickly.

The use of wild genetic resources of rice has received more attention because of the importance of CWRs for food and nutrition security and the creation of modern cultivars employing modern biotechnologies. The genetic diversity preserved in wild species appears to have far more potential for crop enhancement than previously thought. It will be facilitated to bridge genotype-phenotype gaps. Several QTL and SNPs related to agronomically and ecologically important characteristics in wild species have been discovered using linkage analysis, GWAS, and combined analysis of "omics" techniques and linkage mapping.

Further, allele mining may be viewed in the context of the tremendous potential and application of adequate genetic and genomic resources in modern rice breeding. It is critical to developing novel and efficient "mining" strategies to keep up with the increasing sequence data in GenBank and the growing rice gene banks. The current issue is integrating data from whole-genome sequencing, genome-scale SNP assays, and focused gene mapping investigations with phenotypic trait variation to better use genetic and genomic resources for crop enhancement. There is also a need to close the gap between channeling information regarding genes and gene expression to develop novel productive cultivars, which are an essential component of future food production.

Undoubtedly, *Oryza* genetic resources include a plethora of characteristics that have the potential to enhance cultivated rice. Because such resources are underrepresented in global germplasm banks and due to the challenges they face in the wild, coordinated regional and global efforts must ensure that they are collected and conserved correctly. Targeted CWRs may be collected and conserved in geographic areas with the most remarkable taxonomic diversity. Advanced biotechnologies are constantly being developed, such as genome editing and cisgenesis/intragenesis, which will speed up the protection and application of genetic diversity preserved in CWRs, leading to agricultural sustainability.

References

Addanki KR, Balakrishnan D, Yadavalli VR, Surapaneni M, Mesapogu S, Beerelli K (2019) Neelamraju S. Swarna × *Oryza nivara* introgression lines: a resource for seedling vigour traits in rice. Plant Genet 17:12–23

Akram R, Fahad S, Masood N, Rasool A, Ijaz M, Ihsan MZ, Maqbool MM, Ahmad S, Hussain S, Ahmed M, Kaleem S (2019) Plant growth and morphological changes in Rice under abiotic stress. In: Advances in Rice research for abiotic stress tolerance. Springer, Cham, pp 69–85

Allender C (2010) The second report on the state of the World's plant genetic resources for food and agriculture. Food and Agriculture Organization of the United Nations, Experimental Agriculture, Rome, Italy

Aluko G, Martinez C, Tohme J, Castano C, Bergman C, Oard J (2004) QTL mapping of grain quality traits from the interspecific cross *Oryza sativa* × *O. glaberrima*. Theor Appl Genet 109: 630–639

Amante-Bordeos A, Sitch LA, Nelson R, Dalmacio RD, Oliva NP, Aswidinnoor H, Leung H (1992) Transfer of bacterial blight and blast resistance from the tetraploid wild rice *Oryza minuta* to cultivated rice, *Oryza sativa*. Theoret Appl Genet 84:345–354

Angeles ER, Khush GS (2000) Genetics of resistance to green leafhopper in five cultivars of rice, *Oryza sativa* L. SABRAO J Breed Genet 32:1–4

Arunakumari K, Durgarani CV, Satturu V, Sarikonda KR, Chittoor PDR, Vutukuri B, Laha GS, Nelli APK, Gattu S, Jamal M, Prasadbabu A (2016) Marker-assisted pyramiding of genes conferring resistance against bacterial blight and blast diseases into Indian rice variety MTU1010. Rice Sci 23:306–316

Ashkani S, Rafi MY, Shabanimofrad M, Miah G, Sahebi M, Azizi P, Tanweer FA, Akhtar MS, Nasehi A (2015) Molecular breeding strategy and challenges towards improvement of blast disease resistance in rice crop. Front Plant Sci 6:886

Babujee L, Gnanamanickham SS (2000) Molecular tools for characterization of rice blast pathogen (*Magnaporthe grisea*) population and molecular marker assisted breeding for disease resistance. Curr Sci 78:24–25

Bailey-Serres J, Parker JE, Ainsworth EA, Oldroyd GE, Schroeder JI (2019) Genetic strategies for improving crop yields. Nature 575:109–118

Bakti C, Tanaka J (2019) Detection of dominant QTLs for stigma exsertion ratio in rice derived from *Oryza rufipogon* accession' W0120'. Breed Sci 69:143–150

Balakrishnan D, Surapaneni M, Yadavalli VR, Addanki KR, Mesapogu S, Beerelli K, Neelamraju S (2020) Detecting CSSLs and yield QTLs with additive, epistatic and QTL× environment interaction effects from *Oryza sativa* × *O. nivara* IRGC81832 cross. Sci Rep 10:1–17

Balkunde S, Le HL, Lee HS, Kim DM, Kang JW, Ahn SN (2013) Fine mapping of a QTL for the number of spikelets per panicle by using near-isogenic lines derived from an interspecific cross between *Oryza sativa* and *Oryza minuta*. Plant Breed 132:70–76

Banaticla-Hilario MC (2012). An ecogeographic analysis of *Oryza* series Sativae in Asia and the Pacific

Bao Y, Xu S, Jing X, Meng L, Qin Z (2015) De novo assembly and characterization of Oryza officinalis leaf transcriptome by using RNA-seq. Biomed Res Int 2015:982065

Barbier P (1989) Genetic variation and ecotypic differentiation in the wild rice species *Oryza rufipogon*. I. Population differentiation in life-history traits and isozymic loci. Jpn J Genet 64: 259–271

Bellard C, Bertelsmeier C, Leadley P, Thuiller W, Courchamp F (2012) Impacts of climate change on the future of biodiversity. Ecol Lett 15:365–377

Bentur JS, Pasalu IC, Sarma NP, Prasada Rao U, Mishra B (2003) Gall midge resistance in rice. Directorate of Rice Research, Hyderabad, India

Bernier J, Kumar A, Ramaiah V, Spaner D, Atlin G (2007) A large-effect QTL for grain yield under reproductive-stage drought stress in upland rice. Crop Sci 47:507–516

Bessho-Uehara K, Furuta T, Masuda K, Yamada S, Angeles-Shim RB, Ashikari M, Takashi T (2017) Construction of rice chromosome segment substitution lines harboring *Oryza barthii* genome and evaluation of yield-related traits. Breed Sci 67:408–415

Bhasin H, Bhatia D, Raghuvanshi S, Lore JS, Sahi GK, Kaur B, Vikal Y, Singh K (2012) New PCR-based sequence-tagged site marker for bacterial blight resistance gene Xa38 of rice. Mol Breed 30:607–611

Bhatia D, Wing RA, Yu Y, Chougule K, Kudrna D, Lee S, Rang A, Singh K (2018) Genotyping by sequencing of rice interspecific backcross inbred lines identifies QTLs for grain weight and grain length. Euphytica 214:1–6

Bierschenk B, Tagele MT, Ali B, Ashrafuzzaman MD, Wu LB, Becker M, Frei M (2020) Evaluation of rice wild relatives as a source of traits for adaptation to iron toxicity and enhanced grain quality. PLoS One 15:e0223086

Biradar SK, Sundaram RM, Thirumurugan T, Bentur JS, Amudhan S, Shenoy VV, Mishra B, Bennett J, Sarma NP (2004) Identification of flanking SSR markers for a major rice gall midge resistance gene Gm1 and their validation. Theor Appl Genet 109:1468–1473

Birky CW Jr (2001) The inheritance of genes in mitochondria and chloroplasts: laws, mechanisms, and models. Annu Rev Genet 35:125–148

Bonilla P, Dvorak J, Mackell D, Deal K, Gregorio G (2002) RFLP and SSLP mapping of salinity tolerance genes in chromosome 1 of rice (*Oryza sativa* L.) using recombinant inbred lines. Philippine Agricultural Scientist

Brar DS, Khush GS (2002) 14 transferring genes from wild species into Rice. CABI, Cambridge, UK

Brar DS, Khush GS (2006) Cytogenetic manipulation and germplasm enhancement of rice (*Oryza sativa* L.). Genetic Resour Chromosome Engineering Crop Improv 2:115–158

Brar DS, Khush GS (2018) Wild relatives of rice: a valuable genetic resource for genomics and breeding research. In: The wild *Oryza* genomes. Springer, Cham, pp 1–25

Breseghello F, Coelho AS (2013) Traditional and modern plant breeding methods with examples in rice (*Oryza sativa* L.). J Agric Food Chem 61:8277–8286

Brondani C, Rangel P, Brondani R, Ferreira M (2002) QTL mapping and introgression of yield-related traits from *Oryza glumaepatula* to cultivated rice (*Oryza sativa*) using microsatellite markers. Theor Appl Genet 104:1192–1203

Brozynska M, Furtado A, Henry RJ (2016) Genomics of crop wild relatives: expanding the gene pool for crop improvement. Plant Biotechnol 14:1070–1085

Brozynska M, Omar ES, Furtado A, Crayn D, Simon B, Ishikawa R, Henry RJ (2014) Chloroplast genome of novel rice germplasm identified in northern Australia. Trop Plant Biol 7:111–120

Brush SB (2000) The issues of *in-situ* conservation of crop genetic resources. Genes Field: On-Farm Conserv Crop Diversity 2000:3–26

Busungu C, Taura S, Sakagami JI, Ichitani K (2016) Identification and linkage analysis of a new rice bacterial blight resistance gene from XM14, a mutant line from IR24. Breed Sci 66:636–645

Cao Z, Li Y, Tang H, Zeng B, Tang X, Long Q, Wu X, Cai Y, Yuan L, Wan J (2020) Fine mapping of the qHTB1-1 QTL, which confers heat tolerance at the booting stage, using an *Oryza rufipogon* Griff. Introgression line. Theor Appl Genet 133:1161–1175

Chang TT (1976) The origin, evolution, cultivation, dissemination, and diversification of Asian and African rices. Euphytica 25:425–441

Chen G, Liu Y, Ma J, Zheng Z, Wei Y, McIntyre CL, Zheng YL, Liu C (2013) A novel and major quantitative trait locus for *fusarium* crown rot resistance in a genotype of wild barley (*Hordeum spontaneum* L.). PLoS One 8:e58040

Chen J, Huang DR, Wang L, Liu GJ, Zhuang JY (2010) Identification of quantitative trait loci for resistance to white-backed planthopper, *Sogatella furcifera*, from an interspecific cross *Oryza sativa* × *O. rufipogon*. Breed Sci 60:153–159

Chen LN, Yang Y, Yan CQ, Wang XM, Yu CL, Zhou J, Zhang WL, Cheng Y, Cheng XY, Chen JP (2012) Identification of quantitative trait loci for bacterial blight resistance derived from *Oryza meyeriana* and agronomic traits in recombinant inbred lines of *Oryza sativa*. Phytopathology 160:461–468

Cheng X, Wu Y, Guo J, Du B, Chen R, Zhu L, He G (2013) A rice lectin receptor-like kinase that is involved in innate immune responses also contributes to seed germination. Plant J 76:687–698

Choudhary M, Singh V, Muthusamy V, Wani SH (2017) Harnessing crop wild relatives for crop improvement. Int J Life Scin 6:73–85

Chu Z, Yuan M, Yao J, Ge X, Yuan B, Xu C, Li X, Fu B, Li Z, Bennetzen JL, Zhang Q (2006) Promoter mutations of an essential gene for pollen development result in disease resistance in rice. Genes Dev 20:1250–1255

Cohen SP, Leach JE (2019) Abiotic and biotic stresses induce a core transcriptome response in rice. Sci Rep 9:1–11

Comai L, Henikoff S (2006) TILLING: practical single-nucleotide mutation discovery. Plant J 45: 84–94

Comai L, Young K, Till BJ, Reynolds SH, Greene EA, Codomo CA, Enns LC, Johnson JE, Burtner C, Odden AR, Henikoff S (2004) Efficient discovery of DNA polymorphisms in natural populations by Ecotilling. Plant J 37:778–786

Das A, Soubam D, Singh PK, Thakur S, Singh NK, Sharma TR (2012) A novel blast resistance gene, Pi54rh cloned from wild species of rice, *Oryza rhizomatis* confers broad spectrum resistance to *Magnaporthe oryzae*. Funct Integr Genomics 12:215–228

Das G, Rao GJ (2015) Molecular marker assisted gene stacking for biotic and abiotic stress resistance genes in an elite rice cultivar. Front Plant Sci 6:698

Das G, Rao GJ, Varier M, Prakash A, Prasad D (2018) Improved Tapaswini having four BB resistance genes pyramided with six genes/QTLs, resistance/ tolerance to biotic and abiotic stresses in rice. Sci Rep 8:2413

Deen R, Ramesh K, Gautam SK, Rao YK, Lakshmi VJ, Viraktamath BC, Brar DS, Ram T (2010) Identification of new gene for BPH resistance introgressed from *O. rufipogon*

Dempewolf H, Baute G, Anderson J, Kilian B, Smith C, Guarino L (2017) Past and future use of wild relatives in crop breeding. Crop Sci 57:1070–1082

Dempewolf H, Hodgins KA, Rummell SE, Ellstrand NC, Rieseberg LH (2012) Reproductive isolation during domestication. Plant Cell 24:2710–2717

Des Marais DL, Hernandez KM, Juenger TE (2013) Genotype-by-environment interaction and plasticity: exploring genomic responses of plants to the abiotic environment. Annu Rev Ecol Evol Syst 44:5–29

Devi SR, Singh K, Umakanth B, Vishalakshi B, Rao KV, Suneel B, Sharma SK, Kadambari GK, Prasad MS, Senguttvel P, Syamaladevi DP (2020) Identification and characterization of a large effect QTL from *Oryza glumaepatula* revealed Pi68 (t) as putative candidate gene for rice blast resistance. Rice 13:1–3

Dilla-Ermita CJ, Tandayu E, Juanillas VM, Detras J, Lozada DN, Dwiyanti MS, Cruz CV, Mbanjo EG, Ardales E, Diaz MG, Mendioro M (2017) Genome-wide association analysis tracks bacterial leaf blight resistance loci in rice diverse germplasm. Rice 10:1–7

Dokku P, Das KM, Rao GJN (2013a) Genetic enhancement of host plant-resistance of the Lalat cultivar of rice against bacterial blight employing marker-assisted selection. Biotechnol Lett 35: 1339–1348

Dokku P, Das KM, Rao GJN (2013b) Pyramiding of four resistance genes of bacterial blight in Tapaswini, an elite rice cultivar, through marker-assisted selection. Euphytica 192:87–96

Dong L, Liu S, Kyaing MS, Xu P, Tharreau D, Deng W, Li X, Bi Y, Zeng L, Li J, Zhou J (2020) Identification and fine mapping of Pi69 (t), a new gene conferring broad-spectrum resistance against *Magnaporthe oryzae* from *Oryza glaberrima* Steud. Front Plant Sci 11:1190

Du B, Zhang W, Liu B, Hu J, Wei Z, Shi Z, He R, Zhu L, Chen R, Han B, He G (2009) Identification and characterization of Bph14, a gene conferring resistance to brown planthopper in rice. PNAS 106:22163–22168

Dwivedi SL, Upadhyaya HD, Stalker HT, Blair MW, Bertioli DJ, Nielen S, Ortiz R (2008) Enhancing crop gene pools with beneficial traits using wild relatives. Plant Breed Rev 30: 179–230

Edwards MA, Henry RJ (2011) DNA sequencing methods contributing to new directions in cereal research. J Cereal Sci 54:395–400

Eizenga GC, Neves PC, Bryant RJ, Agrama HA, Mackill DJ (2016) Evaluation of a M-202 × *Oryza nivara* advanced backcross mapping population for seedling vigor, yield components and quality. Euphytica 208:157–171

Eizenga GC, Prasad B, Jackson AK, Jia MH (2013) Identification of rice sheath blight and blast quantitative trait loci in two different *O. sativa*/*O. nivara* advanced backcross populations. Mol Breed 31:889–907

Eizenga GC, Sanchez PL, Jackson AK, Edwards JD, Hurwitz BL, Wing RA, Breeding KD (2017) Genetic variation for domestication-related traits revealed in a cultivated rice, Nipponbare (*Oryza sativa* ssp. *japonica*) × ancestral rice, *O. nivara*, mapping population. Mol 37:1–22

Ellstrand NC (2003) Dangerous liaisons?: when cultivated plants mate with their wild relatives. JHU Press, Baltimore, MD

Ellstrand NC, Prentice HC, Hancock JF (1999) Gene flow and introgression from domesticated plants into their wild relatives. Annu Rev Ecol Evol Syst 30:539–563

Fahad S, Bajwa AA, Nazir U, Anjum SA, Farooq A, Zohaib A, Sadia S, Nasim W, Adkins S, Saud S, Ihsan MZ (2017) Crop production under drought and heat stress: plant responses and management options. Front Plant Sci 8:1147

Feng B, Chen K, Cui Y, Wu Z, Zheng T, Zhu Y, Ali J, Wang B, Xu J, Zhang W, Li Z (2018) Genetic dissection and simultaneous improvement of drought and low nitrogen tolerances by designed QTL pyramiding in rice. Front Plant Sci 9:306

Ferrer MC, Duldulao MD, Caguiat XG, Mananghaya TE, Newingham MC, Nombrere JM, Castro JR, Alfonso DO, Regalario JB, Alvarino JB, Pacada IG (2020) PhilRice Genebank: recent developments in managing and sharing the Philippine rice germplasm. In IOP Conference Series: Environ Earth Sci 482:012010

Fiume E, Christou P, Gianì S, Breviario D (2004) Introns are key regulatory elements of rice tubulin expression. Planta 218:693–703

Fogliatto S, Serra F, Patrucco L, Milan M, Vidotto F (2019) Effect of different water salinity levels on the germination of imazamox-resistant and sensitive weedy rice and cultivated rice. Agron J 9:658

Fu Q, Zhang P, Tan L, Zhu Z, Ma D, Fu Y, Zhan X, Cai H, Sun C (2010) Analysis of QTLs for yield-related traits in Yuanjiang common wild rice (*Oryza rufipogon* Griff.). JGG 37:147–157

Fujita D (2004) Introgression of a resistance gene for green rice leafhopper from *Oryza nivara* into cultivated rice, *Oryza sativa* L. Rice Genet Newsl 21:64–66

Fujita D, Doi K, Yoshimura A, Yasui H (2006) Molecular mapping of a novel gene, Grh5, conferring resistance to green rice leafhopper (*Nephotettix cincticeps Uhler*) in rice, *Oryza sativa* L. Theor Appl Genet 113:567–573

Fujita D, Doi K, Yoshimura A, Yasui H (2010) A major QTL for resistance to green rice leafhopper (Nephotettix cincticeps Uhler) derived from African rice (Oryza glaberrima Steud.). Breed Sci 60:336–341

Gaikwad KB, Singh N, Bhatia D, Kaur R, Bains NS, Bharaj TS, Singh K (2014) Yield-enhancing heterotic QTL transferred from wild species to cultivated rice *Oryza sativa* L. PLoS One 9: e96939

Gaikwad KB, Singh N, Bhatia D, Kaur R, Bains NS, Bharaj TS, Singh K (2019) Reinventing heterosis phenomenon through deployment of alien introgression lines in rice (*Oryza sativa* L.). Plant Breed 138:277–289

Gaikwad KB, Singh N, Bhatia D, Sharma N, Bains NS, Bharaj TS, Singh K (2018) Heterotic response of genomic regions derived from *Oryza rufipogon* and *O. nivara* in improving grain morphology and quality of *indica* rice (*Oryza sativa* L.). Indian J Genet 78:155–165

Gaikwad KB, Singh N, Kaur P, Rani S, Babu HP, Singh K (2021) Deployment of wild relatives for genetic improvement in rice (Oryza sativa). Plant Breed 140:23–52

Garg R, Verma M, Agrawal S, Shankar R, Majee M, Jain M (2014) Deep transcriptome sequencing of wild halophyte rice, *Porteresia coarctata*, provides novel insights into the salinity and submergence tolerance factors. DNA Res 21:69–84

Gichuhi E, Himi E, Takahashi H, Maekawa M (2016) Characterization and QTL analysis of *Oryza longistaminata* introgression line, pLIA-1, derived from a cross between *Oryza longistaminata* and *Oryza sativa* (Taichung 65) under non-fertilized conditions. J Rice Res 4:174

Godfray HC, Beddington JR, Crute IR, Haddad L, Lawrence D, Muir JF, Pretty J, Robinson S, Thomas SM, Toulmin C (2010) Food security: the challenge of feeding 9 billion people. Sci 327:812–818

Grillo MA, Li C, Fowlkes AM, Briggeman TM, Zhou A, Schemske DW, Sang T (2009) Genetic architecture for the adaptive origin of annual wild rice, *Oryza nivara*. Evolution 63:870–883

Gu K, Tian D, Yang FW, Wu L, Sreekala C, Wang D, Wang GL, Yin Z (2004) High-resolution genetic mapping of Xa27 (t), a new bacterial blight resistance gene in rice, *Oryza sativa* L. Theor Appl Genet 108:800–807

Guo S, Zhang D, Lin X (2010) Identification and mapping of a novel bacterial blight resistance gene Xa35 (t) originated from *Oryza minuta*. Sci Agric Sin 43:2611–2618

Guttikonda H, Thummala SR, Agarwal S, Mangrauthia SK, Ramanan R, Neelamraju S (2020) Genome-wide transcriptome profile of rice hybrids with and without *Oryza rufipogon* introgression reveals candidate genes for yield. Sci Rep 10:1–15

Han X, Yang Y, Wang X, Zhou J, Zhang W, Yu C, Cheng C, Cheng Y, Yan C, Chen J (2014) Quantitative trait loci mapping for bacterial blight resistance in rice using bulked segregant analysis. Int J Mol Sci 15:11847–11861

Harini AS, Lakshmi SS, Kumar SS, Sivaramakrishnan S, Kadirvel P (2010) Validation and fine-mapping of genetic locus associated with resistance to brown planthopper [Nilaparvata lugens (Stal.)] in rice (*Oryza sativa* L.). Asian J Biol Sci 5:32–37

Haritha G, Swamy BP, Naik ML, Jyothi B, Divya B, Malathi S, Sarla N (2018) Yield traits and associated marker segregation in elite introgression lines derived from *O. sativa* × *O. nivara*. Rice Sci 25:19–31

Harlan JR, De Wet JM (1971) Toward a rational classification of cultivated plants. Taxon 20:509–517

Hasan MM, Rafii MY, Ismail MR, Mahmood M, Rahim HA, Alam MA, Ashkani S, Malek MA, Latif MA (2015) Marker-assisted backcrossing: a useful method for rice improvement. Biotechnol Biotechnol Equip 29:237–254

Hay FR, Hamilton NR, Furman BJ, Upadhayaya HD, Reddy KN, Singh SK (2010) Conservation of tropical plant species. In: Plant taxonomy. Springer, New York

He C, Xiao Y, Yu J, Li J, Meng Q, Qing X, Xiao G (2019) Pyramiding Xa21, Bph14, and Bph15 genes into the elite restorer line Yuehui9113 increases resistance to bacterial blight and brown planthopper in rice. J Crop Prot 115:31–39

He N, Wu R, Pan X, Peng L, Sun K, Zou T, Zhu H, Zeng R, Liu Z, Liu G, Wang S (2017) Development and trait evaluation of chromosome single-segment substitution lines of *O. meridionalis* in the background of *O. sativa*. Euphytica 213(281)

Henry RJ (2014) Sequencing of wild crop relatives to support the conservation and utilization of plant genetic resources. Plant Genet Resour 12(S1):S9–S11

Henry RJ, Rice N, Waters DL, Kasem S, Ishikawa R, Hao Y, Dillon S, Crayn D, Wing R, Vaughan D (2010) Australian *Oryza*: utility and conservation. Rice 3:235–241

Heywood V, Casas A, Ford-Lloyd B, Kell S, Maxted N (2007) Conservation and sustainable use of crop wild relatives. Agric Ecosyst Environ 121:245–255

Hirabayashi H (1998) Identification of brown planthopper resistance gene derived from *O. officinalis* using molecular markers in rice. Breed Sci:48:82

Hirabayashi H, Ideta O, Sato H, Takeuchi Y, Ando I, Nemoto H, Imbe T, Brar DS, Ogawa T (2004) Identification of a resistance gene to brown planthopper derived from *Oryza minuta* in rice. Breed Res 6:285

Hirosawa S, Takumi S, Ishii T, Kawahara T, Nakamura C Mori N (2004) Chloroplast and nuclear DNA variation in common wheat: insight into the origin and evolution of common wheat. Genes Genet Syst 79:271–282

Hou LY, Ping YU, Qun XU, Yuan XP, Yu HY, Wang YP, Wang CH, Guo WA, Tang SX, Peng ST, Wei XH (2011) Genetic analysis and preliminary mapping of two recessive resistance genes to brown planthopper, *Nilaparvata lugens* Stål in rice. Rice Sci 18:238–242

Hu BL, Xie JK, Wan Y, Zhang JW, Zhang FT, Li X (2016) Mapping QTLs for fertility restoration of different cytoplasmic male sterility types in rice using two *Oryza sativa* × *O. rufipogon* backcross inbred line populations. BioMed Res Int 2016:9236573

Hu J, Xiao C, Cheng MX, Gao GJ, Zhang QL, He YQ (2015) A new finely mapped *Oryza australiensis*-derived QTL in rice confers resistance to brown planthopper. Gene 561:132–137
Hu Z, Cao L, Sun X, Zhu Y, Zhang T, Jiang L, Liu Y, Dong S, Sun D, Yang J, He H (2018) Fine mapping of a major quantitative trait locus, qgnp7 (t), controlling grain number per panicle in African rice (*Oryza glaberrima* S.). Breed Sci 68:606–613
Huang D, Qiu Y, Zhang Y, Huang F, Meng J, Wei S, Li R, Chen B (2013) Fine mapping and characterization of BPH27, a brown planthopper resistance gene from wild rice (*Oryza rufipogon* Griff.). Theor Appl Genet 126:219–229
Huang N, Angeles ER, Domingo J, Magpantay G, Singh S, Zhang G, Kumaravadivel N, Bennett J, Khush GS (1997) Pyramiding of bacterial blight resistance genes in rice: marker-assisted selection using RFLP and PCR. Theor Appl Genet 95:313–320
Huang P, Schaal BA (2012) Association between the geographic distribution during the last glacial maximum of Asian wild rice, *Oryza rufipogon* (Poaceae), and its current genetic variation. Am J Bot 99:1866–1874
Huang X, Kurata N, Wang ZX, Wang A, Zhao Q, Zhao Y, Liu K, Lu H, Li W, Guo Y, Lu Y (2012) A map of rice genome variation reveals the origin of cultivated rice. Nature 490:497–501
Hussain K, Yingxing ZH, Anley W, Riaz A, Abbas A, Rani MH, Hong WA, Xihong SH, Liyong CA, Shihua CH (2020) Association mapping of quantitative trait loci for grain size in introgression line derived from *Oryza rufipogon*. Rice Sci 27:246–254
Iftekharuddaula KM, Ahmed HU, Ghosal S, Moni ZR, Amin A, Ali MS (2015) Development of new submergence tolerant rice variety for Bangladesh using marker-assisted backcrossing. Rice Sci 22:16–26
Ishii T, Brar DS, Multani DS, Khush GS (1994) Molecular tagging of genes for brown planthopper resistance and earliness introgressed from *Oryza australiensis* into cultivated rice, *O sativa*. Genome 37:217–221
Isshiki M, Morino K, Nakajima M, Okagaki RJ, Wessler SR, Izawa T, Shimamoto K (1998) A naturally occurring functional allele of the rice waxy locus has a GT to TT mutation at the 5′ splice site of the first intron. Plant J 15:133–138
Jackson MT, Ford-Lloyd B, Parry ML (1990) Climatic change and plant genetic resources. Belhaven Press, London, pp 1–17
Jacquemin J, Bhatia D, Singh K, Wing RA (2013) The international *Oryza* map alignment project: development of a genus-wide comparative genomics platform to help solve the 9 billion-people question. Curr Opin Plant Biol 16:147–156
Jamwal G, Bijral IS (2020) Comparative genomics for exploring new genes and traits for crop improvement. In: Rediscovery of genetic and genomic resources for future food security. Springer, Cham, pp 251–261
Jena KK (1994) Production of intergeneric hybrid between *Oryza sativa* L. and *Porteresia coarctata* T. Curr Sci:744–746
Jena KK (2010) The species of the genus *Oryza* and transfer of useful genes from wild species into cultivated rice, *O sativa*. Breed Sci 60:518–523
Jena KK, Hechanova SL, Verdeprado H, Prahalada GD, Kim SR (2017) Development of 25 near-isogenic lines (NILs) with ten BPH resistance genes in rice (*O. sativa* L.): production, resistance spectrum, and molecular analysis. Theoret Appl Genet 130:2345–2360
Jeung JU, Heu SG, Shin MS, Vera Cruz CM, Jena KK (2006) Dynamics of *Xanthomonas oryzae pv. Oryzae* populations in Korea and their relationship to known bacterial blight resistance genes. Phytopathology 96:867–875
Ji SD, Yuan PR, Xiao L, Lee HS, Ahn SN (2012) Characterization of quantitative trait loci for number of primary branches in near-isogenic lines from a cross between the *Oryza* sativa cultivar 'Hwayeongbyeo' and the wild relative *Oryza rufipogon*. Plant Breed 131:48–53
Ji H, Kim SR, Kim YH, Suh JP, Park HM, Sreenivasulu N, Misra G, Kim SM, Hechanova SL, Kim H, Lee GS (2016a) Map-based cloning and characterization of the BPH18 gene from wild rice conferring resistance to brown planthopper (BPH) insect pest. Sci Rep 6:1–4

Ji ZJ, Yang SD, Zeng YX, Liang Y, Yang CD, Qian Q (2016b) Pyramiding blast, bacterial blight and brown planthopper resistance genes in rice restorer lines. J Integr Agric 15:1432–1440

Jin FX, Kim DM, Ju HG, Ahn SN (2009) Mapping quantitative trait loci for awnness and yield component traits in isogenic lines derived from an *Oryza sativa* /*O. rufipogon* cross. J Crop Sci Biotechnol 12:9–15

Joseph M, Gopalakrishnan S, Sharma RK, Singh VP, Singh AK, Singh NK, Mohapatra T (2004) Combining bacterial blight resistance and basmati quality characteristics by phenotypic and molecular marker-assisted selection in rice. Mol Breed 13:377–387

Kaladhar K, Swamy BP, Babu AP, Reddy CS, Sarla N (2008) Mapping quantitative trait loci for yield traits in BC2F2 population derived from Swarna x *O. nivara* cross. Rice Genet Newslett 24:34–36

Kane N, Sveinsson S, Dempewolf H, Yang JY, Zhang D, Engels JM, Cronk Q (2012) Ultra-barcoding in cacao (*Theobroma* spp.; Malvaceae) using whole chloroplast genomes and nuclear ribosomal DNA. Am J Bot 99:320–329

Kang YJ, Shim KC, Lee HS, Jeon YA, Kim SH, Kang JW, Yun YT, Park IK, Ahn SN (2018) Fine mapping and candidate gene analysis of the quantitative trait locus gw8. 1 associated with grain length in rice. Genes and genomics 40:389–397

Kaur A, Sidana K, Bhatia D, Neelam K, Singh G, Sahi GK, Gill BK, Sharma P, Yadav IS, Singh K (2018) A novel QTL qSPP2. 2 controlling spikelet per panicle identified from *Oryza longistaminata* (a. Chev. et Roehr.), mapped and transferred to *Oryza sativa* (L.). Mol Breed 38:1–3

Khoury C, Laliberté B, Guarino L (2010) Trends in *ex-situ* conservation of plant genetic resources: a review of global crop and regional conservation strategies. Genet Resour Crop Evol 57:625–639

Khush GS (1997) Origin, dispersal, cultivation and variation of rice. Plant Mol Biol 35:25–34

Khush GS (2005) What it will take to feed 5.0 billion rice consumers in 2030. Plant Mol Biol 59: 1–6

Khush GS, Jena KK (2009) Current status and future prospects for research on blast resistance in rice (*Oryza sativa* L.). Adv Genet Genomics Control Rice Blast Dis 1:1–10

Khush GS, Ling KC (1974) Inheritance of resistance to grassy stunt virus and its vector in rice. J Hered 65:135–136

Koseki M, Kitazawa N, Yonebayashi S, Maehara Y, Wang ZX, Minobe Y (2010) Identification and fine mapping of a major quantitative trait locus originating from wild rice, controlling cold tolerance at the seedling stage. Mol Genet Genomics 284:45–54

Kottapalli KR, Narasu ML, Jena KK (2010) Effective strategy for pyramiding three bacterial blight resistance genes into fine grain rice cultivar, samba Mahsuri, using sequence tagged site markers. Biotechnol Lett 32:989–996

Kubota S, Konno I, Kanno A (2012) Molecular phylogeny of the genus asparagus (Asparagaceae) explains interspecific crossability between the garden asparagus (*A. officinalis*) and other asparagus species. Theor Appl Genet 124:345–354

Kumar A, Sandhu N, Dixit S, Yadav S, Swamy BP, Shamsudin NA (2018a) Marker-assisted selection strategy to pyramid two or more QTLs for quantitative trait-grain yield under drought. Rice 11:1–6

Kumar K, Sarao PS, Bhatia D, Neelam K, Kaur A, Mangat GS, Brar DS, Singh K (2018b) High-resolution genetic mapping of a novel brown planthopper resistance locus, Bph34 in *Oryza sativa* L. × *Oryza nivara* (Sharma and Shastry) derived interspecific F 2 population. Theor Appl Genet 131:1163–1171

Kumar PN, Sujatha K, Laha GS, Rao KS, Mishra B, Viraktamath BC, Hari Y, Reddy CS, Balachandran SM, Ram T, Madhav MS (2012) Identification and fine-mapping of Xa33, a novel gene for resistance to *Xanthomonas oryzae* pv. *Oryzae*. Phytopathology 102:222–228

Kumar R, Venuprasad R, Atlin GN (2007) Genetic analysis of rainfed lowland rice drought tolerance under naturally-occurring stress in eastern India: heritability and QTL effects. Field Crops Res 103:42–52

Kumar VA, Balachiranjeevi CH, Naik SB, Rekha G, Rambabu R, Harika G, Pranathi K, Hajira SK, Anila M, Kousik M, Kale R (2017) Marker-assisted pyramiding of bacterial blight and gall midge resistance genes into RPHR-1005, the restorer line of the popular rice hybrid DRRH-3. Mol Breed 37:86

Lafitte HR, Ismail A, Bennett J (2004) Abiotic stress tolerance in rice for Asia: progress and the future. In Proceeding of 4th International Crop Science Congress, Brisbane, Australia

Latha R, Rubia L, Bennett J, Swaminathan MS (2004) Allele mining for stress tolerance genes in *Oryza* species and related germplasm. Mol Biotechnol 27:101–108

Lee SJ, Oh CS, Suh JP, McCouch SR, Ahn SN (2005) Identification of QTLs for domestication-related and agronomic traits in an *Oryza sativa* × *O. rufipogon* BC1F7 population. Plant Breed 124:209–219

Lei D, Tan L, Liu F, Chen L, Sun C (2013) Identification of heat-sensitive QTL derived from common wild rice (*Oryza rufipogon* Griff.). Plant Sci 201:121–127

Lesk C, Rowhani P, Ramankutty N (2016) Influence of extreme weather disasters on global crop production. Nature 529:84–87

Li J, Xiao J, Grandillo S, Jiang L, Wan Y, Deng Q, Yuan L, McCouch SR (2004) QTL detection for rice grain quality traits using an interspecific backcross population derived from cultivated Asian (*O. sativa* L.) and African (*O. glaberrima* S.) rice. Genome 47:697–704

Li L, Chen H, Mao D (2019) Pyramiding of rapid germination loci from *Oryza Sativa* cultivar 'Xieqingzao B' and cold tolerance loci from Dongxiang wild rice to increase climate resilience of cultivated rice. Mol Breed 39:1–1

Li R, Li L, Wei S, Wei Y, Chen Y, Bai D, Yang L, Huang F, Lu W, Zhang X, Li X (2010) The evaluation and utilization of new genes for brown planthopper resistance in common wild rice (*Oryza rufipogon* Griff.). Mol Entomol 4:365–371

Liang Y, Yan C, Zheng J, Nan W, Qin X, Zhang H (2019) Locating QTL associated with spike traits of Dongxiang wild rice (*Oryza rufipogon* Griff.). Euphytica 215:26

Linh LH, Hang NT, Jin FX, Kang KH, Lee YT, Kwon SJ, Ahn SN (2008) Introgression of a quantitative trait locus for spikelets per panicle from *Oryza minuta* to the *O. sativa* cultivar Hwaseongbyeo. Plant Breed 127:262–267

Liu C, Ding S, Zhang A, Hong K, Jiang H, Yang S, Ruan B, Zhang B, Dong G, Guo L, Zeng D (2020) Development of nutritious rice with high zinc/selenium and low cadmium in grains through QTL pyramiding. J Integr Plant Biol 62:349–359

Liu R, Zheng XM, Zhou L, Zhou HF, Ge S (2015) Population genetic structure of *Oryza rufipogon* and *Oryza nivara*: implications for the origin of *O. nivara*. Mol Ecol 24:5211–5228

Liyanage AS, Senanayake G (2010) The atlas of selected crop wild relatives in Sri Lanka. Colombo: Department of Agriculture, Sri Lanka 11–73

Lobell DB, Burke MB, Tebaldi C, Mastrandrea MD, Falcon WP, Naylor RL (2008) Prioritizing climate change adaptation needs for food security in 2030. Sci 319:607–610

Long W, Dan D, Yuan Z, Chen Y, Jin J, Yang W, Zhang Z, Li N, Li S (2020) Deciphering the genetic basis of lodging resistance in wild rice *Oryza longistaminata*. Front Plant Sci 11:628

Lowry DB, Modliszewski JL, Wright KM, Wu CA, Willis JH (2008) The strength and genetic basis of reproductive isolating barriers in flowering plants. Philosophical Transactions of the Royal Society B: Biol Sci 363:3009–3021

Luo X, Wu S, Tian F, Xin X, Zha X, Dong X, Fu Y, Wang X, Yang J, Sun C (2011) Identification of heterotic loci associated with yield-related traits in Chinese common wild rice (*Oryza rufipogon* Griff.). Plant Sci 181:14–22

Luo XD, Jun ZHAO, Dai LF, Zhang FT, Zhou Y, Yong WAN, Xie JK (2016) Linkage map construction and QTL mapping for cold tolerance in *Oryza rufipogon* Griff. at early seedling stage. J Integr Agric 15:2703–2711

Ma X, Chen X, Zhao J, Wang S, Tan L, Sun C, Liu F (2019) Identification of QTLs related to cadmium tolerance from wild rice (*Oryza nivara*) using a high-density genetic map for a set of introgression lines. Euphytica 215:1–2

Ma X, Fu Y, Zhao X, Jiang L, Zhu Z, Gu P, Xu W, Su Z, Sun C, Tan L (2016) Genomic structure analysis of a set of *Oryza nivara* introgression lines and identification of yield-associated QTLs using whole-genome resequencing. Sci Rep 6:1–2

Mandal N, Gupta S (1997) Anther culture of an interspecific rice hybrid and selection of fine grain type with submergence tolerance. Plant Cell Tiss Org 51:79–82

Marri PR, Sarla N, Reddy LV, Siddiq EA (2005) Identification and mapping of yield and yield related QTLs from an Indian accession of *Oryza rufipogon*. BMC Genet 26:1–4

Maxted N, Dulloo E, Ford-Lloyd V, Iriondo JM, Jarvis A (2008a) Gap analysis: a tool for complementary genetic conservation assessment. Divers Distrib 14:1018–1030

Maxted N, Kell S, Ford-Lloyd B, Dulloo E, Toledo Á (2012) Toward the systematic conservation of global crop wild relative diversity. Crop Sci 52:774–785

Maxted N, Kell S, Toledo Á, Dulloo E, Heywood V, Hodgkin T, Hunter D, Guarino L, Jarvis A, Ford-Lloyd B (2010) A global approach to crop wild relative conservation: securing the gene pool for food and agriculture. Kew Bull 65:561–576

Maxted N, Kell SP (2009) Establishment of a global network for the in-situ conservation of crop wild relatives: status and needs, CGRFA. FAO, Paris, France

Maxted N, Kell SP, Ford-Lloyd BV, Iriondo JM, Dulloo E, Turok J (2008b) Crop wild relative conservation and use: establishing the context. In: Crop wild relative conservation and use. CABI, New York, pp 3–30

McCouch SR, Sweeney M, Li J, Jiang H, Thomson M, Septiningsih E, Edwards J, Moncada P, Xiao J, Garris A, Tai T (2007) Through the genetic bottleneck: *O. rufipogon* as a source of trait-enhancing alleles for *O. sativa*. Euphytica 154:317–339

McPherson H, Van der Merwe M, Delaney SK, Edwards MA, Henry RJ, McIntosh E, Rymer PD, Milner ML, Siow J, Rossetto M (2013) Capturing chloroplast variation for molecular ecology studies: a simple next generation sequencing approach applied to a rainforest tree. BMC Ecol 13:1–11

Molla KA, Debnath AB, Ganie SA, Mondal TK (2015) Identification and analysis of novel salt responsive candidate gene based SSRs (cgSSRs) from rice (*Oryza sativa* L.). BMC Plant Biol 15(122)

Moncada P, Martinez CP, Borrero J, Châtel M, Gauch H Jr, Guimaraes E, Tohmé J, McCouch SR (2001) Quantitative trait loci for yield and yield components in an *Oryza sativa* × *Oryza rufipogon* BC2F2 population evaluated in an upland environment. Theor Appl Genet 102:41–52

Morishima H, Oka HI (1960) The pattern of interspecific variation in the genus *Oryza*: its quantitative representation by statistical methods. Evolution:153–165

Morishma H, Sano Y, Oka H (1984) Differentiation of perennial and annual types due to habitat conditions in the wild rice *Oryza perennis*. Plant Syst Evol 144:119–135

Mukamuhirwa A, Persson Hovmalm H, Bolinsson H (2019) Concurrent drought and temperature stress in rice a possible result of the predicted climate change: effects on yield attributes, eating characteristics, and health promoting compounds. Int J Environ Res Public Health 16:1043

Muthu V, Abbai R, Nallathambi J, Rahman H, Ramasamy S, Kambale R, Thulasinathan T, Ayyenar B, Muthurajan R (2020) Pyramiding QTLs controlling tolerance against drought, salinity, and submergence in rice through marker assisted breeding. PLoS One 15:e0227421

Nair S, Kumar A, Srivastava MN, Mohan M (1996) PCR-based DNA markers linked to a gall midge resistance gene, Gm4t, has potential for marker-aided selection in rice. Theor Appl Genet 92:660–665

Ndjiondjop MN, Manneh B, Cissoko M, Drame NK, Kakai RG, Bocco R, Baimey H, Wopereis M (2010) Drought resistance in an interspecific backcross population of rice (*Oryza* spp.) derived from the cross WAB56-104 (*O. sativa*) × CG14 (*O. glaberrima*). Plant Sci 179:364–373

Neelam K, Mahajan R, Gupta V, Bhatia D, Gill BK, Komal R, Lore JS, Mangat GS, Singh K (2020) High-resolution genetic mapping of a novel bacterial blight resistance gene xa-45 (t) identified from *Oryza glaberrima* and transferred to *Oryza sativa*. Theor Appl Genet 133:689–705

Ngu MS, Thomson MJ, Bhuiyan MAR, Ho C, Wickneswari R (2014) Fine mapping of a grain weight quantitative trait locus, qGW6, using near isogenic lines derived from *Oryza rufipogon* IRGC105491 and *Oryza sativa* cultivar MR219. Genet Mol Res 13:9477–9488

Nguyen BD, Brar DS, Bui BC, Nguyen TV, Pham LN, Nguyen HT (2003) Identification and mapping of the QTL for aluminum tolerance introgressed from the new source, *Oryza rufipogon* Griff., into *indica* rice (*Oryza sativa* L.). Theor Appl Genet 106:583–593

Nock CJ, Waters DL, Edwards MA, Bowen SG, Rice N, Cordeiro GM, Henry RJ (2011) Chloroplast genome sequences from total DNA for plant identification. Plant Biotechnol J 9: 328–333

Oka HI (1988) Origin of cultivated rice. Elsevier, Amsterdam, pp 1–254

Pang Y, Chen K, Wang X, Wang W, Xu J, Ali J, Li Z (2017) Simultaneous improvement and genetic dissection of salt tolerance of rice (*O. sativa* L.) by designed QTL pyramiding. Front. Plant Sci 8:1275

Perrino EV, Perrino P (2020) Crop wild relatives: know how past and present to improve future research, conservation and utilization strategies, especially in Italy: a review. Genet Resour Crop Evol 67:1067–1105

Pettersson E, Lundeberg J, Ahmadian A (2009) Generations of sequencing technologies. Genomics 93:105–111

Phi CN, Fujita D, Yamagata Y, Yoshimura A, Yasui H (2019) High-resolution mapping of GRH6, a gene from *Oryza nivara* (Sharma et Shastry) conferring resistance to green rice leafhopper (*Nephotettix cincticeps* Uhler). Breed Sci 2019:19029

Pradhan SK, Nayak DK, Mohanty S, Behera L, Barik SR, Pandit E, Lenka S, Anandan A (2015) Pyramiding of three bacterial blight resistance genes for broadspectrum resistance in Deepwater rice variety. Jalmagna Rice 8:19

Qi L, Ding Y, Zheng X, Xu R, Zhang L, Wang Y, Wang X, Zhang L, Cheng Y, Qiao W, Yang Q (2018) Fine mapping and identification of a novel locus qGL12. 2 control grain length in wild rice (*Oryza rufipogon* Griff.). Theor Appl Genet 131:1497–1508

Qi L, Sun Y, Li J, Su L, Zheng X, Wang X, Li K, Yang Q, Qiao W (2017) Identify QTLs for grain size and weight in common wild rice using chromosome segment substitution lines across six environments. Breed Sci 67:472–482

Qiao W, Qi L, Cheng Z, Su L, Li J, Sun Y, Ren J, Zheng X, Yang Q (2016) Development and characterization of chromosome segment substitution lines derived from *Oryza rufipogon* in the genetic background of *O. sativa* spp. *indica* cultivar 9311. BMC Genet 17:1–2

Qing D, Dai G, Zhou W, Huang S, Liang H, Gao L, Gao J, Huang J, Zhou M, Chen R, Chen W (2019) Development of molecular marker and introgression of Bph3 into elite rice cultivars by marker-assisted selection. Breed Sci 69:40–46

Qiu Y, Guo J, Jing S, Zhu L, He G (2012) Development and characterizationof *japonica* rice lines carrying the brown planthopper-resistance genes BPH12 and BPH6. Theor Appl Ge Genetics 124:485–494

Qu S, Liu G, Zhou B, Bellizzi M, Zeng L, Dai L, Han B, Wang GL (2006) The broad-spectrum blast resistance gene Pi9 encodes a nucleotide-binding site – leucine-rich repeat protein and is a member of a multigene family in rice. Genetics 172:1901–1914

Rahman L, Khanam S, Roh JH, Koh HJ (2011) Mapping of QTLs involved in resistance to rice blast (*Magnaporthe grisea*) using *Oryza minuta* introgression lines. Czech J Genet Plant Breed 47:85–94

Rahman M, Chu SH, Choi MS, Qiao YL, Jiang W, Piao R, Khanam S, Cho YI, Jeung JU, Jena KK, Koh HJ (2007) Identification of QTLs for some agronomic traits in rice using an introgression line from Oryza minuta. In: Molecules and cells. Springer, Cham

Rahman ML, Jiang W, Chu SH, Qiao Y, Ham TH, Woo MO, Lee J, Khanam MS, Chin JH, Jeung JU, Brar DS (2009) High-resolution mapping of two rice brown planthopper resistance genes, Bph20 (t) and Bph21 (t), originating from *Oryza minuta*. Theor Appl Genet 119:1237–1246

Ram T, Deen R, Gautam SK, Ramesh K, Rao YK, Brar DS (2010) Identification of new genes for brown planthopper resistance in rice introgressed from *O. glaberrima* and *O. minuta*. Rice Genet Newsl 25:67–69

Ramos JM, Furuta T, Uehara K, Chihiro N, Angeles-Shim RB, Shim J, Brar DS, Ashikari M, Jena KK (2016) Development of chromosome segment substitution lines (CSSLs) of *Oryza longistaminata* a. Chev. and Röhr in the background of the elite *japonica* rice cultivar, Taichung 65 and their evaluation for yield traits. Euphytica 210:151–163

Rangan L, Constantino S, Khush GS, Swaminathan MS, Bennett J (1999) The feasibility of PCR-based allele mining for stress tolerance genes in rice and related germplasm. Rice Gene Newslett 16:43

Rangel PN, Brondani RP, Rangel PH, Brondani C (2008) Agronomic and molecular characterization of introgression lines from the interspecific cross *Oryza sativa* (BG90-2) × *Oryza glumaepatula* (RS-16). Genet Mol Res 7:184–195

Ranjan RU, Yadav R, Pandey R, Jain N, Bainsla NK, Gaikwad KB, Singh AM (2019) Variation in wheat (*Triticum aestivum*) advance lines and released cultivars for traits associated with nitrogen use efficiency under N limiting environment. Indian J Agric Sci 89:99–104

Renganayaki K, Fritz AK, Sadasivam S, Pammi S, Harrington SE, McCouch SR, Kumar SM, Reddy AS (2002) Mapping and progress toward map-based cloning of brown planthopper biotype-4 resistance gene introgressed from *Oryza officinalis* into cultivated rice, *O sativa*. Crop Sci 42:2112–2117

Roth M, Florez-Rueda AM, Griesser S, Paris M, Städler T (2018) Incidence and developmental timing of endosperm failure in post-zygotic isolation between wild tomato lineages. Ann Bot 121:107–118

Ruan HH, Yan CQ, An DR, Liu RH, Chen JP (2008) Identifying and mapping new gene xa32 (t) for resistance to bacterial blight (*Xanthomonas oryzae* pv *oryzae*, Xoo) from *Oryza meyeriana* L. Acta Agric Bor Sin 17:170–174

Sahebi M, Hanafi MM, Rafii MY, Mahmud TM, Azizi P, Osman M, Abiri R, Taheri S, Kalhori N, Shabanimofrad M, Miah G (2018) Improvement of drought tolerance in rice (*Oryza sativa* L.): genetics, genomic tools, and the WRKY gene family. Biomed Res Int 7

Saikumar S, Varma CMK, Saiharini A (2016) Grain yield responses to varied level of moisture stress at reproductive stage in an interspecific population derived from Swarna /*O. glaberrima* introgression line. J Life Sci 78:111–122

Salamini F, Özkan H, Brandolini A, Schäfer-Pregl R, Martin W (2002) Genetics and geography of wild cereal domestication in the near east. Nat Rev Genet 3:429–441

Samadder P, Sivamani E, Lu J, Li X, Qu R (2008) Transcriptional and post-transcriptional enhancement of gene expression by the 5′ UTR intron of rice rubi3 gene in transgenic rice cells. Mol Genet Genomics 79:429–439

Sanchez PL, Wing RA, Brar DS (2013) The wild relative of rice: genomes and genomics. In: Genetics and genomics of rice. Springer, Cham, pp 9–25

Sandamal S, Tennakoon A, Meng QL, Marambe B, Ratnasekera D, Melo A, Ge S (2018a) Population genetics and evolutionary history of the wild rice species *Oryza rufipogon* and *O. nivara* in Sri Lanka. Ecol Evol 8:12056–12065

Sandamal S, Tennakoon A, Ratnasekera D (2018b) Habitat fragmentation and isolation-by-distance driven population divergence of Oryza rufipogon in Sri Lanka. Forestry and Environment symposium. University of Sri Jayewardenepura, Sri Lanka

Sandhu N, Dixit S, Swamy BP, Raman A, Kumar S, Singh SP, Yadaw RB, Singh ON, Reddy JN, Anandan A, Yadav S (2019) Marker assisted breeding to develop multiple stress tolerant varieties for flood and drought prone areas. Rice 12:8

Sandhu N, Kumar A (2017) Bridging the rice yield gaps under drought: QTLs, genes, and their use in breeding programs. Agron 7:27

Sang T, Ge S (2007) Genetics and phylogenetics of rice domestication. Curr Opin Genet Dev 17: 533–538

Sarao PS, Sahi GK, Neelam K, Mangat GS, Patra BC, Singh K (2016) Donors for resistance to brown planthopper Nilaparvata lugens (Stål) from wild rice species. Rice Sci 23:219–224

Saski C, Lee SB, Fjellheim S, Guda C, Jansen RK, Luo H, Tomkins J, Rognli OA, Daniell H, Clarke JL (2007) Complete chloroplast genome sequences of Hordeum vulgare, Sorghum bicolor and Agrostis stolonifera, and comparative analyses with other grass genomes. Theor Appl Genet 115:571–590

Septiningsih EM, Trijatmiko KR, Moeljopawiro S, McCouch SR (2003) Identification of quantitative trait loci for grain quality in an advanced backcross population derived from *the Oryza sativa* variety IR64 and the wild relative *O. rufipogon*. Theor Appl Genet 107:1433–1441

Shinada H, Iwata N, Sato T, Fujino K (2014) QTL pyramiding for improving of cold tolerance at fertilization stage in rice. Breed Sci 63:483–488

Shrivastava P, Kumar R (2015) Soil salinity: a serious environmental issue and plant growth promoting bacteria as one of the tools for its alleviation. Saudi J Biol Sci 22:123–131

Singh RK, Prasad A, Muthamilarasan M, Parida SK, Prasad M (2020) Breeding and biotechnological interventions for trait improvement: status and prospects. Planta 252:1–8

Singh S, Sidhu JS, Huang N, Vikal Y, Li Z, Brar DS, Dhaliwal HS, Khush GS (2001) Pyramiding three bacterial blight resistance genes (xa5, xa13 and Xa21) using marker-assisted selection into *indica* rice cultivar PR106. Theoret Appl Genet 102:1011–1015

Sitch LA, Amante AD, Dalmacio RD, Leung H (1989) *Oryza minuta*, a source of blast and bacterial blight resistance for rice improvement. In: In 2. International symposium on genetic manipulation in crops, El Batan (Mexico)

Song Z, Li BO, Chen J, Lu BR (2005) Genetic diversity and conservation of common wild rice (*Oryza rufipogon*) in China. Plant Species Biol 20:83–92

Spooner D (2005) Molecular markers for genebank management. Biodivers Int J

Subbaiyan GK, Waters DL, Katiyar SK, Sadananda AR, Vaddadi S, Henry RJ (2012) Genome-wide DNA polymorphisms in elite *indica* rice inbreds discovered by whole-genome sequencing. Plant Biotechnol J 10:623–634

Suh JP, Cho YC, Won YJ, Ahn EK, Baek MK, Kim MK, Kim BK, Jena KK (2015) Development of resistant gene-pyramided *japonica* rice for multiple biotic stresses using molecular marker-assisted selection. Plant Breed Biotech 3:333–435

Sun D, Wang W, Xin X, Cao L, Sun X, Hu Z, Jiang L, Dong S, Liu Y, Yang J, Luo X (2019) Construction and evaluation of introgression lines and fine mapping of ehd8 from Jinghong common wild rice (*Oryza rufipogon*). Plant Breed 138:163–173

Surapaneni M, Balakrishnan D, Mesapogu S, Addanki KR, Yadavalli VR, Tripura Venkata VG, Neelamraju S (2017) Identification of major effect QTLs for agronomic traits and CSSLs in rice from Swarna/*Oryza nivara* derived backcross inbred lines. Front Plant Sci 8:1027

Swamy BM, Kaladhar K, Reddy GA, Viraktamath BC, Sarla N (2014) Mapping and introgression of QTL for yield and related traits in two backcross populations derived from *Oryza sativa* cv. Swarna and two accessions of *O. nivara*. J Genet 93:643–654

Swamy BM, Kaladhar K, Shobha Rani N, Prasad GS, Viraktamath BC, Reddy GA, Sarla N (2012) QTL analysis for grain quality traits in 2 BC2F2 populations derived from crosses between *Oryza sativa* cv Swarna and 2 accessions of *O. nivara*. J Hered 103:442–452

Swathi G, Rani CVD, Md J, Madhav MS, Vanisree S, Anuradha C, Kumar NR, Kumar NAP, Kumari KA, Bhogadhi SC, Ramprasad E (2019) Marker-assisted introgression of the major bacterial blight resistance genes, Xa21 and xa13 , and blast resistance gene, Pi54, into the popular rice variety, JGL1798. Mol Breed 39:58

Tan GX, Ren X, Weng QM, Shi ZY, Zhu LL, He GC (2004) Mapping of a new resistance gene to bacterial blight in rice line introgressed from *Oryza officinalis*. Acta genetica Sinica 31:724–729

Tan L, Liu F, Xue W, Wang G, Ye S, Zhu Z, Fu Y, Wang X, Sun C (2007) Development of *Oryza rufipogon* and *O. sativa* introgression lines and assessment for yield-related quantitative trait loci. J Integr Plant Biol 49:871–884

Tan Q, Zou T, Zheng M, Ni Y, Luan X, Li X, Yang W, Yang Z, Zhu H, Zeng R, Liu G (2020) Substitution mapping of the major quantitative trait loci controlling stigma exsertion rate from *Oryza glumaepatula*. Rice 13:1–10

Tennakoon A, Sandamal S, Ratnasekera D (2018) Genetic structure of common wild rice (Oryza nivara) populations from Sri Lanka. 4th international research symposium. Rajarata University of Sri Lanka Sri Lanka

Tennakoon A, Sandamal S, Ratnasekera D (2019) Habitat preference of common wild rice species in Sri Lanka by species distribution modelling techniques. International Conference on Climate Change and its Impact on Global Food Security and Sustainability of Agriculture, Institute of Sciences, Banaras Hindu University, Varanasi

Thalapati S, Batchu AK, Neelamraju S, Ramanan R (2012) Os11Gsk gene from a wild rice, *Oryza rufipogon* improves yield in rice. Funct Integr Genomics 12:277–289
Thi Lang N, Chi Buu B (2003) Genetic and physical maps of gene Bph10 controlling brown planthopper resistance in rice (*Oryza sativa* L.). Omonrice 11:35–41
Thomson MJ, de Ocampo M, Egdane J, Rahman MA, Sajise AG, Adorada DL, Tumimbang-Raiz E, Blumwald E, Seraj ZI, Singh RK, Gregorio GB (2010) Characterizing the Saltol quantitative trait locus for salinity tolerance in rice. Rice 3:148–160
Thomson MJ, Tai TH, McClung AM, Lai XH, Hinga ME, Lobos KB, Xu Y, Martinez CP, McCouch SR (2003) Mapping quantitative trait loci for yield, yield components and morphological traits in an advanced backcross population between *Oryza rufipogon* and the *Oryza sativa* cultivar Jefferson. Theor Appl Gen 107:479–493
Thormann I, Parra-Quijano M, Endresen DTF, Rubio-Teso, ML, Iriondo JM, Maxted N (2014) Predictive characterization of crop wild relatives and landraces: Technical guidelines version 1
Tian F, Fu Q, Zhu ZF, Fu YC, Wang XK, Sun CQ (2006) Construction of introgression lines carrying wild rice (*Oryza rufipogon* Griff.) segments in cultivated rice (*Oryza sativa* L.) background and characterization of introgressed segments associated with yield-related traits. Theor Appl Genet 112:570–580
Till BJ, Reynolds SH, Greene EA, Codomo CA, Enns LC, Johnson JE, Burtner C, Odden AR, Young K, Taylor NE, Henikoff JG (2003) Large-scale discovery of induced point mutations with high-throughput TILLING. Genome Res 13:524–530
Turelli M, Barton NH, Coyne JA (2001) Theory and speciation. Trends Ecol Evol 16:330–343
Ullah I, Jamil S, Iqbal MZ, Shaheen HL, Hasni SM, Jabeen S, Mehmood A, Akhter M (2012) Detection of bacterial blight resistance genes in basmati rice landraces. Genet Mol Res 11:1960–1966
United Nations (2019) World population prospects 2019: highlights. United Nations Department of Economic and Social Affairs, New York
Van Oort PA, Zwart SJ (2018) Impacts of climate change on rice production in Africa and causes of simulated yield changes. Glob Chang Biol 24:1029–1045
Vaughan D, Morishima H, Kadowaki K (2003) Diversity in the *Oryza* genus. Curr Opin Plant Biol 6:139–146
Vaughan DA (1989) Genus *Oryza* L. Current status of taxonomy. IRRI Research Paper Series (Philippines)
Vaughan DA, Chang TT (1992) *In-situ* conservation of rice genetic resources. Econ Bot 46:368–383
Vaughan DA, Lu BR, Tomooka N (2008) The evolving story of rice evolution. Plant Sci 174:394–408
Venuprasad R, Dalid CO, Del Valle M, Zhao D, Espiritu M, Cruz MS, Amante M, Kumar A, Atlin GN (2009) Identification and characterization of large-effect quantitative trait loci for grain yield under lowland drought stress in rice using bulk-segregant analysis. Theor Appl Genet 120:177–190
Vikram P, Swamy BM, Dixit S, Ahmed HU, Cruz MT, Singh AK, Kumar A (2011) qDTY 1.1, a major QTL for rice grain yield under reproductive-stage drought stress with a consistent effect in multiple elite genetic backgrounds. BMC Genet 12:1–5
Wambugu PW, Furtado A, Waters DL, Nyamongo DO, Henry RJ (2013) Conservation and utilization of African *Oryza* genetic resources. Rice 6:1–3
Wambugu PW, Mathenge PW, Auma EO, VanRheenen HA (2012) Constraints to on-farm maize (*Zea mays* L.) seed production in western Kenya: plant growth and yield. Int Sch Res Notices 2012:153412
Wang E, Wang J, Zhu X, Hao W, Wang L, Li Q, Zhang L, He W, Lu B, Lin H, Ma H (2008) Control of rice grain-filling and yield by a gene with a potential signature of domestication. Nat Genet 40:1370–1374
Wang H, Gao Y, Mao F, Xiong L, Mou T (2019) Directional upgrading of brown planthopper resistance in an elite rice cultivar by precise introgression of two resistance genes using genomics-based breeding. Plant Sci 288:110211

Wang K, Wambugu PW, Zhang B, Wu AC, Henry RJ, Gilbert RG (2015) The biosynthesis, structure and gelatinization properties of starches from wild and cultivated African rice species (*Oryza barthii* and *Oryza glaberrima*). Carbohydr Polym 29:92–100

Wang Y, Jiang W, Liu H, Zeng Y, Du B, Zhu L, He G, Chen R (2017) Marker assisted pyramiding of Bph6 and Bph9 into elite restorer line 93-11 and development of functional marker for Bph9. Rice 10:51

Warschefsky EJ, Rieseberg LH (2021) Laying the groundwork for crop wild relative conservation in the United States. PNAS 2021:118

Waters DL, Henry RJ (2014) Australian wild rice reveals pre-domestication origin of polymorphism deserts in rice genome. PLoS One 9:e98843

Waters DL, Nock CJ, Ishikawa R, Rice N, Henry RJ (2012) Chloroplast genome sequence confirms distinctness of Australian and Asian wild rice. Ecol Evol 2:211–217

Wei Y, Li X, Huang K, Chen Y, Gao G, Deng G, Guo S (2015) Mapping of yield-related quantitative trait locus of *Oryza minuta*. J Southern Agric 46:958–963

Wing RA, Ammiraju JS, Luo M, Kim H, Yu Y, Kudrna D, Goicoechea JL, Wang W, Nelson W, Rao K, Brar D (2005) The *Oryza* map alignment project: the golden path to unlocking the genetic potential of wild rice species. Plant Mol Biol 59:53–62

Wu X (2009) Prospects of developing hybrid rice with super high yield. Agron J 101:688–695

Wu Y, Xiao N, Yu L, Pan C, Li Y, Zhang X, Liu G, Dai Z, Pan X, Li A (2015) Combination patterns of major R genes determine the level of resistance to the *M. oryzae* in rice (*Oryza sativa* L.). PLoS One 10:0126130

Xiao J, Grandillo S, Ahn SN, McCouch SR, Tanksley SD, Li J, Yuan L (1996) Genes from wild rice improve yield. Nature 384:223–224

Xiao J, Li J, Grandillo S, Ahn SN, Yuan L, Tanksley SD, McCouch SR (1998) Identification of trait-improving quantitative trait loci alleles from a wild rice relative, *Oryza rufipogon*. Genetics 150:899–909

Xie J, Agrama HA, Kong D, Zhuang J, Hu B, Wan Y, Yan W (2010) Genetic diversity associated with conservation of endangered Dongxiang wild rice (*Oryza rufipogon*). Genet Resour Crop Evol 57:597–609

Xie X, Jin F, Song MH, Suh JP, Hwang HG, Kim YG, McCouch SR, Ahn SN (2008) Fine mapping of a yield-enhancing QTL cluster associated with transgressive variation in an *Oryza sativa* × *O. rufipogon* cross. Theor Appl Genet 116:613–622

Xiong LZ, Liu KD, Dai XK, Xu CG, Zhang Q (1999) Identification of genetic factors controlling domestication-related traits of rice using an F2 population of a cross between *Oryza sativa* and *O. rufipogon*. Theor Appl Genet 98:243–251

Xu K, Mackill DJ (1996) A major locus for submergence tolerance mapped on rice chromosome 9. Mol Breed 2:219–224

Xu P, Dong L, Zhou J, Li J, Zhang Y, Hu F, Liu S, Wang Q, Deng W, Deng X, Tharreau D (2015) Identification and mapping of a novel blast resistance gene Pi57 (t) in *Oryza longistaminata*. Euphytica 205:95–102

Xuwei J, Chunlian W, Qing Y (2007) Breeding of near-isogenic line CBB30 and molecular mapping of Xa30 (t), a new resistance gene to bacterial blight in rice. Sci Agric Sin 40:1094–1100

Yang H, You A, Yang Z, Zhang F, He R, Zhu L, He G (2004) High-resolution genetic mapping at the Bph15 locus for brown planthopper resistance in rice (*Oryza sativa* L.). Theor Appl Genet 110:182–191

Yang L, Li RB, Li YR, Huang FK, Chen YZ, Huang SS, Huang LF, Liu C, Ma ZF, Huang DH, Jiang JJ (2012) Genetic mapping of bph20 (t) and bph21 (t) loci conferring brown planthopper resistance to Nilaparvata lugens Stål in rice (*Oryza sativa* L.). Euphytica 183:161–171

Yang M, Lin J, Cheng L, Zhou H, Chen S, Liu F, Li R, Hu Y (2020) Identification of a novel planthopper resistance gene from wild rice (*Oryza rufipogon* Griff.). Crop J 8:1057–1070

Yasmin S, Hafeez FY, Mirza MS, Rasul M, Arshad HM, Zubair M, Iqbal M (2017) Biocontrol of bacterial leaf blight of rice and profiling of secondary metabolites produced by rhizospheric *Pseudomonas aeruginosa* BRp3. Front Microbiol 8:1895

Yinglun F, Xuewei C, Chunlian W (2006) Mapping the rice bacterial blight resistance gene Xa23 with RFLP markers and converting RFLP to STS marker. Acta Agron Sin 32:931–935

Yoon DB, Kang KH, Kim HJ, Ju HG, Kwon SJ, Suh JP, Jeong OY, Ahn SN (2006) Mapping quantitative trait loci for yield components and morphological traits in an advanced backcross population between *Oryza grandiglumis* and the *O. sativa japonica* cultivar Hwaseongbyeo. Theor Appl Genet 112:1052–1062

Yuan PR, Kim HJ, Chen QH, Ju HG, Ji SD, Ahn SN (2010) Mapping QTLs for grain quality using an introgression line population from a cross between *Oryza sativa* and *O. rufipogon*. J Crop Sci Biotechnol 13:205–212

Yuan R, Zhao N, Usman B, Luo L, Liao S, Qin Y, Nawaz G, Li R (2020) Development of chromosome segment substitution lines (CSSLs) derived from Guangxi wild rice (*Oryza rufipogon* Griff.) under rice (*Oryza sativa* L.) background and the identification of QTLs for plant architecture, agronomic traits and cold tolerance. Genes 11:980

Yun YT, Chung CT, Lee YJ, Na HJ, Lee JC, Lee SG, Lee KW, Yoon YH, Kang JW, Lee HS, Lee JY (2016) QTL mapping of grain quality traits using introgression lines carrying *Oryza rufipogon* chromosome segments in *japonica* rice. Rice 9:1–4

Zair W, Maxted N, Brehm JM, Amri A (2021) *Ex-situ* and in-situ conservation gap analysis of crop wild relative diversity in the Fertile Crescent of the Middle East. Genet Resour Crop Evol 68: 693–709

Zamir D (2001) Improving plant breeding with exotic genetic libraries. Nat Rev Genet 2:983–989

Zhang F, Xie J (2014) Genes and QTLs resistant to biotic and abiotic stresses from wild rice and their applications in cultivar improvements. Rice-Germplasm Genetics Improv 23:59–78

Zhang JW, Long Y, Xue MD, Xiao XG, Pei XW (2017) Identification of microRNAs in response to drought in common wild rice (*Oryza rufipogon* Griff.) shoots and roots. PLoS One 12:e0170330

Zhang QJ, Zhu T, Xia EH, Shi C, Liu YL, Zhang Y, Liu Y, Jiang WK, Zhao YJ, Mao SY, Zhang LP (2014) Rapid diversification of five *Oryza* AA genomes associated with rice adaptation. PNAS 111:E4954–E4962

Zhao H, Sun L, Xiong T, Wang Z, Liao Y, Zou T, Zheng M, Zhang Z, Pan X, He N, Zhang G (2019) Genetic characterization of the chromosome single-segment substitution lines of *O. glumaepatula* and *O. barthii* and identification of QTLs for yield-related traits. Mol Breed 39:1–19

Zhao Y, Huang J, Wang Z, Jing S, Wang Y, Ouyang Y, Cai B, Xin XF, Liu X, Zhang C, Pan Y (2016) Allelic diversity in an NLR gene BPH9 enables rice to combat planthopper variation. PNAS 113:12850–12855

Zheng CK, Chun-Lian WA, Yuan-Jie YU, Liang YT, Kai-Jun ZH (2009) Identification and molecular mapping of Xa32 (t), a novel resistance gene for bacterial blight (*Xanthomonas oryzae* pv. *Oryzae*) in rice. Acta Agron Sin 35:1173–1180

Zheng W, Wang Y, Wang L, Ma Z, Zhao J, Wang P, Zhang L, Liu Z, Lu X (2016) Genetic mapping and molecular marker development for Pi65 (t), a novel broad-spectrum resistance gene to rice blast using next-generation sequencing. Theoret Appl Genet 12:1035–1104

Zhu Q, Ge S (2005) Phylogenetic relationships among A-genome species of the genus *Oryza* revealed by intron sequences of four nuclear genes. New Phytol 167:249–265

Zou T, Zhao H, Li X, Zheng M, Zhang S, Sun L, He N, Pan X, Liu Z, Fu X (2020) QTLs detection and pyramiding for stigma exsertion rate in wild rice species by using the single-segment substitution lines. Mol Breed 40:1–4

Conservation and Determination of Nutritional Composition of *Mondia whitei* In Vitro

22

Ponnusamy Baskaran, Subashen Naidu, and Johannes Van Staden

Abstract

An efficient in vitro micropropagation protocol from stored synthetic seeds and the macro- and micronutrient composition and its relative distribution are described for *Mondia whitei* (Hook.f.) using shoot apical meristems (SAMs). Higher numbers of shoots were achieved in MS media supplemented with 2.5 μM benzyladenine (BA) and 5 μM thidiazuron (TDZ). A combination of 5 μM phloroglucinol (PG) and 2.5 μM indole-3-butyric acid (IBA) produced a greater number of roots per shoot derived from combined treatments of TDZ and PG or indole-3-acetic acid (IAA) or 6-(2-hydroxy-3-methylbenzylamino) purine (PI55). A 100% survival rate was observed on acclimatization in greenhouse. A scanning electron microscope (SEM) including energy-dispersive X-ray analysis was used to determine macro- and micronutrient composition in SAMs of *M. whitei* in vitro shoots. These shoots yielded elements such as aluminum (Al, 0.20%), calcium (Ca, 0.46%), magnesium (Mg, 0.18%), phosphorus (P, 0.52%), potassium (K, 2.51%), and sulfur (S, 0.17%). The elements (Mg, P, K, and S) were detected in all treatments, while 15 μM phloroglucinol (PG) and 25 μM 6-(2-hydroxy-3-methylbenzylamino) purine (PI55) treatments also contained micronutrients, iron and silicon, respectively. The reported protocol for *M. whitei* here showed the

P. Baskaran (✉)
Biotechnology, Dami OPRS, NBPOL, Dami, Kimbe, Papua New Guinea
e-mail: baskaran.ponnusamy@nbpol.com

S. Naidu
Microscopy and Microanalysis Unit (MMU), University of KwaZulu-Natal, Pietermaritzburg, Scottsville, South Africa

J. Van Staden
Research Centre for Plant Growth and Development, School of Life Sciences, University of KwaZulu-Natal, Pietermaritzburg, Scottsville, South Africa

S. Ramamoorthy et al. (eds.), *Plant Genetic Resources, Inventory, Collection and Conservation*, https://doi.org/10.1007/978-981-16-7699-4_22

significance of the impacts of hormones, PI55 and PG, on micropropagation and elemental composition. The developed protocol offers germplasm conservation and large-scale clonal propagation, and provides indications that different elements were detected in the shoots. Thus, the material can serve as a potent bioresource in traditional medicine and other commercial applications.

Keywords

Elemental composition · SEM-EDX analysis · Micropropagation · Phloroglucinol · 6-(2-Hydroxy-3-methylbenzylamino) purine (PI55)

22.1 Introduction

Trace elements or micronutrients play an important role in the physiological processes of plants and animals (Thangarajan et al. 2012; Yashvanth et al. 2013). They are involved in several physiological functions of the human body and are essential components of a healthy lifestyle. The human body requires minute quantities of these nutrients to aid in enhancing the immune system, defending against pathogens, and promoting a quicker recovery from serious infections. Essential microelements are thus engaged in several enzymatic reactions and immunologic processes and in overall metabolism (Chaturvedi et al. 2004; Selvaraju et al. 2009; Rajan et al. 2014). In addition, micronutrients also play a pivotal role in the biosynthesis of plant secondary metabolites. These chemical components are mainly accountable for the medicinal properties of plants (Adebajo et al. 2013). Plant secondary metabolites have become important avenues for drug lead discovery and development in the pharmaceutical industry. Micropropagation of plant material has been accepted widely as a feasible means for phytopharmaceutical production. As such, the analysis of micronutrient composition in micropropagated plants becomes essential to authenticate and regulate the processes of micropropagation for pharmaceutical benefit. Higher plants differ in their ability to take up micronutrients while different parts of a plant tend to accumulate specific quantities of certain nutrients (Pilon-Smits et al. 2009; Rout et al. 2013; Guerrero-Molina et al. 2014). The external growth factors like hormones might have a central part in in vitro culture and should be explored for their effect on micronutrient composition and correlation with physiological aspects for the overall growth and progress of plants. Interestingly, it was found out that PI-55 could regulate sensing of cytokinins (Spíchal et al. 2009). It acted as an antagonist of cytokinins, particularly at the receptor CRE1/AHK4. This receptor was elucidated as a negative regulator of root growth in *Arabidopsis* (Mähönen et al. 2000; Riefler et al. 2006). Consequently, it is important to investigate in detail the influence of external administration of PI-55 on the determination of macro- and micronutrient composition.

Various methods have been used for elemental detection in medicinal plants (Rout et al. 2013; Yashvanth et al. 2013; Guerrero-Molina et al. 2014); however, energy-dispersive X-ray analysis (EDX) together with SEM is a relatively fast and effective method to qualitatively and quantitatively determine the nutritional composition. This method also provides a means for comparison and characterization of plant tissues and materials (Goldstein et al. 2003). It is also employed in plant studies to examine the bioaccumulation, bioavailability, metal immobilization, and biodegradation (Guerrero-Molina et al. 2014). Accordingly, the application of the EDX technique in phytopharmaceuticals aids in the detection and measurement of valuable micronutrients.

The endangered medicinal plant, *Mondia whitei* (Apocynaceae), contains various bioactive compounds and is used to treat various diseases (SANBI 2015; Baskaran et al. 2017). Based on the present status of *M. whitei* from the Red Data List of South African plants, the species requires the development of an alternative in vitro micropropagation protocol. However, in order to endorse the use of micropropagated plants as substitutes for wild populations, the macro- and micronutrient composition of micropropagated plants needs to be examined. The aims of this study were to develop a method for the production of synthetic seeds from shoot tips and to optimize an easy and efficient, robust in vitro micropropagation protocol from stored synthetic seeds for *M. whitei* using different plant growth regulators (PGRs), PG and PI55. In addition, the macro- and micronutrient composition of shoot apical meristems (SAMs) from the shoot apical buds of *M. whitei* produced by different treatments of shoot regeneration media was determined with EDX techniques and correlated to physiological aspects for in vitro development of *M. whitei*.

22.2 Materials and Methods

22.2.1 Micropropagation from Stored Synthetic Seeds

The stored synthetic seeds (25 ± 2 °C for 10 days in darkness) of somatic embryo (liquid MS (Murashige and Skoog 1962) medium plus sucrose (30 g L^{-1}), *meta*-topolin riboside (1 μM), and naphthaleneacetic acid (NAA) (0.5 μM) of *Mondia whitei* were established earlier (Baskaran et al. 2017). A micropropagation protocol for *M. whitei* was developed using the conserved seeds in this study. The synthetic seeds were germinated on MS medium with sucrose (30 g L^{-1}) and agar (8 g L^{-1}). Shoot tips (approximately 0.5–1.0 cm) of *M. whitei* were excised from 1-month-old germinated plants and were encapsulated (Baskaran et al. 2017) with 3% (w/v) sodium alginate (3%) and calcium chloride (100 mM). Subsequently shoot tips were stored in darkness at 25 ± 2 °C for 10 days according to a previously standardized protocol (Baskaran et al. 2017) and were then moved into MS medium with sucrose (30 g L^{-1}) and agar (8 g L^{-1}). Various concentrations of benzyladenine (BA), *meta*-topolin (*m*T), phloroglucinol (PG), thidiazuron (TDZ), 6-(2-hydroxy-3-methylbenzylamino) purine (PI55), and indole-3-acetic acid (IAA) were added in combinations for shoot regeneration for 8 weeks. The shoots of around 2–4 cm long formed from combined treatments (TDZ and BA/IAA/*m*T/PG/PI55) of shoot regeneration media which were then placed in root-inducing medium of MS supplemented with NAA (2.5 μM) or IBA alone and combinations with 2.5 μM IBA and 5 μM PG or PI55 for 6 weeks to evaluate effective rooting of shoots, as specified in Table 22.2. In all experiments, the MS medium without PGR was used as a control. The pH of every media was made to 5.8 by using 0.1 N NaOH and/or 0.1 N HCl followed by autoclaving at 121 °C for 20 min. The temperature maintained for growth was 25 ± 2 °C whereas the intensity of light was 40 μmol m^{-2} s^{-1} through cool white fluorescent light (OSRAM L 58 W/740, South Africa) for 16-h photoperiod. Analytical graded chemicals were used (Biolab, South Africa; Sigma, USA; and Oxford, England).

22.2.2 Acclimatization

After a time period of 6 weeks, the regenerated plantlets (>4 cm length) from every root-inducing medium and after 8 weeks well-developed rooted shoots from certain shoot regeneration media (5 μM PG, 25 μM PI55, and 5 μM TDZ plus 2.5 μM *m*T or PI55) were harvested. The plantlets were moved to terracotta pots of 95 × 120 mm with vermiculite for 4 weeks and later transferred to a mixture of soil:vermiculite (1: 1). All plantlets were watered with tap water on each third day. Afterwards, plantlets were acclimatized by keeping in the greenhouse at a temperature of 25 ± 2 °C in natural photoperiod along with midday PPF of 950 ± 50 μmol m^{-2} s^{-1}.

22.2.3 Nutritional Evaluation

The 8-week-old shoot apical buds of *M. whitei* produced by control and different treatments of shoot regeneration media containing shoots and rooted shoots (Table 22.1) were used for the analysis of elemental composition (cobalt (Co), iron (Fe), chlorine (Cl), magnesium (Mg), phosphorus (P), potassium (K), calcium (Ca), silicon (Si), aluminum (Al), sodium (Na), and sulfur (S)) using energy-dispersive X-ray (EDX) methods. The shoot apical meristems (SAMs) were removed from the shoot apical buds and cut in half with a single-edged stainless steel blade (GEM, Agar Scientific). This material was subsequently mounted, with the cut-exposed surface facing upwards, onto aluminum scanning electron microscope (SEM) stubs with double-sided carbon tape (Advanced Lab Services, Agar Scientific). The EDX analysis was performed using a Zeiss EVO LS 15 VP SEM (Zeiss, Germany), under variable pressure mode at an EHT of 20 kV. The Oxford X-Max 80 mm silicon drift detector (SDD) was used to extract and process the EDX data. The central meristem region was scanned to determine its macro- and micronutrient composition.

22.2.4 Statistical Analysis

The experiments were performed 3 times with 25 replicates for each treatment. Analysis of variance (ANOVA) studies were carried out and the results are shown

Table 22.1 Effect of plant growth regulators on shoot multiplication from shoot-tip explants of *M. whitei* "stored synthetic seeds" after 8 weeks of culture

PGR (μM) in MS medium	Shoot explant^{-1} (n)	Shoot length (cm)	Root explant^{-1} (n)	Root length (cm)
Control	1.0 ± 0.00 e	2.39 ± 0.19 e	1.2 ± 0.12 d	2.17 ± 0.16 cd
5 BA	1.8 ± 0.37 d	3.18 ± 0.40 de	0	0
15 BA	2.4 ± 0.25 cd	3.36 ± 0.53 de	0	0
5 *m*T	1.2 ± 0.20 e	3.76 ± 0.61 de	0	0
15 *m*T	2.4 ± 0.27 cd	4.90 ± 0.52 cd	0	0
5 TDZ	2.6 ± 0.25 cd	7.16 ± 0.77 ab	0	0
15 TDZ	3.0 ± 0.32 bc	7.36 ± 0.85 ab	0	0
5 PG	1.0 ± 0.00 e	5.08 ± 0.69 cd	7.0 ± 0.58 b	3.72 ± 0.56 ab
15 PG	1.0 ± 0.00 e	7.06 ± 0.54 ab	5.6 ± 0.25 cd	1.64 ± 0.28 d
25 PG	1.0 ± 0.00 e	4.04 ± 0.38 de	4.8 ± 0.14 cd	1.52 ± 0.21 d
5 PI55	1.0 ± 0.00 e	3.66 ± 0.53 de	4.0 ± 0.45 cd	2.06 ± 0.35 cd
15 PI55	1.0 ± 0.00 e	3.40 ± 0.36 de	5.2 ± 0.40 cd	3.38 ± 0.55 ab
25 PI55	1.0 ± 0.00 e	4.30 ± 0.41 cd	6.0 ± 0.45 bc	3.52 ± 0.53 ab
5 TDZ + 2.5 BA	4.6 ± 0.60 a	5.94 ± 0.36 c	2.4 ± 0.34 e	1.70 ± 0.19 d
5 TDZ + 2.5 *m*T	2.2 ± 0.37 cd	4.08 ± 0.44 de	10.2 ± 1.02 a	4.68 ± 0.56 a
5 TDZ + 2.5 IAA	3.2 ± 0.32 b	2.92 ± 0.41 e	3.4 ± 0.25 d	1.52 ± 0.17 d
5 TDZ + 2.5 PG	2.4 ± 0.37 cd	8.26 ± 0.83 a	3.6 ± 0.27 d	2.58 ± 0.29 cd
5 TDZ + 2.5 PI55	2.6 ± 0.26 cd	8.08 ± 0.69 a	7.6 ± 0.45 b	3.36 ± 0.55 ab

PGR plant growth regulator. Means followed by same letters in each column are not significantly different ($P = 0.05$) using Duncan's multiple range test

as mean ± standard error. Duncan's multiple range test and Tukey's multiple comparison test at 5% probability level were used to separate treatment means. Whole statistical part of studies was executed in SPSS (Windows v 24) (SPSS Inc., Chicago, IL, USA).

22.3 Results

22.3.1 Production of Adventitious Shoots

Stored (10 days; 25 ± 2 °C in darkness) "synthetic seeds" of shoot tips of *M. whitei* were grown in medium of various combinations of *m*T, TDZ, BA, IAA, PG, and PI55 in different concentrations. Shoot regeneration responded differently to treatments as depicted in Table 22.1 and Fig. 22.1a, b. Of the different concentrations of individual PGRs used, the most effective PGR for shoot regeneration was observed to be TDZ (15 μM), but this did not differ significantly from 5 μM TDZ, 15 μM BA, or *m*T (Table 22.1). TDZ treatments induced green compact basal callus at the bottom of shoots, whereas *m*T caused formation of broad leaves (4.0 × 2.0 cm) (Fig. 22.1c). In the study, different concentrations (5–25 μM) of PG and PI55 reduced shoot regeneration in *M. whitei* (Table 22.1). On the other hand, PG (15 μM) elongated the shoot similar to 5 or 15 μM TDZ treatments (Table 22.1; Fig. 22.1d). Interestingly, 5–25 μM PG or PI55 treatments induced

Fig. 22.1 Micropropagation of *M. whitei* using synthetic seeds of shoot-tip explants with different plant growth regulator treatments. Development of shoot from synthetic seeds in 15 μM TDZ after 1 week (**a**) and 3 weeks of culture (**b**). (**c**) Shoot regeneration of basal callus in TDZ (15 μM). (**d**) Shoot elongation in 5 μM TDZ after 8 weeks of culture. (**e**) Leaf chlorosis of shoot in 25 μM PG. (**f**) Prolific shoot in TDZ (5 μM) along with BA (2.5 μM). (**g**) Shoots rooted in combination of TDZ and *m*T. Acclimatized plants of *M. whitei* in the vermiculite:soil mixture (1:1, v/v) after 4 weeks (**h**) and after 2 months (**i**) in the greenhouse

roots on shoots after 4 weeks of culture. Rooting (7.0 roots plus 3.72 cm long) increased significantly with 5 μM PG followed by 15 μM PI55 (Table 22.1). However, after 8 weeks, leaves showed chlorosis followed by necrosis at the highest PG concentration of 25 μM (Fig. 22.1e). Prolific shoots were initiated rapidly when explants were cultured on the medium containing combined treatments of TDZ plus BA or *m*T or IAA or PG or PI55 (Table 22.1). Amalgamation of TDZ and IAA or BA was most effective in shoot regeneration. However, the highest number of shoots was obtained via a combined treatment of TDZ (5 μM) and BA (2.5 μM) as depicted in Table 22.1 and Fig. 22.1f. In this study, regenerated shoots formed green compact basal callus and roots; however, the roots significantly varied in number and length (Table 22.1). On comparison to the rest of the treatments, the root growth (10.2 roots and 4.68 cm long) was significantly higher with TDZ and *m*T combinations (Table 22.1; Fig. 22.1g). It was understood from the current research that concentrations of PG, PI55, and TDZ and their combinations played a vital role in rooting of shoots.

22.3.2 Rooting of Shoots and Acclimatization

Roots developed on all shoots derived from shoot regeneration media when transferred to rooting media (Table 22.2). However, the root number, root length, and shoot length varied amongst the control and rooting treatments (Table 22.2). These parameters could have been affected by the shoot regeneration medium prior to

Table 22.2 Effect of plant growth regulators, PG and PI55, on shoot rooting of *M. whitei* after 6-week culture on rooting medium

PGR (μM) in shoot multiplication medium	PGR (μM) in rooting medium	Root-shoot^{-1} (n)	Root length (cm)	Shoot length (cm)
5 TDZ + 2.5 BA	Control	3.4 ± 0.51 jk	4.3 ± 0.60 cd	5.6 ± 0.53 de
5 TDZ + 2.5 *m*T	2.5 IBA	3.2 ± 0.74 jk	3.4 ± 0.59 gh	5.2 ± 0.41 de
5 TDZ + 2.5 IAA	2.5 NAA	2.0 ± 0.32 k	2.2 ± 0.36 j	4.7 ± 0.58 ef
5 TDZ + 2.5 PG	2.5 IBA +	6.6 ± 0.75 de	4.9 ± 0.33 bc	6.5 ± 0.42 cd
5 TDZ + 2.5 PI55	5 PG	6.2 ± 0.37 ef	2.1 ± 0.29 j	5.2 ± 0.49 de
5 TDZ + 2.5 BA	2.5 IBA +	6.0 ± 0.71 fg	4.9 ± 0.76 bc	5.6 ± 0.48 de
5 TDZ + 2.5 *m*T	5 PI55	4.4 ± 0.51 ij	4.5 ± 0.43 cd	5.8 ± 0.27 de
5 TDZ + 2.5 IAA		8.6 ± 0.51 bc	4.7 ± 0.39 bc	8.2 ± 0.41 ab
5 TDZ + 2.5 PG		8.4 ± 0.68 bc	5.6 ± 0.27 ab	7.2 ± 0.39 bc
5 TDZ + 2.5 PI55		7.8 ± 0.80 cd	3.2 ± 0.24 hi	6.1 ± 0.39 cd
5 TDZ + 2.5 BA		4.4 ± 0.75 ij	5.6 ± 0.40 ab	6.2 ± 0.17 cd
5 TDZ + 2.5 *m*T		6.4 ± 0.93 de	6.0 ± 0.59 ab	6.6 ± 0.19 cd
5 TDZ + 2.5 IAA		6.2 ± 0.37 ef	3.6 ± 0.34 fg	6.8 ± 0.41 cd
5 TDZ + 2.5 PG		5.8 ± 0.42 fg	4.7 ± 0.26 cd	5.5 ± 0.31 de
5 TDZ + 2.5 PI55		5.2 ± 0.49 gh	2.7 ± 0.30 ij	4.9 ± 0.37 ef
5 TDZ + 2.5 BA		4.3 ± 0.31 ij	4.0 ± 0.33 ef	8.0 ± 0.42 ab
5 TDZ + 2.5 *m*T		4.4 ± 0.51 ij	2.3 ± 0.39 j	5.3 ± 0.31 de
5 TDZ + 2.5 IAA		9.6 ± 1.12 ab	5.1 ± 0.47 ab	9.1 ± 0.33 a
5 TDZ + 2.5 PG		10.4 ± 0.87 a	6.5 ± 0.37 a	8.2 ± 0.48 ab
5 TDZ + 2.5 PI55		9.6 ± 0.68 ab	5.3 ± 0.26 ab	7.7 ± 0.52 bc
5 TDZ + 2.5 BA		6.0 ± 0.45 fg	6.1 ± 0.59 ab	6.7 ± 0.23 cd
5 TDZ + 2.5 *m*T		5.0 ± 0.55 gh	4.1 ± 0.62 ef	4.3 ± 0.41 f
5 TDZ + 2.5 IAA		5.2 ± 0.86 gh	3.9 ± 0.47 ef	6.0 ± 0.32 cd
5 TDZ + 2.5 PG		7.0 ± 0.71 de	5.8 ± 0.36 ab	6.2 ± 0.61 cd
5 TDZ + 2.5 PI55		8.0 ± 1.34 cd	4.5 ± 0.24 cd	6.7 ± 0.53 cd

PGR plant growth regulator. Rooting medium, MS medium containing sucrose (15 g L^{-1}), 8 g L^{-1} agar, and PGR (μM). Means followed by same letters in each column are not significantly different ($p = 0.05$) using Duncan's multiple range test.

rooting. Combined treatments of TDZ and PG or PI55 produced shoots which rooted best in the control treatment (Table 22.2). In the current investigation, IBA was found to be most efficient in inducing root formation in shoots rather than NAA treatments. Combined treatments of TDZ and IAA or PG or PI55 that produced shoots significantly increased root and shoot growth with IBA application (Table 22.2). However, shoots developed with the combination of TDZ and PG produced a higher number of roots (10.4) with longer roots (6.5 cm) and shoots (8.2 cm) in IBA (2.5 μM) and PG treatments (5 μM), while root number and lengths of root and shoot were not significantly different to certain rooting treatments (Table 22.2). The rooting experiments showed clearly that shoot regeneration medium strongly influenced the rooting ability of shoots. The rooted plantlets from rooting media and certain shoot regeneration media (5 μM PG, 25 μM PI55, and 5 μM TDZ plus 2.5 μM *m*T or PI55) were harvested and acclimatized in the greenhouse conditions (Fig. 22.1h, i). A variation is observed in the rate of survival (20–100%) of plantlets in shoot regeneration and rooting media. The plantlets from shoot regeneration media containing 5 μM PG and 25 μM PI55 showed 20% survival, while the survival rate was increased to 30% on application of TDZ (5 μM) including PI55 (2.5 μM) and 45% on TDZ (5 μM) with *m*T (2.5 μM) treatments. However, the highest survival rate (100%) of plantlets was obtained with all rooting media except the control (25%).

22.3.3 Nutritional Composition in Regenerated Shoots of *M. whitei*

Contents of macro- and microelement in shoot apical meristems of *M. whitei* differ significantly depending on the medium composition (Table 22.3). These results are shown in Table 22.3 and Fig. 22.2a–h. The EDX spectra of SAMs expressed qualitative and quantitative (expressed as weight %) elements (cobalt (Co), iron (Fe), chlorine (Cl), magnesium (Mg), phosphorus (P), potassium (K), calcium (Ca), silicon (Si), aluminum (Al), sodium (Na), and sulfur (S)) from different treatments (Table 22.1). Mg, P, K, and S were detected in SAMs in control and all other treatments (Table 22.3; Fig. 22.2a, b, c, d). There were no significant differences observed with elemental weight (%) of Mg, P, and K between control and PGR, PG, and PI55 treatments (Table 22.3); however, S content was significantly higher in SAMs derived from a combination of TDZ and PI55 compared to control and other treatments (Table 22.3). Elements Al, Ca, Cl, Co, and Na occurred in certain treatments and the control only and concentrations of these varied significantly amongst these treatments (Table 22.3; Fig. 22.2e, f). Interestingly, Fe and Si were only found with 15 μM PG and 25 μM PI55, respectively (Table 22.3; Fig. 22.2 g, h).

22.4 Discussion

Micropropagation from synthetic seeds containing shoot-tip explants of *M. whitei* can be helpful for the storage and transportation for germplasm conservation. Effective shoot regeneration of *M. whitei* was dependent on TDZ and IAA or BA combinations. Similarly, in some Apocynaceae plants, the effects of cytokinin-cytokinin or cytokinin-auxin were reported (Kanungo et al. 2012; Vaiyapuri et al. 2014; Alatar 2015). In this study, different morphogenesis was observed, while broad leaves in *m*T treatment and basal callus as well as root regeneration in certain individual (TDZ, PG, and PI55) and combined (TDZ and *m*T or PG or PI55) treatments were exhibited. Similarly, different morphogenesis with plant growth regulator treatments was stated in some plants (Vaiyapuri et al. 2014; Baskaran et al. 2016). As per the current investigation, it was observed that a higher concentration of TDZ (15 μM) was effective in shoot regeneration; however, this was via induced basal callus at the base of shoots. TDZ influences various morphogenetic processes, mainly basal callus of shoots in vitro (Murthy et al. 1998; Vaiyapuri et al. 2014; Baskaran et al. 2016). PG and PI55 treatments inhibited shoot regeneration but promoted root growth.

Table 22.3 Elemental composition on shoot apical meristem of *M. whitei* shoot regeneration in response to PGR after 8 weeks of culture

Treatments	Elemental weight (%) in shoot apical meristem										
PGR (μM) in MS medium	Aluminum (Al)	Calcium (Ca)	Chlorine (Cl)	Cobalt (Co)	Iron (Fe)	Magnesium (Mg)	Phosphorus (P)	Potassium (K)	Silicon (Si)	Sodium (Na)	Sulfur (S)
Control	0.23 ± 0.03 cd	0	0	0	0	0.07 ± 0.04 de	0.58 ± 0.29 ab	2.08 ± 0.62 bc	0	0	0.24 ± 0.17 bc
15 BA	0.08 ± 0.01 d	0.17 ± 0.08 cd	0	0	0	0.24 ± 0.05 bc	0.73 ± 0.15 a	1.27 ± 0.23 cd	0	0	0.26 ± 0.08 bc
15 *m*T	0	0	0	0.02 ± 0.01 ab	0	0.08 ± 0.01 de	0.28 ± 0.02 ab	1.09 ± 0.02 cd	0	0.11 ± 0.01 b	0.13 ± 0.06 bc
15 TDZ	0.12 ± 0.04 cd	0.37 ± 0.06 c	0	0.03 ± 0.01 a	0	0.20 ± 0.03 c	0.19 ± 0.01 b	0.53 ± 0.01 d	0	0	0.14 ± 0.05 bc
5 PG	0.31 ± 0.01 bc	0.24 ± 0.03 cd	0.66 ± 0.04 a	0	0	0.36 ± 0.01 ab	0.25 ± 0.02 ab	1.98 ± 0.13 bc	0	0	0.12 ± 0.01 bc
15 PG	0.63 ± 0.25 ab	0.23 ± 0.01 cd	0	0	0.15 ± 0.15 a	0.17 ± 0.03 cd	0.51 ± 0.16 ab	1.09 ± 0.11 cd	0	0	0.24 ± 0.01 bc
25 PG	0.70 ± 0.03 a	2.95 ± 0.12 a	0.27 ± 0.02 b	0	0	0.12 ± 0.01 cd	0.41 ± 0.06 ab	4.86 ± 0.06 a	0	0.90 ± 0.29 a	0.28 ± 0.04 bc
5 PI55	0	0	0	0	0	0.22 ± 0.01 c	0.42 ± 0.02 ab	1.01 ± 0.05 cd	0	0	0.21 ± 0.01 bc
15 PI55	0.21 ± 0.01 cd	0.40 ± 0.07 c	0	0	0	0.25 ± 0.03 bc	0.48 ± 0.08 ab	2.51 ± 0.03 bc	0	0	0.38 ± 0.06 b

(continued)

Table 22.3 (continued)

Treatments	Elemental weight (%) in shoot apical meristem										
PGR (μM) in MS medium	Aluminum (Al)	Calcium (Ca)	Chlorine (Cl)	Cobalt (Co)	Iron (Fe)	Magnesium (Mg)	Phosphorus (P)	Potassium (K)	Silicon (Si)	Sodium (Na)	Sulfur (S)
25 PI55	0	0.19 ± 0.02 cd	0.06 ± 0.01 de	0.02 ± 0.01 ab	0	0.19 ± 0.02 cd	0.35 ± 0.02 ab	1.15 ± 0.03 cd	0.09 ± 0.02 a	0	0.17 ± 0.01 bc
5 TDZ + 2.5 BA	0.20 ± 0.03 cd	0.46 ± 0.72 bc	0	0	0	0.18 ± 0.05 cd	0.52 ± 0.11 ab	2.51 ± 0.04 bc	0	0	0.17 ± 0.01 bc
5 TDZ + 2.5 *m*T	0.37 ± 0.02 ab	0.73 ± 0.17 b	0.11 ± 0.01 cd	0	0	0.48 ± 0.02 a	0.40 ± 0.07 ab	1.96 ± 0.63 b	0	0	0.19 ± 0.01 bc
5 TDZ + 2.5 IAA	0.26 ± 0.02 cd	0	0	−0.01 ± 0.00 c	0	0.11 ± 0.02 cd	0.45 ± 0.02 ab	1.06 ± 0.45 cd	0	0	0.21 ± 0.03 bc
5 TDZ + 2.5 PG	0.46 ± 0.06 ab	0.37 ± 0.01 c	0.17 ± 0.04 c	0	0	0.22 ± 0.03 c	0.45 ± 0.06 ab	1.85 ± 0.07 cd	0	0	0.16 ± 0.01 bc
5 TDZ + 2.5 PI55	0	0	0	0	0	0.13 ± 0.02 cd	0.63 ± 0.01 ab	3.73 ± 0.79 ab	0	0	0.77 ± 0.21 a

PGR plant growth regulator. Data are expressed as mean ± SE. Means followed by same letters in each column are not significantly different ($p = 0.05$) using Tukey's multiple comparison test.

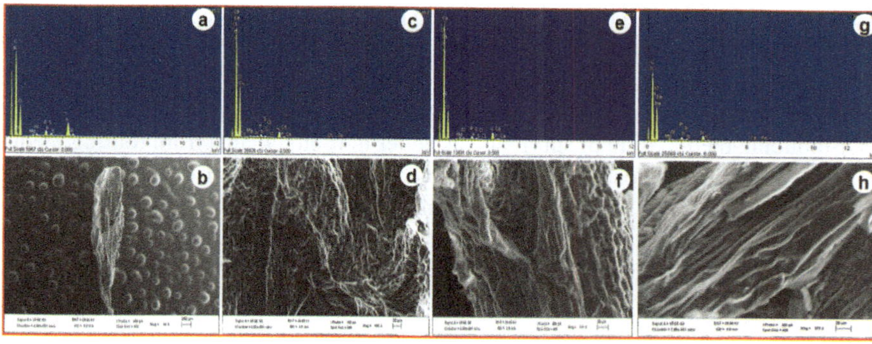

Fig. 22.2 SEM-EDX spectra and map for trace elemental composition of shoot apical meristems of *M. whitei* with different plant growth regulator treatments. The EDX spectrum (**a**, **c**) and SEM map (**b**, **d**) showing Mg, P, K, and S elements in SAMs of shoots in control and 25 μM PI55, respectively. The elements of Al, Ca, and Cl extracted from EDX spectrum (**e**) and SEM map (**f**) with the combination of TDZ and PG treatment. EDX spectrum (**g**) and SEM map (**h**) showing trace elemental composition with Fe from 15 μM PG treatment

Phloroglucinol is a benzenetriol which appears in phenolic rings of various phenolic compounds like flavonoid (A ring). This PGR is also employed for the growth and development of in vitro plants (Baskaran et al. 2016). PG has also been reported as a genotype-reliant auxin promoter (Teixeira da Silva et al. 2013).

Generally, PI55 obstructs the CRE1/AHK4 receptor found in *Arabidopsis thaliana* (Spíchal et al. 2009). These receptors are principally expressed in plant roots (Nishimura et al. 2004). A higher concentration (25 μM) of PG exhibited chlorosis and necrosis of leaves in *M. whitei* in vitro. PG exhibited activities similar to auxin and led to different types of morphogenesis in some other plants (Tallón et al. 2012; Baskaran et al. 2016). It was found that the combinations of TDZ and BA were more efficient in regenerating shoots, which indicates that these combinations are essential for mass shoot regeneration of *M. whitei*. Similar observations were made in other Apocynaceae species (Vaiyapuri et al. 2014). Combinations of TDZ and *m*T decreased shoot regeneration but improved rooting of shoots, which revealed that the combination impacted the formation of roots on shoots of plantlets. The effectiveness of TDZ and BA or other cytokinin combinations on root differentiation of shoots has been demonstrated in a number of Apocynaceae and other family members (Beena et al. 2003; Guo et al. 2011; Baskaran and Van Staden 2013; Vaiyapuri et al. 2014). Our investigation indicates that *m*T, PG, and the chemical compound PI55 mechanism of action needs to be investigated.

Efficient rooting of shoots is essential for effective hardening and varies with shoot regeneration, rooting media composition, and plant species. Synergistic effects in the rooting of shoots are dependent on plant growth regulators in shoot regeneration and rooting media in *M. whitei* in vitro. These effects have been observed in many other plants (Faisal and Anis 2003; Alatar 2015; Baskaran et al. 2016). Shoots produced from shoot regeneration medium containing combinations of TDZ and IAA or PG or PI55 were efficient in rooting with IBA treatments; however, rooting was most effective with combined application of IBA (2.5 μM) and PG (5 μM). The role of IBA in inducing root and its development was documented in various other species of Apocynaceae (Faisal and Anis 2003; Alatar 2015). A variation was observed in the survival rate of plantlets acclimatized from shoot regeneration and root-inducing media. Rate of survival was better in rooted plants from rooting media compared to shoot regeneration media. The promontory effect of PGR and other conditions in vitro on rooting for successful acclimatization has been testified in many other plants (Hazarika et al. 2006). The current results suggest that combined treatments of shoot regeneration media that produced shoots and IBA or IBA plus PG in rooting media can effectively be used for improved acclimatization of *M. whitei*.

The EDX revealed different macro- and micronutrient compositions and its relative distribution in SAMs of *M. whitei*. In this study, nutritional composition varied in SAMs of shoots developed amongst control and treatments. This indicated that the PGR, PG, and PI55 treatments may affect the uptake of macro- and micronutrients in the meristematic regions of the shoots. The elements Mg, P, K, and S accumulated in all treatments, which were expressed as essential elements in SAMs of *M. whitei*. No significant difference was observed with Mg, P, and K elements amongst the control and treatments. However, Al, Ca, Cl, Co, and Na elements were exhibited in certain treatments but varied quantitatively (weight %). The higher accumulation of Na along with K was observed in the SAMs of shoots with 25 μM PG treatment. This may have initiated the chlorosis and necrosis of leaves in *M. whitei*. Sodium is categorized as an advantageous micronutrient (Epstein 1972) but can produce chlorotic and necrotic plants in excessively high concentrations, even in combination with a high concentration of K in the plants (Broadley et al. 2011). Trace elements Fe and Si were observed only in 15 μM PG and 25 μM PI55, respectively. Si plays a pivotal role in plant growth and development in *M. whitei*. This investigation indicated that quantitative and qualitative accumulation of macro- and micronutrients in SAMs of shoots does not affect shoot regeneration of *M. whitei*. On the other hand, all the essential trace elements do play vital roles in pharmacological, physiological, and biochemical functions in *M. whitei*. Similarly, pharmacological action of medicinal plant, *Ichnocarpus frutescens*, and other species has also been well documented with trace elements (Thangarajan et al. 2012; Rajan et al. 2014).

22.5 Conclusions

An effective and simple in vitro micropropagation protocol and nutritional composition analysis of shoot apical meristems of *M. whitei* were established for stored synthetic seeds of shoot-tip explants. The shoot and root regeneration was effective with specific concentrations and combinations of PGR and PG or PI55 during the in vitro phases, whereas PG and PI55 were required for effective rooting and acclimatization of *M. whitei*. Also, improved shoot and root regeneration and acclimatization were established from specific shoot regeneration and rooting media. Moreover, the trace elemental composition of shoot apical meristems of shoots in vitro varied in the treatments but did not affect shoot regeneration of *M. whitei*. Established micropropagation offers potential protocols for large-scale clonal production, germplasm conservation, and transportation, thus serving as an effective bioresource in conventional medicine and for further industrial purposes.

Acknowledgements The financial funding by National Research Foundation (NRF), Pretoria, and the University of KwaZulu-Natal, Pietermaritzburg, is thankfully acknowledged.

Author Contribution Statement Conceived and designed the experiments: PB, JVS, and SN. Performed experiments, analyzed the data obtained, and wrote the manuscript: PB and SN. All authors read and approved the final manuscript.

References

Adebajo AC, Ayoola MD, Odediran SA, Aladesanmi AJ, Schmidt TJ, Verspohl EJ (2013) Evaluation of ethnomedical claims III: antihyperglycaemic activities of *Gongronema latifolium* root and stem. J Diabetes 5:336–343

Alatar AA (2015) Thidiazuron induced efficient in vitro multiplication and ex vitro conservation of *Rauwolfia serpentina* - a potent antihypertensive drug producing plant. Biotechnol Biotechnol Equip 29:489–497

Baskaran P, Kumari A, Naidoo D, Van Staden J (2016) *In vitro* propagation and ultrastructural studies of somatic embryogenesis of *Ledebouria ovatifolia*. In Vitro Cell Dev Biol Plant 50: 271–275

Baskaran P, Kumari A, Van Staden J (2017) Rapid propagation of *Mondia whitei* by embryonic cell suspension culture *in vitro*. S Afr J Bot 108:281–286

Baskaran P, Staden V (2013) Rapid *in vitro* micropropagation of *Agapanthus praecox*. S Afr J Bot 86:46–50

Beena MR, Martin KP, Kirti PB, Hariharan M (2003) Rapid *in vitro* propagation of medicinally important *Ceropegia candelabrum*. Plant Cell Tissue Org Cult 72:285–289

Broadley M, Brown P, Cakmak I, Feng Ma J, Rengel Z, Zhao F (2011) Beneficial elements. In: Marschner P (ed) Mineral nutrition of higher plants. Elsevier, Amsterdam, The Netherlands, pp 249–268

Chaturvedi UC, Shrivastava R, Upreti RK (2004) Viral infections and trace elements: a complex interaction. Curr Sci 87:1536–1554

Epstein E (1972) Mineral nutrition of plants: principles and perspectives. Wiley, New York

Faisal M, Anis M (2003) Rapid mass propagation of *Tylophora indica* Merrill via leaf callus culture. Plant Cell Tissue Org Cult 75:125–129

Goldstein J, Newbury DE, Joy DC, Lyman CE, Echlin P, Llfshin E, Sawyer L, Michael JR (2003) Scanning electron microscopy and X-ray microanalysis. Springer, New York

Guerrero-Molina MF, Lovaisa NC, Salazar SM, Diaz-Ricci JC, Pedraza RO (2014) Elemental composition of strawberry plants inoculated with the plant growth-promoting bacterium *Azospirillum brasilense* REC3, assessed with scanning electron microscopy and energy dispersive X-ray analysis. Plant Biol 16:726–731

Guo B, Abbasi BH, Zeb A, Xu LL, Wei YH (2011) Thidiazuron: a multi-dimensional plant growth regulator. Afr J Biotechnol 10:8984–9000

Hazarika BN, Teixeira da Silva JA, Talukdar A (2006) Floriculture, ornamental and plant biotechnology: advances and topical issues. In: Effective acclimatization of in vitro cultured plants: methods, physiology and genetics. Global Science Books, East Sussex, UK, pp 427–438

Kanungo S, Pradhan C, Sahoo SL, Sahu RK (2012) Role of auxins in the *in vitro* rooting and micropropagation of *Holarrhena antidysenterica* Wall., a woody aromatic medicinal plant, through nodal explants from mature trees. J Med Plant Res 6:4660–4666

Mahonen AP, Bonke M, Kauppinen L, Riikonen M, Benfey PN, Helariutta Y (2000) A novel two-component hybrid molecule regulates vascular morphogenesis of the *Arabidopsis* root. Genes Dev 14:2938–2943

Murashige T, Skoog F (1962) A revised medium for rapid growth and bioassays with tobacco tissue cultures. Physiol Plant 15:473–497

Murthy BNS, Murch SJ, Saxena PK (1998) Thidiazuron: a potent regulator of *in vitro* plant morphogenesis. In Vitro Cell Dev Biol Plant 34:267–275

Nishimura CH, Ohashi Y, Sato S, Kato T, Tabata S, Ueguchi CH (2004) Histidine kinase homologs that act as cytokinin receptors possess overlapping functions in the regulation of shoot and root growth in *Arabidopsis*. Plant Cell 16:1365–1377

Pilon-Smits EAH, Quinn CF, Tapken W, Malagoli M, Schiavon M (2009) Physiological functions of beneficial elements. Curr Opin Plant Biol 12:267–274

Rajan JP, Singh KB, Kumar S, Mishra RK (2014) Trace elements contents in the selected medicinal plants traditionally used for curing skin diseases by the natives of Mizoram, India. Asian Pac J Trop Med 7:410–414

Riefler M, Novak O, Strnad M, Schmulling T (2006) *Arabidopsis* cytokinin receptor mutants reveal functions in shoot growth, leaf senescence, seed size, germination, root development, and cytokinin metabolism. Plant Cell 18:40–54

Rout OP, Acharya R, Gupta R, Mishra SK, Sahoo R (2013) Analysis of some essential elements in medicinal plants used in Ayurveda. Int J Appl Biol Pharm Technol 4:335–338

SANBI (2015) Statistics: Red List of South African Plants version 2015.1. http://redlist.sanbi.org/stats.php

Selvaraju R, Raman RG, Narayanaswamy R, Valliappan R, Baskaran R (2009) Trace element analysis in hepatitis B affected human blood serum by inductively coupled plasma-atomic emission spectroscopy (ICP-AES). Rom J Biophys 19:35–42

Spichal L, Werner T, Popa I, Riefler M, Schmulling T, Strnad M (2009) The purine derivative PI-55 blocks cytokinin action via receptor inhibition. FEBS J 276:244–253

Tallon CI, Porras I, Perez-Tornero O (2012) Efficient propagation and rooting of three citrus rootstocks using different plant growth regulators. In Vitro Cell Dev Biol Plant 48:488–499

Teixeira da Silva JA, Dobranszki J, Ross S (2013) Phloroglucinol in plant tissue culture. In Vitro Cell Dev Biol Plant 49:1–16

Thangarajan S, Paramasivam R, Chinthamony AR, Palanisamy CP, Velliyur KG (2012) Element and functional group analysis of *Ichnocarpus frutescens R. Br.* (Apocynaceae). Int J Pharm Pharm Sci 4:343–345

Vaiyapuri M, Raju K, Karuppusamy S (2014) An efficient *in vitro* micropropagation of *Secamone emetica* (Retz.) R. Br. (Apocynaceae) – a medicinally important plant. South Pac J Technol Sci 2:325–331

Yashvanth S, Rani SS, Madhavendra SS (2013) Microchemical (elemental) analysis of *Leucas aspera* (Willd) link employing SEM-EDAX. Int J Pharma Sci Drug Res 5:32–35

23 Social Benefits, Development, and the Pertinence of Laws: The Case of Commercialization of Plant Genetic Resources in the Global Markets

Usha Swaminathan and Jananipriya Rameshbabu

Abstract

Plant genetic resources (PGR) have always been playing a very prominent role from time immemorial, especially in crop cultivation, other agricultural purposes, and health and nutrition. In the past few decades, PGR has become the main perception of food securities. Fostering of PGR is for and by individuals for a sustainable economic growth as India is rich in biological resources, which is a contributor for the international market. Plant genetics is the key tool to develop new plant genes and diversity, and including it, farmers earn much from crop varieties followed by a sustainable environment for the future generations. This chapter reviews the findings towards the benefits in terms of agriculture, management and economic value, technology, women empowerment and their sustainability through Access and Benefit Sharing (ABS) programs, and laws and policies in protecting the growth and development of agriculture, food security, farmers rights, and plant varieties. From the international perspective, the International Union for the Protection of New Varieties of Plants (UPOV) and the Agreement on Trade-Related Aspects of Intellectual Property Rights (TRIPS) are formed to safeguard plant variety intellectual property, whereas the Food and Agriculture Organization (FAO) and the Convention on Biological Diversity (CBD) are taking the steps to protect farmer rights and benefit sharing and from the Indian perspective the Biological Diversity Act (BDA), the State Biodiversity Boards (SBBs), and the National Biodiversity Authority (NBA) and Intellectual Property (IP) laws and others have taken the lead and are discussed with case studies.

U. Swaminathan (✉) · J. Rameshbabu
School of Social Sciences and Languages, Vellore Institute of Technology, Vellore, India
e-mail: susha@vit.ac.in

S. Ramamoorthy et al. (eds.), *Plant Genetic Resources, Inventory, Collection and Conservation*, https://doi.org/10.1007/978-981-16-7699-4_23

Keywords

Plant genetic resources · Commercialization · Social benefits · Development · Technology · Law and policy

23.1 Introduction to PGR in India

The significance of PGR lies in crop cultivation, other agricultural purposes, and health and nutrition. Fostering of PGR is for and by individuals for a sustainable economic growth as India is rich in biological resources, which is a contributor for the international market. PGR gets more valuable and viable for raw materials in furtherance of crop improvement and future desires for farmers and the environment apart from food diversity for humans for a healthy living. Such PGR traditional knowledge (TK) will provide women, especially from the rural, an economic and social empowerment through various in situ and ex situ management systems (IGNFA, 2002). It is a key tool to develop new plant genes and diversity. PGR is the mainstay for sustainable development though agriculture and other forms of cultivation for export, especially among the developing countries (Ogwu et al. 2014). Agriculture biodiversity is defined as "all crops and cattle, their wild relatives, and all pollinators, symbionts, pests, parasites, predators, and competitors that interact among themselves" (Qualset et al. 1995). PGR is a plant genetic material that aids in determining the characteristics and ability of a new genetic plant to adapt and survive (Hammer 2003). PGR refers to the biological materials found in present and future plant species, as well as their economic, scientific, and social value. The materials that are constructed and applicable in cytogenetic, evolutionary, biochemical, pathological, phylogenetic, physiological, and both cultivated crop and wild varieties are included in the ecological study and breeding (Ulukan 2011). Mainly, PGR is developed to satisfy the food requirements for the growing population. International programmers work to slow down the pace of population growth, but it is a great challenge for worldwide food security and is a significant test for people in the future (Salgotra et al. 2015a; Hoisington et al. 1999). PGR stores basic crop materials to secure for future food demands. It is well established that the biodiversity in PGR field and vegetable crops in India's Northwest Himalaya is rich (Wani et al. 2003), with several crop plants and a huge range of sweet-smelling rice mainly grown in Kashmir, notably in areas of Sangam, Khudwani, Larnoo, and Kokernag, thereby developing these places as hot spots for growing Basmati rice (Salgotra and Gupta 2015; Gupta et al. 2009). Genetic resources' nexus with scientific methods adorns the development of wealth of society, environment, agriculture, crop cultivation, etc. In the last two decades in India, abundant attention has been drawn towards domestically customized endemic cultivars, especially as a result of the helpful genetic variation for the present and future plant breeding and also the speedy rate at which they are disappearing through replacement by high-yielding varieties. The Indian National Programme on Genetic Resources Conservation aims to discover and collect this natural heritage, classify it, investigate it, protect it, and document it for the current and future use. Plant genetic resources in India are essential measures for farmers to deal with future challenges, a feature stressed by the FAO commission on genetic resources for food and agriculture. The practice of PGR in crop improvement programs rests on identification of promising accessions. The germplasm that has been collected or introduced is characterized and evaluated in order to assess its potential by recording knowledge on scientific discipline traits such as yield, quality, and tolerance to organic phenomena and abiotic stresses.

23.2 Social Benefits of PGR

- *Agriculture*: Plant genetic resource is very useful for the agriculture sector. It helps to develop new varieties of the food crop to meet the future requirement of foods for growing population. It makes a staggering contribution to achieving the global goals of food security and sustainability, economic conditions, protection, and property development. PGR promotes spherical development to yield scientific discipline characteristics, weakness tolerance, and improvements in the form and structure of the plant species. Indian germplasm has aided in the discovery of diverse contributors with high yield potential for tolerance to organic phenomena and abiotic stresses. The use of introduced materials in crossbreeding programs has greatly aided sorghum and cereal production in India and other developing countries. PGR is the backbone for growing agriculture and mainly to programs for crop variety production and improvement. Disease and insect pest tolerance and yield and quality enhancement are all possible, thanks to PGR variety (Salgotra et al. 2015b).
- *Management and economic value:* Plant genetic resource management includes exotic as well as indigenous collection, conservation, characterization, evaluation, and documentation as essential steps. To maintain genetic diversity and PGR function is possible for the most part through synchronizing among these activities (Nass and Paterniani 2000). Plant genetic resource management entails using natural mechanisms such as open pollinated reproductive systems, introgression from wild relatives, and mutations to generate genetic diversity across a wide range of environments and time (Mercer and Perales 2010). Low usage of PGR results in an exiguous genetic base of commercial hybrids (Lu et al. 2009). The genomic era has provided various technologies including sequencing and re-sequencing platforms, trait-associated markers, and cost-effective management of PGR. By adding management to the PGR, it is more effective and efficient, but still there are some harnessing the full potential of genomic tools (Singh et al. 2019). Nunes and van den Bergh (2001) discovered the economic value of gene and species interactions with ecosystems and other biological functions, and a recent example has also been reviewed. The benefit extracted from crop genetic resources can be divided into two main categories: use benefit and nonuse benefit (Bragdon 2004). Nonuse benefit, also known as existence values, represents the pleasure that individuals or communities can derive simply from knowing that something exists, regardless of whether it is used (Binod et al. 2016; Krutilla

1967). The market prices for a commodity or commodities, such as grain or seed, during export would be used to assess both existing and potential values. We may use "hedonic analysis" to determine the current value of crop genetic capital embodied in crop varieties for productivity enhancement (Binod et al. 2016; Evenson et al. 1998).

- *Technology:* Plant technologies for energy generation and interpreting huge amounts of environmental, phenotypic, and genotypic conditions are being created at a rapid rate, as are techniques and practices for replicating genetic materials (Policy Science for Environment 2016). The development of technologies for genome sequencing that is strongly parallel has come after numerous approaches for calculating the molecular biological state of cells and species and aids in the automated detection of conventional phenotypes (Furbank and Tester 2011; Halewood et al. 2018). The advancement of modern technologies allowed the biological system discovery, manipulation, and synthesis. It contributes to the generation, study, and utilization of large volumes of data (Halewood et al. 2018). In technology, the National Bureau of Plant Genetic Resources (NBPGR) has been developing mobile apps to find information about the PGR and their activities through mobile phones. Two apps, "GenBank" and "PGR Map," are developed by the NBPGR (ICAR, 2020). These two apps are the first mobile apps in the world gene bank. These apps are applicable for both Android and iOS. The PGR Map app has three benefits—"What's around me": it helps users to get quick access to the genes collected and converted around them in Indian locations; "Search the map": this helps to identify the location of collected and conserved gene bank from the Indian location; and "Search for species": it helps collect the specific crop from the map (Singh et al. 2020).
- *Women empowerment:* Indian women have the most interest and actively participate in agriculture. More than 65% of total agricultural labor is accomplished by Indian women, and involvement ranges from agricultural processing to domestic consumption (Dashora et al. 2009). The cultivating women formulated their concept based on their experience of pests, enhancement of soil, and increase of plant and animal biological resources (FAO 1999; Sood et al. 2015). According to the analysis, women gather the great majority of species of plants for local consumption. Afterward, these plant species are cultivated as broad varieties of landrace crops, and some of the crops are used in cloning processes to improve new varieties (Addison Posey 1999). "Women were responsible for food production processes like winnowing, seed selection, shelling, threshing, drying, pounding, and cooking. Men were more interested in construction than women. Despite their lack of physical and decision-making power in societies, women have played an important role in food storage for future food security" (Matsa and Mukoni 2013; Sood et al. 2015). The majority of the food is produced by the female. Mostly they produce staple foods, either intentionally or unintentionally; in most developed countries, women contribute more than 60% of global food production (FAO 1995). Women are preferred in order to amass a range of plant species associated with medical plants, domestic-use plants which are related to food, and woods and easily firing plants to cook. Empowerment of women

farmers is a method to enhance the effectiveness of agricultural research for development and cut back gender-based disadvantages. It benefits women farmers in terms of better governance regime, appropriate regulation for access to agricultural practices. It is recognized that incentives are given to support and encourage them to conserve plant resources and associated knowledge and practices. Women prefer culinary traditions and preferences, as well as the postharvest processes that are needed so that the supply of edible and generally accepted food has a marked influence on data, selection, and use and conservation of PGR.

23.3 Case Studies: I

1. Amrita Devi was the first person in the world to sacrifice her life to protect the "Khejri," a medium-sized tree that grows in North India (Rajasthan). Jodhpur king Abhay Singh had constructed a palace to collect the wood. He sent his soldiers to cut the trees and collect the wood. Bishnois protested and argued for the tree cut. Amrita Devi requested that her head be severed rather than a tree be cut down. The brutal soldier cut her head as well as her three daughters' heads, and they slaughtered 363 Bishnois; the old Bishnois hung themselves in the tree, but the soldiers collected the woods for the palace. This incident is known as the "Chipko Movement" (Jain 2011). To honor Amrita Devi, the Ministry of Environment and Forests (MoEF), the Government of India, introduced the award "Amrita Devi Bishnoi Wild Life Protection Award" in the year 2001. This award is given to one person every year who protects the environment better (Saini et al. 2019).
2. (A) Bachni Devi from India's Garhwal Himalayas used this to effectively protect and conserve trees and other plant genetic resources. She organized a bunch of local women and encouraged them to adorn trees with sacred thread. The tree's Raksha Bandhan was thus completed. This contributed to the protection of trees and other natural resources. Nowadays, women in all sectors achieve a lot same as in PGR. Women are the backbone to cultivate more food varieties for the society. It increases the standard of living in women's lifestyle. (B) Kamala Pujari is a tribal woman from Koraput in Orissa. She received a Padma Shri from the Honorable President of India Ram Nath Kovind for preserving hundreds of varieties of paddy seeds and collecting rare types of seeds like turmeric and mahakanta. Over 1800 rice landscape residences are located in the Orissa district of Koraput. It is regrettable that only 15 species with high yield are currently cultivated. Landraces are far more suitable to the environmental conditions and therefore less disease-prone than high-rise species. Typically, landraces are also highly valuable as a method for adapting to climate change. Landraces are also a resource with different useful genes, which can be an essential method to strengthen crops. Too much focus has been put on the high yield truths of land erosion. Kamala Pujari has taken over the conservation of rice terraces with the aid of some of her district's other women. Now they are cultivated along with

high-yield species in several areas of Orissa. Mrs. Pujari's contribution was also recognized on the international forum and she was awarded in 2002 at the Johannesburg World Summit on Sustainable Development. (C) In Kerala, India, women from villages such as Panangofe, Pampuchathamana, and Bharathnoor formed a committee and received training to protect medicinal plants in a scientific manner. Women were taught how to collect, preserve, and disseminate herbal medicines. These women built a nexus to educate other village women, which is now heavily involved in the preservation of PGR and TK diversity. (D) Illegal tree felling is a significant issue that has far-reaching negative consequences for the forest. Women in the Nayagarh district of Orissa in India take this issue seriously and acknowledged the challenge of forest conservation. Women organized and founded "the Maa Ghodadei Mahila Samiti (MGMS)" party, specifically designed to control the illicit fall of forest trees. The Samiti people used to patrol the forest area even at night and capture the perpetrators. Women's forces have had such an impact that fearful offenders do not wage to join for illicit activities in the forest now. Women's commitment to the protection of the forest has also helped (Sood et al. 2015).

23.3.1 Access and Benefit Sharing

ABS refers to the genetic resources that are accessed by business, scientists, etc. and where the benefits that arise from their use are shared or exchanged by people or countries that use such resources.

- The effective and honest implementation of mechanisms, supported by acceptable policies and laws, can ultimately be the foremost necessary assessment issue of success of any access and profit-sharing regime.
- The mechanisms embody practices helpful for documentation, adding worth, sustainable use, and conservation, providing access and sharing of the accumulated advantages fairly and equitably.
- Another key space of activity is experimentation and analysis on novel institutional structures at the community and national levels that may perform functions to confirm farmers' rights.
- The essential adoption strategy is associated with the implementation to guide benefit sharing for crop genetic resources toward adaptation efforts and to confirm expedited and cost-effective access to crop genetic resources for adaptation functions.

Article 8(j) - (CBD 2021) states with respect to in situ conservation that

> "*Each contracting Party shall, as far as possible and as appropriate: Subject to national legislation, respect, preserve and maintain knowledge, innovations and practices of indigenous and local communities embodying traditional lifestyles relevant for the conservation and sustainable use of biological diversity and promote their wider application with the*

approval and involvement of the holders of such knowledge, innovations and practices and encourage the equitable sharing of the benefits arising from the utilization of such knowledge innovations and practices".

23.4 Case Study: II

The Kani Tribe live in the tropical forest of the Western Ghats of Kerala, India. The Kani tribes are rich in traditional plant cultivation for health reasons. The Kani tribe and the Jawaharlal Nehru Tropical Botanic Garden and Research Institute (JNTBGRI) had an agreement to cultivate the "Arogyapacha" (scientific name, *Trichopus zeylanicus*) medical plant. JNTBGRI's researchers discovered that not only does the plant (especially the fruit and leaves) have calming and immunological properties, but it also increases endurance, reduces fatigue, aids in tumor control, and stimulates the body's natural defenses and cellular immune system. JNTBGRI named it "Jeevani," which means "giver of life." The Kani traditionally consumed Arogyapacha in the form of its fruit. But JNTBGRI identified that crushing the leaves of the plant is the more effective path to get the 12 compounds. Since JNTBGRI is a research institute, it is unable to commercialize any products derived from its Jeevani creation. As a result, it approved the licensing of Jeevani manufacturing technology to interest groups. JNTBGRI formed a committee to decide which organizations would be the best fit for licensing. The committee selected Arya Vaidya Pharmacy Ltd. (AVP), which is one of India's leading herbal pharmacies, as the main producer and in 1995 AVP agreed to sign a 7-year licensing deal with JNTBGRI and charged a US$50,000 licensing fee. According to the agreement, 2% of royalties on sales of Jeevani products should be given to the JNTBGRI. The access benefit-sharing agreement between Kani tribes and the JNTBGRI has become a model for the world (WIPO 2015).

Article 15 - (CBD 2021) states with respect to access to genetic resources that

"*1. Recognizing the sovereign rights of States over their natural resources, the authority to determine access to genetic resources rests with the national governments and is subject to national legislation.*

2. Each Contracting Party shall endeavour to create conditions to facilitate access to genetic resources for environmentally sound uses by other Contracting Parties and not to impose restrictions that run counter to the objectives of this Convention.

3. For the purpose of this Convention, the genetic resources being provided by a Contracting Party, as referred to in this Article".

23.5 Case Studies: III

1. Seaweed cultivation and exportation: From the export to countries such as Indonesia, Malaysia, and the Philippines to the tune of 2000 metric tonnes from fisherman societies in Tamil Nadu coastal regions such as Tuticorin, Ramanathapuram, Tanjore, and Pudukkottai, a royalty of Rs. 39.09 lakhs (US

$58,721.68) has been earned which is for the distribution to 754 families in the coastal districts taken care by the PepsiCo India Holdings Private Ltd. and NBA agreement (Varma 2017).

2. Bio-India Biologicals Neem Case: The NBA received approximately USD 924 from Bio-India Biologicals (BIB) for the export of 2000 kilograms of neem to Japan. An agreement by the people from Amarchinta, Andhra Pradesh, with BIB to bundle and dry the leaves before handing them over to BIB for export has led to a royalty which is transferred by the NBA to the local biodiversity body in Amarchinta for planting neem saplings and raising awareness on biodiversity and its conservation to Amarchinta's local biodiversity body (Sanghera et al. 2015).

23.5.1 Nagoya Protocol

Article 5 - (CBD 2021) provides for fair and equitable benefit sharing:

> *"1. In accordance with Article 15, paragraphs 3 and 7 of the Convention, benefits arising from the utilization of genetic resources as well as subsequent applications and commercialization shall be shared in a fair and equitable way with the Party providing such resources that is the country of origin of such resources or a Party that has acquired the genetic resources in accordance with the Convention. Such sharing shall be upon mutually agreed terms.*
>
> *2. Each Party shall take legislative, administrative or policy measures, as appropriate, with the aim of ensuring that benefits arising from the utilization of genetic resources that are held by indigenous and local communities, in accordance with domestic legislation regarding the established rights of these indigenous and local communities over these genetic resources, are shared in a fair and equitable way with the communities concerned, based on mutually agreed terms.*
>
> *3. To implement paragraph 1 above, each Party shall take legislative, administrative or policy measures, as appropriate.*
>
> *4. Benefits may include monetary and non-monetary benefits, including but not limited to those listed in the Annex.*
>
> *5. Each Party shall take legislative, administrative or policy measures, as appropriate, in order that the benefits arising from the utilization of traditional knowledge associated with genetic resources are shared in a fair and equitable way with indigenous and local communities holding such knowledge. Such sharing shall be upon mutually agreed terms."*

23.6 Case Studies: IV (NBA 2012)

Anti-snake venom: Dr. Geeta Pandurang Pawar, an Ayurvedic doctor from Pune, India, submitted the application for the production of an Ayurvedic anti-snake venom. This anti-venom "pinak" tablet serves as a temporary cure for patients of snakebites immediately prior to taking the victim to the hospital. NBA determined the 2% of the total profits or gross revenues of the commodity received from the exploitation of biological resources. Royalty was paid upon commercial viability of the patent drug to share the benefits.

Red Sanders case: A worldwide e-auction was conducted for the sale of Red Sanders by the Andhra Pradesh Government involving Indian and foreign bidders. After successful purchase, the buyers had to give 5% to the NBA/SBB and 95% of the benefits to the local Biodiversity Management Committee (BMC). This benefitted the state by sharing the benefits to indigenous people, tribal people, and forest dwellers.

23.7 Challenges in ABS

1. Lack of knowledge and awareness among the farmers and common public: many farmers lack in the knowledge regarding profit sharing.
2. It leads to conflict between the stockholders while dividing the interest.
3. Lack of sufficient laws and regulations for ABS.

23.8 Scope of Development

The scope of development of PGR is to be a future analysis program that should be used to develop a sound body comparative analysis on socioeconomic conditions, gender, and PGR which can link the global and native modification processes. Genetic resources have a great scope in the fields of agriculture and country's economic value. PGR intensifies international exchange and sharing of benefits arising from it.

Legal Revolution

International: During 1980, the apex court in the USA allowed modification among living organisms and determined that patents should cover living transformed organisms. Later, IP rights were extended across the globe through treaties such as the International Convention for the Protection of New Varieties of Plants and the World Trade Organization's (WTO) 1994 Agreement on Trade-Related Aspects of Intellectual Property (TRIPS) having acquired similar features due to the contractual agreements between sovereign countries.

According to the Article 27.3(b) of the TRIPS agreement, *"which requires plant variety protection either by patents or an effective sui generics system making TRIPS and WTO member countries responsible under international law for IP protection of plant varieties."* In the 1950s–1970s, the Green Revolution, also known as the Third Agricultural Revolution, was introduced to agricultural production. Agriculture production increased globally during this time span. Plant varieties grown by IARCs such as the International Maize and Wheat Improvement Center and the International Rice Research Institute raised food production in underdeveloped and developing countries. A commission was established by the FAO during 1983 to embrace the notion of farmers' rights describing as "*rights arising from the past, present and future contributions of farmers in conserving, improving, and making available plant genetic resources, particularly those in the centres of origin/diversity*" (Sullivan, 2004).

The CBD, signed on the 5th of June, 1992, at Rio de Janeiro, Brazil, came to effect on December 29, 1993. As part of the international law, the CBD gave recognition to biodiversity conservation, termed the problem as "common concern of humankind," and stated that it is fundamental in the development process, covering all habitats, plants, and genetic resources. This further builds a nexus between traditional conservation practices and the economic goal by means of biological wealth sustainably. There can be advancement in equity principles and distribution of the benefits ethically, attained through the uses of genetic resources, especially those having economic value (Louafi and Morin 2004).

Article 1 - (CBD 2021) states with respect to its objectives that

"The objectives of this Convention to be pursued in accordance with its relevant provisions are, the conservation of biological diversity, the sustainable use of its components and the fair and equitable sharing of the benefits arising out of the utilization of genetic resources, including by appropriate access to genetic resources and by appropriate transfer of relevant technologies, taking into account all rights over those resources and to technologies, and by appropriate funding."

On November 3, 2001, in Rome, the FAO Commission on Genetic Resources and its subsidiary bodies approved a new International Treaty on Plant Genetic Resources for Food and Agriculture (ITPGRFA) and it entered into force on June 29, 2004. The ITPGRFA is the 31st conference of the FAO, represented by 116 nations except the USA and Japan. The two countries raised concerns regarding the lacuna in clarity of the TRIPS. Concentration was laid towards multilateral system, farmers' rights, benefit sharing, and use of biological components (Sullivan, 2004).

The Treaty aims at identifying the huge involvement of farmers in the diversity of crops that feed the world; launching a comprehensive system to offer farmers, plant breeders, and scientists an access to plant genetic materials; and assuring to share benefits that the recipients get from the use of genetic materials with the countries where they have been originated (International Treaty on Plant Genetic Resources for Food and Agriculture, 2001).

In 2010, Nagoya Protocol on ABS was established with an aim to fulfil the basic three objectives of CBD.

India: Biological Diversity Act, 2002, established by the Indian Parliament was to protect the biological diversity in India and provide equitable sharing of benefits with the help of SBBs and NBA in mediating and monitoring the benefits being shared through royalty and commission. Biodiversity has been defined under Sect. 2 (b) in the Biological Diversity Act 2002, of the Act as *"the variability among living organisms from all sources and the ecological complexes of which they are part, and includes diversity within species or between species and of ecosystems."* The Act also defines under Sect. 2(C) in the Biological Diversity Act 2002, biological resources as *"plants, animals and micro-organisms or parts thereof, their genetic material and by-products (excluding value added products) with actual or potential use or value, but does not include human genetic material."*

The features of the Act are:

- To control access to the country's biological resources with the aim of ensuring equal distribution of benefits resulting from the use of biological resources, as well as related information relating to biological resources
- Biological diversity must be conserved and used in a sustainable manner
- To respect and conserve local communities' biodiversity-related awareness
- To ensure value sharing with local citizens as biological resource conservers and holders of expertise and information about biological resource use
- To protect and establish biological diversity-sensitive areas by designating them as biological diversity heritage sites
- To conserve and rehabilitate endangered species in the international and national framework on ABS
- To include agencies of state governments in the broad scheme of Biological Diversity Act implementation through the establishment of committees (Kumar et al. 2017)

The NBA is managed by the Indian Forest Service (IFS) and governed by the Ministry of Environment, Forest and Climate Change (MoEFCC). The BDA was adopted and signed by India in 2003, after having agreed and signed the Biological Diversity Convention in 1992. Its functions are to come up with regulations for prohibited acts, impart advices to the government on various strategies on conservation techniques and selecting the biological heritage sites, and take appropriate measures in opposing the grant of any IP rights in foreign countries rising from the use of biological resources or associated traditional knowledge.

SBBs aim is at advising the state governments concerning any guiding principles issued by the central government involving conservation of biodiversity, sustainable use of genetic resources, and access sharing equally. SBBs are established under Sect. 22 of the Act, and accordingly in all the states SBBs have been established. The main functions of SBBs are to advise on biodiversity conservation and sustainable use and equitable sharing of benefits arising from commercialization of biological resources; regulate by granting approvals; and perform those functions that are necessary as per the Act or state government (NBA 2018).

BMCs are local government institutions formed at the local community for discussions, sustainable use, biodiversity reporting, and recording of conventional biodiversity information (Rana 2012).

Similarly, for exclusive protection of plant varieties and the rights of farmers and plant breeders, to support the expansion of new varieties, the Government of India established the Protection of Plant Varieties and Farmers' Rights Act, 2001, adopting a sui generis system.

Section 2(1) in The Protection of Plant Varieties and Farmers' Rights Act, 2001 defines *"farmer" as any person who*

> *"(i) cultivates crops by cultivating the land himself; or*
>
> *(ii) cultivates crops by directly supervising the cultivation of land through any other person; or*

(iii) conserves and preserves, severally or jointly, with any person any wild species or traditional varieties, or adds value to such wild species or traditional varieties through selection and identification of their useful properties."

Its objectives are to launch a system for active protection of plant species and the rights of producers and breeders of plants and promotion of the growth of new plant species; to consider and preserve farmers' rights with regard to their contributions to the conservation, improvement, and provision of plant genetic resources for the production of new plant varieties at all times; protect plant breeders' rights in order to accelerate farm production in the region and stimulate public and private research and development investment in new crop development; and facilitate seed sector development in the country by ensuring high-quality seed and planting material availability to the farmers (Protection of Plant Varieties and Farmers' Rights Act, 2001).

Another important leap in the biodiversity protection is the ICAR-NBPGR, established during 1976 by the Indian Council of Agriculture Research (ICAR) headquartered at New Delhi. Its objectives are to concentrate on the exploration and collection of indigenous and exotic plant genetic resources, conserve and commercialize in collaboration with national organizations, exchange and quarantine, improve information network, conduct research, undertake teaching and training, develop guidelines, and create public awareness on plant genetic resources.

A 2016 global analysis of over 1000 crop wild relative organisms identified 70% as major priority for additional gathering to increase their coverage in seed banks, while the remaining 29% were completely absent from ex situ conservation (Castañeda-Alvarez et al. 2016). According to a 2020 report, 93.3% of crop wild relatives specific to the USA are underrepresented in ex situ conservation reservoirs, while 93.1% are under-conserved in their natural habitats (Khoury et al. 2020).

23.9 Conclusion

PGR is the key for developing new varieties of crops. It helps to fulfil the requirement of food scarcity for the growing population, to grow in the economic value of the countries. PGR is more useful in the modern world; it helps to grow the excess of crops in a limited time. It creates a new idea for cultivating crops in a small test tube. By the help of PGR nowadays many women are cultivating food crops at home. These plant genetic resources are used not only for humans but also for domestic animals. Due to a lot of new varieties of crops an increase in economic development can be foreseen in terms of sales and in the agriculture sector it helps to increase the crop cultivation such as rice, maize, and wheat. Additional preventive measures for plant genetic resources are needed. PGR is helping women to earn on their own; it increases women empowerment and saves wild trees. This is more useful to gain economic value in developing and underdeveloped countries.

References

Addison Posey D (1999) Cultural and spiritual values of biodiversity. Martyn Bramwell Publication, Surrey, UK

Binod PL, Krishna HG, Bal KJ, Ryu KY, Sung JS, Rhee J et al (2016) Current status of plant genetic resources, their research and Management in the Genebank of Nepal. The Journal of the Korean Society of International Agriculture 28(2):143–151

Bragdon, S. (2004). *International law of relevance to plant genetic resources: a practical review for scientists and other professionals working with plant genetic resources*. p. 87–108

Castañeda-Álvarez NP, Khoury CK, Achicanoy HA, Bernau V, Dempewolf H, Eastwood RJ, Guarino L, Harker RH, Jarvis A, Maxted N, Müller JV, Ramirez-Villegas J, Sosa CC, Struik PC, Vincent H, Toll J (2016) Global conservation priorities for crop wild relatives. Nat Plants 2(4):16022

CBD (2021) Access and benefit-sharing meaning referred from https://www.cbd.int/abs/infokit/brochure-en.pdf

Dashora K, Gupta A, Gupta V (2009) The hand that rocks the cradle, feeds the nation. Curr Sci 96(8):1016

Evenson RE, Gollin D, Santaniello V (1998) Agricultural values of plant genetic resources. In: Symposium on economics of valuation and conservation of genetic resources for agriculture may 1996 Roma (Italia) (no. 631.52 S989 1996). CAB International

FAO (1995) A fairer future for rural women. FAO, Rome, Italy

FAO (1999) Women – users, preservers and managers of agrobiodiversity. Food and Agriculture Organization, Rome, Italy

Furbank RT, Tester M (2011) Phenomics–technologies to relieve the phenotyping bottleneck. Trends Plant Sci 16(12):635–644

Gupta, B. B., Salgotra, R. K., & Bali, A. S. (2009). Status paper on rice in Jammu and Kashmir. Rice knowledge management portal. Directorate of Rice Research, Rajendranagar, Hyderabad, 500030

Halewood M, Chiurugwi T, Sackville Hamilton R, Kurtz B, Marden E, Welch E et al (2018) Plant genetic resources for food and agriculture: opportunities and challenges emerging from the science and information technology revolution. New Phytol 217(4):1407–1419

Hammer K (2003) A paradigm shift in the discipline of plant genetic resources. Genet Resour Crop Evol 50(1):3–10

Hoisington D, Khairallah M, Reeves T, Ribaut JM, Skovmand B, Taba S, Warburton M (1999) Plant genetic resources: what can they contribute toward increased crop productivity? Proc Natl Acad Sci 96(11):5937–5943

ICAR-National Bureau of Plant Genetic Resources (2020). http://www.nbpgr.ernet.in/About_NBPGR/Mandate.aspx

IGNFA (2002) Biodiversity and Access and Benefit Sharing in India, https://www.ignfa.gov.in/document/biodiversity-cell-ntfp-related-issues1.pdf

International Treaty on Plant Genetic Resources for Food and Agriculture, (2001) AIMS. http://www.fao.org/plant-treaty/overview/en/

Jain P (2011) Dharma and ecology of Hindu communities: sustenance and sustainability, 1st edn. Routledge

Khoury CK, Carver D, Greene SL, Williams KA, Achicanoy HA, Schori M et al (2020) Crop wild relatives of the United States require urgent conservation action. Proc Natl Acad Sci 117(52):33351–33357

Krutilla JV (1967) Conservation reconsidered. Am Econ Rev 57(4):777–786

Kumar A, Singh TP, Gautam S (2017) International and National Framework on access and benefit sharing. In: Biodiversity for sustainable development. Springer, Cham, pp 3–15

Louafi S, Morin J-F (2004) International governance of biodiversity: involving all the users of genetic resources. IDDRI, Paris, France

Lu Y, Yan J, Guimaraes CT, Taba S, Hao Z, Gao S et al (2009) Molecular characterization of global maize breeding germplasm based on genome-wide single nucleotide polymorphisms. Theor Appl Genet 120(1):93–115

Matsa W, Mukoni M (2013) Traditional science of seed and crop yield preservation: exploring the contributions of women to indigenous knowledge systems in Zimbabwe. Int J Humanit Soc Sci 3(4):1–12

Mercer KL, Perales HR (2010) Evolutionary response of landraces to climate change in centers of crop diversity. Evol Appl 3(5-6):480–493

Nass LL, Paterniani E (2000) Pre-breeding: a link between genetic resources and maize breeding. Sci Agric 57(3):581–587

NBA (2012) Access and Benefit Sharing Experience from India. http://nbaindia.org/uploaded/pdf/ABS_Factsheets_1.pdf

NBA (2018). http://nbaindia.org/text/13/StateBiodiversityBoards.html#:~:text=The%20Functions%20of%20State%20Biodiversity,of%20utilization%20of%20biological%20resources

Nunes PA, van den Bergh JC (2001) Economic valuation of biodiversity: sense or nonsense? Ecol Econ 39(2):203–222

Ogwu MC, Osawaru ME, Ahana CM (2014) Challenges in conserving and utilizing plant genetic resources (PGR). Int J Genet Mol Biol 6(2):16–23

Policy Science for Environment (2016) Synthetic biology and biodiversity. Future brief 15. Bristol, UK: For the European Commission DG Environment by the Science Communication Unit, UWE

Protection of Plant Varieties and Farmers' Rights Act (2001). http://www.plantauthority.gov.in/

Qualset C, McGuire P, Warburton M (1995) In California: agrobiodiversity key to agricultural productivity. Calif Agric 49(6):45–49

Rana RS (2012) Accessing plant genetic resources and sharing the benefits: experiences in India. Indian J Plant Genetic Resour 25(1):31–51

Saini I, Kaushik P, Saini BL (2019) Role of Indian women in conservation of biodiversity. Int J Curr Microbiol App Sci 8(4):63–68

Salgotra RK, Gupta BB (2015) Plant genetic resources and traditional/indigenous knowledge: potentials and challenges. In: Plant genetic resources and traditional knowledge for food security. Springer, Singapore, pp 1–21

Salgotra RK, Gupta BB, Bhat JA, Sharma S (2015a) Genetic diversity and population structure of basmati rice (*Oryza sativa* L.) germplasm collected from North Western Himalayas using trait linked SSR markers. Plos one 10(7):e0131858

Salgotra RK, Gupta BB, Sood M (2015b) Biotechnological interventions and their role in sustainable hill agriculture. J Plant Sci Res 2(1):1–8

Sanghera GS, Bhatia D, Thind KS (2015) Access and benefit sharing on the use of indigenous traditional knowledge. In: Plant genetic resources and traditional knowledge for food security. Springer, Singapore, pp 163–181

Section 2(b) in The Biological Diversity Act (2002). https://www.indiacode.nic.in/bitstream/123456789/2046/1/200318.pdf

Section 2(c) of Biological Diversity Act (2002). https://www.indiacode.nic.in/bitstream/123456789/2046/1/200318.pdf

Section 2(l) in The Protection of Plant Varieties and Farmers Rights Act (2001). https://www.indiacode.nic.in/show-data?actid=AC_CEN_23_31_00002_200153_1517807322692&sectionId=1830&sectionno=2&orderno=2

Singh K, Gupta K, Tyagi V, Rajkumar S (2020) Plant genetic resources in India: management and utilization. Vavilovskii Zhurnal Genet Selektsii 24(3):306–314

Singh K, Kumar S, Kumar SR, Singh M, Gupta K (2019) Plant genetic resources management and pre-breeding in genomics era. Indian J. Genet 79(1 Suppl 117):130

Sood M, Gupta V, Jan A (2015) Role of women in conserving plant genetic resources and related traditional knowledge for food security. In: Plant genetic resources and traditional knowledge for food security. Springer, Singapore, pp 237–253

Sullivan SN (2004) Plant genetic resources and the law: past, present, and future. Plant Physiol 135(1):10–15

Ulukan H (2011) Plant genetic resources and breeding: current scenario and future prospects. Int J Agric Biol 13:447–454

WIPO (2015) Using traditional knowledge to revive the body and a community. Accessed at https://www.wipo.int/ipadvantage/en/details.jsp?id=2599

Varma RV (2017) Access and benefit sharing in India: challenges ahead. In: Biodiversity for sustainable development. Springer, Cham, pp 87–96

Wani SH, Wani SA, Ishfaq Abidi DZ, Lone A, Gull Zaffar KM (2003) Plant genetic resources of Kashmir region – prospect for crop improvement. In *National Conference on status and conservation of biodiversity in India*. Kashmir University, Srinagar, India

The manufacturer's authorised representative in the EU is Springer
Nature Customer Service Centre GmbH, Europaplatz 3, 69115 Heidelberg,
Germany. If you have any concerns regarding our products, please
contact ProductSafety@springernature.com

Printed and bound by CPI Group (UK) Ltd, Croydon, CR0 4YY
30/09/2026
02225245-0017